Recent Titles in This Series

159 **William J. Heinzer, Craig L. Huneke, and Judith D. Sally, Editors,** Commutative algebra: Syzygies, multiplicities, and birational algebra, 1994

158 **Eric M. Friedlander and Mark E. Mahowald, Editors,** Topology and representation theory, 1994

157 **Alfio Quarteroni, Jacques Periaux, Yuri A. Kuznetsov, and Olof B. Widlund, Editors,** Domain decomposition methods in science and engineering, 1994

156 **Steven R. Givant,** The structure of relation algebras generated by relativizations, 1994

155 **William B. Jacob, Tsit-Yuen Lam, and Robert O. Robson, Editors,** Recent advances in real algebraic geometry and quadratic forms, 1994

154 **Michael Eastwood, Joseph Wolf, and Roger Zierau, Editors,** The Penrose transform and analytic cohomology in representation theory, 1993

153 **Richard S. Elman, Murray M. Schacher, and V. S. Varadarajan, Editors,** Linear algebraic groups and their representations, 1993

152 **Christopher K. McCord, Editor,** Nielsen theory and dynamical systems, 1993

151 **Matatyahu Rubin,** The reconstruction of trees from their automorphism groups, 1993

150 **Carl-Friedrich Bödigheimer and Richard M. Hain, Editors,** Mapping class groups and moduli spaces of Riemann surfaces, 1993

149 **Harry Cohn, Editor,** Doeblin and modern probability, 1993

148 **Jeffrey Fox and Peter Haskell, Editors,** Index theory and operator algebras, 1993

147 **Neil Robertson and Paul Seymour, Editors,** Graph structure theory, 1993

146 **Martin C. Tangora, Editor,** Algebraic topology, 1993

145 **Jeffrey Adams, Rebecca Herb, Stephen Kudla, Jian-Shu Li, Ron Lipsman, and Jonathan Rosenberg, Editors,** Representation theory of groups and algebras, 1993

144 **Bor-Luh Lin and William B. Johnson, Editors,** Banach spaces, 1993

143 **Marvin Knopp and Mark Sheingorn, Editors,** A tribute to Emil Grosswald: Number theory and related analysis, 1993

142 **Chung-Chun Yang and Sheng Gong, Editors,** Several complex variables in China, 1993

141 **A. Y. Cheer and C. P. van Dam, Editors,** Fluid dynamics in biology, 1993

140 **Eric L. Grinberg, Editor,** Geometric analysis, 1992

139 **Vinay Deodhar, Editor,** Kazhdan-Lusztig theory and related topics, 1992

138 **Donald St. P. Richards, Editor,** Hypergeometric functions on domains of positivity, Jack polynomials, and applications, 1992

137 **Alexander Nagel and Edgar Lee Stout, Editors,** The Madison symposium on complex analysis, 1992

136 **Ron Donagi, Editor,** Curves, Jacobians, and Abelian varieties, 1992

135 **Peter Walters, Editor,** Symbolic dynamics and its applications, 1992

134 **Murray Gerstenhaber and Jim Stasheff, Editors,** Deformation theory and quantum groups with applications to mathematical physics, 1992

133 **Alan Adolphson, Steven Sperber, and Marvin Tretkoff, Editors,** p-Adic methods in number theory and algebraic geometry, 1992

132 **Mark Gotay, Jerrold Marsden, and Vincent Moncrief, Editors,** Mathematical aspects of classical field theory, 1992

131 **L. A. Bokut′, Yu. L. Ershov, and A. I. Kostrikin, Editors,** Proceedings of the International Conference on Algebra Dedicated to the Memory of A. I. Mal′cev, Parts 1, 2, and 3, 1992

130 **L. Fuchs, K. R. Goodearl, J. T. Stafford, and C. Vinsonhaler, Editors,** Abelian groups and noncommutative rings, 1992

129 **John R. Graef and Jack K. Hale, Editors,** Oscillation and dynamics in delay equations, 1992

(Continued in the back of this publication)

CONTEMPORARY MATHEMATICS

159

Commutative Algebra: Syzygies, Multiplicities, and Birational Algebra

AMS-IMS-SIAM Summer Research Conference
on Commutative Algebra
July 4–10, 1992
Mount Holyoke College

William J. Heinzer
Craig L. Huneke
Judith D. Sally
Editors

American Mathematical Society
Providence, Rhode Island

The AMS-IMS-SIAM Summer Research Conference in the Mathematical Sciences on Commutative Algebra: Syzygies, Multiplicities, and Birational Algebra was held at Mount Holyoke College, South Hadley, Massachusetts, from July 4 to July 10, 1992 with support from the National Science Foundation Grant No. DMS 8918200-02 and the National Security Agency Grant No. MDA 904-92-H-3059.

1991 *Mathematics Subject Classification.* Primary 13–06; Secondary 13–02, 14–02, 14–06.

Library of Congress Cataloging-in-Publication Data

Commutative algebra: syzygies, multiplicities, and birational algebra/William J. Heinzer, Craig L. Huneke, Judith D. Sally, editors.

p. cm. — (Contemporary mathematics, ISSN 0271-4132; v. 159)

"The conference Cummutative Algebra: Syzygies, Multiplicities, and Birational Algebra was one of the 1992 AMS-IMS-SIAM Summer Research Conferences in the Mathematical Sciences which were held at Mount Holyoke College in South Hadley, Massachusetts"—CIP pref.

Includes bibliographical references.

ISBN 0-8218-5188-8

1. Commutative algebra—Congresses. I. Heinzer, William J. II. Huneke, C. (Craig) III. Sally, Judith D. IV. Series: Contemporary mathematics (American Mathematical Society); v. 159.

QA251.3.C657 1994 93–45007

512′.24—dc20 CIP

♾ The paper used in this book is acid-free and falls within the guidelines
established to ensure permanence and durability.
♻ Printed on recycled paper.

All articles in this volume were printed from copy prepared by the authors.
Some articles were typeset using $\mathcal{A}\mathcal{M}\mathcal{S}$-TEX or $\mathcal{A}\mathcal{M}\mathcal{S}$-LATEX,
the American Mathematical Society's TEX macro systems.

10 9 8 7 6 5 4 3 2 1 99 98 97 96 95 94

Contents

Preface vii

Grothendieck"s localization problem
LUCHEZAR L. AVRAMOV AND HANS-BJØRN FOXBY 1

A simple proof of Grothendieck's theorem on the parafactoriality of local rings
FREDERICK CALL AND GENNADY LYUBEZNIK 15

Resolutions with a given Hilbert function
HARA CHARALAMBOUS AND E. GRAHAM EVANS, JR. 19

Complete ideals in algebra and geometry
STEVEN DALE CUTKOSKY 27

On the Cohen-Macaulay type of perfect ideals
J. ELIAS, A. V. GERAMITA, AND G. VALLA 41

On the Gorensteinness of graded rings associated to ideals of analytic deviation one
SHIRO GOTO AND YUKIO NAKAMURA 51

Prime ideals in birational extensions of polynomial rings
WILLIAM J. HEINZER, DAVID LANTZ, AND SYLVIA M. WIEGAND 73

On the index of a homogeneous Gorenstein ring
JÜRGEN HERZOG 95

Solid closure
MELVIN HOCHSTER 103

Tight closure in equal characteristic, big Cohen-Macaulay algebras, and solid closure
MELVIN HOCHSTER 173

Indecomposable canonical modules and connectedness
MELVIN HOCHSTER AND CRAIG HUNEKE 197

Multiplicities in graded rings I: The general theory
D. KIRBY AND D. REES 209

Pfaffian identities, with applications to free resolutions, DG-algebras, and algebras with straightening law
ANDREW R. KUSTIN 269

Proximity inequalities for complete ideals in two-dimensional regular local rings
JOSEPH LIPMAN 293

Cohomological annihilators and Castelnuovo-Mumford regularity
UWE NAGEL AND PETER SCHENZEL 307

Local-global principle for annihilation of local cohomology
K. N. RAGHAVAN 329

Multiplicities and Chern classes
PAUL C. ROBERTS 333

A computation of local cohomology
PAUL C. ROBERTS 351

Algebra structures for graded free resolutions
HEMA SRINIVASAN 357

Primary decompositions of powers of ideals
IRENA SWANSON 367

Artin-Nagata properties and reductions of ideals
BERND ULRICH 373

Hilbert functions, analytic spread, and Koszul homology
WOLMER V. VASCONCELOS 401

Infinite cyclic covers of strongly F-regular rings
KEI-ICHI WATANABE 423

Torsion in Picard groups of affine rings
ROGER WIEGAND 433

PREFACE

The conference Comutative Algebra: Syzygies, Multiplicities and Birational Algebra was one of the 1992 AMS-IMS-SIAM Summer Research Conferences in the Mathematical Sciences which were held at Mount Holyoke College in South Hadley, Massachusetts. The scientific program of the conference centered around a series of invited one-hour lectures on recent advances in commutative algebra and the corresponding interactions with related fields such as algebraic geometry, representation theory and combinatorics. The majors themes of the conference were tight closure, Hilbert functions, birational algebra, free resolutions and the homological conjectures, Rees algebras and local cohomology.

The papers in this volume are refereed research papers on these themes. We would like to thank the referees for their help. We would also like to thank the American Mathematical Society, the Society for Industrial and Applied Mathematics, and the Institute of Mathematical Statistics for sponsoring the conference, and the National Science Foundation and the National Security Agency for support. Our thanks also go to Conference Coordinator Carole Kohanski and to Donna Harmon, who coordinated these proceedings.

Contemporary Mathematics
Volume **159**, 1994

Grothendieck's Localization Problem

LUCHEZAR L. AVRAMOV AND HANS–BJØRN FOXBY

ABSTRACT. The singularity of a fiber of a flat homomorphism of noetherian rings $\varphi\colon R \to S$ at a prime ideal $\mathfrak{p} \in \operatorname{Spec} R$ is controlled by the singularity of the fiber of φ at any specialization of $\mathfrak{p}$ and by the singularities of the formal fibers of R at $\mathfrak{p}$.

Introduction

Let $\varphi\colon R \to S$ be a homomorphism of commutative rings. For a prime ideal $\mathfrak{p} \in \operatorname{Spec} R$, the residue field of the localization of R at $\mathfrak{p}$ is denoted $k(\mathfrak{p})$, and the ring $k(\mathfrak{p}) \otimes_R S$ is called the *fiber* of φ at $\mathfrak{p}$. The fibers of the canonical maps from R to its completions in the $\mathfrak{p}$–adic topologies are known as the *formal fibers* of R.

For various properties $\mathcal{P}$ of commutative rings, Grothendieck [**9**, (7.5)] considers the following Localization Problem: Suppose that φ is a flat homomorphism of noetherian rings and that the formal fibers of R have $\mathcal{P}$; if the fibers of φ at the primes of R contracted from maximal ideals of S have $\mathcal{P}$, is it then true that all the fibers of φ have $\mathcal{P}$? We obtain positive answers for properties related to "complete intersection," "Gorenstein," and "Cohen–Macaulay" in Section 4 and compare them to earlier ones in Section 5. However, the main thrust of this paper is to investigate the more general thesis formulated in the Abstract.

To make precise statements we describe some invariants which measure the complexity of the singularity of a local ring S with maximal ideal $\mathfrak{n}$ and residue field ℓ. The formula $\operatorname{cid} S = \operatorname{rank}_\ell \mathrm{H}_1(\mathbf{K}) - \operatorname{rank}_\ell \mathfrak{n}/\mathfrak{n}^2 + \dim S$, where $\mathbf{K}$ denotes the Koszul complex on a minimal set of generators of $\mathfrak{n}$, defines the *complete intersection defect* of S (introduced under the name *deviation* and denoted $d(S)$ in [**12**]). This integer is non-negative, and vanishes if and only if the $\mathfrak{n}$–adic

1991 *Mathematics Subject Classification.* 13H10, 14D40.
L.L.A. was partly supported by NSF Grant No. DMS-9102951.
H.B.F. was partly supported by the Danish Research Council.
"This paper is in final form and no version of it will be submitted for publication elsewhere."

completion $\widehat{S}$ is the quotient of a regular local ring by a regular sequence, that is, if S is a *complete intersection*, cf. also (1.2.a) below.

The *Cohen–Macaulay defect* of S is the non-negative integer $\operatorname{cmd} S = \dim S - \operatorname{depth} S$ (in [**8**, (16.4.9)] it is named *codepth* and denoted $\operatorname{coprof} S$). An equality $\operatorname{cmd} S = 0$ characterizes the *Cohen–Macaulay* rings S. In such S any two ideals generated by systems of parameters require the same number of irreducible ideals for their irredundant primary decompositions, [**16**, Theorem 3], cf. also (1.2.c). This integer is called the *type* of S and denoted $\operatorname{type} S$. *Gorenstein* rings may be described as the Cohen–Macaulay rings of type 1.

Each of the invariants introduced above is extended to arbitrary noetherian rings A by taking the supremum of its values on $A_{\mathfrak{p}}$, when $\mathfrak{p}$ ranges over the maximal ideals of A, or – equivalently, cf. (1.4) – over its prime ideals.

Now we can state our main result, to be proved in Section 3, based on the existence [**6**] of Cohen factorizations for local homomorphisms to complete local rings. The necessary preparation is carried out in Section 2.

MAIN THEOREM. *Let $\varphi\colon (R, \mathfrak{m}) \to (S, \mathfrak{n})$ be a flat local homomorphism of local rings.*

For any prime ideal $\mathfrak{q}$ in S and its inverse image $\mathfrak{p} = \mathfrak{q} \cap R$ in R there are inequalities:

(a) $$\operatorname{cid} S_{\mathfrak{q}}/\mathfrak{p}S_{\mathfrak{q}} + \operatorname{cid}(k(\mathfrak{q}) \otimes_S \widehat{S}) \leq \operatorname{cid} S/\mathfrak{m}S + \operatorname{cid}(k(\mathfrak{p}) \otimes_R \widehat{R}) .$$

(b) $$\operatorname{cmd} S_{\mathfrak{q}}/\mathfrak{p}S_{\mathfrak{q}} + \operatorname{cmd}(k(\mathfrak{q}) \otimes_S \widehat{S}) \leq \operatorname{cmd} S/\mathfrak{m}S + \operatorname{cmd}(k(\mathfrak{p}) \otimes_R \widehat{R}) .$$

(c) *If $S/\mathfrak{m}S$ and the formal fibers of R at $\mathfrak{p}$ are Cohen–Macaulay, then $S_{\mathfrak{q}}/\mathfrak{p}S_{\mathfrak{q}}$ and the formal fibers of S at $\mathfrak{q}$ are Cohen–Macaulay, and there is an inequality:*

$$\operatorname{type}(S_{\mathfrak{q}}/\mathfrak{p}S_{\mathfrak{q}}) \cdot \operatorname{type}(k(\mathfrak{q}) \otimes_S \widehat{S}) \leq \operatorname{type}(S/\mathfrak{m}S) \cdot \operatorname{type}(k(\mathfrak{p}) \otimes_R \widehat{R}) .$$

The inequalities of the Main Theorem are themselves limit cases of more general results. Suitably modified, the thesis stated in the Abstract is valid for many properties of *homomorphisms of finite flat dimension*, cf. [**5**] for extensions of (b) and (c). In fact, it is consideration of this situation which led us to the present point of view, but the proof of the Main Theorem in the flat case is technically less involved than that of the general result. It is also simpler than earlier proofs of Localization Theorems for $\mathcal{P}$ = "complete intersection" or "Gorenstein," and provides the first such theorem working in full generality for $\mathcal{P}$ = "Cohen–Macaulay."

1. Defects and type

In this section $(S, \mathfrak{n}, \ell)$ denotes a local ring. All the results below are known, but some arguments are sketched so as to make the text independent of the theory of dualizing complexes and canonical modules.

(1.1) COMPLETIONS. Complete intersection defect, Cohen–Macaulay defect, and type of Cohen–Macaulay rings are invariant under maximal-ideal-adic completions of local rings.

(1.2) REGULAR PRESENTATIONS. Assume that $S \cong Q/\mathfrak{b}$ with Q a regular local ring. If the ideal $\mathfrak{b}$ is minimally generated by $\nu_Q\mathfrak{b}$ elements, then

(a) $$\operatorname{cid} S = \nu_Q\mathfrak{b} - \operatorname{grade}_Q S \ :$$

the equivalent expression $\nu_Q\mathfrak{b} - \operatorname{ht}\mathfrak{b}$ is obtained for $\operatorname{cid} S$ in [**12**, Satz 1], cf. also [**14**, (21.1.iii)]; here and below grade denotes the maximal length of a regular sequence in the annihilator of a module, and ht stands for the height of an ideal.

From the Auslander–Buchsbaum Equality for the projective dimension $\operatorname{pd}_R S$, and from the Cohen–Macaulayness of regular rings, one easily sees that

(b) $$\operatorname{cmd} S = \operatorname{pd}_Q S - \operatorname{grade}_Q S\,.$$

It follows from (b) that S is Cohen–Macaulay if and only if $\mathfrak{b}$ is perfect, that is, when $\operatorname{pd}_R S = \operatorname{grade}_R S$. When this is the case, there is an equality

(c) $$\operatorname{type} S = \operatorname{rank}_\ell \operatorname{Tor}_p^Q(\ell, S) \ \text{ for } \ p = \operatorname{pd}_Q S\,.$$

Indeed, as the ring Q is regular the Tor in (c) is the last non-vanishing homology group of the Koszul complex on a set of generators of $\mathfrak{n}$. This group is isomorphic to the vector space $((\mathbf{x}) : \mathfrak{n})/(\mathbf{x})$, where $\mathbf{x}$ denotes a maximal S–regular sequence, cf. [**2**, (1.8)]. Assuming that S is Cohen–Macaulay, $\mathbf{x}$ is a system of parameters, and we see that $\operatorname{rank}_\ell((\mathbf{x}) : \mathfrak{n})/(\mathbf{x})$ does not depend on the choice of the system. On the other hand, this dimension is well known and easily seen to be equal to the number of irreducible $\mathfrak{n}$–primary components in an irredundant primary decomposition of $(\mathbf{x})$, that is, to the type of S.

Note that the preceding argument also establishes the invariance of the number of irreducible ideals in irredundant primary decomposition of parameter ideals in a Cohen–Macaulay local ring.

(1.3) FLAT EXTENSIONS. If $\varphi\colon (R, \mathfrak{m}) \to (S, \mathfrak{n})$ is a flat local homomorphism, then the following hold:

(a) [**4**, (3.6)] $$\operatorname{cid} S = \operatorname{cid} R + \operatorname{cid} S/\mathfrak{m}S\,.$$

(b) [**9**, (6.3.2)] $$\operatorname{cmd} S = \operatorname{cmd} R + \operatorname{cmd} S/\mathfrak{m}S\,.$$

(c) [**11**, (1.24)] The ring S is Cohen–Macaulay if and only if R and $S/\mathfrak{m}S$ have this property, and when this is the case there is an equality:

$$\operatorname{type} S = (\operatorname{type} R) \cdot (\operatorname{type} S/\mathfrak{m}S)\,.$$

Alternative references are provided by [**14**, (15.1.ii) and (23.3)] for (b); [**14**, (23.4.Proof)] for (c).

(1.4) LOCALIZATIONS. For a prime ideal $\mathfrak{q}$ in S there are equalities:

(a) [**4**, (3.8)] $\operatorname{cid} S_{\mathfrak{q}} \leq \operatorname{cid} S\,.$

(b) [**9**, (6.11.5)] $\operatorname{cmd} S_{\mathfrak{q}} \leq \operatorname{cmd} S\,.$

(c) [**11**, (6.16)] If S is Cohen–Macaulay, then so is $S_{\mathfrak{q}}$ and

$$\operatorname{type} S_{\mathfrak{q}} \leq \operatorname{type} S\,.$$

When S is complete use Cohen's Structure Theorem to present it in the form $Q/\mathfrak{b}$ with a regular local ring Q. The inequalities in (a) and (b) then follow because the right–hand sides of the first two formulas in (1.2) do not increase upon localization; some more care should be exercised in connection with (c), cf. (2.2). For a general ring S, note that by (1.1) its invariants are equal to the corresponding ones of $\widehat{S}$, then choose by faithful flatness a prime $\mathfrak{q}^*$ in $\widehat{S}$ lying over $\mathfrak{q}$, and apply (1.3) to the flat local homomorphism $S_{\mathfrak{q}} \to \widehat{S}_{\mathfrak{q}^*}$.

All local rings of the formal fibers of S are obtained as closed fibers of such homomorphisms by varying $\mathfrak{q}^*$ in $\operatorname{Spec} \widehat{S}$. Thus, when S is a complete intersection (respectively: Gorenstein, Cohen–Macaulay), its formal fibers have the same property, and this yields the more general statement below.

(1.5) FORMAL FIBERS. If S is a quotient of a complete intersection (respectively: Gorenstein, Cohen–Macaulay) local ring, then its formal fibers are complete intersections (respectively: Gorenstein, Cohen–Macaulay).

In particular, the formal fibers of a complete local ring are complete intersections, and hence also Gorenstein and Cohen–Macaulay. Of course, they are even geometrically regular, cf. e.g. [**14**, (32.3)], but this will not be needed here.

2. Deviation, imperfection, and canonical number

Let $(S, \mathfrak{n}, \ell)$ be a local noetherian ring. If $S \cong Q/\mathfrak{b}$ for some local ring Q, and the projective dimension $\operatorname{pd}_Q S$ is finite, then motivated by (1.2) we consider the following numbers associated with the Q–module S:

- the *deviation* $\operatorname{dev}_Q S = \operatorname{rank}_\ell \operatorname{Tor}_1^Q(\ell, S) - \operatorname{grade}_Q S$; due to the canonical isomorphism $\operatorname{Tor}_1^Q(\ell, S) \cong \ell \otimes_Q \mathfrak{b}$, it can also be defined as $\nu_Q \mathfrak{b} - \operatorname{grade}_Q S$.
- the *imperfection* $\operatorname{imp}_Q S = \operatorname{pd}_Q S - \operatorname{grade}_Q S$;
- the *canonical number* $\operatorname{cnn}_Q S = \operatorname{rank}_\ell \operatorname{Tor}_p^Q(\ell, S)\,,$ where $p = \operatorname{pd}_Q S$.

We start with a couple of elementary remarks.

(2.1) REMARK. $\operatorname{dev}_{\widehat{Q}} \widehat{S} = \operatorname{dev}_Q S$, $\operatorname{imp}_{\widehat{Q}} \widehat{S} = \operatorname{imp}_Q S$, and $\operatorname{cnn}_{\widehat{Q}} \widehat{S} = \operatorname{cnn}_Q S$.

(2.2) REMARK. If $\mathfrak{q}$ is a prime ideal of S, and $\mathfrak{q}' = \mathfrak{q} \cap Q$ is its inverse image in Q, then

(a) $\operatorname{dev}_{Q_{\mathfrak{q}'}} S_{\mathfrak{q}} \leq \operatorname{dev}_Q S\,.$

(b) $\operatorname{imp}_{Q_{\mathfrak{q}'}} S_{\mathfrak{q}} \leq \operatorname{imp}_Q S\,.$

(c) If the Q–module S is perfect, then so is the $Q_{\mathfrak{q}'}$–module $S_{\mathfrak{q}}$, and

$$\operatorname{cnn}_{Q_{\mathfrak{q}'}} S_{\mathfrak{q}} \leq \operatorname{cnn}_Q S\,.$$

Indeed, projective dimension does not go up and grade does not go down under localization: this establishes (b), and shows that if S is a perfect module of grade p over Q, then $S_{\mathfrak{q}}$ has the same property over $Q_{\mathfrak{q}'}$. If $\mathbf{F}$ is a minimal free resolution of the Q–module S, then $\mathbf{F}_{\mathfrak{q}'}$ is free resolution of the $Q_{\mathfrak{q}'}$–module $S_{\mathfrak{q}}$; since $\operatorname{Tor}_n^Q(\ell, S) \cong \ell \otimes_Q F_n$ for $n \in \mathbb{Z}$, there are (well known) inequalities:

$$\operatorname{rank}_{k(\mathfrak{q}')} \operatorname{Tor}_n^{Q_{\mathfrak{q}'}}(k(\mathfrak{q}'), S_{\mathfrak{q}}) \leq \operatorname{rank}_\ell \operatorname{Tor}_n^Q(\ell, S)$$

which now take care of (a) and the remaining part of (c)

Next we show that these numbers are invariant under deformation.

(2.3) PROPOSITION. *Let $P \xrightarrow{\pi} Q \xrightarrow{\varkappa} S$ be surjective homomorphisms of local rings, such that* $\operatorname{Ker}\pi$ *is generated by a P–regular sequence.*

When $\operatorname{pd}_Q S$ *is finite there are equalities:*

(a) $$\operatorname{dev}_P S = \operatorname{dev}_Q S\,.$$

(b) $$\operatorname{imp}_P S = \operatorname{imp}_Q S\,.$$

(c) $$\operatorname{cnn}_P S = \operatorname{cnn}_Q S\,.$$

PROOF. It suffices to establish the equalities when $Q = P/(x)$ for a non zero-divisor x in the maximal ideal $\mathfrak{r}$ of P.

Consider the standard change of rings spectral sequence:

$${}^2\mathrm{E}_{pq} = \operatorname{Tor}_p^Q(\ell, \operatorname{Tor}_q^P(Q, S)) \Rightarrow \operatorname{Tor}_{p+q}^P(\ell, S)\,.$$

Since $\operatorname{Tor}_q^P(Q, S)$ vanishes for $q \neq 0, 1$, and is isomorphic to S for $q = 0, 1$, it shows that $\operatorname{pd}_P S = p + 1$ when $p = \operatorname{pd}_Q S$, and also provides a "corner" isomorphism $\operatorname{Tor}_{p+1}^P(\ell, S) \cong \operatorname{Tor}_p^Q(\ell, S)$, which settles (c).

The obvious equality $\operatorname{grade}_P S = \operatorname{grade}_Q S + 1$ now yields (b).

To establish (a), we must show that the minimal number of generators of $\operatorname{Ker}(\varkappa\pi)$ is one more than that of $\operatorname{Ker}(\varkappa)$. If this is not the case, then the non zero-divisor x is contained in $\mathfrak{r}\operatorname{Ker}(\varkappa\pi)$. In such a situation [**20**, Theorem 5] yields isomorphisms:

$$\operatorname{Tor}_n^Q(\ell, S) \cong \bigoplus_{i \geq 0} \operatorname{Tor}_{n-2i}^P(\ell, S) \quad \text{for } n \in \mathbb{Z}\,,$$

which can only hold when $\operatorname{pd}_Q S = \infty$.

However, this is ruled out by our hypothesis. □

The module invariants considered above enter the proof of the Main Theorem through their links with the ring invariants of fibers of flat homomorphisms. The bridge is provided by Cohen factorizations of local homomorphisms, whose definition we recall next.

A local homorphism is said to be *weakly regular* if it is flat and its closed fiber is a regular ring. It is proved in [**6**, (1.1)] that for any local homomorphism $\varphi\colon R \to S$ the composition $\dot{\varphi}\colon R \to \widehat{R} \xrightarrow{\widehat{\varphi}} \widehat{S}$ factors as:

$$\begin{array}{ccccc} & & R' & & \\ & \dot{\varphi} \nearrow & & \searrow \varphi' & \\ R & & \xrightarrow[\dot{\varphi}]{} & & \widehat{S}\,, \end{array}$$

with $(R', \mathfrak{m}')$ a complete local ring, $\dot{\varphi}$ a weakly regular homomorphism, and φ' a surjective homomorphism.

Such a decomposition is called a *Cohen factorization* of $\dot{\varphi}$.

The next proposition is a crucial ingredient in the proof of the Main Theorem. In [**6**, (2.8) and (3.6)] its part (b) is derived from more general results; the proof given below avoids, in particular, an indirect use of the New Intersection Theorem.

(2.4) PROPOSITION. *Let $\varphi\colon (R, \mathfrak{m}) \to (S, \mathfrak{n})$ be a flat local homomorphism.*

For any Cohen factorization $R \to R' \xrightarrow{\varphi'} \widehat{S}$ of $\dot{\varphi}$ the projective dimension of the R'-module $\widehat{S}$ is finite, and the following hold:

(a) $$\operatorname{cid}(S/\mathfrak{m}S) = \operatorname{dev}_{R'} \widehat{S}\,.$$

(b) $$\operatorname{cmd}(S/\mathfrak{m}S) = \operatorname{imp}_{R'} \widehat{S}\,.$$

(c) *The ring $S/\mathfrak{m}S$ is Cohen–Macaulay if and only if the R'-module $\widehat{S}$ is perfect, and then*

$$\operatorname{type}(S/\mathfrak{m}S) = \operatorname{cnn}_{R'} \widehat{S}\,.$$

PROOF. By (1.1) we may assume S is complete. We write S instead of $\widehat{S}$, set $k = R/\mathfrak{m}$, and denote by $\overline{\varphi}'$ the surjective homomorphism $k \otimes_R \varphi'$.

Let $\mathbf{F}$ be an R'-free resolution of S. Using first the flatness of R' over R, and then the flatness of S over R, we get: $\mathrm{H}_i(k \otimes_R \mathbf{F}) = \operatorname{Tor}_i^R(k, S) = 0$ for $i > 0$. Thus, the complex $k \otimes_R \mathbf{F}$ is a free resolution of $\overline{S} = S/\mathfrak{m}S$ over $\overline{R}' = R'/\mathfrak{m}R'$. It follows that for any integer n there are isomorphisms:

(2.4.1) $$\operatorname{Tor}_n^{R'}(\ell, S) = \mathrm{H}_n(\ell \otimes_{R'} \mathbf{F}) \cong \mathrm{H}_n(\ell \otimes_{\overline{R}'} (k \otimes_R \mathbf{F})) = \operatorname{Tor}_n^{\overline{R}'}(\ell, \overline{S})\,.$$

As a first consequence, we obtain:

(2.4.2) $$\operatorname{pd}_{R'} S = \operatorname{pd}_{\overline{R}'} \overline{S}\,.$$

Since the ring $\overline{R}'$ is regular, the latter number is finite, say equal to p.

Choose next in $\operatorname{Ker} \overline{\varphi}'$ an $\overline{R}'$-regular sequence of length $g = \operatorname{grade}_{\overline{R}'} \overline{S}$. As $\operatorname{Ker} \overline{\varphi}' = (\operatorname{Ker} \varphi')\overline{R}'$, it can be lifted to a length g sequence in $\operatorname{Ker} \varphi'$. Since R'

is R–flat, this sequence is necessarily R'–regular by [**14**, (22.5.Corollary)]. Thus, $\operatorname{grade}_{R'} S \geq g = \operatorname{grade}_{\overline{R}'} \overline{S}$. On the other hand, there are relations:

$$\begin{aligned}\operatorname{grade}_{R'} S &\leq \dim R' - \dim S \\ &= (\dim R + \dim \overline{R}') - (\dim R + \dim \overline{S}) \\ &= \dim \overline{R}' - \dim \overline{S} \\ &= \operatorname{grade}_{\overline{R}'} \overline{S}\,,\end{aligned}$$

where the first equality stems from the flatness of R' and S over R, and the third one from the Cohen–Macaulayness of the regular local ring $\overline{R}'$. Thus, we have:

$$\operatorname{grade}_{R'} S = \operatorname{grade}_{\overline{R}'} \overline{S}\,. \tag{2.4.3}$$

By comparing (2.4.3) with (2.4.1) for $n = 1$, one sees that $\operatorname{dev}_{R'} S = \operatorname{dev}_{\overline{R}'} \overline{S}$. Since the ring $\overline{R}'$ is regular, (1.2.a) shows that $\operatorname{dev}_{\overline{R}'} \overline{S} = \operatorname{cid} \overline{S}$, and (a) follows.

Similarly, for (b) compare (2.4.2) and (2.4.3) to obtain $\operatorname{imp}_{R'} S = \operatorname{imp}_{\overline{R}'} \overline{S}$, and then apply (1.2.b) to get $\operatorname{imp}_{\overline{R}'} \overline{S} = \operatorname{cmd} \overline{S}$.

It follows from (b) that the perfection of the R'–module S is equivalent to the Cohen–Macaulayness of the ring $\overline{S}$, which we now assume. In this case (2.4.1) with $n = p$ yields $\operatorname{cnn}_{R'} S = \operatorname{cnn}_{\overline{R}'} \overline{S}$, and then (1.2.c) shows that the last number equals $\operatorname{type} \overline{S}$, as required for (c).

The proof of the proposition is now complete. □

A consequence of (a) will be needed at an early stage of the proof of the Main Theorem, so we note it here explicitly:

(2.5) Corollary. *If $\varphi\colon (R, \mathfrak{m}) \to (S, \mathfrak{n})$ is a flat local homomorphism of complete local rings, and the closed fiber of φ is a complete intersection, then in any Cohen factorization $R \to R' \xrightarrow{\varphi'} S$ the kernel of φ' is generated by a regular sequence.* □

3. Proof of the Main Theorem

For this section we fix the following notation: $\varphi\colon (R, \mathfrak{m}) \to (S, \mathfrak{n})$ is a flat local homomorphism of local rings, $\mathfrak{q}$ is a prime ideal of S, $\mathfrak{p} = \mathfrak{q} \cap R$ is the inverse image of $\mathfrak{q}$ in R, and $\varphi_\mathfrak{q}\colon R_\mathfrak{p} \to S_\mathfrak{q}$ is the induced local homomorphism.

First we establish a special case of the Localization Theorem for complete intersections.

Step 1. *If φ is weakly regular and both rings R and S are complete, then the closed fiber of $\varphi_\mathfrak{q}$ is a complete intersection.*

Proof. If both R and S are regular, then the closed fiber of $\varphi_\mathfrak{q}$ is the quotient of the regular local ring $S_\mathfrak{q}$ by the extension of the maximal ideal of the regular local ring $R_\mathfrak{p}$. The latter is generated by a regular sequence. Due to the flatness of $\varphi_\mathfrak{q}$, so is the former. Thus, the closed fiber of $\varphi_\mathfrak{q}$ is a complete intersection.

In the general case, choose by Cohen's Structure Theorem a surjective homomorphism $\pi\colon P \to R$ from a regular local ring P. By [**6**, (1.6)] there is a commutative diagram of local homomorphisms:

$$\begin{array}{ccc} P & \xrightarrow{\zeta} & Q \\ {\scriptstyle\pi}\downarrow & & \downarrow{\scriptstyle\varkappa} \\ R & \xrightarrow[\varphi]{} & S \end{array}$$

where Q is complete, ζ is weakly regular, $\varkappa$ is surjective, and the induced map $R \otimes_P Q \to S$ is an isomorphism.

It follows in particular that the ring Q is regular, cf. [**14**, (23.7.ii)], and that the rings $k(\mathfrak{q} \cap P) \otimes_P Q$ and $k(\mathfrak{q} \cap R) \otimes_R S$ are canonically isomorphic. This isomorphism induces one of the closed fiber of $\zeta_{\mathfrak{q}\cap Q}$ with that of $\varphi_{\mathfrak{q}}$, so that the assertion of the lemma follows from the special case already settled above. □

Next we prove the theorem in the complete case, when the invariants of the formal fibers are all trivial by (1.5).

STEP 2. *If R and S are complete local rings, then there are inequalities:*

(a) $$\operatorname{cid} S_{\mathfrak{q}}/\mathfrak{p}S_{\mathfrak{q}} \le \operatorname{cid} S/\mathfrak{m}S\,.$$

(b) $$\operatorname{cmd} S_{\mathfrak{q}}/\mathfrak{p}S_{\mathfrak{q}} \le \operatorname{cmd} S/\mathfrak{m}S\,.$$

(c) *If furthermore $S/\mathfrak{m}S$ is Cohen–Macaulay, then so is $S_{\mathfrak{q}}/\mathfrak{p}S_{\mathfrak{q}}$ and*

$$\operatorname{type} S_{\mathfrak{q}}/\mathfrak{p}S_{\mathfrak{q}} \le \operatorname{type} S/\mathfrak{m}S\,.$$

PROOF. The arguments will be given for Cohen–Macaulay defects. The other two assertions are obtained by manual or electronic changes in notation, accompanied by the corresponding switches to parts (a) or (c) of the quoted results.

Take a Cohen factorization $\varphi\colon R \xrightarrow{\dot\varphi} R' \xrightarrow{\varphi'} S$, set $\mathfrak{p}' = \mathfrak{q} \cap R$, and consider the induced factorization

$$\varphi_{\mathfrak{q}}\colon R_{\mathfrak{p}} \xrightarrow{\dot\varphi_{\mathfrak{p}'}} R'_{\mathfrak{p}'} \xrightarrow{\varphi'_{\mathfrak{q}}} S_{\mathfrak{q}}\,.$$

Writing $\widetilde{R}$, $\widetilde{R}'$, and $\widetilde{S}$ for the completions of the local rings $R_{\mathfrak{p}}$, $R'_{\mathfrak{p}'}$, and $S_{\mathfrak{q}}$ in their respective maximal-ideal-adic topologies, we obtain flat local homomorphisms:

$$\widetilde{\varphi}\colon \widetilde{R} \xrightarrow{\psi} \widetilde{R}' \xrightarrow{\vartheta} \widetilde{S}\,.$$

Denoting by $\widetilde{\mathfrak{p}}$ the maximal ideal of $\widetilde{R}$ and remarking that the closed fiber of $\widetilde{\varphi}$ is the completion of that of $\varphi_{\mathfrak{q}}$, we record the equality:

$$\operatorname{cmd} S_{\mathfrak{q}}/\mathfrak{p}S_{\mathfrak{q}} = \operatorname{cmd} \widetilde{S}/\widetilde{\mathfrak{p}}\widetilde{S}\,.$$

Next we take a Cohen factorization

$$\widetilde{R} \xrightarrow{\dot\psi} \widetilde{R}'' \xrightarrow{\psi'} \widetilde{R}'$$

of ψ, and note that $(\vartheta\psi')\dot{\psi}$ is one of $\widetilde{\varphi}$. Thus, (2.4.b) yields:

$$\operatorname{cmd} \widetilde{S}/\widetilde{\mathfrak{p}}\widetilde{S} = \operatorname{imp}_{\widetilde{R}''} \widetilde{S}.$$

The closed fiber $\widetilde{R}'/\widetilde{\mathfrak{p}}\widetilde{R}'$ of $\dot{\psi}$ is the completion of that of $\dot{\varphi}_{\mathfrak{p}'}$. By Step 1 the latter is a complete intersection, hence so is the former. By (2.5) the kernel of ψ' is generated by an $\widetilde{R}''$-regular sequence. Applications of (2.3.b), (2.1), and (2.2.b) yield:

$$\operatorname{imp}_{\widetilde{R}''} \widetilde{S} = \operatorname{imp}_{\widetilde{R}'} \widetilde{S} = \operatorname{imp}_{R'_{\mathfrak{p}'}} S_{\mathfrak{q}} \leq \operatorname{imp}_{R'} S.$$

Referring again to (2.4.b) we get:

$$\operatorname{imp}_{R'} S = \operatorname{cmd} S/\mathfrak{m}S.$$

To obtain the assertion of Step 2(b) it only remains to concatenate the (in)equalities above. □

Finally, we deal with the general case.

STEP 3. *For any φ there are inequalities:*

(a) $\operatorname{cid} S_{\mathfrak{q}}/\mathfrak{p}S_{\mathfrak{q}} + \operatorname{cid}(k(\mathfrak{q}) \otimes_S \widehat{S}) \leq \operatorname{cid} S/\mathfrak{m}S + \operatorname{cid}(k(\mathfrak{p}) \otimes_R \widehat{R}).$

(b) $\operatorname{cmd} S_{\mathfrak{q}}/\mathfrak{p}S_{\mathfrak{q}} + \operatorname{cmd}(k(\mathfrak{q}) \otimes_S \widehat{S}) \leq \operatorname{cmd} S/\mathfrak{m}S + \operatorname{cmd}(k(\mathfrak{p}) \otimes_R \widehat{R}).$

(c) *If $S/\mathfrak{m}S$ and the formal fibers of R at $\mathfrak{p}$ are Cohen–Macaulay, then $S_{\mathfrak{q}}/\mathfrak{p}S_{\mathfrak{q}}$ and the formal fibers of S at $\mathfrak{q}$ are Cohen–Macaulay, and there is an inequality:*

$$\operatorname{type}(S_{\mathfrak{q}}/\mathfrak{p}S_{\mathfrak{q}}) \cdot \operatorname{type}(k(\mathfrak{q}) \otimes_S \widehat{S}) \leq \operatorname{type}(S/\mathfrak{m}S) \cdot \operatorname{type}(k(\mathfrak{p}) \otimes_R \widehat{R}).$$

PROOF. Here again, we write down the argument for (b) only, and refer to the beginning of the preceding proof for details on the treatment of the other two assertions.

Choose by faithful flatness a prime ideal $\mathfrak{q}^* \in \operatorname{Spec} \widehat{S}$ which lies over $\mathfrak{q} \in \operatorname{Spec} S$, and set $\mathfrak{p}^* = \mathfrak{q}^* \cap \widehat{R}$. By the additivity (1.3.b) of Cohen–Macaulay defects on local flat extensions we now have the following simple computation:

$$\begin{aligned}\operatorname{cmd} S_{\mathfrak{q}}/\mathfrak{p}S_{\mathfrak{q}} + \operatorname{cmd} \widehat{S}_{\mathfrak{q}^*}/\mathfrak{q}\widehat{S}_{\mathfrak{q}^*} &= (\operatorname{cmd} S_{\mathfrak{q}} - \operatorname{cmd} R_{\mathfrak{p}}) + (\operatorname{cmd} \widehat{S}_{\mathfrak{q}^*} - \operatorname{cmd} S_{\mathfrak{q}}) \\ &= \operatorname{cmd} \widehat{S}_{\mathfrak{q}^*} - \operatorname{cmd} R_{\mathfrak{p}} \\ &= (\operatorname{cmd} \widehat{R}_{\mathfrak{p}^*} - \operatorname{cmd} R_{\mathfrak{p}}) + (\operatorname{cmd} \widehat{S}_{\mathfrak{q}^*} - \operatorname{cmd} \widehat{R}_{\mathfrak{p}^*}) \\ &= \operatorname{cmd} \widehat{R}_{\mathfrak{p}^*}/\mathfrak{p}\widehat{R}_{\mathfrak{p}^*} + \operatorname{cmd} \widehat{S}_{\mathfrak{q}^*}/\mathfrak{p}\widehat{S}_{\mathfrak{q}^*}.\end{aligned}$$

Because $\widehat{R}_{\mathfrak{p}^*}/\mathfrak{p}\widehat{R}_{\mathfrak{p}^*}$ is a localization of $k(\mathfrak{p}) \otimes_R \widehat{R}$, we get from (1.4.b) an inequality:

$$\operatorname{cmd} \widehat{R}_{\mathfrak{p}^*}/\mathfrak{p}\widehat{R}_{\mathfrak{p}^*} \leq \operatorname{cmd}(k(\mathfrak{p}) \otimes_R \widehat{R}).$$

On the other hand, the result of Step 2 and (1.1) show that:

$$\operatorname{cmd} \widehat{S}_{\mathfrak{q}^*}/\mathfrak{p}\widehat{S}_{\mathfrak{q}^*} \leq \operatorname{cmd} \widehat{S}/\widehat{\mathfrak{m}}\widehat{S} = \operatorname{cmd} S/\mathfrak{m}S.$$

Using these formulas to replace the corresponding terms of the last sum above, we obtain:

$$\operatorname{cmd} S_{\mathfrak{q}}/\mathfrak{p}S_{\mathfrak{q}} + \operatorname{cmd} \widehat{S}_{\mathfrak{q}^*}/\mathfrak{q}\widehat{S}_{\mathfrak{q}^*} \leq \operatorname{cmd} S/\mathfrak{m}S + \operatorname{cmd}(k(\mathfrak{p}) \otimes_R \widehat{R}).$$

By taking the supremum over the prime ideals $\mathfrak{q}^* \in \operatorname{Spec} \widehat{S}$ which lie over $\mathfrak{q}$, we see that part (b) of the Main Theorem follows from the last inequality. □

4. Localization Theorems

We first note a very special case of the Main Theorem.

(4.1) THEOREM. *Let $\varphi\colon (R, \mathfrak{m}) \to (S, \mathfrak{n})$ be a flat local homomorphism of local rings. Assume that the formal fibers of R and the closed fiber $S/\mathfrak{m}S$ of φ have one of the following properties:*

(CI) *complete intersection.*
(G) *Gorenstein.*
(CM) *Cohen–Macaulay.*

All fibers of φ and formal fibers of S then have the corresponding property. □

(4.2) REMARK. Assume that the formal fibers of R have property (CI) (respectively: (G), (CM)) above. The Main Theorem shows that if for some integer n the closed fiber $S/\mathfrak{m}S$ has complete intersection defect at most n (respectively: is Cohen–Macaulay of type at most n, has Cohen–Macaulay defect at most n), then the latter condition holds for all fibers of φ and all formal fibers of S.

Next we refine the Localization Theorem (4.1) by involving Serre's condition (S_n) and its variants, like Ischebeck's condition (G_n). Let A be a noetherian ring. For a fixed integer n, consider the properties:

(CI_n) if $\mathfrak{p} \in \operatorname{Spec} A$ has $\operatorname{depth} A_{\mathfrak{p}} < n$, then $A_{\mathfrak{p}}$ is a complete intersection.
(G_n) if $\mathfrak{p} \in \operatorname{Spec} A$ has $\operatorname{depth} A_{\mathfrak{p}} < n$, then $A_{\mathfrak{p}}$ is Gorenstein.
(S_n) if $\mathfrak{p} \in \operatorname{Spec} A$ has $\operatorname{depth} A_{\mathfrak{p}} < n$, then $A_{\mathfrak{p}}$ is Cohen–Macaulay.

Clearly, A has the absolute property $\mathcal{P}$ = "complete intersection" (respectively: "Gorenstein," "Cohen–Macaulay") precisely when it has the partial property $\mathcal{P}_n$ = "(CI_n)" (respectively: "(G_n)," "(S_n)") for all $n \in \mathbb{Z}$. Rings satisfying (G_n) are known as *n–Gorenstein* [**18**]; by restriction of language, we say A is *n–complete intersection* if it has (CI_n).

We need information on the behavior of the partial properties $\mathcal{P}_n$ under flat homomorphisms. It is derived from the following facts about the corresponding absolute properties $\mathcal{P}$, which are limit cases of the equalities (1.3).

(4.3) If $\varphi\colon (R, \mathfrak{m}) \to (S, \mathfrak{n})$ is a flat local homomorphism of local rings, then:

(1) If R and $S/\mathfrak{m}S$ have $\mathcal{P}$ then S has $\mathcal{P}$.
(2) If S has $\mathcal{P}$ then so does R.

When $\mathcal{P}_n$ = "(S_n)" the argument for the next lemma is indicated in [**9**, (7.3.4), (7.3.8)]; it is noted in [**17**, Proposition 3] that the same pattern applies to $\mathcal{P}_n$ = "(G_n)."

(4.4) LEMMA. *Let $\alpha\colon A \to B$ and $\beta\colon B \to C$ be flat homomorphisms of noetherian rings, and let $\mathcal{P}_n$ denote one of the conditions* (CI_n), (G_n), *or* (S_n). *The following then hold:*

(1) *If the fibers of α and β have $\mathcal{P}_n$ then so do the fibers of $\beta\alpha$.*

(2) *If β is faithfully flat and the fibers of $\beta\alpha$ have $\mathcal{P}_n$ then so do those of α.*

PROOF. If k is the quotient field of a localization of A at a prime ideal, the induced homomorphism $k \otimes_A B \to k \otimes_A C$ is flat (respectively, faithfully flat) along with β, and its fibers are among those of β. Thus, it suffices to show the following, proved in [**9**, (6.4.1)] for $\mathcal{P}_n =$"(S_n)" by an argument which applies equally well to the other cases, cf. [**18**, Proposition 1] or [**17**, Proposition 1] for $\mathcal{P}_n =$"(G_n)."

(1') If B and the fibers of β satisfy $\mathcal{P}_n$ then C satisfies $\mathcal{P}_n$.

(2') If β is faithfully flat and C satisfies $\mathcal{P}_n$ then so does B.

For $\mathfrak{q} \in \operatorname{Spec} C$ with $\operatorname{depth} C_\mathfrak{q} < n$, set $\mathfrak{p} = \mathfrak{q} \cap B$ and consider the induced flat local homomorphism $B_\mathfrak{p} \to C_\mathfrak{q}$. The additivity of depth on flat extensions,

$$\operatorname{depth} B_\mathfrak{p} + \operatorname{depth} C_\mathfrak{q}/(\mathfrak{p})C_\mathfrak{q} = \operatorname{depth} C_\mathfrak{q}\,,$$

cf. e.g. [**14, (23.3)**], implies that the depth $B_\mathfrak{p}$ and $C_\mathfrak{q}/(\mathfrak{p})C_\mathfrak{q}$ is smaller than n, hence both rings satisfy $\mathcal{P}$. By (4.3.1) then so does $C_\mathfrak{q}$, and this proves (1').

To prove (2'), start from a given $\mathfrak{p} \in \operatorname{Spec} R$ with $\operatorname{depth} B_\mathfrak{p} < n$ and choose $\mathfrak{q} \in \operatorname{Spec} C$ minimal over $\mathfrak{p}$. Using again the additivity of depth, one sees that $\operatorname{depth} C_\mathfrak{q} = \operatorname{depth} B_\mathfrak{p} < n$, hence $C_\mathfrak{q}$ has $\mathcal{P}$. By (4.3.2) so does $B_\mathfrak{p}$. □

Next we derive the refined form of the Localization Theorem from its primitive form (4.1) by the argument of [**15**, (2.4.Proof)], repeated in [**7**, (3.3.Proof)] to obtain a partial Localization Theorem for $\mathcal{P} =$"(S_n)" from one for $\mathcal{P} =$"Cohen–Macaulay." We reproduce it here for completeness.

(4.5) THEOREM. *Let $\varphi\colon (R, \mathfrak{m}) \to (S, \mathfrak{n})$ be a flat local homomorphism of local rings. If for some integer n one of the conditions* (CI_n), (G_n), *or* (S_n) *holds for the formal fibers of R and the closed fiber of φ, then it holds for all fibers of φ.*

PROOF. Consider first the case when R is complete. Since its formal fibers then have $\mathcal{P}$ by (1.5), if the Localization Theorem fails there is a prime $\mathfrak{p} \subsetneqq \mathfrak{m}$ such that $k(\mathfrak{p}) \otimes_R S$ does not satisfy $\mathcal{P}_n$. By Noetherian induction choose $\mathfrak{p}$ maximal among such primes. After replacing φ by the induced flat local homomorphism $R/\mathfrak{p} \to S/\mathfrak{p}S$, we change notation, and assume that R is a domain and all the fibers of φ but the generic one enjoy $\mathcal{P}_n$.

Thus, there is a prime $\mathfrak{q}$ in S lying over (0) in R, with $\operatorname{depth} S_\mathfrak{q} < n$ and for which the ring $(R_{(0)} \otimes_R S)_\mathfrak{q}$ does not have $\mathcal{P}$. Choose a non-zero element $x \in \mathfrak{m}$, and then a prime $\mathfrak{q}'$ in S, minimal over $(\mathfrak{q} + Sx)$. Set $\mathfrak{p}' = \mathfrak{q}' \cap R$, and consider the induced flat local homomorphism $\varphi_{\mathfrak{q}'}\colon R_{\mathfrak{p}'} \to S_{\mathfrak{q}'}$. Note that $\operatorname{depth} R_{\mathfrak{p}'} \geq 1$ as $x \in \mathfrak{p}'$, and hence:

$$\operatorname{depth} S_{\mathfrak{q}'}/\mathfrak{p}'S_{\mathfrak{q}'} = \operatorname{depth} S_{\mathfrak{q}'} - \operatorname{depth} R_{\mathfrak{p}'} \leq \operatorname{depth} S_\mathfrak{q} + \operatorname{ht}(\mathfrak{q}'/\mathfrak{q}) - \operatorname{depth} R_{\mathfrak{p}'} < n\,.$$

Because $S_{\mathfrak{q}'}/\mathfrak{p}'S_{\mathfrak{q}'}$ is a localization of the non-generic fiber $k(\mathfrak{p}') \otimes_R S$ of φ, the depth estimate above implies that it satisfies $\mathcal{P}$. Applying (4.1) we see that all the fibers of $\varphi_{\mathfrak{q}'}$ then have $\mathcal{P}$. This contradicts the assumption on its generic fiber, and thus finishes the argument for complete R.

For arbitrary R, consider the composition $R \xrightarrow{\rho} \widehat{R} \xrightarrow{\varphi^*} S^*$, where S^* denotes the $(\mathfrak{m}S)$–adic completion of S. The closed fiber of φ^* is canonically isomorphic with that of φ, hence all the fibers of φ^* satisfy $\mathcal{P}_n$ by the case already settled above. The fibers of ρ satisfy $\mathcal{P}_n$ by assumption. It follows from (4.4.1) that $\mathcal{P}_n$ holds for the fibers of $\varphi^*\rho = \sigma\varphi$, where $\sigma \colon S \to S^*$ is the completion map. As σ is faithfully flat, (4.4.2) implies that the fibers of φ have the same property, as was to be proved. □

5. Concluding remarks

Grothendieck's original approach to the Localization Problem begins with a careful analysis in [**9**, (7.3)] of the behavior under base change of the properties $\mathcal{P}$ under consideration. He then proves Localization Theorems assuming the induced residue field extension $R/\mathfrak{m} \to S/\mathfrak{n}$ is finite, [*idem*, (7.5.2)], or assuming the existence of resolution of singularities, [*idem*, (7.9.8)], cf. also [**13**, (2.1)].

The properties considered in (4.1), (4.2), and (4.5) satisfy Grothendieck's conditions, hence these results are known in the residually finite and characteristic zero cases.

The first unrestricted positive answer to the Localization Problem has been obtained by André [**1**] for "geometrically regular," using the André–Quillen homology theory of commutative algebras. Nishimura [**15**] has applied this result to prove Localization Theorems for $\mathcal{P}$ ="geometrically reduced" and $\mathcal{P}$ ="geometrically normal," and his method has been extended by Brezuleanu and Ionescu [**7**] to higher versions of normality, like "geometrically $(\mathrm{R}_n) + (\mathrm{S}_{n+1})$."

The Localization Problem for $\mathcal{P}$ ="complete intersection" is settled by Tabaâ [**19**], sharpening the method of [**13**, (4.6)]. Both papers rely on André–Quillen homology and the fact that the analog of (4.4) holds for the complete intersection property, cf. [**3**]. The proof of (4.1.CI) does not use André–Quillen homology.

The Localization Theorem for $\mathcal{P}$ ="Gorenstein" is reduced by Marot [**13**, (3.2)] to the complete case, established by Hall and Sharp [**10**, (3.3)] with the help of dualizing complexes. The proof of (4.1.G) does not depend on that theory.

For $\mathcal{P}$ ="Cohen–Macaulay" and its generalizations $\mathcal{P}$ ="(S_n)" and $\mathcal{P}$ ="of Cohen–Macaulay defect at most n," the only previous advance beyond the cases originally settled by Grothendieck is in small dimensions: When the dimension of R is at most 4 the original argument can be adapted to use "macaulayfications" in place of resolutions of singularities, cf. [**7**, (3.1), (3.3)]. The results of (4.1.CM), (4.2), and $(4.5.\mathrm{S}_n)$ provide the first unrestricted Localization Theorems for these properties.

References

1. M. André, *Localisation de la lissité formelle*, manuscripta math. **13** (1974), 297–307.
2. M. Auslander and D. A. Buchsbaum, *Codimension and multiplicity*, Ann. of Math. (2) **68** (1958), 625–657.
3. L. L. Avramov, *Flat morphisms of complete intersections*, Dokl. Akad. Nauk. SSSR **225** (1975), 11–14 (Russian); English transl.: Soviet Math. Dokl. **16** (1975), 1413–1417.
4. ———, *Homology of local flat extensions and complete intersection defects*, Math. Ann. **228** (1977), 27–37.
5. ——— and H.-B. Foxby, *Cohen–Macaulay properties of local homomorphisms*, in preparation.
6. ———, ———, and B. Herzog, *Structure of local homomorphisms*, J. Algebra (to appear).
7. A. Brezuleanu and C. Ionescu, *On the localization theorems and completions of* **P***-rings*, Rev. Roumaine Math. Pures Appl. **29** (1984), 371–380.
8. A. Grothendieck, *Éléments de géométrie algébrique.* Chapitre IV, Première Partie, Publ. Math. IHES **20** (1964).
9. ———, *Éléments de géométrie algébrique.* Chapitre IV, Seconde Partie, Publ. Math. IHES **24** (1965).
10. J. E. Hall and R. Y. Sharp, *Dualizing complexes and flat homomorphisms of commutative noetherian rings*, Math. Proc. Cambridge Philos. Soc. **84** (1978), 37–45.
11. J. Herzog and E. Kunz, *Der kanonische Modul eines Cohen–Macaulay Rings*, Lecture Notes in Math., vol. 238, Springer-Verlag, Berlin–New York, 1971.
12. R. Kiehl and E. Kunz, *Vollstañdige Durchschnitte und p-Basen*, Arch. Math. (Basel) **16** (1965), 348–362.
13. J. Marot, **P***-rings and* **P***-homomorphisms*, J. Algebra **87** (1984), 136–149.
14. H. Matsumura, *Commutative ring theory*, Cambridge Univ. Press, Cambridge, 1986.
15. J. Nishimura, *On ideal adic completions of noetherian rings*, J. Math. Kyoto Univ. (1981), 153–169.
16. D. G. Northcott, *On irreducible ideals in local rings*, J. London Math. Soc. **32** (1957), 82–88.
17. M. Paugam, *La condition G_q de Ischebeck*, C. R. Acad. Sci. Paris Sér. A **276** (1973), 109–112.
18. I. Reiten and R. Fossum, *Commutative n-Gorenstein rings*, Math. Scand. **31** (1972), 33–48.
19. M. Tabaâ, *Sur les homomorphismes d'intersection complète*, C. R. Acad. Sci. Paris Sér. I **298** (1984), 437–439.
20. J. Tate, *Homology of Noetherian rings and of local rings*, Illinois J. Math. **1** (1957), 14–27.

Department of Mathematics, Purdue University, West Lafayette, Indiana 47907

E-mail address: avramov@math.purdue.edu

Matematisk Institut, Københavns Universitet, Universitetsparken 5, DK–2100 København Ø, Denmark

E-mail address: foxby@math.ku.dk

Contemporary Mathematics
Volume **159**, 1994

A SIMPLE PROOF OF GROTHENDIECK'S THEOREM ON THE PARAFACTORIALITY OF LOCAL RINGS

FREDERICK CALL AND GENNADY LYUBEZNIK

All rings in this paper are commutative. Recall that a local ring is defined to be parafactorial, if its depth is ≥ 2 and the Picard group of its punctured spectrum is trivial. Grothendieck proved the following result on the parafactoriality of local rings [1,XI, 3.13ii]:

THEOREM. *Let A be a local ring of dimension ≥ 4 and let $\hat{A}$ be its completion with respect to the maximal ideal. If $\hat{A}$ is a complete intersection, then A is parafactorial.*

Grothendieck's proof is quite hard. It uses sophisticated techniques like formal schemes, etc. The purpose of this paper is to present a simple proof of this theorem. Our proof is, basically, Grothendieck's original proof stripped of unnecessary complications. The only somewhat less elementary ingredient is local duality for Gorenstein rings (LDGR), which we state here for the reader's convenience.

LDGR. If A is a local Gorenstein ring, $\mathcal{M}$ is the maximal ideal of A and M is a finitely generated A-module, then $H^i_{\mathcal{M}}(M) = Ext_A^{dim A - i}(M, A)^*$.

Here $H^i_{\mathcal{M}}(M)$ is the ith local cohomology module of M with support in $\mathcal{M}$ and $(-)^* = Hom_A(-, E)$, where E is the injective hull of the residue field of A in the category of A-modules. It should be pointed out that M has finite length in the category of A-modules if and only if M^* does.

LEMMA. *Let A be a local ring with maximal ideal $\mathcal{M}$.*

(i) A is parafactorial if and only if depth $A \geq 2$ and every ideal I of A, such that

(i_1) I_P is principal for every prime $P \neq \mathcal{M}$,

(i_2) I is not contained in any associated prime of A,

(i_3) $\mathcal{M}$ is not associated to I,

1991 Mathematics Subject Classification. Primary 13F15. Secondary 13D45

Second author supported by the NSF.

This paper is in final form and no version of it will be submitted for publication elsewhere.

is principal.

(ii) Let $A \to B$ be a flat local homomorphism with $\dim(B/\mathcal{M}B) = 0$. *If B is parafactorial, so is A.*

PROOF. (i) If A is parafactorial, then depth $A \geq 2$. Assume that depth $A \geq 2$. Then $H^i_{\mathcal{M}}(A) = 0$ for $i \leq 1$. Let X be the punctured spectrum of A. The exact sequence $0 \to H^0_{\mathcal{M}}(A) \to A \to H^0(X, \mathcal{O}_X) \to H^1_{\mathcal{M}}(A) \to 0$ implies that $H^0(X, \mathcal{O}_X) = A$. Let F be an invertible sheaf on X. Then $F = \tilde{M}|X$, where M is a suitable finitely generated submodule of $H^0(X, F)$. Let S be the set of the non-zero-divisors of A. Since depth $A \geq 2$, S contains elements of the maximal ideal of A. Hence every maximal ideal of A_S defines a point of X. Since A_S is a semilocal ring and $(M_S)_P = (A_S)_P$ for every maximal ideal P of A_S, we conclude that $M_S = A_S$. Let $\varphi : M \to A$ be a homomorphism that induces an isomorphism $M_S \to A_S$. Let $J \subset A$ be the image of φ. Then $\tilde{J}|X$ is a subsheaf of $\mathcal{O}_X$. If $I \subset A$ is an ideal, then $\tilde{I}|X$ is invertible if and only if I satisfies (i_1) and (i_2). The maximal ideal is not associated to I if and only if $I = H^0(X, \tilde{I}|X)$. Set $I = H^0(X, \tilde{J}) \subset H^0(X, \mathcal{O}_X) \simeq A$. Then $I \supset J$ is an ideal of A with $\tilde{I}|X \simeq \tilde{J}|X$ and $\mathcal{M}$ is not an associated prime of I. Since $J_P = I_P$ is principal for every prime $P \neq \mathcal{M}$ and $J \subset I$ does not belong to any associated prime of A, we conclude that $\tilde{I}|X$ is an invertible sheaf. Hence φ induces an isomorphism $F \to \tilde{I}|X$. We have proven that a sheaf F on X is invertible if and only if it is of the form $\tilde{I}$, where I is an ideal of A satisfying (i_1), (i_2) and (i_3). The ring A is parafactorial iff $\tilde{I} = \mathcal{O}_X$ for every such I. But $\tilde{I} = \mathcal{O}_X$ iff $H^0(X, \tilde{I}) = I = H^0(X, \mathcal{O}_X) = A$, that is, iff I is principal. This finishes the proof of (i).

(ii) Since B is flat over A, an A-regular sequence of elements of A is B-regular, so depth $B \geq$ depth A. If $a_1, \ldots, a_r$ is a maximal A-regular sequence of elements of A, then there exists an injection $A/\mathcal{M} \to A/(a_1, \ldots a_r)$. Since B is flat over A, there exists an injection $B/\mathcal{M}B \to B/(a_1, \ldots, a_r)B$. Since $\dim B/\mathcal{M}B = 0$, the B-regular sequence $a_1, \ldots, a_r$ is maximal. So, depth $B =$ depth $A \geq 2$, and all we have to prove is that every ideal I of A that satisfies the conditions of (i) is principal. This follows from the fact that IB also satisfies the conditions of (i), hence it is principal. But I and IB have the same minimal number of generators because, by flatness, $IB = I \otimes_A B$. This proves (ii) and the lemma.

PROOF OF THE THEOREM. By part (ii) of the lemma it is enough to prove that $\hat{A}$ is parafactorial, so we can assume that A is complete with respect to the maximal ideal. Let A be a quotient of a regular local ring C by a regular sequence of length s. Clearly, we can assume C complete as well. We use induction on s. If $s = 0$, then A is regular, hence A is a factorial, so part(i) of the lemma implies that A is parafactorial. This finishes the case where $s = 0$. Assume $s > 0$ and let B be the quotient of C by the ideal generated by the first $(s-1)$ elements. Then B is complete and $A = B/(a)$, where $a \in B$ is a regular element. Set $A_i = B/(a^i)$. Denote by X_i the punctured spectrum of A_i and by $\mathcal{O}^*_{X_i}$ the multiplicative sheaf of units of $\mathcal{O}_{X_i}$. Reduction modulo (a) gives a surjection $\mathcal{O}^*_{X_{i+1}} \to \mathcal{O}^*_{X_i}$ whose kernel is $a^i\mathcal{O}_{X_{i+1}} = \mathcal{O}_{X_1}$ with its usual additive structure. The corresponding short exact sequence gives the exact sequence $H^1(\mathcal{O}_{X_1}) \to PicX_{i+1} \to PicX_i \to H^2(\mathcal{O}_{X_1})$. But $H^i(\mathcal{O}_{X_1}) = H^{i+1}_{\mathcal{M}}(A) = 0$ for $0 < i < 3$, as depth $A > 3$. So reduction modulo (a) induces an isomorphism

$PicX_{i+1} \to PicX_i$.

To prove the theorem all we have to show is that every invertible sheaf F_1 on X_1 is isomorphic to $\mathcal{O}_{X_1}$. Let F_i be the unique invertible sheaves on $X_i (i > 1)$ with $F_i/aF_i = F_{i-1}$. We claim that $M_i = H^0(F_i)$ is a finite A_i-module. Indeed, let L_i be a finite submodule of M_i with $\tilde{L}_i|X_i \equiv \tilde{M}_i|X_i \equiv F_1$. By LDGR $H^1_{\mathcal{M}}(L_i) \equiv Ext_A^{dimA_i-1}(L_i, A_i)^*$. Since $\dim A_i - 1 > 0$ and L_i is free at every non-maximal prime, the Ext in question vanishes at every non-maximal prime, so it has finite length. Thus $H^1_{\mathcal{M}}(L_i)$ has finite length. Now the claim follows from the exact sequence $L_i \to H^0(X_i, \tilde{L}_i|X_i)(\equiv M_i) \to H^1_{\mathcal{M}}(L_i)$. As $H^1(F_1) \equiv H^2_{\mathcal{M}}(L_1) \equiv Ext_A^{\dim A-2}(L_1, A)^*$ and $\dim A - 2 > 0$, the same argument shows that $H^1(F_1)$ is a finite A-module.

Let $\tilde{\varphi}_{i,j} : F_{i+j} \to F_i$ be the reduction modulo a^j and let $\tilde{\psi}_{i,j} : F_i \to F_{i+j}$ be the composition of the isomorphism $F_i \equiv a^j F_{i+j}$ with the natural injection $a^j F_{i+j} \to F_{i+j}$. Let $\varphi_{i,j} : M_{i+j} \to M_i$ and $\psi_{i,j} : M_i \to M_{i+j}$ be the induced maps. Note that $\tilde{\varphi}_{i,j} \circ \tilde{\psi}_{i,j} : F_i \to F_i$ and $\varphi_{i,j} \circ \psi_{i,j} : M_i \to M_i$ are multiplications by a^j. Our next claim is that the inverse system $\ldots M_3 \to M_2 \to M_1$ satisfies the Mittag-Leffler condition. Set $B' = A \oplus (a)/(a^2) \oplus (a^2)/(a^3) \ldots$. Let H^1 be the graded B'-module $H^1(F_1) \oplus H^1(aF_2) \oplus H^1(a^2F_3) \ldots$. The exact sequence $0 \to a^i F_{i+1} \to F_{i+1} \to F_i \to 0$ gives the exact sequence $M_{i+1} \overset{\varphi_{i,1}}{\to} M_i \to H^1(a^i F_{i+1})$. Set $G_i = coker\varphi_{i,1} \subset H^1(a^i F_{i+1})$. Since a is B-regular, B' is the polynomial ring $A[t]$. The maps $\psi_{i+1,j} : M_{i+1} \to M_{i+1+j}$ and $\psi_{i,j} : M_i \to M_{i+j}$ induce a map $G_i = coker(\varphi_{i,1} : M_{i+1} \to M_i) \to coker(\varphi_{i+j,1} : M_{i+1+j} \to M_{i+j}) = G_{i+j}$, which we call the multiplication by t^j. This makes $G = G_1 \oplus G_2 \oplus \ldots$ into a graded submodule of H^1. Since $F_1 \equiv a^i F_{i+1}$, the multiplication by t^i induces an isomorphism $H^1(F_1) \equiv H^1(a^i F_{i+1})$. Since $H^1(F_1)$ is a finite A-module, H^1 is a finite B'-module. Since B' is Noetherian, G also is finite. Assume G is generated by elements of degree $< k$. Then for $i > k$ the multiplication by $t^j : G_i \to G_{i+j}$ is onto. Hence $M_{i+j} \to G_{i+j}$ maps $im\psi_{i,j}$ onto G_{i+j}. So M_{i+j} is generated by $ker(M_{i+j} \to G_{i+j+1}) = im\varphi_{i+j,1}$ and $im\psi_{i,j}$. If $j > i$, then $\varphi_{i,j} \circ \psi_{i,j} = 0$, as M_i is annihilated by a^j. So for $j > max\{i,k\}$, $im\varphi_{i,j} = im\varphi_{i,j}(im\varphi_{i+j,1}) = im\varphi_{i,j+1}$. This proves the Mittag-Leffler condition.

Let N_i be the stable submodule of M_i. Set $N = \underset{\leftarrow}{\lim} N_i = \underset{\leftarrow}{\lim} M_i$. Since B is complete with respect to the maximal ideal, it is also complete with respect to the ideal (a) so $B = \underset{\leftarrow}{\lim} A_i$. The A_i-module structures on N_i make N a B-module. For a suffficiently large i the exact sequence $0 \to F_j \to F_{i+j} \to F_i \to 0$ gives the exact sequence $0 \to M_j \to M_{i+j} \to N_i \to 0$. Since $\underset{\leftarrow}{\lim}$ is left-exact, we get the exact sequence $0 \to N \to N \to N_i \to 0$, where exactness on the right follows from the Mittag-Leffler condition. Here $N \to N$ is the multiplication by a^i since it is induced by $\varphi_{-,j} \circ \varphi_{-,j}$. Hence the kernel of the natural surjection $N \to N_i$ is $a^i N$.

It follows that N is (a)-adically separated. Since $N/aN \equiv N_1$ is finite, N is finite as well. Hence the set $X' \subset SpecB$ on which N is invertible is open. Since $\tilde{\varphi}_{i,j}$ are surjective, $\tilde{N}_i|X_i \equiv \tilde{M}_i|X_i \equiv F_i$. If $Q \supset (a)$ is a non-maximal prime of B, then $(N/a^j N)_Q = (N_j)_Q = (F_j)_Q = (A_j)_Q = (B/a^j B)_Q$ for all j, so $N_Q \equiv B_Q$, i.e. $Q \in X'$. If P is a prime of B with $\dim B/P > 1$, then $N_Q \equiv B_Q$ for any non-maximal prime $Q \supset P + (a)$, so $N_P \equiv (N_Q)_P \equiv B_P$, i.e. $P \in X'$. Hence $X - X'$, where X is the punctured spectrum of B, is a closed (i.e. finite) set of primes $P_1, P_2, \cdots \notin V(a)$ with $\dim B/P_i = 1$. Let $F' = N|X'$ and let F be the pushforward of F' to all of X. The stalk of F at P_j is $H^0(X_{P_i}, F'|X_{P_i})$ where $X_{P_i} = SpecB_{P_i} - \{P_i\}$. Since B is complete, it is catenary, so $\dim B_{P_i} \geq 4$. By induction on s, B_{P_i} is parafactorial, so $F'|X_{P_i} \equiv \mathcal{O}_{X_{P_i}}$. In the exact sequence $H^0_{P_i}(B_{P_i}) \to B_{P_i} \to H^0(\mathcal{O}_{X_{P_i}}) \to H^1_{P_i}(B_{P_i})$ the end terms are 0, as depth $B_{P_i} > 1$. So the stalk of F at P_i is B_{P_i}, and F is invertible on all of X. By induction on s, B is parafactorial, so $F \equiv \mathcal{O}_X$. So $F_1 \equiv \tilde{N}_1|X_1 \equiv (\tilde{N}/a\tilde{N})|X_1 \equiv F/aF$ (since $P_i \notin V(a)) \equiv \mathcal{O}_X/a\mathcal{O}_X \equiv \mathcal{O}_{X_1}$. Q.E.D.

For the reader's convenience, we include a standard consequence of the above result [1,XI,3.14]:

COROLLARY (P. Samuel's conjecture). Let A be a local ring and $\hat{A}$ its completion with respect to the maximal ideal. If $\hat{A}$ is a complete intersection and for each prime ideal P of A of height ≤ 3, A_P is factorial, then A is factorial.

PROOF. A_P is factorial if dim $A_P = 1$, hence A has R_1. Since $\hat{A}$ is Cohen-Macaulay, so is A, that is, A has S_2. Now by Serre's criterion A is normal, hence a domain. So, it is enough to prove that every height one prime of A is principal. We use induction on dim A, the case dim $A \leq 3$ being trivial.

Assume dim $A \geq 4$. Let $Q \subset A$ be a height one prime. We claim Q is locally principal on the punctured spectrum. Indeed, let $P \supset Q$ be a nonmaximal prime. If ht $P \leq 3$, then A_P is factorial, so Q_P is principal. If ht $P \geq 4$, let $P' \subset \hat{A}$ be a minimal prime of $P\hat{A}$. Then $\hat{A}_{P'}$ is a complete intersection and $\dim \hat{A}_{P'} = \dim A_P \geq 4$. By the preceding theorem, $\hat{A}_{P'}$ is parafactorial. By part (ii) of the lemma, A_P also is parafactorial, so by part (i) of the lemma, Q_P is principal. This proves the claim. Since by the preceding theorem and part (ii) of the lemma A is parafactorial, it follows from part (i) of the lemma, that Q is principal. Q.E.D.

REFERENCES

1. A. Grothendieck, *Cohomologie locale des faisceaux cohérents et théorèmes de Lefschetz locaux et globaux* (SGA2), North-Holland, Amsterdam, 1968.

Department of Mathematics, Clark University, Worcester, MA 01610
Department of Mathematics, University of Minnesota, Minneapolis, MN 55455

Contemporary Mathematics
Volume **159**, 1994

Resolutions With A Given Hilbert Function

HARA CHARALAMBOUS AND E. GRAHAM EVANS, JR.

In his remarkable 1890 paper Hilbert [**Hi**] used the projective resolution of a graded module M, over $k[x_1, \dots, x_n]$ to create its Hilbert function. Recently there has been some progress in the other direction. That is, we fix a Hilbert function for a graded module and then survey the possible projective resolutions that occur. In general there are many different possible resolutions. If

$$0 \longrightarrow \oplus_d R[-d]^{n_d} \longrightarrow \cdots \longrightarrow \oplus_d R[-d]^{0_d} \longrightarrow M \longrightarrow 0$$

is a typical minimal resolution, the graded betti numbers of M denoted by $\beta_{m\,d}$ are m_d. We partially order two resolutions F^α and F^β with graded betti numbers $\alpha_{m\,d}$ and $\beta_{m\,d}$ respectively by $F^\alpha \geq F^\beta$ if $\alpha_{m\,d} \geq \beta_{m\,d}$ for all m and d.

We concentrate on the cyclic module case. If the dimension of the ring is two then the situation is completely understood. We show that there are some rather simple restrictions on the degrees of the generators and relations in a resolution which must be satisfied (Theorem 1) and if we have a set of degrees satisfying those relations how to produce an ideal generated by monomials that has such a resolution (Theorem 2). This enables us to find all possible sequences of degrees of generators and relations that can exist for a given Hilbert function. These results are related to work of Briançon [**B**] and Iarrobino [**I**] on the Hilbert scheme of $C[x, y]$. Their work considers rings which are not graded but then examines the Hilbert series of the associated graded ring using the grading coming from powers of the ideal generated by x and y. These results are implicit in the paper of Campanella [**Cp**] and in the paper by Sauer [**Sa**]. We present our version here as it is shorter and more direct. It also gives an algorithm for finding the various possible resolutions with a given Hilbert function and for creating explicit ideals I so that R/I has a given resolution type. For dimension three or

---

1991 Mathematics Subject Classification. Primary 18G10. Secondary 13D10, 13D40, 14C05.

Project partially sponsored by the National Security Agency under Grant Number MDA904-92- H-3017. The United States Government is authorized to reproduce and distribute reprints notwithstanding any copyright notation hereon.

This paper is in final form and no version of it will be submitted for publication elsewhere.

greater the situation is not well understood. In particular in those cases there are always possible Hilbert functions of cyclic modules so that there are incomparable minimal sets of graded betti numbers. We show the existence of such examples in Theorem 3. The general question of determining the smallest possible resolutions of finite length modules is one that has attracted our attention for some time. The smallest known resolutions come from maximal R sequences. These can only arise if the generating function of the Hilbert function factors as a product of terms $1 + t + t^2 + \ldots + t^n$. What kinds of smallest resolution can occur for more general Hilbert functions seems to be an interesting question.

We would like to thank the organizers of the conference at Mt. Holyoke for the stimulating series of lectures they assembled.

We will temporarily restrict ourselves to the two dimensional case. That is $R = k[a,\ b]$ where k is a field and a and b are indeterminates of degree one. Let R/I be a finite length graded module and

$$0 \longrightarrow R[-t_{21}] \oplus \cdots \oplus R[-t_{2n}] \longrightarrow R[-t_{10}] \oplus \cdots \oplus R[-t_{1n}] \longrightarrow R \longrightarrow R/I \longrightarrow 0$$

be a minimal graded projective resolution of R/I. This gives us two lists of numbers, the t_{2i} where i goes from 1 to n and the t_{1i} where i goes from 0 to n. We order the t_{1i} and the t_{2i} so that they are nondecreasing as i increases.

THEOREM 1. *Let $R = k[a,\ b]$ be a two dimensional graded regular ring and R/I a finite length graded module with minimal resolution and lists of t_{2i} and t_{1i} positive integers as above. Then 1) $\sum_{i=0}^{n} t_{1i} = \sum_{i=1}^{n} t_{2i}$ and 2) $t_{1i} < t_{2i}$ for all i.*

PROOF. In order to show the first claim we use the Hilbert polynomial rather than the Hilbert series. Or perhaps more simply we compute the dimension over k of the dth graded piece of R/I from the resolution. Since R is dimension two, the Hilbert polynomial of R evaluated at d is simply $1 + d$, the number of monomials in R of degree d. Similarly the Hilbert polynomial of $R[-n]$ is $1+d-n$. Thus the Hilbert polynomial of R/I can be computed from the resolution above as

$$1 + d - \sum_{i=0}^{n}(1 + d - t_{1i}) + \sum_{i=1}^{n}(1 + d - t_{2i}).$$

But R/I has finite length so eventually the dth graded piece of R/I is zero. Thus the Hilbert polynomial is simply 0. Expanding the sum above shows that $\sum_{i=0}^{n} t_{1i} = \sum_{i=1}^{n} t_{2i}$ as desired.

If the second claim were false, then $t_{1i} \geq t_{2i}$ for some i. Since our resolution is a minimal one, no nonzero entry of the matrices giving the maps can have degree 0 or less. Thus there are i syzygies with degrees $t_{21}, \ldots, t_{2i}$ that are relations on at most the first i generators of I that have degree $t_{10}, \ldots, t_{1(i-1)}$. But in a dimension two ring an ideal generated by i elements has exactly $i - 1$ minimal relations. This gives the desired contradiction. $\square$

Needless to say this argument is very special to the two dimensional case. We know of no similar elementary criteria for the higher dimensional cases. However in this case the criteria are not only necessary but are sufficient to yield an ideal I generated by monomials with the resolution of R/I having the desired terms.

THEOREM 2. *Let $R = k[a, b]$ and let positive integers t_{1i} and t_{2i} be given so that 1) $\sum_{i=0}^{n} t_{1i} = \sum_{i=1}^{n} t_{2i}$ and 2) $t_{1i} < t_{2i}$ for all i. Then there is an ideal I generated by monomials in a and b so that the minimal resolution of R/I has terms exactly the $R[-t_{1i}]$ in the first position and $R[-t_{2i}]$ in the second position. The resulting R/I will have finite length.*

PROOF. We let the zeroth generator, m_0, of I be $a^{t_{10}}$. The next generator, m_1, must have degree t_{11} and have the relation on the two generators have degree t_{21}. If we let $m_1 = a^{t_{10}-(t_{21}-t_{11})}b^{t_{21}-t_{10}}$, then m_1 has degree t_{11} and the relation $b^{t_{21}-t_{10}}m_0 - a^{t_{21}-t_{10}}m_1$ is a minimal relation of degree t_{21}. We continue this pattern of lowering the exponent on a and raising it on b in order to make the degrees of the next generator and the relation it introduces be correct. If the ith generator is $m_i = a^{\alpha}b^{\beta}$, then the $i+1$st generator is $m_{i+1} = a^{\alpha-(t_{2i+1}-t_{1i+1})}b^{\beta-(t_{1i}-t_{2i+1})}$. It is clear that the degree of the new generator is t_{1i+1} and that the relation introduced has degree t_{2i+1}. Since $t_{2i+1} > t_{1i+1} \geq t_{1i}$, the degree in a does go down while the degree of b goes up. Condition 1) assures that the degree of a in m_n, the last generator, is 0. This assures that R/I has finite length as desired. □

COROLLARY 1. *Let $R = k[a, b]$ and let t_{1i} and t_{2i} be sets of numbers satisfying the conditions of theorem 1. Then*

1) if $t_{1i} = t_{2j}$ for some i and j and if we form new lists by deleting t_{1i} and t_{2j} from the old lists, these new lists satisfy the conditions of theorem 1 and thus yield an ideal whose resolution has the desired terms.

2) if there is an i and a q so that $t_{1i} < q < t_{2i}$ and if we insert q in both lists, then the new lists satisfy the conditions of the theorem and yield an ideal whose resolution has the desired terms.

PROOF. One only needs to check that the conditions of the theorem are still satisfied. Condition 1 is immediate since we have deleted or inserted the same term in both sums. Condition 2 is also easy. For part 1) we note that if $t_{1i} = t_{2j} = q$ then $i > j$. Thus when delete q from both lists, the new correspondence is between elements of the first list t_{1k} and elements of the second list t_{2s} where $s \geq k$. Since $t_{1k} < t_{2k} \leq t_{2s}$ we are done. For part 2) we observe that when we reorder we find j so that $t_{1j} < q \leq t_{1j+1}$ and $t_{2k-1} \leq q < t_{2k}$. We insert q in the two lists at those points. Thus we get the string of inequalities:

$$t_{1k} \leq \ldots \leq t_{1i} \leq \ldots \leq t_{1j} < q < t_{2k} \leq \ldots \leq t_{2i} \leq \ldots \leq t_{2j}.$$

Since the only changes in the comparisons is that we will compare t_{1k} with q, t_{1l+1} with t_{2l} for l from k to $j-1$, and, finally, q with t_{2j}, the desired inequalities hold. □

This corollary allows us to move around the possible graded betti numbers for a fixed Hilbert series. In particular we can delete common terms until the two lists are disjoint or insert terms until each $t_{2i} = t_{1i} + 1$ to get the smallest or biggest possible graded betti numbers.

COROLLARY 2. *Let $R = k[a,\ b]$ and let $H(t)$ be a Hilbert series for a cyclic module. Then there is a unique smallest set of graded betti numbers that yields that series and there is a unique biggest set of graded betti numbers that yields that series. In both cases the resulting ideal can be selected to be generated by monomials in a and b.*

PROOF. By Macaulay [**M**] there is a monomial ideal that has Hilbert series $H(t)$. We note that the Hilbert series is unchanged when we add (or delete) a generator of the ideal and a relation on those generators of the same degree. The corollary follows from the above. □

Perhaps a simple example will clarify the process. In ring dimension 2, $H(t) = (1 - \sum \beta_{1d} t^d + \sum \beta_{2d} t^{2d})/(1-t)^2$. If one takes as Hilbert series the polynomial $1 + 2t + 3t^2 + t^3$, then after multiplying by $(1-t)^2$ to see the numerator in the description of the Hilbert series we get $1 - 3t^3 + t^4 + t^5$. Thus one sequence of t_{1i} is (3, 3, 3) while the sequence of t_{2i} is (4, 5). This yields the ideal $I = (a^3,\ a^2b,\ b^3)$. The Hilbert series could also have come from the numerator polynomial being $1 - (3t^3 + t^4) + 2t^4 + t^5$ which yields lists (3, 3, 3, 4) and (4, 4, 5). This yields the ideal $I = (a^3,\ a^2b,\ ab^2,\ b^4)$. It is interesting to note that the presenting matrices for the ideals that we create this way have a very similar structure. That is, they all have powers of b down the main diagonal and powers of a down the diagonal just below the main diagonal. This makes it easy to check that the ideal generated by the minors of maximal size has both a power of a and a power of b and thus generates an ideal of height two [as it must]. In fact this enables us to see that the relations on any subset consisting of the first n generators also has such a presentation. We used this remark implicitly in the proof of theorem 2 to insure that the new relation we added really did give us, along with the earlier relations, a full set of relations on the generators. The matrices of relations for our two example ideals are:

$$\begin{pmatrix} -b & 0 \\ a & -b \\ 0 & a^2 \end{pmatrix} \text{ and } \begin{pmatrix} -b & 0 & 0 \\ a & -b & 0 \\ 0 & a & -b^2 \\ 0 & 0 & a \end{pmatrix} \text{ respectively.}$$

We note that if all the powers of a are just a itself, then we are in the case of the biggest resolution. If also all the powers of b are just b itself, then we are resolving a power of the maximal ideal.

Going from the lists (3, 3, 3) and (4, 5) to the lists (3, 3, 3, 4) and (4, 4, 5) as in the above example shows part 2) of Corollary 2 at work while going in the opposite direction shows part 1) of Corollary 2 at work.

Now we pass to the dimension three or higher case. Here the situation is much more confusing. Essentially we are working on the Hilbert scheme of modules with a given Hilbert series. This scheme can have many different components. On each component there will be a "generic" resolution type which will be smallest. Unfortunately there may be no relation between the smallest resolutions as you go from component to component. On the other hand as you go from component to component these smallest resolutions can degenerate in many different ways. Surprisingly there is a worst degeneration leading to the biggest resolution.

There is a well understood procedure for going from the Hilbert function to information about the betti numbers. Stanley [**St**] and Avramov and Buchweitz [**A-B**] have exploited it in some detail for finite length modules. Briefly the relation between the generating Hilbert function $f(t) = \sum_d \dim_k M_d\, t^d$ and the graded betti numbers for modules over $R = k[x_1, \ldots, x_n]$ is given by $f(t) = (\sum_{m=0}^{n}(-1)^m \beta_{m\,d} t^d)/(1-t)^n$. If M has finite length then $f(t)$ is just a polynomial. In that case we can multiply it by $(1-t)^n$ to get the numerator. This gives information on the graded betti numbers. Unfortunately one can only see the sum over all m of the $(-1)^m \beta_{m\,d}$ for each d. We remark that this is the same as taking the alternating sum in the betti diagrams along the diagonals that go from the upper right to the lower left. These sums are the same for any resolution with the same Hilbert function. So there can be lots of cancellation and even if there is no cancellation, one cannot tell how the sum distributes over the possible places except if the sum is negative then the terms belong in odd indexed places and if the sum is positive then the terms belong in even indexed ones. Nevertheless this procedure is enough to show that if the length of M is odd then the sum of the betti numbers must be at least 2^n. This is consistent with the Buchsbaum-Eisenbud conjectures [**B-E**] which ask that the betti numbers in the resolution of a finite length module over a regular local ring of dimension n be at least as large as those of the residue field. That is the ith betti number should be at least $\binom{n}{i}$.

THEOREM 3. *Let $R = k[x_1, \ldots, x_n]$ with n at least 3. Then there is a Hilbert series for a cyclic finite length R module and two incomparable smallest sets of graded betti numbers for that Hilbert series.*

PROOF.. First we begin with the dimension three case. We let $R = k[x_1, x_2, x_3]$ and consider the Hilbert function, $f(t)$ where $f(0) = 1$, $f(1) = 3$, $f(2) = 4$, $f(3) = 2$, $f(4) = 1$ and $f(d) = 0$ for all d greater than 4. Let $I_1 = (x_2x_3, x_1x_3, x_1^3, x_2^3, x_3^3)$ and $I_2 = (x_3^2, x_2^2, x_1^2x_2, x_1^2x_3, x_1^5)$. As is easily checked, R/I_1 and R/I_2 have the above Hilbert function. We claim that the minimal projective resolutions of R/I_1 and R/I_2 are incomparable ones, in the sense that there is no smaller resolution sitting under both of them. Indeed the betti diagram of R/I_1 is:

total	1	5	6	2
0 :	1	–	–	–
1 :	–	2	1	–
2 :	–	3	4	1
3 :	–	–	–	–
4 :	–	–	1	1.

while that of R/I_2 is:

total	1	5	6	2
0 :	1	–	–	–
1 :	–	2	–	–
2 :	–	2	4	–
3 :	–	–	–	1
4 :	–	1	2	1.

We remark that the notation for the betti diagram is that of the computer program Macaulay. The entries in the ith column correspond to ith syzygies. An entry in the ith column and jth row is of degree $i+j$ where we begin counting with zero for both the rows and the columns. So according to this diagram $\beta_{m\,d}$ is the entry in the mth column and $d-m$th row.

The two sequences of graded betti numbers are smallest for that Hilbert series. The proof of this is somewhat tedious. Instead we will prove a slightly weaker fact, that there is no resolution smaller than both of them. Since one can pass to any minimal element under the resolutions of R/I_1 and R/I_2 in the partial order on resolutions with the above Hilbert function, this is enough to prove our claim that there are incomparable smallest resolutions corresponding to the same Hilbert series.

Suppose that the finite length module R/I has the same Hilbert function as R/I_1, but a smaller resolution than R/I_1 and R/I_2. The following remark will be useful in the higher dimensional case: in R/I_1, the images of $x_1^2x_2^2$ and x_3^2 generate the socle. While for R/I_2 the socle is generated by the images of x_1^4 and $x_1x_2x_3$. Thus in the first case there are third syzygies of degree 4+3 and 2+3 while in the second there are third syzygies of degree 4+3 and 3+3. In general in dimension three an element of degree d annihilated by the maximal ideal gives a third syzygy of degree d+3. Since R/I has finite length, it must have a third syzygy. The only way the resolution of R/I can be smaller than the two we are considering is if there were only one third syzygy (in degree 7). This implies that the ideal is Gorenstein, because it is Cohen-Macaulay and has highest betti number one. But the Hilbert function isn't symmetric. Since Hilbert functions of Gorenstein modules of finite length are symmetric, see for example Stanley's book **[St, page 59]**, we get a contradiction. Thus there is no smallest graded betti series for this Hilbert function.

For the higher dimensional case one case use the same Hilbert series. Thus if the dimension of R is $n = m + 3$ and R/I is a module with this Hilbert series, then I will have to have m linearly independent forms in degree one which we can take to be $x_4, \dots, x_n$. We now form new ideals I_1 and I_2 by taking the old ideals and adjoining the variables $x_4, \dots, x_n$. Once again the first ideal will have elements annihilated by the maximal ideal in degrees 4 and 2 while the second will have them in degree 4 and 3. These will give nth syzygies in degrees $4+n$ and $2+n$ in the first case and in degrees $4+n$ and $3+n$ in the second. Thus to have an ideal with a smaller resolution than both of these one must have only one nth syzygy, forcing the ideal to be Gorenstein and contradicting the lack of symmetry in the Hilbert function. □

Even though we cannot expect to have a resolution with minimal graded betti numbers for fixed Hilbert function, the case is quite different as mentioned for the resolution with maximal graded betti numbers. The next results are known in characteristic zero. Given the Hilbert function of a cyclic module of finite length, independently Hulett [**Hu**] and Bigatti [**Bi**] showed that there is a resolution with maximal graded betti numbers. More generally we expect that there is a biggest resolution for all finitely generated modules. That is, once you are told that M is the image of a free module with fixed number of generators (of fixed degrees), then there is a resolution which is greater than or equal to any other resolution with the given Hilbert function. Hulett [**Hu1**] is working on this as part of her Ph. D. thesis at Illinois. The discrepancy between the smallest and maximal cases could be interpreted as M. Green pointed out as follows. One can make a "generic" change of variables and then take the initial ideals. In terms of betti numbers this has the effect of increasing them. This is because the change of variables has no effect on the Hilbert scheme while taking initial ideals corresponds to a deformation and the betti numbers can only go up. So this procedure leads to a degenerate point on the same component of the Hilbert scheme where the largest behaviour is found. The various components meet at a point where the largest overall behaviour is found. We remark that these "generic initial ideals" in case of characteristic 0 are easy to resolve (see Eliahou and Kervaire [**E-K**] for example) and play a critical role in the study of the biggest resolutions.

In our example of noncomparable minimal resolutions the betti numbers, the sum of the graded betti numbers, are the same. We believe though that this is not the general case, and that there should be examples of Hilbert functions with incomparable minimal betti numbers.

In response to remarks by Eisenbud and Buchweitz this summer we wrote a short program that treated various polynomials in one variable as the generating functions of Hilbert functions of finite length modules. The program then computed the forced twists in the resolution and the sum of the forced betti numbers coming from those twists. We used the program to treat over 50 million polynomials as generating functions in dimension 6. Of these only 22 had the sum of their betti numbers less than the desired $2^6 = 64$. Unfortunately it is very difficult to find the smallest possible betti sequences for a given Hilbert function as an earlier part of this article points out. However the surprisingly small number of possible "bad" Hilbert functions seems to indicate that there is a stronger result waiting to be proved.

REFERENCES

[A-B] L. Avramov and R.-O. Buchweitz, *Lower bounds for betti numbers*, preprint of University of Toronto Department of Mathematics.

[Bi] A. M. Bigatti, *Upper Bounds for the Betti Numbers of a Given Hilbert Function*, to appear, Comm. in Alg.

[B] J. Briançon, *Description de* $Hilb^n C\{x,y\}$, Invent. Math. **41** (1977), 45-89.

[B-E] D. Buchsbaum and D. Eisenbud, *Algebra structures for finite free resolutions and some structure theorems for ideals of codimension 3*, American J. of Mathematics **99, 3** (1977), 447–485.

[Cp] G. Campanella, *Standard Bases of Perfect Homogeneous Polynomial Ideals of Height 2*, J. of Algebra **101** (1986), 47-60.

[E-K] S. Eliahou and M. Kervaire, *Minimal resolutions of some monomial ideals*, J. of Alg. **129** (1990), 1-25.

[Hi] D. Hilbert, *Über die Theorie der algebraischen Formen*, Math. Annalen **36** (1890), 473-534.

[Hu] H. A. Hulett, *Maximal betti numbers with a given Hilbert function*, to appear, Comm. in Alg.

[Hul] H. A. Hulett, *thesis*, Univ. of Illinois, Urbana Champaign.

[I] A. Iarrobino, *Deforming Complete Intersection Artin Algebras. Appendix: Hilbert Functions of* $C[x,y]$, Proc. Symp. Pure Math. **40, 1** (1983), 593-608.

[M] F.S.Macaulay, *Some Properties of Enumeration in the Theory of Modular Systems*, Proc. London Math. Soc. **26** (1927), 531-555.

[Sa] T. Sauer, *Smoothing Projectively Cohen-Macaulay Space Curves*, Math. Ann. **272** (1985), 83-90.

[St] R. Stanley, *Combinatorics and Commutative Algebra*, Birkhaüser, Boston, 1983.

DEPARTMENT OF MATHEMATICS, SUNY AT ALBANY, ALBANY, NY 12222

DEPARTMENT OF MATHEMATICS, UNIVERSITY OF ILLINOIS, URBANA, IL 61801

Contemporary Mathematics
Volume **159**, 1994

Complete Ideals in Algebra and Geometry

STEVEN DALE CUTKOSKY

ABSTRACT. We begin by summarizing the theory of factorization of complete ideals in normal local rings of dimension two. We then develop a framework in higher dimensional local rings in which factorization and related questions can be studied. A number of examples are given showing the complexity of complete ideal theory in higher dimensions.

The theory of complete ideals was begun by Zariski in 1938 [**Z1**] as the arithmetic analogue of complete linear systems in projective geometry. There is a very good theory for normal local rings of dimension two. Zariski proved that unique factorization of complete ideals into products of indecomposables holds in regular local rings of dimension two. For normal local rings of dimension two there are sufficent conditions for a good theory of factorization of complete ideals. To obtain necessary and sufficient conditions for a good factorization theory conditions such as complete or with algebraically closed residue field must be imposed. In sections one and two, which are expository, we present this theory. We also give some related results on Hilbert functions of valuations. The natural extensions of these Theorems to dimension three are false.

In section three, we develop a framework in which complete ideals may be studied in higher dimensions. Complete ideals and their factorizations can be analyzed by the characteristic cone of a morphism. This is essentially the technique used by Lipman [**L1**] in dimension two. However, there is extra structure available in dimension two which can not be used in higher dimensions. Because of this, developing a theory in higher dimension requires new techniques and the results are quite different. The advantage of this approach is that understanding the myriad interesting new phenomena occuring in higher dimensional regular local rings R can be reduced to understanding the possible shapes of cones in

---

1991 *Mathematics Subject Classification*. Primary 13C, 14B.

Partially supported by NSF

This paper is in final form, and no version of it will be submitted for publication elsewhere

finite dimensional real vector spaces which can occur as characteristic cones of birational projective morphisms onto spec(R).

For instance, the notion of semi-unique factorization of complete ideals (Definition 7) can be easily understood in this language. A cone is semi-factorial if it is rational polyhedral, with linearly independent vertices. Semi-unique factorization of complete ideals in R is equivalent to all characteristic cones of birational projective morphisms onto spec(R) being semi-factorial (Theorem 18). In the case of dimension two, this specializes to give Lipman's Theorem that the characteristic cone of a projective birational morphism from a normal scheme to a regular local ring of dimension two is always semi-factorial. However, we exhibit examples of characteristic cones showing that semi-unique factorization of complete ideals fails in dimension larger than two. Some of these cones deviate wildly from being semi-factorial.

Our theory in this paper seeks to explain why the situation in dimension three is so much different from dimension two, and to give methods which allow one to understand what kinds of new behavior one can expect in higher dimensions. As a demonstration of this line of thought, we give an example of infinitely many normal ideals in a regular local ring of dimension three which have the same Rees valuations, but distinct blowups (Example 2 of section 3). This answers a question of Heinzer ([**H**], page 21).

1. Complete ideals and Linear systems with base conditions

We begin with definitions which can be found in [**ZS**] and [**L1**]. Suppose that (R, m) is a normal local ring, with quotient field K.

DEFINITION 1. ([**ZS**], Appendix 4) Suppose that $I \subset R$ is an ideal. $z \in K$ is integrally dependent on I if it satisfies an equation of the form:

$$z^q + a_1 z^{q-1} + \ldots + a_q = 0,\ a_i \in I^i.$$

The set of $z \in K$ which are integrally dependent on I form an R ideal $\overline{I}$ called the integral closure of I, or the completion of I.

Complete ideals are the local analogue of linear systems with base conditions in projective geometry. Zariski apparently came up with the idea of complete ideals as a method of understanding such linear systems. Linear systems with base conditions can be constructed as follows. Let V be a nonsingular projective variety, D an effective divisor on V, $f : T \to V$ a projective birational morphism with T nonsingular. A divisor E on T is called exceptional if every irreducible component P of E satisfies codim$(f(P), V) \geq 2$. Suppose that E is an effective exceptional divisor on T. The "base condition" is

$$f_*\mathcal{O}_T(-E) \subset \mathcal{O}_V.$$

This induces an inclusion

$$\Gamma(V, f_*\mathcal{O}_T(-E) \otimes \mathcal{O}_V(D)) \subset \Gamma(V, \mathcal{O}_V(D))$$

The zero loci of the sections in $\Gamma(V, f_*\mathcal{O}_T(-E) \otimes \mathcal{O}_V(D))$ form a sub linear system of $\mid D \mid$. This is a "linear system with base conditions".

Computing this linear system can be very difficult. However, Zariski showed that the local analogue has a very nice solution on a nonsingular surface.

The local analogue of this problem is to localize, and consider the case where $V = \text{spec}(R)$. Then $\Gamma(T, \mathcal{O}_T(-E)) \subset R$ is a height two ideal. Zariski found that these ideals are exactly the (height ≥ 2) complete ideals. In fact, a further analysis ([L1]) shows that if R is normal, $f : X \longrightarrow \text{spec}(R)$ is a birational projective morphism with X normal, and $I\mathcal{O}_Y$ is invertible, then $\Gamma(Y, I\mathcal{O}_Y) = \overline{I}$.

The product of m-primary complete ideals is in general not complete. There is a *-product $I * J = \overline{IJ}$. If $f : X \longrightarrow \text{spec}(R)$ is a birational projective morphism with X normal, and $I\mathcal{O}_Y$, $J\mathcal{O}_Y$ are invertible, then $\Gamma(Y, IJ\mathcal{O}_Y) = I * J$.

2. Theory of complete ideals in dimension two

In this section, we summarize the theory of factorization of complete ideals in dimension two.

Let R be a normal local ring with maximal ideal m, $m(R)$ be the semi-group of m-primary complete ideals with *-product.

There is a very strong factorization Theorem by Zariski for complete ideals in regular local rings of dimension two.

Zariski used classical results on birational geometry in dimension two and the translation of local geometry into ideal theory outlined in section one, to prove the following theorem.

THEOREM 2. *(Zariski,* **[Z1]**, *1938, Theorems 7.1, 12.1;* **[ZS]**, *1960, Appendix 5) Suppose that R is a regular local ring of dimension two. Then*

(1) *products of complete ideals are complete.*
(2) *$m(R)$ has unique factorization (into indecomposables).*

This theorem has been generalized by Lipman to rational singularities.

THEOREM 3. *(Lipman* **[L1]**, *1969, Theorems 7.1, 20.1, section 25) Suppose that R is a normal local domain of dimension two. Then*

(1) *If R is a rational singularity, then products of complete ideals are complete.*
(2) *If R is complete and has algebraically closed residue field, then $m(R)$ has unique factorization if R is a UFD.*

Characterization of R satisfying 1) and 2) of Zariski's Theorem. The converse to Lipman's Theorems are true.

THEOREM 4. *(***[C4]***, Theorem 1) Suppose that (R, m) is an analytically normal local domain of dimension two, and $k = R/m$ is algebraically closed. Then R has a rational singularity iff the product of any two complete ideals is complete.*

The if part of Theorem 4 is false if k is not algebraically closed. An example is given in **[C4]**.

THEOREM 5. *([**C2**], Theorem 5, [**C3**], Corollary to Theorem 4) Suppose that (R, m) is a complete normal local domain of dimension two with algebraically closed residue field. Then $m(R)$ has unique factorization iff R is a UFD.*

The only if part of Theorem 5 is true with no restriction on the residue field [**C3**]. An example showing that an algebraically closed residue field is necessary for the if part of Theorem 5 to be true is given in [**C3**].

Condition (N) and Semi-factoriality. Muhly and Sakuma [**MS**] considered a less stringent condition than factoriality for $m(R)$, which they called condition (N). This is the weakest condition ensuring a good factorization theory of complete ideals (in dimension two). We first introduce some notation, and then give a definition of condition (N) which is equivalent to Muhly and Sakuma's definition.

Given a normal local ring R, and a discrete valuation v of R, let

$$I_n(v) = \{g \in R \mid v(g) \geq n\}.$$

A discrete valuation v of R is called a prime divisor of the second kind if the center of v on R is m, and if the residue field of the valuation ring of v has transcendence degree $\dim(R) - 1$ over R/m.

DEFINITION 6. (Muhly and Sakuma) A normal local two dimensional ring R satisfies condition (N) if $\oplus_{n \geq 0} I_n(v)$ is a finitely generated R algebra for all prime divisors of the second kind v.

Göhner [**G**] later determined the proper notion of factoriality in $m(R)$ which is equivalent to condition (N).

DEFINITION 7. Let G be a commutative semi-group with cancellation law. $g \in G$ with $g \neq 0$ is called extremal if g has no inverse in G, and if a factorization $ng = a + b$ implies that $sa = qg$ and $tb = pg$ for suitable integers a, b, p, q. Two extremal elements x and y are called equivalent, $x \sim y$, if there are positive integers m and n such that $nx = my$. G is semi-factorial if to each $g \in G$ with $g \neq 0$, and such that g has no inverse, there is an integer $n > 0$ such that ng is a sum of extremal elements, and this factorization is unique in the following sense: If $ng \sim a_1 + \ldots + a_s$, a_i extremal $a_i \not\sim a_j$ for $i \neq j$, and $mg = b_1 + \ldots + b_r$, b_i extremal, $b_i \not\sim b_j$ if $i \neq j$, then $r = s$ and $a_i \sim b_i$ after reindexing.

THEOREM 8. *(Göhner [**G**], 1975, Theorem 3.10) Suppose that (R, m) is a m-adically complete normal local domain of dimension two. Then R satisfies condition (N) if and only if $m(R)$ is semi-factorial.*

Göhner conjectured (in the introduction to [**G**]) that a complete normal local domain R of dimension two satisfies condition (N) if and only if the divisor class group of R, $\mathrm{Cl}(R)$, is a torsion group. Göhner proved $\Leftarrow$ of this conjecture ([**G**], Theorem 4.6). We recently gave a proof of $\Rightarrow$ to prove the conjecture.

THEOREM 9. *(Theorem 4* **[C3]***) Suppose that R is a complete normal local domain of dimension 2. Then R satisfies condition (N)* $\iff$ *Cl(R) is a torsion group.*

This Theorem gives a surprising connection between height one and height two ideals.

A Hilbert function type problem. Theorem 9 shows that for a normal local ring R of dimension two and a prime divisor v of R, $\oplus_{n\geq 0} I_n(v)$ is in general not a finitely generated R algebra. As such, we might expect that the function

$$\text{length}_R(R/I_n(v))$$

is not very well behaved for large n.

However, at least in equicharacteristic zero, $\oplus_{n\geq 0} I_n(v)$ has a Hilbert function as if it were finite generated, which is a consequence of the following Theorem.

THEOREM 10. *(Corollary to Theorem 9* **[CS]***) Let R be an excellent two dimensional equicharacteristic zero normal local ring, and let v be a prime divisor of the second kind on R. Then there exists a quadratic polynomial P(n) and a periodic function* $\lambda(n)$ *such that* $\ell(R/I_n(v)) = P(n) + \lambda(n)$ *for all* $n >> 0$.

Theorem 10 follows from Theorem 11 below, which is the local analogue of our solution with Srinivas to the Riemann Roch problem for projective surfaces. To obtain Theorem 10 from Theorem 11, we need only consider a resolution of singularities S of spec(R) such that the center of v on S is a divisor E. Then take $D = -E$.

THEOREM 11. *(Theorem 9* **[CS]***) Let R be an excellent two dimensional equicharacteristic zero normal local ring. Let* $f : S \longrightarrow \text{spec}(R)$ *be a resolution of singularities, and let* $D \neq 0$ *be an exceptional divisor on S. Then there is a quadratic polynomial P(n) and a periodic function* $\lambda(n)$ *such that*

$$\textit{length}(R/\Gamma(S, \mathcal{O}_S(nD))) = P(n) + \lambda(n)$$

for all $n >> 0$.

In **[CS]** a counterexample is given to the conclusions of Theorem 10 when the residue field has positive characteristic, of positive transcendence degree over the prime field.

3. Complete ideals in higher dimensions and characteristic cones

All of the Theorems in section two about complete ideals in dimension two are false in dimension three.

Counterexamples to unique and semi-unique factorization of complete m-primary ideals are in **[C1]**, **[C2]**. An example of a product of complete ideals which is not complete in a three dimensional regular local ring is given in **[Hu]**. A counterexample to the existence of a good Hilbert function for a prime divisor of the second kind in a three dimensional normal local ring is given in **[CS]**.

In this section we give a framework in which complete ideals and their factorization in higher dimensions can be studied.

The method is to study linebundles which are generated by global sections. This generalizes the approach of Lipman [**L1**] in dimension two.

The following definitions are a minor adaption from Kleiman's paper [**K**]. Our schemes are relative to a local ring instead of a field. The characteristic cone of a morphism was first defined by Hironaka [**Hi**].

For simplicity of exposition, we will assume throughout this section that (R, m) is a normal local ring, essentially of finite type over an algebraically closed field k. Let $S = \text{spec}(R)$. Let X be a normal S-scheme and $f : X \longrightarrow S$ a birational projective map.

Let $C(R)$ be the semi-group of complete ideals in R of height larger than one, with $*$-product. Let $m(R)$ be the semi group of m-primary complete ideals in R with $*$-product.

DEFINITION 12.

(1) Given $\mathcal{L} \in \text{Pic}(X)$, $\mathcal{L}$ is numerically equivalent to zero if $(\mathcal{L} \cdot C) = 0$ for all closed integral curves $C \subset f^{-1}(m)$. Define

$$\text{Pic}^n(X/S) = \{\mathcal{L} \in \text{Pic}(X) \mid \mathcal{L} \text{ is numerically equivalent to zero}\}.$$

(2) Define

$$A^1_{\mathbf{Q}}(X/S) = (\text{Pic}(X)/\text{Pic}^n(X/S)) \otimes \mathbf{Q}, \text{ and}$$
$$A^1(X/S) = A^1_{\mathbf{Q}}(X/S) \otimes \mathbf{R}.$$

(3) Define the ample cone $\overset{o}{P}(X/S)$ to be the cone in $A^1(X/S)$ generated by ample line bundles.

(4) Define the characteristic cone of f, $\tilde{P}(X/S) \subset A^1(X/S)$ to be the cone generated by the images of the invertible sheaves on X which are generated by their global sections. Let $\tilde{P}_{\mathbf{Q}}(X/S) = \tilde{P}(X/S) \cap A^1_{\mathbf{Q}}(X/S)$.

When T is a projective scheme over spec(k), we can define a linebundle $\mathcal{L}$ on T to be numerically equivalent to zero if $(\mathcal{L} \cdot C) = 0$ for every projective curve C on T. Then we can define $A^1(X/S)$, $\overset{o}{P}(T/k)$, $\tilde{P}(T/k)$ as in definition 12.

Let $\rho(X/S) = \dim_{\mathbf{R}}(A^1(X/S)$. By definition, we have an inclusion

$$A^1(X/S) \rightarrow A^1(f^{-1}(m)/k).$$

This vector space has finite dimension by IV.4.3 [**K**]. Hence $\rho(X/S) < \infty$.

We recall two more definitions from [**K**]. A boundary component of a cone K is a maximal cone contained in K − interior(K). The cells of K are defined by descending induction on dim K. A cone C is a cell of K if $C = K$ or if C is a boundary component of some bigger cell of K.

A Stein factor of f is a factorization $X \xrightarrow{g} W \longrightarrow S$ such that $g_*\mathcal{O}_X = \mathcal{O}_W$. The map g^* on Pic induces a natural inclusion $g^*(\tilde{P}(W/S)) \longrightarrow \tilde{P}(X/S)$.

THEOREM 13. *([**K**]) Suppose that S and X are as in Definition 1. Then*

(1) *interior*$(\tilde{P}(X/S)) = \overset{o}{P}(X/S)$.
(2) $\dim_{\mathbf{R}} \tilde{P}(X/S) = \dim_{\mathbf{R}} A^1(X/S) = \rho(X/S)$.
(3) *The map* $W \mapsto g^*(\overset{o}{P}(W/S))$ *is an order isomorphism of the isomorphism classes of Stein factors of f ordered by domination onto the cells of $\tilde{P}(X/S)$ ordered by inclusion.*

PROOF. This follows as in IV.5.1 [**K**].

The fundamental locus of f is the reduced subscheme of S on which f is not an isomorphism.

LEMMA 14. *$\tilde{P}(X/S)$ is generated as a cone by the classes of $I\mathcal{O}_X$ such that $I \subset R$ is a complete ideal and $I\mathcal{O}_X$ is invertible. If R is factorial, R/I can be taken to be supported on the fundamental locus of f.*

PROOF. Suppose that $\mathcal{L} \in \operatorname{Pic}(X)$ is generated by global sections. Choose a nonzero section $\sigma \in \Gamma(X, \mathcal{L}^{-1})$. σ induces an inclusion $\mathcal{L} \longrightarrow \mathcal{O}_X$. Let $I = f_*\mathcal{L} \subset f_*\mathcal{O}_X = R$.

In the case where R and X are regular, $A^1(X/S)$ is quite simple, as the next Lemma shows. The characteristic cone $\tilde{P}(X/S)$ is also very simple in the case where R has dimension two (Theorem 19). However, the cone becomes extremely complicated when R has dimension larger than two (Theorem 20 and Examples 1-3).

LEMMA 15. *Suppose that R and X are regular, $n = \rho(X/S)$, and $E_1, \ldots, E_n$ are the prime exceptional divisors for f. Then $\{E_1, \ldots, E_n\}$ generate Pic(X) and $\{E_1, \ldots, E_n\}$ is a $\mathbf{R}$-basis for $A^1(X/S)$.*

PROOF. Given $\mathcal{L} \in \operatorname{Pic}(X)$, we have prime divisors D_j on X which are not exceptional, and integers a_i, b_j such that $\mathcal{L} \cong \mathcal{O}_X((\sum_i a_iE_i) + (\sum_j b_jD_j))$. Let $D'_j = f(D_j)$. Note that $\mathcal{O}_S(\sum_j b_jD'_j) \cong R$. $\mathcal{L} \cong \mathcal{L} \otimes f^*(\mathcal{O}_S(-\sum_j b_jD'_j)) = \mathcal{O}_X(\sum_i c_iE_i)$ for some integers c_i. This shows that the E_i generate Pic(X) and $A^1(X/S)$.

It remains to show that there is no dependence relation on the E_i. Let $d = \dim(R)$. Suppose that $\sum_i a_iE_i$ is numerically equivalent to zero. Let $\mathcal{L}$ be a very ample line bundle on X, and let $T_1, \ldots, T_{d-2}$ be the zero locuses of $d-2$ independent general global sections of $\Gamma(X, \mathcal{L})$. Let $W = T_1 \cdots T_{d-2}$. Then W is nonsingular of dimension two, and each $E_i \cdot W$ is an integral curve C_i by Bertini's Theorem. But $\sum a_iC_i$ is numerically equivalent to zero on W, and this divisor is the exceptional locus for $f \mid W$, so that the intersection matrix $(C_i \cdot C_j)$ is negative definite (c.f. [**Mu**] or Lemma 14.1 [**L1**]), and thus each $a_i = 0$.

Note that $\tilde{P}_{\mathbf{Q}}(X/S)$ is a commutative semi-group with cancellation law. The only invertible element in $\tilde{P}_{\mathbf{Q}}(X/S)$ is zero.

THEOREM 16. *The characteristic cone $\tilde{P}_{\mathbf{Q}}(X/S)$ has semi-unique factorization if and only if there are elements $v_1, \ldots, v_n \in \tilde{P}_{\mathbf{Q}}(X/S)$ with $n = \rho(X/S)$*

such that

$$\tilde{P}_{\mathbf{Q}}(X/S) = \sum_{i=1}^{n} \mathbf{Q}_+ v_i$$

PROOF. Suppose that $\tilde{P}_{\mathbf{Q}}(X/S) = \sum_{i=1}^{n} \mathbf{Q}_+ v_i$. Then the extremal elements are the nonzero elements of the rays $\mathbf{Q}_+ v_i$, and $\tilde{P}_{\mathbf{Q}}(X/S)$ is semi-factorial, since $\{v_1, \ldots, v_n\}$ is a basis for $A^1_{\mathbf{Q}}(X/S)$.

Conversely, suppose that $\tilde{P}_{\mathbf{Q}}(X/S)$ is semi-factorial. Let $v_1, \ldots, v_m$ be a set of inequivalent extremal elements. Suppose that $v_1, \ldots, v_m$ are linearly dependent in $A^1_{\mathbf{Q}}(X/S)$. We will derive a contradiction. After reindexing the v_i, there is a relation $a_1 v_1 = a_2 v_2 + \ldots + a_m v_m$ for some integers a_i. If $a \geq \max\{| a_2 |, \ldots, | a_m |\}$, then $a_1 v_1 + a v_2 + \ldots + a v_m = (a_2 + a) v_2 + \ldots + (a_m + a) v_m$, a contradiction to the assumption that $\tilde{P}_{\mathbf{Q}}(X/S)$ is semi-factorial. Since the extremal elements of $\tilde{P}_{\mathbf{Q}}(X/S)$ generate $\tilde{P}(X/S)$, and hence generate $A^1_{\mathbf{Q}}(X/S)$ as a rational vector space, there are exactly n inequivalent extremal elements.

LEMMA 17. *Suppose that R is regular and $I \in C(R)$ ($I \in m(R)$). The following are equivalent.*

(1) *I is extremal in $C(R)$ (in $m(R)$).*
(2) *$I\mathcal{O}_Y$ is extremal in $\tilde{P}_{\mathbf{Q}}(Y/S)$ for every normal Y and projective morphism $g : Y \longrightarrow S$ such that $I\mathcal{O}_Y$ is invertible (and Y is the blow up of an m-primary ideal).*
(3) *$I\mathcal{O}_Y$ is extremal in $\tilde{P}_{\mathbf{Q}}(Y/S)$, whenever Y is the normalization of the blowup of I.*

PROOF. We will first show 1) $\Rightarrow$ 2). By Lemma 14, $I\mathcal{O}_Y$ not extremal in some $\tilde{P}_{\mathbf{Q}}(Y/S)$ implies $I^n\mathcal{O}_Y = JK\mathcal{O}_Y$ for some $n > 0$ and complete ideals J and K which are not equivalent to I. Hence $\overline{I^n} = J * K$ and I is not extremal.

We will next show 2) $\Rightarrow$ 1). If I is not extremal, then $\overline{I^n} = J * K$ for some $n > 0$, and ideals J, K not equivalent to I. If Y is normal such that $J\mathcal{O}_Y$ and $K\mathcal{O}_Y$ are invertible, we have $I^n\mathcal{O}_Y = JK\mathcal{O}_Y$, which implies that $I\mathcal{O}_Y$ is not extremal in $\tilde{P}_{\mathbf{Q}}(Y/S)$.

2) $\Leftrightarrow$ 3) follows from Theorem 13, 3).

THEOREM 18. *Suppose that R is regular. $C(R)$ has semi-unique factorization ($m(R)$ has semi-unique factorization) if and only if $\tilde{P}_{\mathbf{Q}}(Y/S)$ has semi-unique factorization for every projective birational morphism $g : Y \longrightarrow S$ with Y normal (such that Y is the blowup of an m-primary ideal).*

PROOF. This follows from Lemmas 14 and 17 since (high multiples) of extremal elements in $\tilde{P}_{\mathbf{Q}}(Y/S)$ correspond to extremal complete ideals in R, and a factorization of complete ideals lifts to a factorization in $\tilde{P}_{\mathbf{Q}}(Y/S)$ for some Y.

For regular local rings of dimension two, the shape of $\tilde{P}_{\mathbf{Q}}(X/S)$ is very simple. Theorem 19 and it corollaries follow from Theorem 20.1 and preceding material of [**L1**].

THEOREM 19. *Suppose that R is two-dimensional and regular, and X is regular. Let $n = \rho(X/S)$, and $E_1, \dots, E_n$ be the irreducible exceptional curves for f, so that $\{E_i\}$ is a basis for $A^1(X/S)$. Let $D_i \in A^1(X/S)$ be such that $(D_i \cdot -)$ is the dual form δ_{ij} for this basis. Then*

$$\tilde{P}_{\mathbf{Q}}(X/S) = \sum_{i=1}^{n} \mathbf{Q}_+ D_i$$

In particular, $\tilde{P}_{\mathbf{Q}}(X/S)$ has semi-unique factorization. $\tilde{P}_{\mathbf{Q}}(X/S)$ (relative is the class of a line bundle

PROOF. Some multiple mD_i of D_i is (the class of) a linebundle on X. By Theorem 12.1 [**L1**], $\mathcal{O}_X(mD_i)$ is generated by global sections, since R has rational singularities, $f^{-1}(m)$ has dimension one, and D_i has non-negative intersection number with each E_i. Hence each $D_i \in \tilde{P}(X/S)$. Suppose that $\mathcal{L}$ is a line bundle on X. Any exceptional curve C with negative intersection number with $\mathcal{L}$ must be in the zero locus of every global section $\sigma \in \Gamma(X, \mathcal{L})$. By Theorem 12.1 [**L1**], we then have $\mathcal{L} \in \tilde{P}(X/S)$ if and only if $(E_i \cdot \mathcal{L}) \geq 0$ for $1 \leq i \leq n$. Now there exists a unique element $D = \sum_i a_i D_i \in \tilde{P}(X/S)$ such that $(D \cdot E_i) = (\mathcal{L} \cdot E_i)$ for $1 \leq i \leq n$, since the intersection form is negative definite. By Lemma 15, D and $\mathcal{L}$ have the same class in $\tilde{P}(X/S)$.

COROLLARY 1. *Suppose that R is regular of dimension two. Then R has semi-unique factorization of complete ideals.*

PROOF. This is immediate from Theorems 18 and 19.

The arguments of Theorem 19 can be used to prove Zariski's Theorem. To make the basic idea transparent, we will only prove it in the case where R is complete.

COROLLARY 2. *(Zariski) Suppose that R is regular of dimension two. Then $m(R)$ has unique factorization.*

PROOF. (When R is complete) The map

$$\Theta : \mathrm{Pic}(X) \longrightarrow \mathrm{Hom}(\bigoplus \mathbf{Z}E_i, \mathbf{Z})$$

defined by $\Theta(\mathcal{L})(E_i) = (\mathcal{L} \cdot E_i)$ is surjective by proposition 14.4 [**L1**]. In particular, the D_i of Theorem 19 are integral divisors, and generate the semi-group of integral points of $\tilde{P}_{\mathbf{Q}}(X/S)$.

THEOREM 20. *Suppose that R is the localization at (x, y, z) of the polynomial ring $k[x, y, z]$. Let $f : X_n \longrightarrow S$ be obtained by first blowing up m, and then blowing up n closed points in general position on the exceptional divisor. Then $\tilde{P}_{\mathbf{Q}}(X_n/S)$ has semi-unique factorization if and only if $n \leq 2$.*

Let $X_0 \longrightarrow S$ be the blowup of m with exceptional curve E. Let E_0 be the strict transform by f of the exceptional divisor E and let $E_1, \dots, E_n$ be the remaining n prime exceptional divisors of f, obtained by blowing up points p_i on E. $A^1(X_n/S)$ has $\{E_0, E_1, \dots, E_n\}$ as a basis. Let $H = -E_0 - E_1 - E_2 - \ldots - E_n$.

LEMMA 21. *Let notation be as in Theorem 20 and the above paragraph. Then the natural restriction map Pic*(X_n) $\longrightarrow$ *Pic*(E_0) *induces an isomorphism* $\tilde{P}(X_n/S) \longrightarrow \tilde{P}(E_0/k)$

PROOF. Let $h = \mathcal{O}_{E_0} \otimes \mathcal{O}_{X_n}(H)$. Let $e_i = E_i \cdot E_0$. $\{h, e_1, \ldots, e_n\}$ is a basis for $A^1(E_0/k)$, and the restriction map induces an isomorphism on Pic and A^1.

Suppose that $\mathcal{L} \in \mathrm{Pic}(X_n)$ is such that $\mathcal{L} \otimes \mathcal{O}_{E_0}$ is generated by global sections. We will show that the natural map $\Gamma(X_n, \mathcal{L}) \longrightarrow \Gamma(E_0, \mathcal{L} \otimes \mathcal{O}_{E_0})$ is a surjection. Since $\mathcal{L} \otimes \mathcal{O}_{E_0}$ is generated by global sections, we have $\mathcal{L} \cong \mathcal{O}_{X_n}(mH - a_1E_1 - \ldots - a_nE_n)$ with $m > 0$ and each $a_i \geq 0$. Given $\sigma \in \Gamma(E_0, \mathcal{O}_{E_0}(mh - a_1e_1 - \ldots - a_ne_n))$, the zero locus (σ) is $(g^*(\tau)) - \sum a_ie_i$ for some section $\tau \in \Gamma(E, \mathcal{O}_E(m))$ with $\mathrm{mult}_{p_i}(\tau) \geq a_i$ for $1 \leq i \leq n$. There exists a section $\lambda \in \Gamma(X_0, \mathcal{O}_{X_0}(-mE))$, induced by a homogeneous form in $k[x, y, z]$, such that $\lambda \mid E_0 = \tau$. There are homogeneous height two primes in $k[x, y, z]$ whose strict transforms on X_0 are curves l_i in X_0, such that $l_i \cdot E_i = p_i$ and $\mathrm{mult}_{l_i}(\lambda) = \mathrm{mult}_{p_i}(\tau)$. We have

$$\mathrm{mult}_{p_i}(\lambda) \leq \mathrm{mult}_{p_i}(\tau) = \mathrm{mult}_{l_i}(\lambda) \leq \mathrm{mult}_{p_i}(\lambda).$$

so that $\mathrm{mult}_{p_i}(\lambda) = \mathrm{mult}_{p_i}(\tau)$.

Hence $(g^*(\lambda)) - \sum a_iE_i$ is effective, and induces a global section of $\mathcal{L}$ which restricts to σ.

$\mathcal{L}$ is thus generated by global sections in a neighborhood of E_0, so that there are at most a finite number of closed points where $\mathcal{L}$ is not generated by global sections. By a local analogue of Theorem 6.2 [**Z2**], some multiple of $\mathcal{L}$ is generated by global sections.

Proof of Theorem 20. We will first calculate $\tilde{P}(X_3/S)$. Let $NE(E_0) \subset A^1(E_0/k)$ be the cone generated by effective divisors on E_0. Then

$$\begin{aligned} NE(E_0/k) &= \mathbf{R}_+(h - e_1 - e_2) + \mathbf{R}_+(h - e_2 - e_3) \\ &\quad + \mathbf{R}_+(h - e_1 - e_3) + \mathbf{R}_+e_1 + \mathbf{R}_+e_2 + \mathbf{R}_+e_3 \end{aligned}$$

by Theorem 26.2 [**M**]. The cone of (classes in $A^1(E_0/k)$) of linebundles with non-negative intersection number with each of these curves is

$$\mathbf{R}_+(2h - e_1 - e_2 - e_3) + \mathbf{R}_+(h - e_1) + \mathbf{R}_+(h - e_2) + \mathbf{R}_+(h - e_3) + \mathbf{R}_+h,$$

which must contain $\tilde{P}(E_0/k)$. But each of these vertices induce morphisms, so the two cones are the same. By Lemma 21,

$$\begin{aligned} \tilde{P}(X_3/S) &= \mathbf{R}_+H + \mathbf{R}_+(H - E_1) + \mathbf{R}_+(H - E_2) \\ &\quad + \mathbf{R}_+(H - E_3) + \mathbf{R}_+(2H - E_1 - E_2 - E_3). \end{aligned}$$

Since $\tilde{P}(X_3/S)$ is a cell of $\tilde{P}(X_n/S)$ for $n \geq 3$, we have that $\tilde{P}(X_n/S)$ is not semi-factorial for $n \geq 3$.

The Theorem now follows from consideration of appropriate cells of $\tilde{P}(X_3/S)$, which shows that

$$\tilde{P}(X_1/S) = \mathbf{R}_+H + \mathbf{R}_+(H - E_1)$$

$$\tilde{P}(X_2/S) = \mathbf{R}_+H + \mathbf{R}_+(H - E_1) + \mathbf{R}_+(H - E_2).$$

COROLLARY. *R has semi-unique factorization of m-primary complete ideals if and only if R has dimension ≤ 2.*

PROOF. This is immediate from Theorem 19 and its corollary, Theorem 20.

DEFINITION 22. A Cone of the form

$$\sum_{i=1}^{r} \mathbf{R}_+ v_i$$

in a finite dimensional real vector space defined over $\mathbf{Q}$ is called a finite rational polyhedral cone if the v_i are vectors defined over $\mathbf{Q}$. The v_i may be linearly dependent.

For $n \leq 8$, we have that the cone of curves $NE(E_n/k)$ is finite rational polyhedral (Theorem 26.2 [**M**]), and $\tilde{P}(X_n/S) \cong \tilde{P}(E_0/k)$ is finite rational polyhedral.

EXAMPLE 1. $\tilde{P}(X/S)$ is in general not finite rational polyhedral.

Suppose that R is the localization of the polynomial ring $k[x,y,z]$ at (x,y,z). Let $f : X \longrightarrow S$ be the birational projective morphism obtained by first blowing up m, and then blowing up the nine intersection points of two general cubics on the exceptional divisor. Let E_0 be the strict transform of the exceptional divisor of the blowup of m. E_0 has infinitely many -1 curves [**N1**]. These are (rational) curves C with $(C^2) = -1$ and $(C \cdot K_{E_0}) = -1$, where K_{E_0} is the canonical divisor on E_0. Each of these curves can be contracted by a morphism, which extends to a morphism of X by Lemma 21. Since normal projective S-schemes minimal to factoring $X \longrightarrow S$ correspond to minimal cells of $\tilde{P}(X/S)$ (Theorem 13, 3)), There must be infinitely many minimal cells in $\tilde{P}(X/S)$.

We use example 1 to answer a question of Heinzer (Question on page 21 of [**H**]).

EXAMPLE 2. There exist an infinite set of normal ideals in a three dimensional regular local ring such that they have the same Rees valuations, but their blowups are pairwise distinct.

Consider the X and S of example 1. Infinitely many of the cells of $\tilde{P}(X/S)$ must have the same exceptional divisors, and a set of normal ideals in R whose blowups give these schemes have the same Rees valuations, giving the assertion of example 2.

EXAMPLE 3. $\tilde{P}(X/S)$ is in general not a closed cone.

For instance, $\tilde{P}(X_{16}/S)$ is not closed. This follows from a Theorem of Nagata [**N2**] which shows that

$$\tilde{P}(E_0/k) \cap (\mathbf{R}h + \mathbf{R}(e_1 + \ldots + e_{16})) = \{ah - b(e_1 - \ldots - e_{16}) \mid b \geq 0, a > 4b\} \cup \{(0,0)\}$$

Any birational morphism between two dimensional normal schemes which is an isomorphism in codimension one must be an isomorphism by Zariski's main theorem. The really new phenomenon encountered in dimension three is the existence of nontrivial birational morphisms which are isomorphisms in codimension one. This is the explanation for the existence of example 2.

Birational morphisms $f : X \longrightarrow Y$ which are isomorphisms in codimension one are determined by certain non-torsion elements of the group $\mathrm{Cl}(Y)/\mathrm{Pic}(\mathrm{Y})$. This follows from Lemma 23 below. The converse to Lemma 23 is false. For instance, examples of Weil divisors on three dimensional rational singularities such that the blow up of all (non-zero) integral multiples have an exceptional divisor are given in [**C5**].

LEMMA 23. *Suppose that $f : X \longrightarrow Y$ is a birational projective morphism of normal schemes which is an isomorphism in codimension one. Suppose that f is the blow up of an ideal sheaf $I \subset \mathcal{O}_Y$. Then I is a divisorial ideal sheaf.*

PROOF. Let $I \subset \mathcal{O}_Y$ be an ideal sheaf such that its blow up is f. Then $I\mathcal{O}_X = \mathcal{O}_X(-\sum a_i E_i)$ for prime Weil divisors E_i and some positive integers a_i. Let v_i be the natural discrete valuation associated to E_i. Defined $\mathcal{O}_Y$ ideals I_i by

$$I_i = \{g \in \mathcal{O}_Y \mid v_i(g) \geq a_i\}.$$

Each I_i is a divisorial ideal sheaf, since the center of v_i on Y must have codimension one. Hence $I = \cap I_i$ is a divisorial ideal sheaf.

REFERENCES

[C1] S.D. Cutkosky, *Factorization of complete ideals*, J. Algebra **115** (1988), 144-149.

[C2] S.D. Cutkosky, *On unique and almost unique factorization of complete ideals*, American J. Math. **111** (1989), 417-433.

[C3] S.D. Cutkosky, *On unique and almost unique factorization of complete ideals II*, Invent. Math **98** (1989), 59-74.

[C4] S.D. Cutkosky, *A new characterization of rational surface singularities*, Invent. Math. **102** (1990), 157-177.

[C5] S.D. Cutkosky, *Weil divisors and symbolic divisors*, Duke math. J. **57** (1988), 175-183.

[CS] S.D. Cutkosky and V. Srinivas, *On a problem of Zariski on dimensions of linear systems*, Annals of Math (to appear).

[G] H. Göhner, *semi-factoriality and Muhly's condition (N) in two dimensional local rings*, Journal of Algebra **34** (1975), 403-429.

[H] W. Heinzer, *Proceedings of the Midwest/Great Plains workshop in commutative algebra at The University of Nebraska*, 1989.

[HL] W. Heinzer and D. Lantz, *Commutative Algebra, Proceedings of a Microprogram held June 15- July 2, 1987*, Springer-Verlag, New York, 1989.

[Hi] H. Hironaka, *On the theory of birational blowing up*, Thesis, Harvard (1960).

[Hu] C. Huneke, *Integral closures and primary components of ideals in three-dimensional regular local rings*, Math. Ann. **275** (1986), 617-635.

[Hu1] C. Huneke, *Commutative Algebra, Proceedings of a Microprogram held June 15- July 2, 1987*, Springer-Verlag, New York, 1989.

[J] B. Johnston, *A finiteness condition on regular local overrings of a local domain*, Trans, Amer. Math. Soc. **299** (1987), 513-524.

[K] S. Kleiman, *Toward a numerical theory of ampleness*, Annals of Math. **84** (1966), 293-344.

[L1] J. Lipman, *Rational singularities, with applications to algebraic surfaces and unique factorization*, Publ. Math. I.H.E.S. **36** (1969).

[L2] J. Lipman, *Algebraic Geometry and Commutative Algebra Vol. I*, Kinokuniya, Tokyo, 1988.

[M] Y.I. Manin, *Cubic forms: Algebra, Geometry, Arithmetic*, North Holland, London, 1974.

[MS] H. Muhly and M. Sakuma, *Asymptotic factorizations of ideals*, J. London Math. Soc. **38** (1963), 341-350.

[Mu] D. Mumford, *The topology of normal singularities of an algebraic surface*, Publ. Math. Inst. Hautes Etud. Sci. **9** (1961).

[N1] M. Nagata, *On rational surfaces II*, Mem. Coll. Sci. Univ. Kyoto **33** (1960), 271-293.

[N2] M. Nagata, *On the fourteenth problem of Hilbert*, Amer. J. Math. **81** (1959), 766-772.

[R] D. Rees, *Hilbert functions and pseudo-rational local rings of dimension two*, J. London Math. Soc. **24** (1981), 467-479.

[S] J. Sally, *Commutative Algebra: Proceedings of a Microprogram held June 15- July 2, 1987*, Springer-Verlag, New York, 1989.

[Sp] M. Spivakovsky, *Valuations in function fields of surfaces*, Amer. J. Math. **112** (1990), 107-156.

[V] J. Verma, *Rees algebras of contracted ideals in two dimensional regular local rings*, J. Alg. **141,** (1991), 1-10.

[Z1] O. Zariski, *Polynomial ideals defined by infinitely near base points*, Amer. J. Math. **60** (1938), 151-204.

[Z2] O. Zariski, *Theorem of Riemann-Roch for high multiples of an effective divisor on an algebraic surface*, Annals of Math. vol 76 (1962), 560-615.

[ZS] O. Zariski and P. Samuel, *Commutative algebra vol. 2*, Van Nostrand, Princeton, 1960.

Department of Mathematics University of Missouri Columbia, MO 65211

Contemporary Mathematics
Volume **159**, 1994

On the Cohen-Macaulay Type of Perfect Ideals

J.ELIAS AND A.V.GERAMITA AND G.VALLA

ABSTRACT. In this paper we give upper bounds for the Cohen-Macaulay type of rings of the type R/I where R is a polynomial ring and I ranges over the class of either perfect ideals with given codimension and multiplicity or the class of perfect ideals with given codimension, multiplicity and initial degree. We also show that all the bounds we obtain are sharp.

Let I be a perfect homogeneous ideal of a polynomial ring R over a field k. The Cohen-Macaulay type of the ring $A = R/I$ is the last Betti number in the minimal free resolution of A as a R-module; it will be denoted by $r(R/I)$.

This short paper is devoted to finding upper bounds for $r(R/I)$ when: $i)$ I ranges over the class of perfect ideals with given codimension and multiplicity; $ii)$ I ranges over the class of perfect ideals with given codimension, multiplicity and initial degree.

All the bounds we obtain are sharp since they are shown to be attained by suitable monomial ideals.

The result of this paper should be compared with the analogous result found for the minimal number of generators $v(I)$ of I in [**3**].

The result we present here is, in fact, a consequence of the Main Lemma 3.9 in [**3**] and of the result, recently proved independently by Bigatti and Hulett, on upper bounds for the Betti numbers in free resolutions of rings R/I having a given Hilbert function (see [**1**] and [**6**]).

1991 *Mathematics Subject Classification.* Primary 13D02, 13C14, 14M05; Secondary 13C15, 13P10.

Key words and phrases. Betti numbers, codimension 3, Cohen-Macaulay type.

The first author was partially supported by the DGICYT under grand number PB91-0231-C02-01, the second author was partially supported by NSERC (Canada) and the third author was partially supported by M.P.I.(Italy).

This paper is in final form and no version of it will be submitted for publication elsewhere.

Let $R = k[X_1, \ldots, X_n]$ be a polynomial ring over the field k. Let T_R be the monoid of terms in $X_1, \ldots, X_n$; we consider, in T_R, the total ordering defined as follows (and called the *degree-lexicographic* order, or *lex* for short):

$$X_1^{a_1} \cdots X_n^{a_n} > X_1^{b_1} \cdots X_n^{b_n}$$

if the first non zero coordinate of the vector $(\sum_i (a_i - b_i), a_1 - b_1, \ldots, a_n - b_n)$ is positive.

For instance if $n = 3$ the terms of degree 2 of R are ordered as follows:

$$X_1^2 > X_1X_2 > X_1X_3 > X_2^2 > X_2X_3 > X_3^2,$$

and any term of degree 2 is bigger than any term of degree > 2. It therefore makes sense to talk about *lex-segments*, i.e. a totally ordered sequence of terms $t_1, \ldots, t_r$ with the property that any term t for which $t_1 > t > t_r$ is in the given sequence. In this paper we will only use the term lex-segment to refer to a sequence of terms which are all contained in R_t (for some t) and which start with the first term in R_t i.e. start with X_1^t.

Thus, for us, X_1^2, X_1X_2, X_1X_3 is a lex segment in R_2, but X_1X_2, X_1X_3 is not.

DEFINITION 1. *Let $I = \oplus_{t \geq 0} I_t$ be a graded ideal of R. We say that I is a lex-segment ideal if for every $t \geq 0$, I_t is generated (as a k-vector space) by a lex-segment of terms.*

Notice that, with our conventions, if M is a monomial of degree t in a lex-segment ideal I then every monomial M' in I of degree t satisfying $M' \geq M$ is also in I.

Given a standard graded algebra $G := k[X_1, \ldots, X_n]/J = \oplus_{t \geq 0} G_t$, let us consider its Hilbert function

$$H_G(t) := dim_k(G_t).$$

It is proved in **[7]** that if, for every $t \geq 0$, we leave out the smallest $H_G(t)$ monomials in R_t, the remaining monomials form a k-vector base for a monomial ideal I which is, then, a lex-segment ideal with the property that

$$H_{R/I} = H_G.$$

Since the ideal I is uniquely determined by its Hilbert function it is called **the** lex-segment ideal associated to the Hilbert function H_G.

Now let I be any lex-segment ideal in R. We denote by $\overline{I}$ the ideal $(I, X_n)/(X_n)$ in the ring

$$\overline{R} := R/X_nR \simeq k[X_1, \ldots, X_{n-1}].$$

It is clear that $\overline{I}$ is also a lex-segment ideal.

We recall now that if $A = R/I$ is a graded artinian ring then its Cohen-Macaulay type can be computed as the dimension of the socle of R/I where

$$soc(A) := (I : R_1)/I.$$

Furthermore, if A is a standard graded algebra we denote by $e(A)$ the multiplicity of A.

THEOREM 2. *Let I be a lex-segment ideal of R containing no linear forms and such that R/I is an artinian ring. Then*

$$r(R/I) = e(\overline{R}/\overline{I}).$$

i.e. $r(R/I)$ is the number of elements in the canonical base of I which are divisible by X_n.

PROOF. We first claim that

$$I : R_1 = I : X_n.$$

It is clear that $I : R_1 \subseteq I : X_n$. On the other hand, since I is a monomial ideal, if $FX_n \in I$ then all the monomials of FX_n are in I. In other words, $MX_n \in I$ for all monomials M of F. Since $MX_n \leq MX_i$ for every i and I is a lex-segment ideal, the claim follows.

As we noted above, $r(R/I) = \dim((I : R_1)/I)$. Clearly this latter dimension is $\leq \dim((I : F)/I)$ for any $F \in R_1$. Since $\dim((I : F)/I) = e((R/I)/((I, F)/I)) = e(R/(I, F))$ for any F we have, by using the claim and letting $F = X_n$, that $r(R/I) = e(\overline{R}/\overline{I})$, as we wanted to show. (Thanks to Bill Heinzer and Craig Huneke for this simplification of our earlier proof.)

Example. Let $n = 3$ and

$$I = (X_1^2, X_1X_2, X_1X_3^2, X_2^3, X_2^2X_3, X_2X_3^4, X_3^5)$$

be the lex-segment ideal corresponding to the Hilbert function

$$\{1, 3, 4, 2, 2, 1, 0, \dots\}.$$

Then we have $v(I) = 7$. Since $\overline{I} = (X_1^2, X_1X_2, X_2^3)$ we have $v(\overline{I}) = 3$ and so $r(R/I) = 4$. Notice that we now know all the (non-graded) Betti numbers in the minimal free resolution of R/I. They are

$$b(1) = 7, \quad b(2) = 10, \quad b(3) = 4.$$

It is worth noting that in this example we also know all the graded Betti numbers for this homogeneous ideal since we have the degrees of the generators of I, the degrees of the socle generators of R/I and the Hilbert function of R/I. We will not enter into that aspect of the resolution in this paper.

Given the polynomial ring $R = k[X_1, \dots, X_n]$, if we fix a number $e \geq n + 1$, we can consider the *maximal Hilbert function with respect to the multiplicity e*. If $t = t(e)$ is defined as the unique integer such that

$$\binom{n+t-1}{t-1} \leq e < \binom{n+t}{t},$$

the function is defined by the formula

$$H(p) = \begin{cases} \binom{n+p-1}{p} & \text{for } p \le t-1 \\ e - \binom{n+t-1}{t-1} & \text{for } p = t \\ 0 & \text{for } p > t. \end{cases}$$

We denote by $J(e)$ the lex-segment ideal corresponding to this maximal Hilbert function.

In the same manner, given the integer e and the integer $t(e)$ defined as above (which is certainly ≥ 2) choose i to be any integer such that $2 \le i \le t(e)$.

We can consider the *maximal Hilbert function corresponding to the multiplicity e and the initial degree i*. If $s = s(e,i)$ is defined as the unique integer such that

$$\binom{n+s-1}{s-1} - \binom{n+s-i-1}{s-i-1} \le e < \binom{n+s}{s} - \binom{n+s-i}{s-i},$$

this function is defined by the formula

$$H(p) = \begin{cases} \binom{n+p-1}{p} & \text{for } 0 \le p \le i-1 \\ \binom{n+p-2}{p} + \binom{n+p-3}{p-1} + \cdots + \binom{n+p-i-1}{p-i+1} & \text{for } p = i, \ldots, s-1 \\ e - \binom{n+s-1}{s-1} + \binom{n+s-i-1}{s-i-1} & \text{for } p = s \\ 0 & \text{for } p > s. \end{cases}$$

One can show that this function satisfies the necessary conditions in order that it be the Hilbert function of some standard graded algebra (see [**3**]). It is not difficult to see that the multiplicity of any standard graded algebra with this Hilbert function is exactly e.

We denote by $J(e,i)$ the lex-segment ideal corresponding to this maximal Hilbert function.

Using the notation above, we recall the following result (see [**3**], Lemmas 3.9, 4.1, and 4.3.).

THEOREM 3.

(1) *Let e be an integer $e \ge n+1$, and I a lex-segment ideal such that R/I is artinian and $e(R/I) = e$. Then,*

$$e(\overline{R}/\overline{I}) \le e(\overline{R}/\overline{J(e)}).$$

(2) *Let e, i be integers $e \ge n+1$, $2 \le i \le t(e)$, and let I be a lex-segment ideal with initial degree i such that R/I is artinian and $e(R/I) = e$. Then,*

$$e(\overline{R}/\overline{I}) \le e(\overline{R}/\overline{J(e,i)}).$$

We can now prove the main result of the paper. We have decided to formulate the result in the local case since it is more general and no more difficult to prove.

THEOREM 4.

(1) *Let α be a codimension n perfect ideal of the regular local ring (B, m). If $e = e(B/\alpha)$, then*

$$r(B/\alpha) \leq r(R/J(e)).$$

(2) *Let α be a codimension n perfect ideal of the regular local ring (B, m). If $e = e(B/\alpha)$ and $i \geq 2$ is the initial degree of α, then*

$$r(B/\alpha) \leq r(R/J(e, i)).$$

PROOF. It is well known that there exists a minimal reduction $\underline{l} := l_1, \ldots, l_d$ of m modulo α, such that the initial degree of α is the same as the initial degree of $\overline{\alpha} := [\alpha + (\underline{l})]/(\underline{l})$ (see for example [**2**]). The local ring $(\overline{B} := B/(\underline{l}), \overline{m} := m/(\underline{l}))$ is a regular local ring of dimension n. The associated graded ring of the artinian ring $\overline{B}/\overline{\alpha}$ is the artinian standard graded ring

$$G := gr_{\overline{m}/\overline{\alpha}}(\overline{B}/\overline{\alpha}) = k[X_1, \ldots, X_n]/I,$$

where I is the homogeneous ideal of the initial forms of $\overline{\alpha}$.

It is clear that the initial degree of I is the same as that of $\overline{\alpha}$ which, as we remarked before, is the same as that of α. On the other hand since $\underline{l}$ is a regular sequence modulo α, we also have

$$e(B/\alpha) = e(\overline{B}/\overline{\alpha})$$

and

$$r(B/\alpha) = r(\overline{B}/\overline{\alpha}).$$

Now it is well known that passing to the associated graded ring does not change the multiplicity, while the Betti numbers can only increase (see for example [**5**], Corollary 3.2). Hence we get

$$e(B/\alpha) = e(k[X_1, \ldots, X_n]/I)$$

and

$$r(B/\alpha) \leq r(k[X_1, \ldots, X_n]/I).$$

We now use the result of Bigatti-Hulett which says that all the Betti numbers of a given ideal are bounded above by the Betti numbers of the lex-segment ideal with the same Hilbert function (see [**1**] and [**6**]). Hence, if we denote by $L(I)$ the lex-segment ideal with the same Hilbert Function as I, we get

$$r(B/\alpha) \leq r(R/I) \leq r(R/L(I)) = e(\overline{R}/\overline{L(I)}) \leq e(\overline{R}/\overline{J(e)}) = r(R/J(e)).$$

This proves (1) and the analogous argument also proves (2).

In the following Corollary we use $b_j(A)$ to denote the $j-th$ Betti number of a local or graded ring A considered as a $B-$module where B is either a regular local ring or a standard graded polynomial ring.

COROLLARY 5. *Let α be a codimension three perfect ideal of the regular local ring (B, m). Let $R := k[X_1, X_2, X_3]$ and let the initial degree of α be at least two.*

(1) *If $e = e(B/\alpha)$, then*

$$b_j(B/\alpha) \leq b_j(R/J(e))$$

for every $j = 1, 2, 3$.

(2) *If $e = e(B/\alpha)$ and i is the initial degree of α, then*

$$b_j(B/\alpha) \leq b_j(R/J(e, i))$$

for every $j = 1, 2, 3$.

PROOF. By the main result of [**3**] we have

$$b_1(B/\alpha) \leq b_1(R/J(e)).$$

By the theorem above we have

$$b_3(B/\alpha) \leq b_3(R/J(e)).$$

Since $1 + b_2 = b_1 + b_3$, the conclusion follows.

Remark 1. Let us compute the bounds for a perfect codimension three ideal with multiplicity $e = 13$. Since

$$J(13) = (X_1^3, X_1^2X_2, X_1^2X_3, X_1X_2^2, X_1X_2X_3, X_1X_3^2, X_2^3, X_2^2X_3^2, X_2X_3^3, X_3^4),$$

it follows that

$$\overline{J(13)} = (X_1^3, X_1^2X_2, X_1X_2^2, X_2^3).$$

Hence we get

$$b_1(R/J(13)) = 10, \qquad b_2(R/J(13)) = 15, \qquad b_3(R/J(13)) = 6.$$

These are upper bounds for the Betti numbers of any perfect codimension three ideal with multiplicity 13.

Remark 2. The bounds we have found in this paper are sharp since they can all be obtained by monomial ideals. It is also true that we can reach these bounds with radical ideals, in fact with ideals which define zero-dimensional reduced subschemes of $\mathbf{P}^n$.

This follows from a result of Hartshorne which implies that monomial ideals in $k[X_1, \ldots, X_n]$ can be lifted to ideals of distinct points in $\mathbf{P}^n$ with the same graded Betti numbers and hence with the same multiplicity and the same initial degree (see [**4**] for an elementary exposition of this aspect of Hartshorne's result).

Thus the ideal

$$J(13) = (X_1^3, X_1^2X_2, X_1^2X_3, X_1X_2^2, X_1X_2X_3, X_1X_3^2, X_2^3, X_2^2X_3^2, X_2X_3^3, X_3^4)$$

in $R = k[X_1, X_2, X_3]$ can be lifted to the following radical ideal in $k[X_0, X_1, X_2, X_3]$:

$$\begin{aligned} I = (&X_1(X_1 - X_0)(X_1 - 2X_0), X_1(X_1 - X_0)X_2, X_1(X_1 - X_0)X_3, \\ &X_1X_2(X_2 - X_0), X_1X_2X_3, X_1X_3(X_3 - X_0), X_2(X_2 - X_0)(X_2 - 2X_0), \\ &X_2(X_2 - X_0)X_3(X_3 - X_0), X_2X_3(X_3 - X_0)(X_3 - 2X_0), \\ &X_3(X_3 - X_0)(X_3 - 2X_0)(X_3 - 3X_0)). \end{aligned}$$

This is the defining ideal of the following 13 points in $\mathbf{P}^3$:

$$\begin{gathered} P_1 = (1,0,0,0), P_2 = (1,0,0,1), P_3 = (1,0,0,2) \\ P_4 = (1,0,0,3), P_5 = (1,0,1,0), P_6 = (1,0,1,1) \\ P_7 = (1,0,1,2), P_8 = (1,0,2,0), P_9 = (1,0,2,1) \\ P_{10} = (1,1,0,0), P_{11} = (1,1,0,1), P_{12} = (1,1,1,0), P_{13} = (1,2,0,0). \end{gathered}$$

Remark 3. The theorem above does not extend to ideals which are not perfect. Let $R = k[X,Y]$ and $I = (X) \cap (X,Y)^n$. Then

$$r(R/I) = n - 1$$

while $e(R/I) = 1$ does not depend on n.

Remark 4. We can explicitly compute the Cohen-Macaulay type of the rings $R/J(e)$ and $R/J(e,i)$.

We need some additional notation and results about lex-segment ideals as developed in [**3**].

We recall that if p and q are positive integers then p can be uniquely written as

$$p = \binom{p_q}{q} + \binom{p_{q-1}}{q-1} + \cdots + \binom{p_j}{j}$$

where

$$p_q > p_{q-1} > \cdots > p_j \geq j \geq 1.$$

This is called the **q-binomial expansion of p**.

We define for every positive integers p and q:

$$p^{<q>} := \binom{p_q+1}{q+1} + \binom{p_{q-1}+1}{q} + \cdots + \binom{p_j+1}{j+1}$$

and

$$p_{<q>} := \binom{p_q-1}{q} + \binom{p_{q-1}-1}{q-1} + \cdots + \binom{p_j-1}{j}.$$

Notice that the first of these formulas is the $(q+1)$-expansion of $p^{<q>}$ while the second is the q expansion of $p_{<q>}$ if and only if $p_j > j$. The following result is not difficult and can be found in [**3**], Corollary 2.8.

Let J be a lex-segment ideal and let $\overline{J}$ and $\overline{R}$ be as defined above. Then

$$H_{\overline{R}/\overline{J}}(q) = (H_{R/J}(q))_{<q>}$$

for every $q \geq 1$.

THEOREM 6. *Let e be an integer $e \geq n+1$. Let $t = t(e)$ be as defined above and let*

$$c := e - \binom{n+t-1}{t-1}.$$

Then

$$r(R/J(e)) = \binom{n+t-2}{t-1} + c_{<t>}.$$

PROOF. If we now write J instead of $J(e)$ then by Theorem 2, Corollary 2.8 of [**3**] (as explained above) and the definition of the Hilbert Function of R/J, we have

$$r(R/J) = e(\overline{R}/\overline{J}) = \sum_{q \geq 0} H_{\overline{R}/\overline{J}}(q) = \sum_{q \geq 0} (H_{R/J}(q))_{<q>} =$$
$$= \sum_{q=0}^{t-1} \binom{n+q-2}{q} + c_{<t>} = \binom{n+t-2}{t-1} + c_{<t>},$$

which is what we wanted to show.

In the same way, given integers e, i with $e \geq n+1$ and $2 \leq i \leq t(e)$, and $s = s(e,i)$ as defined above, let

$$c := e - \binom{n+s-1}{s-1} + \binom{n+s-i-1}{s-i-1}.$$

Then

$$r(R/J(e,i)) = \binom{n+s-2}{s-1} - \binom{n+s-i-2}{s-i-1} + c_{<s>}.$$

The proof of this last assertion follows the same lines as the proof of Theorem 6. For the sake of simplicity we now write J instead of $J(e,i)$ and as before we get,

$$r(R/J) = e(\overline{R}/\overline{J}) = \sum_{q \geq 0} H_{\overline{R}/\overline{J}}(q) = \sum_{q \geq 0} (H_{R/J}(q))_{<q>} =$$
$$= \sum_{q=0}^{i-1} \binom{n+q-2}{q} + \sum_{q=i}^{s-1} \left(\sum_{k=q-i+1}^{q} \binom{n-3+k}{k} \right) + c_{<s>} =$$
$$= \sum_{q=0}^{i-1} \binom{n+q-2}{q} + \sum_{q=i}^{s-1} \binom{q+n-2}{q} - \sum_{q=i}^{s-1} \binom{q-i+n-2}{n-2} + c_{<s>} =$$
$$= \sum_{q=0}^{s-1} \binom{n+q-2}{q} - \sum_{h=0}^{s-i-1} \binom{h+n-2}{n-2} + c_{<s>} =$$
$$= \binom{n+s-2}{s-1} - \binom{n+s-i-2}{s-i-1} + c_{<s>}.$$

REFERENCES

1. A.M. Bigatti, *Upper Bounds for the Betti Numbers of a Given Hilbert Function*, Comm. in Algebra (to appear).
2. J.Elias, A. Iarrobino, *Extremal Gorenstein Algebras of Codimension Three; the Hilbert Function of a Cohen-Macaulay Local Algebra* **110** (1987), 344-356.
3. J. Elias, L. Robbiano, G. Valla, *Number of Generators of Ideals*, Nagoya Math. J. **123** (1991), 39-76.
4. A.V. Geramita, D. Gregory, L. Roberts, L., *Monomial Ideals and Points in Projective Space*, J. Pure and Applied Alg. **40** (1986), 33-62.
5. J.Herzog, M.E. Rossi, G. Valla, *On the Depth of the Symmetric Algebra*, Trans. A.M.S. **296** (1986), 577-606.
6. H. Hulett, 1992, Comm. in Alg. (to appear).
7. F.S. Macaulay, *Some Properties of Enumeration in the Theory of Modular Systems*, Proc. Lond. Math. Soc. **26** (1927), 531–555.

DEPARTMENT DE ALGEBRA I GEOMETRIA, UNIVERSITAT DE BARCELONA, 08007 BARCELONA, SPAIN

DEPARTMENT OF MATHEMATICS, QUEEN'S UNIVERSITY, KINGSTON, ONTARIO, CANADA

DIPARTIMENTO DI MATEMATICA, UNIVERSITÀ DI GENOVA, VIA L.B.ALBERTI 4, 16132 GENOVA, ITALY

Contemporary Mathematics
Volume **159**, 1994

On the Gorensteinness of graded rings associated to ideals of analytic deviation one

SHIRO GOTO AND YUKIO NAKAMURA

ABSTRACT. Let I be an ideal in a Cohen-Macaulay local ring A with infinite residue class field. We assume that (i) $\mathrm{ht}_A I = s$, (ii) IA_Q is generated by an A_Q-regular sequence of length s for any $Q \in \mathrm{Min}_A A/I$, and (iii) the analytic spread $\lambda(I)$ of I is equal to $s+1$. Let J be a minimal reduction of I. Then, when $\dim A = 1$, the associated graded ring $G(I) = \bigoplus_{n \geq 0} I^n/I^{n+1}$ of I is a Gorenstein ring if and only if (i) A is a Gorenstein ring, (ii) $I^2 = JI$, and (iii) $(I : \mathfrak{m}) \cap (JU : I) = I$, where $\mathfrak{m}$ is the maximal ideal of A and U is the unmixed component of I. As a consequence, we have $I = J$ if $G(I)$ is a Gorenstein ring and A/I is Cohen-Macaulay. The assertions of higher-dimensional cases will be reduced to those of the case where $\dim A = 1$. Besides, a certain criterion for the Rees algebra $R(I) = \bigoplus_{n \geq 0} I^n$ of I to be a Cohen-Macaulay ring shall be given, in the case where the ring A/I is a Cohen-Macaulay and the reduction number $\mathrm{r}_J(I)$ of I with respect to J is at most 2.

1. Introduction

The purpose of this paper is to study the conditions for a certain class of ideals under which Rees algebras and associated graded rings are Cohen-Macaulay or Gorenstein rings. Our results are fundamentally based on the reduction to the case where the base rings have dimension one: a reduction modulo super-regular sequences. For that, the following theorem is one of our key-, hence main-results and we shall prove it in Section 2.

1991 *Mathematics Subject Classification.* Primary 13A30, 13D45, 13H10,14B15; Secondary 14E35,14H20,14H50.

Key words and phrases. Rees algebras, associated graded rings, Cohen-Macaulay rings, Gorenstein rings, analytic deviation, reductions of ideals, a-invariant.

Both authors are partially supported by Grant-in-Aid for Co-operative Research.

This paper is in final form and no version of it will be submitted for publication elsewhere.

THEOREM (1.1). *Let A be a Cohen-Macaulay local ring of* $\dim A = 1$ *and I an ideal in A such that $IA_Q = (0)$ for all $Q \in \mathrm{Min}_A A/I$. Let U denote the unmixed component of I and assume that I contains an element b such that $I^{r+1} = bI^r$ for some $r \geq 0$. Then the following two conditions are equivalent.*

(1) *$G(I) = \bigoplus_{n \geq 0} I^n/I^{n+1}$ is a Gorenstein ring.*

(2) *(i) A is a Gorenstein ring, (ii) $I^2 = bI$, and (iii) $(I : \mathfrak{m}) \cap (bU : I) = I$.*

Here $\mathfrak{m}$ denotes the maximal ideal in A.

In this theorem, if A/I is a Cohen-Macaulay ring, the condition (iii) in (2) is naturally satisfied, and we furthermore have that $G(I)$ is a Gorenstein ring if and only if A is a Gorenstein ring and $I = (b)$. We will prove this rather strong assertion also in Section 2 (Corollary (2.11)). But, before citing further results, we would like to fix some basic notations which we shall maintain throughout this paper.

Let A be a Noetherian local ring of $\dim A = d$ and $\mathfrak{m}$ the maximal ideal in A. Assume that the field $A/\mathfrak{m}$ is infinite. Let I ($\neq A$) be an ideal in A. We put $s = \mathrm{ht}_A I$ and $\mathcal{F} = \mathrm{Min}_A A/I$. Let

$$R(I) = A[It] \subseteq A[t] \quad (t \text{ is an indeterminate over } A) \text{ and}$$
$$G(I) = R(I)/IR(I) \quad (= \bigoplus_{n \geq 0} I^n/I^{n+1})$$

which we call respectively the Rees algebra and the associated graded ring of I. We fix a minimal reduction J of I, so that $J \subseteq I$ and $I^{r+1} = JI^r$ for some $r \geq 0$. Let $\mathrm{r}_J(I) = \min\{r \geq 0 \mid I^{r+1} = JI^r\}$ and call it the reduction number of I with respect to J. Let $\lambda(I)$ denote the analytic spread of I, that is $\lambda(I) = \dim A/\mathfrak{m} \otimes_A G(I)$. Then as is well-known, J is minimally generated by $\lambda(I)$-elements ([**NR**]). Following [**HH1**], we put $\mathrm{ad}(I) = \lambda(I) - s$, and call it the analytic deviation of I. We say that I is generically a complete intersection in A, if IA_Q is, for any $Q \in \mathcal{F}$, generated by an A_Q-regular sequence of length s.

Ideals having analytic deviation 0 are called equimultiple and $\mathfrak{m}$- primary ideals are typical examples. For equimultiple ideals I the Cohen-Macaulay and Gorenstein properties of $R(I)$ are closely investigated by several authors (cf. [**GS1**], [**I1**], [**I2**], [**HIO**] etc.) in connection with the corresponding ring-theoretic properties of $G(I)$. The researches on ideals having positive analytic deviation (except those [**HSV**], [**Br2**], [**GS2**], [**HRZ**] on almost complete intersections) have just started from [**HH1**], [**HH2**], [**GH**]. This paper should be looked upon a succession to them and aims at searching for the conditions under which $R(I)$ and $G(I)$ are Cohen-Macaulay and Gorenstein rings.

In what follows, let us assume that *our base ring A is Cohen-Macaulay and our ideal I is generically a complete intersection in A of* $\mathrm{ad}(I) = 1$. In this situation Huckaba and Huneke [**HH1**] showed $G(I)$ is a Cohen-Macaulay ring if and only if depth$A/I \geq \dim A/I - 1$, provided $\mathrm{r}_J(I) \leq 1$.[1] Thereafter the first author and Huckaba [**GH**] succeeded in a translation of the a-invariant $a(G(I))$ (see [**GW, (3.1.4)**] for definition) in terms of $\mathrm{r}_J(I)$ and showed

PROPOSITION (1.2) [**GH, 2.4**]. *Suppose $G(I)$ is a Cohen-Macaulay ring. Then $a(G(I)) = -s$ if $I = J$ and $a(G(I)) = \mathrm{r}_J(I) - s - 1$ if $I \neq J$.*

They proved, using (1.2), that $R(I)$ is a Cohen-Macaulay ring if and only if $s > 0$, $G(I)$ is a Cohen-Macaulay ring, and $\mathrm{r}_J(I) \leq s$ ([**GH, 2.1**]). (They discussed also the Gorenstein property and essentially showed that in the case where $s \geq 2$, $R(I)$ is a Gorenstein ring if and only if $G(I)$ is a Gorenstein ring and $s = 2$, cf. [**GH, 2.12**]. Later we will back to this result.) Therefore, combining [**HH1, 2.9**] with [**GH, 2.1**], we may directly cover the assertion [HH2, 2.1] that $R(I)$ is a Cohen-Macaulay ring, if $s > 0$, $\mathrm{r}_J(I) \leq 1$, and depth$A/I \geq \dim A/I - 1$. Our purpose is to find similar practical conditions on I, as were given in [**HH1**], [**HH2**], for $R(I)$ and $G(I)$ to be Cohen-Macaulay or Gorenstein rings.

Let us now cite the results in this paper. First we recall that J contains a minimal system $a_1, a_2, \ldots, a_s, b$ of generators such that, with $K = (a_1, a_2, \ldots, a_s)$, $IA_Q = KA_Q$ for all $Q \in \mathcal{F}$ ([**HH1, Proof of 2.2**]). Then as $K : I \not\subseteq Q$ for any $Q \in \mathcal{F}$, we may choose, a system $x_1, x_2, \ldots, x_{d-s-1}, y$ of parameters for the ring A/I, with $y \in K : I$. Let $G = G(I)$ and $\mathbf{M} = \mathfrak{m}G + G_+$. Then it is easy to check that $\mathbf{M} = \sqrt{\mathbf{a}}$ for the ideal $\mathbf{a} = (a_1t, \ldots, a_st, x_1, \ldots, x_{d-s-1}, y + bt)G$ (see Section 3). Thus $a_1t, \ldots, a_st, x_1, \ldots, x_{d-s-1}, y + bt$ is a G-regular sequence, if G is Cohen-Macaulay. We put $\overline{A} = A/(K + (x_1, x_2, \ldots, x_{d-s-1}))$, and the isomorphism

$$G(I\overline{A}) \cong G/(a_1t, \ldots, a_st, x_1, \ldots, x_{d-s-1})G$$

naturally reduces the Gorenstein property of G to that of $G(I\overline{A})$. And, passing to the case of dimension one, we will have the following result, that is a higher-dimensional version of Corollary (2.11).

THEOREM (1.3). *Let I be an ideal with $\mathrm{ad}(I) = 1$ which is generically a complete intersection in a Cohen-Macaulay local ring A. Let J denote a minimal reduction of I. Assume that A/I is a Cohen-Macaulay ring. Then the following three conditions are equivalent.*

(1) *$G(I)$ is a Gorenstein ring.*
(2) *A is a Gorenstein ring and $\mathrm{r}_J(I) \leq 1$.*
(3) *A is a Gorenstein ring and $\mathrm{r}_J(I) = 0$.*

In particular, $\mathrm{r}_J(I) \neq 1$, if A is a Gorenstein ring.

Huckaba and Huneke [**HH1, 2.2**] showed that $\mathrm{r}_J(I) \leq 1$, if $I^{(n)} = I^n$ for all $n \geq 1$. Hence it directly follows from (1.3) that $I = J$, if A is a Gorenstein ring, A/I is a Cohen-Macaulay ring, and $I^{(n)} = I^n$ for all $n \geq 1$ (see Corollary

[1]For ideals I of $\mathrm{ad}(I) = 1$ Huckaba and Huneke [**HH1**], [**HH2**] always assume $s \geq 1$ in their theorems. But their proof still works with slight modifications in the case where $s = 0$ too.

(3.9)). We particularly have the next result, in which $\mu_A(*)$ denotes the number of generators.

COROLLARY (1.4). *Let* $\mathfrak{p}$ *be a prime ideal in a regular local ring* A *and assume* $A/\mathfrak{p}$ *is a Cohen-Macaulay ring of* $\dim A/\mathfrak{p} = 2$. *Then* $\mu_A(\mathfrak{p}) \leq \dim A - 1$, *if* $G(\mathfrak{p})$ *is an integral domain.*

We shall prove in Section 4 the following theorem, which provides a certain class of ideals with a sufficient condition of $R(I)$ being Cohen-Macaulay.

THEOREM (1.5). *Let* I *be an ideal with* $\mathrm{ad}(I) = 1$ *which is generically a complete intersection in a Cohen-Macaulay local ring* A. *Let* A/I *be a Cohen-Macaulay ring and assume that* $\mathrm{r}_J(I) \leq 2$ *for some minimal reduction* J *of* I. *Then the following conditions are equivalent.*

(1) $G(I)$ *is a Cohen-Macaulay ring.*

(2) $\mathrm{depth} A/I^2 \geq \dim A/I - 1$.

Hence $R(I)$ *is a Cohen-Macaulay ring, if* $\dim A \geq 3$ *and* $\dim A/I = 1$.

As was noted before, $R(I)$ is a Gorenstein ring if $G(I)$ is Gorenstein and $s = 2$ ([**GH, 2.12**]). Hence applying (1.3) and (1.5) to space curves, we have the following

COROLLARY (1.6). *Let* A *be a regular local ring of dimension* 3 *and* $\mathfrak{p}$ *a prime ideal in* A *of* $\dim A/\mathfrak{p} = 1$. *Let* J *denote a minimal reduction of* $\mathfrak{p}$. *Then*

(1) $R(\mathfrak{p})$ *is a Cohen-Macaulay ring if and only if* $\mathrm{r}_J(\mathfrak{p}) \leq 2$.

(2) $R(\mathfrak{p})$ *is a Gorenstein ring if and only if* $\mathrm{r}_J(\mathfrak{p}) = 0$.

In particular, if $\mathrm{r}_J(\mathfrak{p}) \leq 1$, $\mathfrak{p} = J$ *and hence* $\mu_A(\mathfrak{p}) \leq 3$.

Part (1) of the corollary is independently proven by [**AH, 7.1**] (see also [**Va, 3.1.3**] for a related assertion).

We will prove Theorem (1.1) in Section 2. The case where $\dim A \geq 2$ will be discussed in Section 3. The rings $G(I)$ can be Gorenstein, even though A/I are *not* Cohen-Macaulay (see [**HH1, Section 4**] for typical examples). But, the authors have not yet obtained any facts extended directly from Theorem (1.1) without the extra assumption that A/I *is Cohen-Macaulay*, which the authors would like to leave to readers as an open problem. Instead, in Section 3 we will explore examples of height 2 prime ideals $\mathfrak{p}$ in a 4-dimensional regular local ring A whose Rees algebras $R(\mathfrak{p})$ are Gorenstein in order to see how our theorem (1.1) works for the analysis of higher dimension. We will prove in Section 4 Theorem (1.5) and its corollaries.

Throughout this paper let $(A, \mathfrak{m})$ stand for a Cohen-Macaulay local ring of $\dim A = d$. For simplicity we always assume that the residue class field $A/\mathfrak{m}$ is infinite. (If necessary, one can delete the assumption, passing to the ideals IB in the ring $B = A[X]_{\mathfrak{m}A[X]}$, where X denotes an indeterminate over A.) Let I $(\neq A)$ be an ideal in A of $\mathrm{ad}(I) = 1$ which is generically a complete intersection. Let $s = \mathrm{ht}_A I$ and $\mathcal{F} = \mathrm{Min}_A A/I$. We fix a minimal reduction J of I and put $r = \mathrm{r}_J(I)$.

2. The case of dimension 1 and Proof of Theorem (1.1)

In this section we assume $\dim A = \lambda(I) = 1$. Hence $s = \mathrm{ht}_A I = 0$. Let $J = (b)$ and let U denote the unmixed component of I. We put $\mathfrak{a} = (0) : b$, the annihilator of J. The main purpose of this section is to prove Theorem (1.1).

To begin with we note

LEMMA (2.1). *(1) $U \cap \mathfrak{a} = (0)$ and hence b is $A/\mathfrak{a}$-regular.*

(2) $\mathfrak{a} = (0) : U = (0) : I$.

(3) $I : \mathfrak{m} \subseteq U$.

(4) $bU : b = \mathfrak{a} + U$.

PROOF. (1) If $U \cap \mathfrak{a} \neq (0)$, we may choose $Q \in \mathrm{Ass} A$ so that $(U \cap \mathfrak{a})A_Q \neq (0)$. Then $b \in Q$ as $\mathfrak{a}A_Q \neq (0)$ and hence $I \subseteq Q$ as $\sqrt{I} = \sqrt{(b)}$, while as $Q \in \mathrm{Ass} A$ we get $\mathrm{ht}_A Q = 0$. Hence $Q \in \mathcal{F}$ and so we have $UA_Q = IA_Q = (0)$, which is absurd. The element b is certainly $A/\mathfrak{a}$-regular, since $(b) \cap \mathfrak{a} = (0)$.
(2) This follows from (1), as $(0) : U \subseteq (0) : I \subseteq \mathfrak{a}$.
(3) We have $I : \mathfrak{m} \subseteq U$, as $U : \mathfrak{m} = U$ (recall A/U is a Cohen-Macaulay ring and $\dim A/U = 1$).
(4) Let $x \in A$ such that $bx \in bU$. Then writing $bx = bu$ with $u \in U$, we get $x - u \in \mathfrak{a}$. Hence $x \in \mathfrak{a} + U$ and so we have $bU : b \subseteq \mathfrak{a} + U$. The opposite inclusion is obvious.

As $\mathfrak{m} = \sqrt{\mathfrak{a} + (b)}$ by (2.1) (1), $\mathfrak{a}$ contains an element y such that $y+b$ is regular. As A/U is a Cohen-Macaulay ring of dimension 1, $y + b$ is regular on A/U too. Hence $U = U : y$, because $y \equiv y + b \mod U$. As $U \subseteq (0) : \mathfrak{a} \subseteq (0) : y \subseteq U : y$, we have $U = (0) : y$. Since $U \cap \mathfrak{a} = (0)$ by (2.1) (1), y is regular on the A-module $\mathfrak{a}$. Thus we have the following

COROLLARY (2.2). *The ideal $\mathfrak{a}$ contains an element y such that y is regular on the A-module $\mathfrak{a}$, $y + b$ is A-regular, and $U = (0) : y$.*

If A is a homomorphic image of a Gorenstein ring, A has the canonical module K_A and $G(I)$ has the *graded* canonical module $K_{G(I)}$. The theory of canonical modules for local rings started from the book **[HK]**. The graded case was thereafter developed by **[GW]**. It has now grown up to be part of a standard theory of graded algebras defined over local rings (cf. **[I2, Section 1]**). In this paper we would like to freely use the results in **[GW]**, **[I2]**, as well as those in **[HK]**, on canonical modules. The readers may consult also **[HIO, Chap. VII]** for details.

We put $R = R(I)$ and $G = G(I)$. For a given graded module X and $n \in \mathbb{Z}$, let X_n denote the homogeneous component of X in degree n. We regard A to be a $\mathbb{Z}$-graded ring whose graduation is given by $A_0 = A$ and $A_n = (0)$ $(n \neq 0)$, so that the canonical projection $\rho : R \to A$ is a homomorphism of graded rings. For a given graded A-module X, let ${}_\rho X$ denote X which is considered, via ρ, to be a graded R-module.

Recall $I \cap \mathfrak{a} = (0)$ (cf. (2.1)(1)), and we have $[\mathfrak{a}G]_n = (0)$ for $n \neq 0$ and $[\mathfrak{a}G]_0 = \mathfrak{a} \cdot (A/I) \cong \mathfrak{a}$. Therefore, regarding $\mathfrak{a}$ as a *trivially* graded A-module, i.e., $\mathfrak{a}_0 = \mathfrak{a}$ and $\mathfrak{a}_n = (0)$ $(n \neq 0)$, we get ${}_\rho\mathfrak{a} \cong \mathfrak{a}G$, so that we have an exact sequence

$$(2.3) \qquad 0 \to {}_\rho\mathfrak{a} \xrightarrow{\varphi} G \xrightarrow{\psi} G/\mathfrak{a}G \to 0$$

of graded R-modules. We note that $G/\mathfrak{a}G \cong G((I+\mathfrak{a})/\mathfrak{a})$ as graded A-algebras, since $I \cap \mathfrak{a} = (0)$.

PROPOSITION (2.4). *Assume that G is a Gorenstein ring. Then A is a Gorenstein ring and $\mathrm{r}_J(I) \leq 1$.*

PROOF. Certainly A is a Gorenstein ring (see, e.g., [**G, Proof of (2.6)**]). As $\mathfrak{a} = (0) : U$, by [**HK, 5.20**] we get $\mathfrak{a} = K_{A/U}$. Therefore by [**HK, 6.1**] we have isomorphisms

$$\mathrm{Hom}_G({}_\rho\mathfrak{a}, K_G) \cong \mathrm{Hom}_G(\mathrm{Hom}_G({}_\rho(A/U), K_G), K_G) \cong {}_\rho(A/U)$$

(recall that A/U is a Cohen-Macaulay ring of $\dim A/U = 1$). And so, taking the K_G-dual of the sequence (2.3), we get an exact sequence

$$(2.3') \qquad 0 \to K_{G/\mathfrak{a}G} \to K_G \to {}_\rho(A/U)$$

of graded R-modules. Let us now assume $r = \mathrm{r}_J(I) \geq 2$. Then by (1.2) we have $a(G) = r - 1$; hence $K_G \cong G(r-1)$ (cf. [**HIO, (36.12)**]) and in particular we have $[K_G]_{1-r} = A/I$. Recall that $[{}_\rho(A/U)]_{1-r} = (0)$, as $1 - r < 0$. Then, considering the homogeneous components in the sequence (2.3') with degree $1-r$, we find $[K_{G/\mathfrak{a}G}]_{1-r}$ is isomorphic to A/I. Hence the ideal $\mathfrak{a}$ kills A/I as it kills $K_{G/\mathfrak{a}G}$, so that we have $\mathfrak{a} \subseteq I$. Therefore by (2.1) (1), $\mathfrak{a} = (0)$. This is impossible, because $\mathrm{ht}_A I = 0$. Thus $\mathrm{r}_J(I) \leq 1$.

We put $f = y + bt$. For each $x \in R$ we denote by $\overline{x}$ the reduction of $x \mod IR$.

LEMMA (2.5). *If $\mathrm{r}_J(I) \leq 1$, the following assertions are true.*

(1) *bt is $G/\mathfrak{a}G$-regular and f is G-regular.*

(2) *$\sum_{i \geq 2} R_i \subseteq fR$.*

(3) *For any element a of I, $\overline{at} \in fG$ if and only if $a \in bU$.*

PROOF. (1) Since b is $A/\mathfrak{a}$-regular (cf. (2.1) (1)) and $I^2 = bI$, by [**VV, 3.1**] we have bt is $G((I+\mathfrak{a})/\mathfrak{a})$-regular; hence both the elements bt and f are $G/\mathfrak{a}G$-regular, as $G/\mathfrak{a}G = G((I+\mathfrak{a})/\mathfrak{a})$ and $f \equiv bt \mod \mathfrak{a}R$. Because bt kills ${}_\rho\mathfrak{a}$ and y is regular on ${}_\rho\mathfrak{a}$ by (2.2), we have f to be regular on ${}_\rho\mathfrak{a}$ too; thus f is regular on G (use the sequence (2.3)).

(2) As $yI = (0)$ by (2.1) (1), we get $I^2t^2 = (y + bt)(It)$; hence $I^2t^2 \subseteq fR$ and so $R_i \subseteq fR$ for $i \geq 2$.

(3) Assume that $\overline{at} \in fG$ and choose $g = \sum_{0 \leq i \leq n} g_i t^i \in R$ $(g_i \in I^i)$ so that $at \equiv fg$ mod IR. Then as $fg = (y + bt)(\sum_{0 \leq i \leq n} g_i t^i) = yg_0 + bg_0 t + bg_1 t^2 + \ldots + bg_n t^{n+1}$

(use the fact that $yI = (0)$), we get $yg_0 \equiv 0 \mod I$ and $a \equiv bg_0 \mod I^2$; hence $g_0 \in U$, as $yg_0 \in I \cap \mathfrak{a} = (0)$ and $U = (0) : y$ by (2.2). Thus we have $a \in bU$, because $a \equiv bg_0 \mod bI$. If $a = bu$ for some $u \in U$, then $at = but = (y + bt)u$, and we get $\overline{at} \in fG$.

For an A-module M let $\mathrm{Socle}_A M$ denote the socle $[(0) :_M \mathfrak{m}]$ of M.

LEMMA (2.6). *Suppose A is a Gorenstein ring. Then $(y + b) : \mathfrak{m} \subseteq (y, b)$.*

PROOF. As $\mathfrak{a} \cap U = (0)$ by (2.1) (1), we get $(y, b) = (y) \oplus (b)$; hence $\mu_A((y, b)) = 2$, while $\mu_A((y+b)) = 1$. Thus $(y, b) \neq (y+b)$. Hence $\mathrm{Socle}_A A/(y+b) \subseteq (y, b)/(y + b)$, because $(y, b)/(y + b) \neq (0)$ and $\mathrm{Socle}_A A/(y + b)$ is generated by a single element.

For a given Cohen-Macaulay local ring B let $\mathrm{r}(B)$ stand for the Cohen-Macaulay type of B, that is $\mathrm{r}(B) = \ell_B(\mathrm{Ext}^n_B(B/\mathfrak{n}, B))$, where $\ell_B(*)$ denotes the length, $n = \dim B$, and $\mathfrak{n}$ is the maximal ideal in B.

We put $\mathbf{M} = \mathfrak{m}G + G_+$, the unique graded maximal ideal in G.

THEOREM (2.7). *Assume A is a Gorenstein ring and $\mathrm{r}_J(I) \leq 1$. Let $S = ((I : \mathfrak{m}) \cap (bU : I))/I$. Then G is a Cohen-Macaulay ring and $\mathrm{r}(G_{\mathbf{M}}) = \ell_A(S) + 1$.*

PROOF. G is a Cohen-Macaulay ring, as f is G-regular (cf. (2.5) (1)). Notice that bt is $G/\mathfrak{a}G$-regular. Then by (2.3) we have an exact sequence

$$(\#) \qquad 0 \to {}_{\rho}(\mathfrak{a}/y\mathfrak{a}) \xrightarrow{\overline{\varphi}} G/fG \xrightarrow{\overline{\psi}} G/(\mathfrak{a}G + btG) \to 0$$

of R-modules, where $\overline{\varphi}$, $\overline{\psi}$ denote the maps induced from φ and ψ, respectively. We put $Z = [fG :_G \mathbf{M}]$ and $L = (I : \mathfrak{m}) \cap (bU : I)$. Then $Z/fG = \mathrm{Socle}_G G/fG$ and $S = L/I$. We have $\mathrm{r}(G_{\mathbf{M}}) = \ell_G(Z/fG)$, as f is G-regular ((2.5) (1)). We will show that $Z/fG \cong L/I \oplus \mathrm{Socle}_A(\mathfrak{a}/y\mathfrak{a})$ as A-modules.

If $u \in L$, by (2.5) (3) we have $It \cdot \overline{u} \subseteq fG$. Hence

CLAIM 1. *$\overline{u} \in Z$, if $u \in L$.*

Let $z \in Z$ and choose $g \in R$ so that $z = \overline{g}$; we may assume $g = g_0 + g_1 t$ ($g_0 \in A$ and $g_1 \in I$), as we are interested in Z/fG and as $\sum_{i \geq 2} G_i \subseteq fG$ by (2.5) (2). Then because $(It)g \equiv (0) \mod fR + IR$ and $(It)g \equiv (Ig_0)t \mod fR$, we have $(Ig_0)t \subseteq fR + IR$, whence $Ig_0 \subseteq bU$ by (2.5) (3). Thus $g_0 \in bU : I$. We write $bg_0 = bu$ with $u \in U$ and put $a = g_0 - u$. Then $a \in \mathfrak{a} = (0) : b$. We furthermore have

CLAIM 2. *(1) $u \in L$.*

(2) $a \in (y)$ and $g_1 \in (b)$.

PROOF OF CLAIM 2. Let x be an element of $\mathfrak{m}$. We write $xg \equiv fh \mod IR$ with $h = \sum_{0 \leq i \leq n} h_i t^i \in R$ ($h_i \in I^i$). Then because $x\{(u+a) + g_1 t\} \equiv yh_0 + bh_0 t + bh_1 t^2 + \dots + bh_n t^{n+1} \mod IR$, comparing the coefficients of t in both sides, we find $xg_1 = bh_0 + bi$ for some $i \in I$, while considering the constant terms, we find

$x(u+a) = yh_0 + j$ for some $j \in I$; hence $xu - j = yh_0 - xa \in U \cap \mathfrak{a} = (0)$, so that we get $xu = j$ and $xa = yh_0$. Thus $u \in I : \mathfrak{m}$. On the other hand, since $I\mathfrak{a} = (0)$ and $g_0 \in bU : I$ as we checked above, we have $u = g_0 - a \in bU : I$; hence $u \in L$. Now recall that $xa = yh_0 = y(h_0 + i)$ and $xg_1 = b(h_0 + i)$. Then as $x(a+g_1) = (y+b)(h_0+i)$, we get $a + g_1 \in \mathrm{Socle}_A A/(y+b)$. Hence $a + g_1 \in (y, b)$ by (2.6). Because $a \in \mathfrak{a}$, $g_1 \in U$, and $(y,b) = (y) \oplus (b) \subseteq \mathfrak{a} + U = \mathfrak{a} \oplus U$, we get $a \in (y)$ and $g_1 \in (b)$, as claimed.

We write $g_1 = bc$ with $c \in A$. Then $g_1 t \equiv -yc \mod fR$, as $g_1 t = (y + bt)c - yc$, and so we get $z \equiv \overline{u + a - yc} \mod fG$. Therefore we may assume $g_0 = u + a$ and $g_1 = 0$. Then as $\overline{u} \in Z$ by Claim 1, we see $\overline{a} \in Z$ too. Hence $a \in [y\mathfrak{a} :_{\mathfrak{a}} \mathfrak{m}]$, because $\overline{a} \mod fG = \overline{\varphi}(a \mod y\mathfrak{a})$ in the above exact sequence (#). Thus we have $Z/fG = \{\overline{u} \mod fG \mid u \in L\} + \overline{\varphi}(\mathrm{Socle}_A(\mathfrak{a}/y\mathfrak{a}))$. As $\overline{\varphi}(\mathrm{Socle}_A(\mathfrak{a}/y\mathfrak{a}) = \mathrm{Im}\overline{\varphi} \cap Z/fG$ and as $\overline{\psi}(Z/fG) \cong S$ by Claim 3 below, we get the required isomorphism $Z/fG \cong S \oplus \mathrm{Socle}_A(\mathfrak{a}/y\mathfrak{a})$. Now recall that $\mathfrak{a} \cong K_{A/U}$ and y is $\mathfrak{a}$-regular (cf. (2.1) and (2.2)). Then we know by **[HK, 6.10]** $\ell_A(\mathrm{Socle}_A(\mathfrak{a}/y\mathfrak{a})) = \mathrm{r}_A(\mathfrak{a}) = 1$, so that $\mathrm{r}(G_{\mathbf{M}}) = \ell_A(S) + 1$ as claimed. This completes the proof of (2.7) modulo the next

CLAIM 3. $\overline{\psi}(Z/fG) \cong S$.

PROOF OF CLAIM 3. We have $\overline{\psi}(Z/fG) \cong (L + \mathfrak{a})/(I + \mathfrak{a})$, while

$$\begin{aligned}(L+\mathfrak{a})/(I+\mathfrak{a}) &\cong L/(L \cap (I + \mathfrak{a}))\\ &= L/(I + (L \cap \mathfrak{a}))\\ &= L/I,\end{aligned}$$

Because $L \cap \mathfrak{a} = (0)$ (cf. (2.1) (1), (3)); hence we get $\overline{\psi}(Z/fG) \cong S$.

Let us note a proof of Theorem (1.1).

PROOF OF THEOREM (1.1). We may assume $\lambda(I) = 1$; otherwise $I = (0)$ and our assertion is obviously true.

(1) $\Rightarrow$ (2) A is Gorenstein and $\mathrm{r}_J(I) \leq 1$ by (2.4). We get $(I : \mathfrak{m}) \cap (bU : I) = I$ by (2.7), as $\mathrm{r}(G_{\mathbf{M}}) = 1$.

(2) $\Rightarrow$ (1) This directly follows from (2.7).

The Gorensteinness of G naturally affects the number $\mu_A(I)$ of generators for I. To see it we need the following

LEMMA (2.8). *Suppose* $\mathrm{r}_J(I) \leq 1$. *Then* $G/\mathfrak{m}G$ *is a Cohen-Macaulay ring, and* bt *is regular on* $G/\mathfrak{m}G$.

PROOF. As $\dim G/\mathfrak{m}G = \lambda(I) = 1$, it suffices to check the second assertion. We identify $G/\mathfrak{m}G = \bigoplus_{n \geq 0} I^n/\mathfrak{m}I^n$. Let $a \in I^n$ $(n \geq 0)$ and assume $ab \in \mathfrak{m}I^{n+1}$. If $n > 0$, then as $\mathfrak{m}I^{n+1} = b\mathfrak{m}I^n$, we get $ab = bc$ for some $c \in \mathfrak{m}I^n$; hence $a = c \in \mathfrak{m}I^n$, as $a - c \in \mathfrak{a} \cap I = (0)$. When $n = 0$, we have $a \in \mathfrak{m}$, since $ab \in \mathfrak{m}I$ and $b \notin \mathfrak{m}I$ (recall that $J = (b)$ is a minimal reduction of I). Hence $a \in \mathfrak{m}I^n$ in any case, which proves bt is $G/\mathfrak{m}G$-regular.

Let $H^i_{\mathbf{M}}(*)$ ($i \in \mathbb{Z}$) denote the i-th local cohomology functor of G relative to $\mathbf{M}$.

THEOREM (2.9). *If G is a Gorenstein ring, then $\mu_A(I) = \ell_A(\mathrm{Socle}_A(A/I)) + 1$.*

PROOF. By (2.4) we have $\mathrm{r}_J(I) \leq 1$. Hence $[G/(\mathfrak{m}G + btG)]_2 = (0)$ and bt is, by (2.8), regular on $G/\mathfrak{m}G$. Consequently, $[G/(\mathfrak{m}G + btG)]_1$ $(= I/((b) + \mathfrak{m}I))$ is contained in the socle $\mathrm{Socle}_G(G/(\mathfrak{m}G + btG))$ of $G/(\mathfrak{m}G + btG)$. Hence $[\mathrm{Socle}_G(G/(\mathfrak{m}G + btG))]_1 = [G/(\mathfrak{m}G + btG)]_1$, so that we have

$$\ell_G([\mathrm{Socle}_G(G/(\mathfrak{m}G + btG))]_1) = \mu_A(I) - 1.$$

Now apply the functors $H^i_{\mathbf{M}}(*)$ to the exact sequence

$$0 \to [G/\mathfrak{m}G](-1) \overset{bt}{\to} G/\mathfrak{m}G \to G/(\mathfrak{m}G + btG) \to 0$$

and consider the derived exact sequence

$$0 \to H^0_{\mathbf{M}}(G/(\mathfrak{m}G + btG)) \to [H^1_{\mathbf{M}}(G/\mathfrak{m}G)](-1) \overset{bt}{\to} H^1_{\mathbf{M}}(G/\mathfrak{m}G) \to 0$$

of local cohomology. Then, because $G/(\mathfrak{m}G + btG) = H^0_{\mathbf{M}}(G/(\mathfrak{m}G + btG))$ and because the socle $\mathrm{Socle}_G(H^1_{\mathbf{M}}(G/\mathfrak{m}G))$ of $H^1_{\mathbf{M}}(G/\mathfrak{m}G)$ is annhilated by bt, we get an isomorphism

$$\mathrm{Socle}_G(H^1_{\mathbf{M}}(G/\mathfrak{m}G)) \cong [\mathrm{Socle}_G(G/(\mathfrak{m}G + btG))](1).$$

Hence $[\mathrm{Socle}_G(H^1_{\mathbf{M}}(G/\mathfrak{m}G))]_0 = [\mathrm{Socle}_G(G/(\mathfrak{m}G + btG))]_1$ and so we have $\ell_G([\mathrm{Socle}_G(H^1_{\mathbf{M}}(G/\mathfrak{m}G))]_0) = \mu_A(I) - 1$; thus $\ell_G([K_{G/\mathfrak{m}G}]_0) = \mu_A(I) - 1$ (cf. [**GW, (2.1.2)**]). On the other hand, because $K_{G/\mathfrak{m}G} \cong \mathrm{Hom}_G(G/\mathfrak{m}G, G)$ by [**GW, (2.2.9)**] (recall that G is a Gorenstein ring of $a(G) = 0$, cf. (1.2)), we get $[K_{G/\mathfrak{m}G}]_0 \cong [(0) :_G \mathfrak{m}]_0 = \mathrm{Socle}_A(A/I)$. Hence $\mu_A(I) = \ell_A(\mathrm{Socle}_A(A/I)) + 1$, as claimed.

As $b \notin \mathfrak{m}I$, we have $\mu_A(I) = 1$ if and only if $I = J$. Hence by (2.9) we get the following, in which the *if* part was first proven by [**HRZ, (4.1)**].

COROLLARY (2.10). *Assume G is a Gorenstein ring. Then A/I is a Cohen-Macaulay ring if and only if $I = J$.*

Recall that the condition (iii) in (2) of Theorem (1.1) is satisfied, once A/I is Cohen-Macaulay. Hence we have the following

COROLLARY (2.11). *Assume A/I is a Cohen-Macaulay ring. Then the following conditions are equivalent.*

(1) *G is a Gorenstein ring.*
(2) *A is a Gorenstein ring and $\mathrm{r}_J(I) \leq 1$.*
(3) *A is a Gorenstein ring and $I = J$.*

3. Higher-dimensional cases and Proof of Theorem (1.3)

In this section we assume $d = \dim A \geq 1$ and $s = \mathrm{ht}_A I \geq 0$. We put $R = R(I)$, $G = G(I)$, and $\mathbf{M} = \mathfrak{m}G + G_+$. Let J be a minimal reduction of I. We put $\mathcal{F} = \mathrm{Min}_A A/I$. The main purpose of this section is to prove Theorem (1.3). To begin with we note the following, whose proof may be found in [**HH1, Proof of 2.2**].

LEMMA (3.1). *J contains a system $a_1, a_2, \ldots, a_s, b$ of generators such that $IA_Q = KA_Q$ for all $Q \in \mathcal{F}$, where $K = (a_1, a_2, \ldots, a_s)$.*

Let $a_1, a_2, \ldots, a_s, b$ be the system of generators for J as is in (3.1). Then, since $K : I \not\subseteq Q$ for any $Q \in \mathcal{F}$ and $\dim A/I = d - s$, we may choose, with $y \in K : I$, a system $x_1, x_2, \ldots, x_{d-s-1}, y$ of parameters for the ring A/I. Let $L = (x_1, x_2, \ldots, x_{d-s-1})$.

LEMMA (3.2). *(1) $a_1, a_2, \ldots, a_s, x_1, x_2, \ldots, x_{d-s-1}, y + b$ is a system of parameters of A.*

(2) *The ideal $(I + L)/(K + L)$ is generically a complete intersection in the ring $A/(K + L)$.*

PROOF. (1) Choose $Q \in \mathrm{Spec} A$ so that $Q \supseteq K + L + (y + b)$. Then, as $y^2 \equiv -yb \mod Q$ and $yb \in K$, we have $y, b \in Q$; hence $Q \supseteq J = K + (b)$. Thus $Q \supseteq I$ as $\sqrt{I} = \sqrt{J}$, and so we have $Q = \mathfrak{m}$, since Q contains the system $x_1, x_2, \ldots, x_{d-s-1}, y$ of parameters for A/I.
(2) Let $Q \in \mathrm{Min}_A A/(I + L)$. Then $Q \neq \mathfrak{m}$, as $\dim A/(I + L) = 1$. Hence $y \notin Q$, so that we have $(I + L)A_Q = (K + L)A_Q$, as $K : I$ is not contained in Q. Hence the assertion follows.

We put $\mathbf{b} = (a_1 t, a_2 t, \ldots, a_s t, x_1, x_2, \ldots, x_{d-s-1}, y + bt)R$.

PROPOSITION (3.3). $\mathbf{M} = \sqrt{\mathbf{b}G}$. *Hence $a_1 t, a_2 t, \ldots, a_s t, x_1, x_2, \ldots, x_{d-s-1}, y + bt$ is a G-regular sequence if and only if G is a Cohen-Macaulay ring.*

PROOF. Choose $Q \in \mathrm{Spec} R$ so that $Q \supseteq \mathbf{b} + IR$. Then, as $y^2 \equiv -y(bt) \mod Q$ and $(yb)t \in Kt$, we get $y, bt \in Q$. Hence $Q \supseteq Jt$, so that Q contains $R_+ = \sum_{n \geq 1} I^n t^n$, while $Q \supseteq \mathfrak{m}R$ as $Q \supseteq I + L + (y)$. Hence $Q \supseteq \mathfrak{m}R + R_+$. Thus $\mathbf{M} = \sqrt{\mathbf{b}G}$, and so the sequence $a_1 t, a_2 t, \ldots, a_s t, x_1, x_2, \ldots, x_{d-s-1}, y + bt$ forms a system of parameters for the local ring $G_{\mathbf{M}}$. Hence the second assertion follows.

We put $\overline{A} = A/(K + L)$, $\overline{I} = I\overline{A}$, and $\overline{J} = J\overline{A}$. Then $\overline{A}$ is a Cohen-Macaulay ring, $\dim \overline{A} = 1$, $\mathrm{ht}_{\overline{A}} \overline{I} = 0$, and $\overline{I}$ is generically a complete intersection in $\overline{A}$ (cf. (3.2)). Furthermore, if G is a Cohen-Macaulay ring, the sequence $a_1 t$, $a_2 t, \ldots$, $a_s t$, x_1, $x_2, \ldots, x_{d-s-1}$ is G-regular by (3.3) and we have a canonical isomorphism

$$G(\overline{I}) \cong G/(a_1 t, \ldots, a_s t, x_1, \ldots, x_{d-s-1})G$$

of graded A-algebras as well as the equality

$$(K + L) \cap I^n = KI^{n-1} + LI^n \tag{3.4}$$

for all $n \in \mathbb{Z}$ (cf. [**VV, 2.3**]). In particular $G(\overline{I})$ is necessarily a Cohen-Macaulay ring. It is a Gorenstein ring if and only if so is the ring G.

LEMMA (3.5). *Assume G is a Cohen-Macaulay ring. Then $\lambda(\overline{I}) = 1$, $\overline{J}$ is a minimal reduction of $\overline{I}$, and $\mathrm{r}_{\overline{J}}(\overline{I}) = \mathrm{r}_J(I)$.*

PROOF. If $\lambda(\overline{I}) \neq 1$, we have $\lambda(\overline{I}) = 0$, so that $\overline{I}^n = (0)$ for some $n > 0$. Then as $I^n \subseteq K+L$, by the equality (3.4) we get $I^n = KI^{n-1}+LI^n$. Thus $I^n = KI^{n-1}$ by Nakayama's lemma. Hence K is a reduction of I, which contradicts our assumption $\lambda(I) = s+1$. Thus $\lambda(\overline{I}) = 1$ and $\overline{J}$ is a minimal reduction of $\overline{I}$. To see that $\mathrm{r}_{\overline{J}}(\overline{I}) \geq \mathrm{r}_J(I)$, let $\overline{I}^{n+1} = \overline{J}\,\overline{I}^n$ for some $n \geq 0$. Then as $I^{n+1} \subseteq JI^n+K+L$, by (3.4) we have $I^{n+1} = JI^n + (K+L) \cap I^{n+1} = JI^n + LI^{n+1}$; so we get $I^{n+1} = JI^n$ and hence $\mathrm{r}_{\overline{J}}(\overline{I}) \geq \mathrm{r}_J(I)$. The opposite inequality is obvious.

The above lemmas (3.2) and (3.5) enable us to generalize several results obtained in Section 2 to those of the case where $d = \dim A \geq 1$ and $s = \mathrm{ht}_A I \geq 0$. Let us state the assertions precisely, although some of them are of the same formulation as were given in Section 2. We begin with the next

PROPOSITION (3.6). *If G is a Gorenstein ring, then A is a Gorenstein ring and $\mathrm{r}_J(I) \leq 1$.*

PROOF. As $G(\overline{I})$ is a Gorenstein ring, by (2.4) we get $\mathrm{r}_{\overline{J}}(\overline{I}) \leq 1$, whence $\mathrm{r}_J(I) \leq 1$ by (3.5).

PROOF OF THEOREM (1.3). G is a Cohen-Macaulay ring if $\mathrm{r}_J(I) \leq 1$, as A/I is Cohen-Macaulay by our assumption (cf. [**HH1, 2.9**]). Because $a_1, \ldots, a_s, x_1, \ldots, x_{d-s-1}$ is an A-regular sequence ((3.2)), $\overline{A}$ is a Gorenstein ring if and only if so is A, while we have $\mathrm{r}_{\overline{J}}(\overline{I}) = \mathrm{r}_J(I)$ by (3.5) if G is Cohen-Macaulay. Thus the equivalence in (1.3) follows from (2.11).

The implication (1) $\Rightarrow$ (2) of the next result is due to [**HRZ, (4.1)**]. We would like to give a brief proof in our context.

COROLLARY (3.7). *Assume that $I = J$. Then the following conditions are equivalent.*

(1) *G is a Gorenstein ring.*
(2) *A is a Gorenstein ring and A/I is Cohen-Macaulay.*

PROOF. (1) $\Rightarrow$ (2) As $G(\overline{I})$ is Gorenstein and $\overline{I} = \overline{J}$, by (2.10) we get $\overline{A}/\overline{I}$ is a Cohen-Macaulay ring. Since $x_1, x_2, \ldots, x_{d-s-1}$ is a G-regular sequence and $G_0 = A/I$, we have it is A/I-regular, too. Hence A/I is a Cohen-Macaulay ring, as $\overline{A}/\overline{I} = A/I + (x_1, x_2, \ldots, x_{d-s-1})$. The assertion that A is Gorenstein is well-known (cf. e.g., [**G, Proof of (2.6)**]).
(2) $\Rightarrow$ (1) This follows from (1.3).

The generalization of Theorem (2.9) is as follows.

THEOREM (3.8). *Suppose that G is a Gorenstein ring. Then we have*

$$\mu_A(I) = \ell_A(\mathrm{Ext}_A^{d-s-1}(A/\mathfrak{m}, A/I)) + s + 1.$$

PROOF. As $G(\overline{I})$ is a Gorenstein ring, we have $\mu_A(\overline{I}) = \ell_A(\mathrm{Socle}_A \overline{A}/\overline{I}) + 1$ by (2.9). Since $\overline{I} = (I+L)/(K+L) \cong I/((K+L)\cap I) = I/(K+LI)$ (cf. (3.4)), we get $\mu_A(\overline{I}) = \mu_A(I/K)$. Hence $\mu_A(\overline{I}) = \mu_A(I) - s$, as the elements $a_1, a_2, ..., a_s$ are part of a minimal system of generators for I. As $x_1, x_2, ..., x_{d-s-1}$ is an A/I-regular sequence, we have isomorphisms

$$\mathrm{Socle}_A \overline{A}/\overline{I} \cong \mathrm{Hom}_A(A/\mathfrak{m}, A/(I+L)) \cong \mathrm{Ext}_A^{d-s-1}(A/\mathfrak{m}, A/I),$$

and we get $\mu_A(I) = \mu_A(\overline{I}) + s = \{\ell_A(\mathrm{Ext}_A^{d-s-1}(A/\mathfrak{m}, A/I)) + 1\} + s$, as claimed.

We put $S = A \setminus \bigcup_{Q\in\mathcal{F}} Q$. Let $I^{(n)}$ ($n \in \mathbb{Z}$) denote the n-th symbolic power of I, that is $I^{(n)} = \{x \in A | sx \in I^n \text{ for some } s \in S\}$.

COROLLARY (3.9). *Assume A is Gorenstein and A/I is Cohen-Macaulay. Then $I = J$, if $I^{(n)} = I^n$ for all $n \in \mathbb{Z}$.*

PROOF. Let $Q \in \mathrm{Spec}A$ such that $Q \supseteq I$ and $\mathrm{ht}_A Q = s+1$. Then as $I^n A_Q = (IA_Q)^{(n)}$, we get $\mathrm{depth} A_Q/I^n A_Q \geq 1$ for all $n \geq 1$; hence by Burch's theorem [**Bu**] we have $\lambda(IA_Q) = s$. Consequently, by [**CN**] IA_Q is generated by an A_Q-regular sequence of length s, so that we have $IA_Q = JA_Q$. Hence the hypothesis of [**HH1, 2.2**] is fulfilled and we get $I^2 = JI$; thus $I = J$ by (1.3).

PROOF OF COROLLARY (1.4). As $\mathfrak{p}^{(n)} = \mathfrak{p}^n$ for all $n \geq 1$, by [**Bu**] we get $d-2 \leq \lambda(\mathfrak{p}) < d = \dim A$. If $\lambda(\mathfrak{p}) = d-2$, then $\mathfrak{p}$ is a complete intersection (cf. [**CN**]), so that $\mu_A(\mathfrak{p}) = d-2$. Assume $\lambda(\mathfrak{p}) = d-1$. Then we have $\mathrm{ad}(\mathfrak{p}) = 1$, so the assertion follows from (3.9).

Kunz [**K**] showed that A/I cannot be a Gorenstein ring for any almost complete intersection I in a regular local ring A. The next result is an analytic deviation one version of his theorem.

COROLLARY (3.10). *Assume that A is a regular local ring and that for any $Q \in \mathrm{Spec}A$ such that $Q \supseteq I$ and $\mathrm{ht}_A Q = s+1$, IA_Q is generated by an A_Q-regular sequence of length s. Then A/I is not a Gorenstein ring.*

PROOF. If A/I were a Gorenstein ring, we get $\mathrm{r}_J(I) \leq 1$ by [**HH1, 2.2**], so $I = J$ by (1.3), whence A/I cannot be Gorenstein by Kunz's theorem [**K, 1.2**]. This is a contradiction.

We would like to devote the rest of this section to the analysis of an example (3.11). The prime ideal $\mathfrak{p}$ in it was chosen among defining ideals of projective space monomial curves. Analyzing their affine pieces, we can easily check $\mathfrak{p}^{(n)} \neq \mathfrak{p}^n$ for any $n \geq 2$ (cf. [**H2, (2.5)**]). In particular $G(\mathfrak{p})$ is not an integral domain.

EXAMPLE (3.11). Let $k[X,Y,Z,W]$ and $k[s,t]$ be polynomial rings over an infinite field k. Let $n \geq 1$ be an integer and let $\varphi : k[X,Y,Z,W] \to k[s,t]$ be the homomorphism of k-algebras defined by $\varphi(X) = s^{7n+6}$, $\varphi(Y) = s^{7n+5}t$, $\varphi(Z) = s^{4n+3}t^{3n+3}$, and $\varphi(W) = t^{7n+6}$. We put $S = k[X,Y,Z,W]$ and $A = S_M$, where $M = (X,Y,Z,W)S$. Let $P = \mathrm{Ker}\varphi$ and $\mathfrak{p} = PA$. Then $R(\mathfrak{p})$ and $G(\mathfrak{p})$ are Gorenstein rings.

PROOF. The ideal P is minimally generated by the following five elements:[2]

$$F_1 = X^{n+1}W - Y^nZ^2, \qquad F_2 = X^nZ^5 - Y^{n+3}W^2,$$
$$F_3 = X^{2n+1}Z^3 - Y^{2n+3}W, \; F_4 = X^{3n+2}Z - Y^{3n+3}, \text{ and}$$
$$F_5 = Z^7 - XY^3W^3.$$

There are relations

$$F_3^2 = X^nY^{n+3}ZF_1^2 + F_2F_4,$$
$$F_3F_5 = XF_2^2 - Y^3Z^3WF_1^2, \text{ and}$$
$$F_4F_5 = XF_2F_3 - Y^3Z(X^{n+1}W + Y^nZ^2)F_1^2.$$

Therefore, letting $J = (F_1, F_2, F_4 - F_5)$, we get $P^2 = JP$. Hence $\mathrm{ad}(\mathfrak{p}) = 1$, and so $G(\mathfrak{p})$ is a Cohen-Macaulay ring (cf. [**HH1, 2.9**]).

CLAIM 1. *(1)* $PS_P = (F_1, F_2)S_P$.

(2) $Y^nZ^5W \in (F_1, F_2) : P$ *and* $Y^nZ^5W \notin P$.

(3) Y^nZ^5W, $X - W$ *is a system of parameters for* $A/\mathfrak{p}$.

PROOF OF CLAIM 1. We have

$$WF_3 = X^nZ^3F_1 + Y^nF_2,$$
$$Y^nF_5 = XWF_2 - Z^5F_1, \text{ and}$$
$$Z^5F_4 = Y^{n+3}Z(X^{n+1}W + Y^nZ^2)F_1 + X^{2n+2}ZF_2.$$

Hence $Y^nZ^5W \in (F_1, F_2) : P$. As $Y^nZ^5W \notin P$, we get $PS_P = (F_1, F_2)S_P$. Thus the assertions (1) and (2) follow. The assertion (3) is clear, because $\varphi(Y^nZ^5W) = s^{7n^2+25n+15}t^{23n+21}$ and $\varphi(X - W) = s^{7n+6} - t^{7n+6}$ (recall $S/P \cong k[s^{7n+6}, s^{7n+5}t, s^{4n+3}t^{3n+3}, t^{7n+6}]$).

We find, by Claim 1 and (3.3), that $F_1t, F_2t, X - W$ forms a $G(\mathfrak{p})$-regular sequence. Let $\overline{A} = A/(F_1, F_2, X - W)$ and $\overline{\mathfrak{p}} = \mathfrak{p}\overline{A}$. Then $\overline{A}$ is a one dimensional Gorenstein local ring, $\overline{\mathfrak{p}}$ is generically a complete intersection of height 0, $\lambda(\overline{\mathfrak{p}}) = 1$, and $\overline{\mathfrak{p}}^2 = (F_4 - F_5)\overline{\mathfrak{p}}$, (see (3.2) and (3.5)). We will show that $\overline{\mathfrak{p}}$ satisfies the condition (iii) in (2) of Theorem (1.1). Then by the reduction technique following (3.3), we have $G(\mathfrak{p})$ is a Gorenstein ring and so by [**GH, 2.12**], $R(\mathfrak{p})$ is necessarily a Gorenstein ring (recall $\mathrm{ht}_A\mathfrak{p} = 2$).

We put $T = k[X, Y, Z]$ and $B = k[X, Y, Z]_{(X,Y,Z)}$. Let us canonically identify $T = S/(X - W)S$ and $B = A/(X - W)A$. Let f_i denote the image of F_i in T; hence

$$f_1 = X^{n+2} - Y^nZ^2, \qquad f_2 = X^nZ^5 - X^2Y^{n+3},$$
$$f_3 = X^{2n+1}Z^3 - XY^{2n+3}, \; f_4 = X^{3n+2}Z - Y^{3n+3}, \text{ and}$$
$$f_5 = Z^7 - X^4Y^3.$$

We put $I = (f_1, f_2, f_3, f_4, f_5)B$, $K = (f_1, f_2)B$, and $b = f_4 - f_5$. Let $\mathfrak{m}$ denote the maximal ideal in B. Then as $I/K = \overline{\mathfrak{p}}$ under the above identification, it suffices to check that $[(bU(I) + K) : I] \cap (I : \mathfrak{m}) = I$, where $U(I) = \bigcap_{Q \in \mathcal{F}} (IA_Q \cap A)$ denotes

[2]One can check the generation by computing a Gröbner basis for the ideal P. See [**BR**] for the detail, where the algorithm is described.

the unmixed component of I. Notice that $I^2 \subseteq bI+K$, as $\overline{\mathfrak{p}}^2 = (F_4 - F_5)\overline{\mathfrak{p}}$; hence $I \subseteq [(bU(I)+K):I] \cap (I:\mathfrak{m})$.

For simplicity, let us assume that $n \geq 2$. (The case where $n = 1$ is similarly proved. Choose $g_1 = f_1$, $g_2 = Z^5 - XY^4$, and $g_3 = Y^5 - X^2Z^3$.) We put $g_1 = f_1$, $g_2 = Y^{n+3} - X^{n-2}Z^5$, $g_3 = f_5$, and $U = (g_1, g_2, g_3)$. Then we have

CLAIM 2. *(1)* $U = U(I)$.

(2) $(g_1, g_3) : g_2 = (X^4, Z^2)$, $(g_1, g_2) : g_3 = (X^{n-2}, Y^n)$, *and* $(g_2, g_3) : g_1 = (Y^3, Z^5)$.

(3) g_1, g_2, g_3 *forms an unconditioned d-sequence in* B *(cf.* [**H1, 1.1**]*).*

PROOF OF CLAIM 2. (1) Notice that the ideal U is generated by the maximal minors of the matrix
$$\begin{bmatrix} Z^5 & X^4 & Y^n \\ Y^3 & Z^2 & X^{n-2} \end{bmatrix}$$
and that the sequence

(a)
$$0 \to B^2 \xrightarrow{\begin{bmatrix} Z^5 & X^4 & Y^n \\ Y^3 & Z^2 & X^{n-2} \end{bmatrix}} B^3 \xrightarrow{\begin{bmatrix} g_1 \\ g_2 \\ g_3 \end{bmatrix}} B$$

is exact (cf. [**BE**]). Then we have B/U is Cohen-Macaulay and $\dim B/U = 1$. Since

(b)
$$\begin{aligned} f_2 &= -X^2 g_2, \\ f_3 &= X^{n-1}Z^3 g_1 - XY^n g_2, \text{ and} \\ f_4 &= X^{n-2}Z(X^{n+2} + Y^n Z^2) g_1 - Y^{2n} g_2, \end{aligned}$$

we get $UB[1/X] = IB[1/X]$. Hence $U = U(I)$, as X is regular on B/U.
(2) Because

(c)
$$Z^5 g_1 + X^4 g_2 + Y^n g_3 = 0 \quad \text{and}$$

(d)
$$Y^3 g_1 + Z^2 g_2 + X^{n-2} g_3 = 0,$$

we have $(X^4, Z^2) \subseteq (g_1, g_3) : g_2$. To check the opposite inclusion, let $y \in (g_1, g_3) : g_2$ and write $yg_2 = -(xg_1 + zg_3)$ with $x, z \in B$. Then by the exactness of the sequence (a), we have
$$[x \quad y \quad z] = [u \quad v] \begin{bmatrix} Z^5 & X^4 & Y^n \\ Y^3 & Z^2 & X^{n-2} \end{bmatrix}$$
for some $u, v \in B$, whence $y \in (X^4, Z^2)$ as claimed. We similarly have $(g_1, g_2) : g_3 = (X^{n-2}, Y^n)$ and $(g_2, g_3) : g_1 = (Y^3, Z^5)$ too.
(3) As X^{n-2}, Y^n, g_3 is a B-regular sequence and as $(g_1, g_2) : g_3 = (X^{n-2}, Y^n)$ by (2), we get $(g_1, g_2) : g_3 = (g_1, g_2) : g_3^2$, which guarantees g_1, g_2, g_3 is a d-sequence (cf. [**H1, p.258**]. Notice that the sequence g_1, g_2 is B-regular.). We similarly have g_1, g_2, g_3 forms a d-sequence in any order.

CLAIM 3. $I : \mathfrak{m} = I + (XY^{n-1}, Y^{2n-1})Zg_2$.

PROOF OF CLAIM 3. Notice that $I = (g_1, g_3) + (X^2, XY^n, Y^{2n}, Z^2)g_2$ (use relations (b) and (d) in the proof of Claim 2). First, let $\alpha \in I : \mathfrak{m}$. Then as $\alpha \in U$ by Claim 2 (1), we have $\alpha = \beta g_2 + \gamma$ for some $\beta \in B$ and $\gamma \in I$. Let $\xi \in \mathfrak{m}$. Then as $\xi\alpha, \xi\gamma \in I$, we have $\xi\beta g_2 \in I$, and so we may write $\xi\beta g_2 = \eta + \zeta g_2$ with $\eta \in (g_1, g_3)$ and $\zeta \in (X^2, XY^n, Y^{2n}, Z^2)$. Then $\xi\beta - \zeta \in (g_1, g_3) : g_2 = (X^4, Z^2)$, so we have $\xi\beta \in (X^2, XY^n, Y^{2n}, Z^2)$. Thus $\beta \in (X^2, XY^n, Y^{2n}, Z^2) : \mathfrak{m}$ and so we get $\beta \in (XY^{n-1}Z, Y^{2n-1})$, because $(X^2, XY^n, Y^{2n}, Z^2) : \mathfrak{m} = (XY^{n-1}Z, Y^{2n-1}Z)$. Thus $\alpha \in (XY^{n-1}Z, Y^{2n-1}Z)g_2 + I$ and we have the inclusion $I : \mathfrak{m} \subseteq I + (XY^{n-1}Z, Y^{2n-1}Z)g_2$. The opposite inclusion is obvious.

We will show $((bU + K) : I) \cap (I : \mathfrak{m}) \subseteq I$. First notice that

$$bU + K = (g_1, X^2g_2, g_3^2, Y^{2n}g_2^2 - g_2g_3),$$

which directly follows from equations (b) and (c). Let $\xi \in ((bU+K) : I) \cap (I : \mathfrak{m})$. Then by Claim 3 we may write, with $\alpha, \beta \in k$, $\xi \equiv \alpha XY^{n-1}Zg_2 + \beta Y^{2n-1}Zg_2$ mod I. Let $\eta = \alpha XY^{n-1}Zg_2 + \beta Y^{2n-1}Zg_2$. Then as $\eta \in (bU + K) : I$ and as $XY^ng_2 \in I$, we get the element $\eta(XY^ng_2) = \alpha X^2Y^{2n-1}Zg_2^2 + \beta XY^{3n-1}Zg_2^2$ belonging to the ideal $bU+K$; hence $\beta XY^{3n-1}Zg_2^2 \in bU+K$, as $X^2g_2 \in bU+K$. Thus $\beta = 0$ by Claim 4 below, and so we have $\eta = \alpha XY^{n-1}Zg_2$. Similarly, as $Y^{2n}g_2 \in I$, we have $\eta(Y^{2n}g_2) = \alpha XY^{3n-1}Zg_2^2 \in bU + K$, and so by Claim 4, $\alpha = 0$ too. Thus $\eta = 0$ and $\xi \in I$. This completes the proof of (3.11) modulo the next

CLAIM 4. $XY^{3n-1}Zg_2^2 \notin bU + K$.

Before entering the proof of the claim, we would like to note here that

(e) $$U^2 \cap (g_1, g_2) = (g_1, g_2)U,$$

(f) $$U^2 \cap (g_1) = g_1U, \quad \text{and}$$

(g) $$R(U) \cong \mathrm{Sym}_A(U), \quad \text{the symmetric algebra of } U$$

(cf. [**H1, 2.1**] and [**HSV, p.488 Remark**]). In particular, the defining ideal of $R(U)$ is generated by linear forms. More precisely, let $B[W_1, W_2, W_3]$ be a polynomial ring and let $\Phi : B[W_1, W_2, W_3] \to R(U)$ denote the homomorphism of B-algebras defined by $\Phi(W_1) = g_1t$, $\Phi(W_2) = g_2t$, and $\Phi(W_3) = g_3t$. Then the kernel of Φ is generated by two linear forms $Z^5W_1 + X^4W_2 + Y^nW_3$ and $Y^3W_1 + Z^2W_2 + X^{n-2}W_3$.

PROOF OF CLAIM 4. Assume $XY^{3n-1}Zg_2^2 \in bU + K$ and write

(h) $$XY^{3n-1}Zg_2^2 = p_1g_1 + p_2X^2g_2 + p_3g_3^2 + p_4(Y^{2n}g_2^2 - g_2g_3)$$

with $p_i \in B$. First we will show that $p_1, p_2 \in U$. In fact, as $p_1g_1 + p_2X^2g_2 \in U^2 \cap (g_1, g_2)$, by (e) we have $p_1g_1 + p_2X^2g_2 = u_1g_1 + u_2g_2$ with $u_1, u_2 \in U$. Hence $X^2p_2 - u_2 \in (g_1) : g_2 = (g_1)$ and so $X^2p_2 \in U$. Since X^2 is B/U-regular, we get

$p_2 \in U$. Moreover, since $p_1 g_1 \in U^2 \cap (g_1) = g_1 U$ by (f), we have $p_1 \in U$. We write $p_1 = q_1 g_1 + q_2 g_2 + q_3 g_3$ and $p_2 = r_1 g_1 + r_2 g_2 + r_3 g_3$ with $q_i, r_i \in B$. Then the relation (h) gives rise to the equality

$$q_1 g_1^2 + (q_2 + X^2 r_1) g_1 g_2 + q_3 g_1 g_3 + (-XY^{3n-1}Z + X^2 r_2 + p_4 Y^{2n}) g_2^2 + (X^2 r_3 - p_4) g_2 g_3 + p_3 g_3^2 = 0$$

and so we get a form of degree 2 in the defining ideal of the Rees algebra $R(U)$. As the ideal is generated by $Z^5W_1 + X^4W_2 + Y^nW_3$ and $Y^3W_1 + Z^2W_2 + X^{n-2}W_3$, we get in $B[W_1, W_2, W_3]$

$$q_1 W_1^2 + (q_2 + X^2 r_1) W_1 W_2 + q_3 W_1 W_3 + (-XY^{3n-1}Z + X^2 r_2 + p_4 Y^{2n}) W_2^2 + (X^2 r_3 - p_4) W_2 W_3 + p_3 W_3^2 = (Z^5W_1 + X^4W_2 + Y^nW_3)(a_1W_1 + a_2W_2 + a_3W_3) + (Y^3W_1 + Z^2W_2 + X^{n-2}W_3)(b_1W_1 + b_2W_2 + b_3W_3)$$

for some $a_i, b_i \in B$. Let us compare the coefficients of W_2^2 in both sides of this equality. Then we have

$$-XY^{3n-1}Z + X^2 r_2 + p_4 Y^{2n} = X^4 a_2 + Z^2 b_2.$$

Hence $b_2 \in (X^2, Y^{2n})$ and $p_4 - XY^{n-1}Z \in (X^2, Z^2)$. On the other hand, comparing the coefficients of W_2W_3 , we have

$$X^2 r_3 - p_4 = X^4 a_3 + Y^n a_2 + Z^2 b_3 + X^{n-2} b_2.$$

Thus $p_4 \in (X^2, Y^n, Z^2)$, whence $XY^{n-1}Z \in (X^2, Y^n, Z^2)$ as $p_4 - XY^{n-1}Z \in (X^2, Z^2)$. This is impossible.

4. Reduction numbers and the Cohen-Macaulay property of $R(I)$

In this section we shall prove Theorem (1.5). Similarly as is in Section 3, we assume $d = \dim A \geq 1$ and $s = \mathrm{ht}_A I \geq 0$. Let $R = R(I)$ and $G = G(I)$. Then, as $I \not\subseteq Q$ for some $Q \in \mathrm{Ass}A$, we have $\dim R = d + 1$ (cf. [**V, 1.6**]). We note the following results due to [**TI**] and [**I2**].

THEOREM (4.1). *(1)([**TI, 1.1**]) The following conditions are equivalent.*

(i) *R is a Cohen-Macaulay ring.*
(ii) *G is a Cohen-Macaulay ring and $a(G) < 0$.*

*(2) ([**I2, 3.1**]) If $s \geq 2$, the following conditions are equivalent.*

(i) *R is a Gorenstein ring.*
(ii) *G is a Gorenstein ring and $a(G) = -2$.*

We divide the proof of Theorem (1.5) into a few steps. Similarly as before, let J be a minimal reduction of I and let $a_1, a_2, ..., a_s, b$ be the system of generators for J given by (3.1); hence $a_1, a_2, ..., a_s$ forms an A-regular sequence (cf. (3.2)). Let $K = (a_1, a_2, ..., a_s)$, so that $J = K + (b)$. We put $\mathcal{F} = \mathrm{Min}_A A/I$. We begin with the following

LEMMA (4.2). *Assume that A/I is a Cohen-Macaulay ring. Then $I^2 \cap K = KI$. Furthermore, if $\mathrm{r}_J(I) \leq 2$, we have $I^n \cap K = KI^{n-1}$ for all $n \in \mathbb{Z}$.*

PROOF. Assume $L = (I^2 \cap K)/KI \neq (0)$ and choose $Q \in \mathrm{Ass}_A A/KI$ so that $L_Q \neq (0)$. Then $Q \supseteq I$ clearly and so, considering the exact sequence $0 \to \bigoplus^s A/I \to A/KI \to A/K \to 0$ (recall $K/KI \cong \bigoplus^s A/I$), we get $Q \in \mathcal{F} = \mathrm{Ass}_A A/I$. Hence $LA_Q = (0)$, as $IA_Q = KA_Q$ (cf. (3.1)). This is impossible. Thus $I^2 \cap K = KI$. To check the second assertion, assume that $n \geq 3$ and that our equality holds for $n-1$. Then $I^n = JI^{n-1}$ as $I^3 = JI^2$, so that we have

$$\begin{aligned} I^n \cap K &= (KI^{n-1} + bI^{n-1}) \cap K \\ &= KI^{n-1} + bI^{n-1} \cap K \\ &= KI^{n-1} + b[(K : b) \cap I^{n-1}]. \end{aligned}$$

Hence, as $(K : b) \cap I = K$ (cf. **[HH1, 2.1]**), we get $I^n \cap K = KI^{n-1} + b(K \cap I^{n-1})$, and the hypothesis on n now yields $I^n \cap K = KI^{n-1} + bKI^{n-2} = KI^{n-1}$, as claimed.

When A/I is Cohen-Macaulay and $\mathrm{r}_J(I) \leq 2$, it follows from (4.2) that a_1t, $a_2t, \ldots, a_st$ is a G-regular sequence and there is an isomorphism $G/(a_1t, \ldots, a_st)G \cong G(I/K)$ of graded A-algebras (cf. **[VV, 2.3]**). Notice that the ideal I/K of A/K is generically a complete intersection, and $\mathrm{ad}(I/K) = 1$ (cf. Proof of (3.5)).

COROLLARY (4.3). *Assume A/I is a Cohen-Macaulay ring and* $\mathrm{depth} A/I^2 \geq \dim A/I - 1$. *Then* $\mathrm{depth} A/(I^2 + K) \geq \dim A/I - 1$.

PROOF. We get $(I^2+K)/I^2 \cong K/KI$, as $I^2 \cap K = KI$ by (4.2), so we have an exact sequence $0 \to \bigoplus^s A/I \to A/I^2 \to A/(I^2+K) \to 0$ (recall $K/KI \cong \bigoplus^s A/I$). Depth Lemma now yields the assertion.

Passing to the ideal I/K in A/K, Lemma (4.2) and Corollary (4.3) will enable us to reduce Theorem (1.5) to that of the case where $s = \mathrm{ht}_A I = 0$. In this case, as $(0) : I \not\subseteq Q$ for any $Q \in \mathcal{F}$, we may choose $y \in (0) : I$ so that $\dim A/(I + (y)) = d - 1$. We furthermore have the following

PROPOSITION (4.4). *Suppose that A/I is a Cohen-Macaulay ring, $\mathrm{ht}_A I = 0$, and $\mathrm{depth} A/I^2 \geq d-1$. Let $\mathrm{r}_J(I) \leq 2$. Then $\mathfrak{m}$ contains a system $x_1, x_2, \ldots, x_{d-1}$, y of parameters for A/I such that $x_1, x_2, \ldots, x_{d-1}$ forms an A/I^n-regular sequence for all $n \geq 1$.*

PROOF. We will show $\mathrm{depth} A/I^n \geq d - 1$ for all $n \geq 1$; this is certainly true for $n = 1, 2$. Let $n \geq 3$ and assume $\mathrm{depth} A/I^{n-1} \geq d - 1$; hence we have $\mathrm{depth}_A I^{n-1} = d$. Therefore, because $I^n = bI^{n-1} \cong I^{n-1}$ (recall that $I \cap ((0) : b) = (0)$, cf. **[HH1, 2.1]**), we have $\mathrm{depth}_A I^n = d$ too, whence $\mathrm{depth} A/I^n \geq d - 1$ as required. We put $\mathcal{H} = \bigcup_{n>0} \mathrm{Ass}_A A/I^n$. Notice that $\mathcal{H}$ is a finite set (cf. **[Br1]**). If $d \geq 2$, as $\mathfrak{m} \notin \mathcal{H}$ and as $\dim A/(I + (y)) > 0$, we may choose $x_1 \in \mathfrak{m}$ so that

$$x_1 \notin \bigcup_{Q \in \mathcal{H}} Q \text{ and } x_1 \notin \bigcup_{Q \in \mathrm{Min}_A A/(I+(y))} Q.$$

Repeat this procedure and we will get the required sequence $x_1, x_2, ..., x_{d-1}$.

We are now ready to prove Theorem (1.5).

PROOF OF THEOREM (1.5). By the remark following (4.2) and Corollary (4.3) we may assume $s = 0$.

(1) $\Rightarrow$ (2) Let $x_1, x_2, ..., x_{d-1}$ be the system of parameters for A/I, which we chose in the remark after (3.1). Then by (3.3) $x_1, x_2, ..., x_{d-1}$ is a G-regular sequence. Hence we have $\mathrm{depth}_A I^n/I^{n+1} \geq d-1$, so that the induction on n yields $\mathrm{depth} A/I^n \geq d-1$ for all $n \geq 1$.

(2) $\Rightarrow$ (1) Let $x_1, x_2, ..., x_{d-1}, y$ be the elements of $\mathfrak{m}$ given by (4.4). We put $L = (x_1, x_2, ..., x_{d-1})$. Then as $L \cap I^n = LI^n$ for any $n \in \mathbb{Z}$, the sequence $x_1, x_2, ..., x_{d-1}$ is G-regular; hence $G/LG \cong G((I+L)/L)$. Because $(I+L)/L$ is generically a complete intersection in A/L (cf. (3.2) (2)), passing to the ideal $(I+L)/L$, we may assume $d = 1$. Let $f = y + bt \in R$. We will show f is G-regular. Let $g = \sum_{0 \leq i \leq n} g_i t^i \in R$ ($g_i \in I^i$) and assume that $fg \equiv 0 \mod IR$. Then, as $fg = yg_0 + bg_0 t + bg_1 t^2 + ... + bg_n t^{n+1} \in IR$, we get $bg_i \in I^{i+2} = bI^{i+1}$ for $i > 0$. Write $bg_i = bs_i$ with $s_i \in I^{i+1}$. Then since $g_i - s_i \in ((0):b) \cap I = (0)$ (cf. (2.1) (1)), we have $g_i = s_i$ if $i > 0$. For $i = 0$, we get $yg_0 \in I$ whence $g_0 \in I$, as y is A/I-regular. Thus $g \in IR$, and so f is G-regular. See the next proposition for a proof of the last assertion.

PROPOSITION (4.5). *Assume* $\dim A \geq 3$ *and* A/I *is a Cohen-Macaulay ring of* $\dim A/I = 1$. *Then* R *is a Cohen-Macaulay ring, if* $\mathrm{r}_J(I) \leq 2$. *When* $\dim A = 3$, *the converse is also true.*

PROOF. If $\mathrm{r}_J(I) \leq 2$, by (1.5) G is Cohen-Macaulay, so that by [**GH, 2.1**] R is a Cohen-Macaulay ring. The second assertion is due to [**GH, 2.1**], since $s = 2$ if $d = 3$.

Let us note an example of prime ideals $\mathfrak{p}$ which satisfy the requirements in (1.5).

EXAMPLE (4.6). Let $S = k[X, Y, Z, W]$ and $T = k[s]$ be polynomial rings over an infinite field k. Let $\varphi : S \to T$ be the homomorphism of k-algebras defined by $\varphi(X) = s^4$, $\varphi(Y) = s^5$, $\varphi(Z) = s^6$, and $\varphi(W) = s^7$. We put $P = \mathrm{Ker}\varphi$, $M = (X, Y, Z, W)S$, and $A = S_M$. Let $\mathfrak{p} = PA$. Then $R(\mathfrak{p})$ is a Cohen-Macaulay ring, but neither $R(\mathfrak{p})$ nor $G(\mathfrak{p})$ are Gorenstein. We have $\mathrm{r}_J(\mathfrak{p}) \geq 2$ for any minimal reduction J of $\mathfrak{p}$.

PROOF. P is minimally generated by the maximal minors of the matrix

$$N = \begin{bmatrix} X & Y & Z & W \\ Y & Z & W & X^2 \end{bmatrix}.$$

Let $[ij]$ denote the minor of N consisting of the i-th and j-th columns. We put $a_1 = [12]$, $a_2 = [13]$, $a_3 = [34]$, $a_4 = [14] + [23]$, $a_5 = [14]$, and $a_6 = [24]$. Then $P = (a_1, a_2, a_3, a_4, a_5, a_6)$ and we have $a_5^2 = a_1a_3 - a_2a_6 + a_4a_5$ and $a_6^2 = a_3a_4 - Xa_1a_5 + Xa_2^2 + Za_1^2 - Ya_1a_2$. A direct checking shows $P^3 = (a_1, a_2, a_3, a_4)P^2$. Hence $J = (a_1, a_2, a_3, a_4)A$ is a minimal reduction of $\mathfrak{p}$. So by (1.5) $R(\mathfrak{p})$ and $G(\mathfrak{p})$ are Cohen-Macaulay rings. As $\mathfrak{p} \neq J$, $G(\mathfrak{p})$ is not Gorenstein by (1.3), whence $R(\mathfrak{p})$ cannot be a Gorenstein ring (cf. (4.1) (2)). The last assertion follows from (1.3).

The Gorensteinness of $R(I)$ is a rather strong restriction on I, as we claim in (4.7) and (4.9). First we will explore the case $s \geq 2$.

THEOREM (4.7). *Assume A/I is a Cohen-Macaulay ring and $s \geq 2$. Then the following conditions are equivalent.*

(1) *R is a Gorenstein ring.*
(2) *A is a Gorenstein ring, $I^2 = JI$, and $s = 2$.*
(3) *A is a Gorenstein ring, $I = J$, and $s = 2$.*

PROOF. (cf. [**GH, 2.12**]) As $s = \operatorname{grade}(I, A) \geq 2$, R is a Gorenstein ring if and only if G is a Gorenstein ring and $a(G) = -2$ (cf. (4.1) (2)). If G is a Gorenstein ring, we have $a(G) = -s$ by (1.2) as $\mathrm{r}_J(I) \leq 1$ by (3.6). Hence R is a Gorenstein ring if and only if G is a Gorenstein ring and $s = 2$. Thus the assertion directly follows from (1.3).

Next we consider the case where s = 1. To do this we need the following

PROPOSITION (4.8) ([**GN, (4.2)**]). *Suppose R is a Gorenstein ring. Then*

(1) *G is a Cohen-Macaulay ring,*
(2) *$K_A = \operatorname{Hom}_A(I, A)$, and*
(3) *there exists an exact sequence*

$$0 \to G(-2) \to K_G \to {}_\rho\mathrm{Ext}_A^1(A/I, A)(-1) \to 0$$

of graded R-modules, where ${}_\rho\mathrm{Ext}_A^1(A/I, A)$ denotes $\mathrm{Ext}_A^1(A/I, A)$ that is considered to be a graded R-module via the canonical projection $\rho : R \to A$.

In the next theorem A/I is not assumed to be Cohen-Macaulay.

THEOREM (4.9). *If $s = 1$, the following conditions are equivalent.*

(1) *R is a Gorenstein ring.*
(2) *$I = J$ and $K_A = \operatorname{Hom}_A(I, A)$.*

PROOF. Let a, b be a system of generators for J such that a is A-regular and $IA_Q = aA_Q$ for any $Q \in \mathcal{F}$.
(2) $\Rightarrow$ (1) As $I = J$, we have $(a) : b = (a) : I$; hence $(a) : b = K_A$, as $(a) : I \cong \operatorname{Hom}_A(I, A)$. The assertion that R is a Gorenstein ring now follows similarly as in the proof of [**GS2, (2) $\Rightarrow$ (1) in (4.4)**].

(1) $\Rightarrow$ (2) As $K_A = \mathrm{Hom}_A(I, A)$ by (4.8), it suffices to show $I = J$. As G is a Cohen-Macaulay ring by (4.1) (1), by (3.3) we may choose elements $x_1, x_2, ..., x_{d-2} \in \mathfrak{m}$ so that $at, x_1, x_2, ..., x_{d-2}$ is a G-regular sequence. As is in Section 3, we put $K = (a)$ and $L = (x_1, x_2, ..., x_{d-2})$. Then since $I^n \cap L = LI^n$ for all $n \in \mathbb{Z}$ (cf. **[VV, 2.3]**), we have an isomorphism $R/LR \cong R((I + L)/L)$; hence $R((I+L)/L)$ is a Gorenstein ring. Thus passing to the ideal $(I+L)/L$ in A/L, we may assume $d = 2$ (cf. (3.2) and (3.5)).

We have $\mathrm{r}_J(I) \leq 1$ by (1.2), as $a(G) = -1$ by (4.8). Let $\mathfrak{a} = K : I$ and $\overline{G} = G((I + \mathfrak{a})/\mathfrak{a})$. Then bt is $\overline{G}$-regular by (2.5)(1). Therefore we have $a(\overline{G}) = a(\overline{G}/bt\overline{G}) - 1$ by **[GW, (3.1.6)]**), whence $a(\overline{G}) \leq 0$ (recall $\overline{G}/bt\overline{G} = G((I + \mathfrak{a})/((b) + \mathfrak{a}))$ and $I^2 \subseteq (b) + \mathfrak{a}$). We will show $a(\overline{G}) \neq 0$. First, consider the natural epimorphism $G(I/K) \to \overline{G}$. Then, applying the functor $\mathrm{Ext}^1_G(*, K_G)$, we have an embedding $K_{\overline{G}} \subseteq K_{G(I/K)}$ of graded modules (use the fact that $\overline{G}$ and $G(I/K)$ are Cohen-Macaulay rings of dimension 1). On the other hand, we have an isomorphism $K_{G(I/K)} \cong [K_G/(at)K_G](1)$ (cf. **[GW, (2.2.10)]**), since at is G-regular (cf. (3.3)) and since $G/(at)G \cong G(I/K)$. Therefore we have $[K_{G(I/K)}]_0 \cong [K_G]_1$, as $a(G) = -1$. Hence $[K_{\overline{G}}]_0 \subseteq [K_{G(I/K)}]_0 = [K_G]_1$, by which we have an embedding

$$H^0_{\mathfrak{m}}([K_{\overline{G}}]_0) \subseteq H^0_{\mathfrak{m}}([K_{G(I/K)}]_0) = H^0_{\mathfrak{m}}([K_G]_1)$$

of A-modules. Now recall the exact sequence

$$0 \to G(-2) \to K_G \to {}_{\rho}\mathrm{Ext}^1_A(A/I, A)(-1) \to 0$$

given in (4.8) and we have $H^0_{\mathfrak{m}}([K_G]_1) = (0)$, since $[K_G]_1 \cong \mathrm{Ext}^1_A(A/I, A)$ and since both G and K_G are Cohen-Macaulay G-modules of dimension 2. Thus we get $H^0_{\mathfrak{m}}([K_{\overline{G}}]_0) = (0)$, while the $\mathfrak{m}$-primary ideal $I + \mathfrak{a}$ kills $K_{\overline{G}}$. Hence $[K_{\overline{G}}]_0 = (0)$ and so we have $a(\overline{G}) \neq 0$. Consequently $a(\overline{G}/bt\overline{G}) = a(\overline{G}) + 1 \leq 0$, so that we have $I + \mathfrak{a} = (b) + \mathfrak{a}$, since $[\overline{G}/bt\overline{G}]_1 \cong (I + \mathfrak{a})/((b) + \mathfrak{a})$. Thus $I = (b) + \mathfrak{a} \cap I = (b) + K = J$ (cf., e.g., **[HH1, 2.1]**), which completes the proof of (4.9).

COROLLARY (4.10). *G is not a Gorenstein ring, if $s = 1$ and if R is Gorenstein.*

PROOF. Assume G is a Gorenstein ring. By (4.9) we have $I = J$, and hence by (3.7) A/I is Cohen-Macaulay. Let a, b be the system of generators for I given in (3.1). Then, as A is a Gorenstein ring, by (4.9) the ideal $(a) : I$ is principal, say $(a) : I = (c)$. We write $a = ce$ and $bc = fa$ with $e, f \in A$. Then $e \in (a) : c$ and $(a) : c = (a) : ((a) : I) = I$ (cf. **[K, 3.1]**), recall A/I is Cohen-Macaulay). Let $e = za + wb$ with $z, w \in A$. Then we have $a = c(za + wb) = acz + afw$, whence $cz + fw = 1$. As the ideal I is not principal, we have $c \in \mathfrak{m}$. Hence $f \notin \mathfrak{m}$, so that $a \in (b)$, and we have $I = (b)$. This is a contradiction.

We conclude this paper with an exploration of the case where the base ring A is regular.

THEOREM (4.11). *Let A be a regular local ring and $\mathfrak{p}$ a prime ideal in A of* $\dim A/\mathfrak{p} = 1$. *Then the following conditions are equivalent.*

(1) *$G(\mathfrak{p})$ is a Gorenstein ring.*

(2) $\mu_A(\mathfrak{p}) \leq \dim A$.

PROOF. Recall $\mathrm{ad}(\mathfrak{p}) \leq 1$. If $\mathrm{ad}(\mathfrak{p}) = 0$, by **[CN]** $\mathfrak{p}$ is a complete intersection, and we have nothing to prove. Assume $\mathrm{ad}(\mathfrak{p}) = 1$. Then by (1.3), $G(\mathfrak{p})$ is a Gorenstein ring if and only if $\mathfrak{p} = J$, where J is a minimal reduction of $\mathfrak{p}$. Since $\mu_A(J) = \lambda(\mathfrak{p}) = d$ and J is generated by part of a minimal system of generators for $\mathfrak{p}$, the assertion (2) is equivalent to saying that $\mathfrak{p} = J$.

To prove Corollary (1.6), we may assume $\mathrm{ad}(\mathfrak{p}) = 1$. Hence the assertion (1) (resp. (2)) now directly follows from (4.5) (resp. (4.7)).

REFERENCES

[AH] I.M. Aberbach and C. Huneke, *An improved Briançon-Skoda theorem with applications to the Cohen-Macaulayness of Rees algebras*, Preprint (1993).

[BE] D. Buchsbaum and D. Eisenbud, *What makes a complex exact*, J. Alg. **25** (1973), 259-268.

[Bu] L. Burch, *Codimension and analytic spread*, Proc. Camb. Phil. Soc. **72** (1972), 369-373.

[Br1] M. Brodmann, *Asymptotic stability of* $\mathrm{Ass}(M/I^nM)$, Proc. Amer. Math. Soc. **74** (1979), 16-18.

[Br2] M. Brodmann, *Rees rings and form rings of almost complete intersections*, Nagoya Math. J. **88** (1982), 1-16.

[BR] H. Brensinsky and B. Renschuch, *Bsaisbestimmung Veronesescher Projektionsideale mit allgemeiner Nullstelle* $(t_0^m, t_0^{m-r}t_1^r, t_0^{m-s}t_1^s, t_1^m)$, Math. Nachr. **96** (1980), 257-269.

[CN] R. C. Cowsik and M. V. Nori On the fibres of blowing up, J. Indian Math. Soc. **40** (1976), 217-222.

[G] S. Goto, *The Cohen-Macaulay symbolic Rees algebras for curve singularities*, Memoirs of the Americal Mathematical Society (to appear).

[GH] S. Goto and S. Huckaba, *On graded rings associated to analytic deviation one ideals*, Amer. J. Math. (to appear).

[GN] S. Goto and K. Nishida, *Filtrations and the Gorenstein property of the associated Rees algebras*, Memoirs of the Americal Mathematical Society (to appear).

[GS1] S. Goto and Y. Shimoda, *On the Rees algebras of Cohen-Macaulay local rings* **68** (1982), Lecture Notes in Pure and Appl. Math., 201-231.

[GS2] S. Goto and Y. Shimoda, *On the Gorensteinness of Rees and form rings of almost complete intersections*, Nagoya Math. J. **92** (1983), 69-88.

[GW] S. Goto and K. Watanabe, *On graded rings I*, J. Math. Soc. Japan **30** (1978), 179-213.

[H1] C. Huneke, *The theory of d-sequences and powers of ideals*, Advances in Math. **46** (1982), 249-279.

[H2] C. Huneke, *The primary components of and integral closures of ideals in 3-dimensional regular local rings*, Math. Ann. **275** (1986), 617-635.

[HH1] S. Huckaba and C. Huneke, *Powers of ideals having small analytic deviation*, Amer. J. Math. **114** (1992), 367-403.

[HH2] S. Huckaba and C. Huneke, *Rees algebras of ideals having small analytic deviation*, Trans. Amer. Math. Soc. (to appear).

[HIO] M. Herrmann, S. Ikeda, and U. Orbanz, *Equimultiplicity and Blowing-up*, Springer-Verlag, Berlin. Heidelberg. New York, 1988.

[HK] J. Herzog and E. Kunz, *Der kanonische Modul eines Cohen-Macaulay-Rings*, Springer L. N. M., 238, Springer-Verlag, Berlin. Heidelberg. New York. Tokyo, 1971.

[HRZ] M. Herrmann, J. Ribbe, and S. Zarzuela, *On Rees and form rings of almost complete intersections*, Preprint.

[HSV] J. Herzog, A. Simis, and W. V. Vasconcelos, *Approximation complexes of blowing-up rings*, J. Alg. **74** (1982), 466-493.

[I1] S. Ikeda, *The Cohen-Macaulayness of the Rees algebras of local rings*, Nagoya Math. J. **89** (1983), 47-63.

[I2] S. Ikeda, *On the Gorensteinness of Rees algebras over local rings*, Nagoya Math. J. **102** (1986), 135-154.

[K] E. Kunz, *Almost complete intersections are not Gorenstein rings*, J. Alg. **28** (1974), 111-115.

[NR] D. G. Northcott and D. Rees, *Reductions of ideals in local rings*, Proc. Camb. Phil. Soc. **50** (1954), 145-158.

[TI] N. V. Trung and S. Ikeda, *When is the Rees algebra Cohen-Macaulay*, Communications in Algebra **17** (1989), 2893-2922.

[V] G. Valla, *Certain graded algebras are always Cohen-Macaulay*, J. Alg. **42** (1976), 537-548.

[VV] P. Valabrega and G. Valla, *Form rings and regular sequences*, Nagoya Math. J. **72** (1978), 93-101.

[Va] W. V. Vasconcelos, *On the equations of Rees algebras*, J. Reine Angew. Math. **418** (1991), 189-218.

DEPARTMENT OF MATHEMATICS, SCHOOL OF SCIENCE AND TECHNOLOGY MEIJI UNIVERSITY, HIGASHIMITA 1-1-1, TAMA-KU KAWASAKI-SHI 214 JAPAN

E-mail address: goto@math.meiji.ac.jp

DEPARTMENT OF MATHEMATICS, TOKYO METROPOLITAN UNIVERSITY, MINAMI-OHSAWA 1-1, HACHIOJI-SHI 192-03 JAPAN

E-mail address: ynakamu@math.metro-u.ac.jp

Contemporary Mathematics
Volume 159, 1994

Prime ideals in birational extensions of polynomial rings

WILLIAM J. HEINZER, DAVID LANTZ, SYLVIA M. WIEGAND

1. Introduction. Our work in this paper concerns the general question: Which partially ordered sets, or equivalently topological spaces, can occur as the prime spectra of certain types of Noetherian rings? In **[HW]**, Heinzer and Wiegand consider the spectrum of $R[x]$ for R a countable semilocal Noetherian domain of dimension one and x an indeterminate over R. In the case where R has a unique maximal ideal it is shown in [**HW, Theorem 2.7**] that there are exactly two possibilities for Spec($R[x]$), one of which occurs when R is Henselian and the other when R is not Henselian. If R has more than one maximal ideal, then the spectrum of $R[x]$ is uniquely determined up to isomorphism by the number of maximal ideals of R. (In this latter case, R cannot be Henselian.) In [**HLW, Theorem 2.3**], in analogy with the above-mentioned result for polynomial rings, it is shown that there are precisely two possibilities for the spectrum of the projective line Proj($R[s,t]$) over a countable local Noetherian domain R of dimension one, the crucial distinction again being whether or not R is Henselian.

In this article we continue an investigation begun in [**HLW**] of the prime spectra of certain birational extensions of the affine line Spec($R[x]$) over a countable semilocal Noetherian domain R of dimension one. We focus our attention on

---

1991 *Mathematics Subject Classification.* 13E05, 13F20, 13G05, 13H99, 13J15.

Heinzer and Wiegand gratefully acknowledge the support of the National Science Foundation. Lantz and Wiegand thank Purdue University for its hospitality while we were working on this research.

This is the final version of this paper.

extensions of the following type: Let $\mathbf{m}_1, \mathbf{m}_2, ..., \mathbf{m}_n$ be the maximal ideals of R, $f \in R[x] - \bigcup_{i=1}^{n} \mathbf{m}_i[x]$, $(f,\ g)$ an $R[x]$-sequence, and $B = R[x, g/f]$. In Theorems 2.6 and 2.8 we describe explicitly the j-spectrum of B.[1] We prove in Theorem 4.1 that if R is Henselian, then Spec(B) is uniquely determined by the j-spectrum of B. We conjecture that when R is not Henselian, there is also only one partially ordered set Spec(B) with a given j-spectrum. This conjecture is equivalent to the assertion that if R is not Henselian and T is a finite set of maximal ideals of B of height two, then there exist infinitely many prime ideals of B of height one that are contained in each maximal ideal in T and in no other maximal ideals of B. In Theorem 3.1 we prove this assertion in the case where T is a singleton set. In Corollary 4.6, we exhibit a large class of examples for which the conjecture is true for all T.

We are grateful to Roger Wiegand for helpful conversations and to the referee for helpful suggestions.

All rings we consider are commutative and contain a multiplicative identity. The terms "local" and "semilocal" include "Noetherian."

1.1 NOTATION. For U a partially ordered set of finite dimension, elements u, w of U, and T a finite subset of U, we set

$$\begin{aligned} \mathrm{G}(u) &= \{w \in U \mid w > u\} \qquad \text{and} \\ \mathrm{L_e}(T) &= \{w \in U \mid w < t \iff t \in T\} \\ &= \{w \in U \mid \mathrm{G}(w) = T\} \quad . \end{aligned}$$

Note that the set called L(T) in **[HW]** is denoted $\mathrm{L_e}(T)$ here. (The notation is chosen to suggest the "exactly-less-than" set.)

Let $\mathcal{M}(U)$ denote the set of maximal elements of U of maximal height.

If $\mathbf{m}_1, \mathbf{m}_2, ..., \mathbf{m}_n$ are the maximal ideals of R, the phrase "B is a *small birational extension* of $R[x]$" will mean that there exist a polynomial $f \in R[x] - \bigcup_{i=1}^{n} \mathbf{m}_i[x]$ and an $R[x]$-sequence (f, g) so that $B = R[x, g/f]$. We write j- Spec(B) for the j-spectrum of B.

[1] The j-spectrum of B is the partially ordered set of prime ideals of B which are intersections of maximal ideals.

In preparation for our analysis of the spectra of small birational extensions of polynomial rings, we reformulate the conclusions of [**HW**] using the j-spectrum.

1.2 THEOREM. *If R is a countable semilocal domain of dimension one with exactly n maximal ideals, then $U = j\text{-}\operatorname{Spec}(R[x])$ has the following properties which characterize it as a partially ordered set:*

(P0) U is countable.

(P1) U has a unique minimal element u_0.

(P2) U has dimension 2.

(P3) U has infinitely many height-one maximal elements.

(P4) U has exactly n height-one nonmaximal elements. We denote these elements $u_1, u_2, \ldots u_n$. They satisfy:

(i) $\mathrm{G}(u_1) \cup \cdots \cup \mathrm{G}(u_n) = \mathcal{M}(U)$,

(ii) $\mathrm{G}(u_i) \cap \mathrm{G}(u_j) = \emptyset$ *for* $i \neq j$, *and*

(iii) $\mathrm{G}(u_i)$ *is infinite for each* i, $1 \leq i \leq n$.

NOTES. (1) To say that certain properties *characterize* a given partially ordered set means that every partially ordered set with those properties is order-isomorphic to the given partially ordered set.

(2) If R is uncountable, then (P1)–(P4) hold, but, obviously, different cardinality assertions are necessary to obtain a characterization of $j\text{-}\operatorname{Spec}(R[x])$.

(3) If D is a domain of dimension two in which each element of the j-spectrum has the same height in the j-spectrum of D as in the entire spectrum of D (so that, in particular, (0) is a j-prime), then $\operatorname{Spec}(D)$ differs from $j\text{-}\operatorname{Spec}(D)$ solely by the additional height-one elements which are not j-primes. (The *j-primes* are the primes which are intersections of maximal ideals of B.)

As Theorem 1.3 (below) indicates, there are just two ways to complete $j\text{-}\operatorname{Spec}(R[x])$ to $\operatorname{Spec}(R[x])$. In order to state the theorem, it is helpful to introduce the *j-subset*, $j\text{-}V$, of a partially ordered set V of finite dimension, defined recursively as follows: $u \in j\text{-}V$ iff either (i) u is maximal in V, or (ii) there are infinitely many covers of u in $j\text{-}V$, where an element w of V is called a *cover* of u if $u < w$ and there is no element v of V for which $u < v < w$.

1.3 THEOREM. **[HW, Theorem 2.7]** *If R is a countable semilocal domain of dimension one with exactly n maximal ideals, then there are at most two possibilities for the partially ordered set $U = \mathrm{Spec}(R[x])$ and they are distinguished by the properties (P6) and (P6′) below:*

(1) In case R is not Henselian, U satisfies

(P6) For each nonempty finite subset T of $\mathcal{M}(U)$, $\mathrm{L_e}(T)$ is infinite.

(1′) In case R is Henselian (and then $n = 1$), U satisfies

(P6′) For each finite subset T of $\mathcal{M}(U)$ of cardinality greater than one, $\mathrm{L_e}(T)$ is empty. For each element t of $\mathcal{M}(U)$, $\mathrm{L_e}(\{t\})$ is infinite.

(2) Let V_1, V_2 be countable partially ordered sets of dimension two with unique minimal elements such that

(i) The subsets j-V_1 and j-V_2 of V_1 and V_2 are order-isomorphic by an order-isomorphism φ,

(ii) For $i = 1, 2$, the height of each element of j-V_i agrees with its height in V_i, and

(iii) For each finite subset T of $\mathcal{M}(V_1)$, $\mathrm{L_e}(T)$ and $\mathrm{L_e}(\varphi(T))$ are either both empty or both infinite.

Then V_1 is order-isomorphic to V_2 by an extension of φ.

(3) Let V be a countable partially ordered set of dimension two with unique minimal element such that

(i) The subset j-V has the properties (P0)–(P4) in 1.2,

(ii) The height of each element of j-V agrees with its height in V, and

(iii) V satisfies (P6) or(P6′) (whichever U satisfies).

Then V is order-isomorphic to U.

NOTE. When we stipulate the j-spectrum and either (P6) or (P6′), we have described all possibilities for $\mathrm{Spec}(R[x])$ given that R is a countable semilocal domain of dimension one with exactly n maximal ideals. However, condition (ii) is necessary in (3), because conditions (i) and (iii) together do not uniquely determine a partially ordered set V. For example, j-V might have dimension two while V has dimension three, or there could be maximal elements of V of height two in V but only of height one in j-V.

PROOF OF 1.3. Part (1) is stated just as in [**HW**] and the proof of part (2) is a simple adaptation of the last part of the proof of [**HW, Theorem 2.6**]. For part (3), we note that, since j-U also satisfies (P0)–(P4) of 1.2, we can adapt the first part of the proof of [**HW, Theorem 2.6**] to establish an order-isomorphism $\varphi : j\text{-}U \to j\text{-}V$. Then by (2), φ extends to an order-isomorphism from U to V. □

In [**HW**], it is shown that if $S = \mathbb{Z} - \bigcup_{i=1}^{n} \mathbf{p}_i$, where $\{\mathbf{p}_i \mid 1 \le i \le n\}$ is a finite set of nonzero primes, and $R = S^{-1}\mathbb{Z}$, then $\operatorname{Spec}(R[x])$ is of the first type from 1.3. A partially ordered set diagram for $\operatorname{Spec}(S^{-1}\mathbb{Z}[x])$ is:

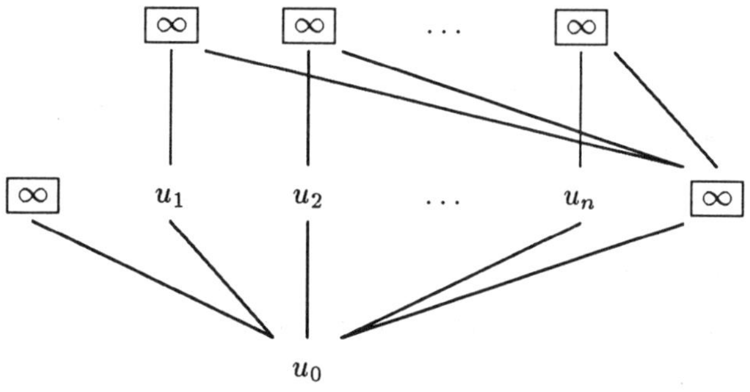

(The relationships involving the lower right boxed section, which are the primes which are not j-primes, are too complicated to display. However, (P6) implies that each finite subset of the top row contains infinitely many of them.)

The second type of partially ordered set in 1.3 is the type of $\operatorname{Spec}(V[x])$, where V is a countable Henselian discrete rank-one valuation domain. A partially ordered set diagram for $\operatorname{Spec}(V[x])$ is:

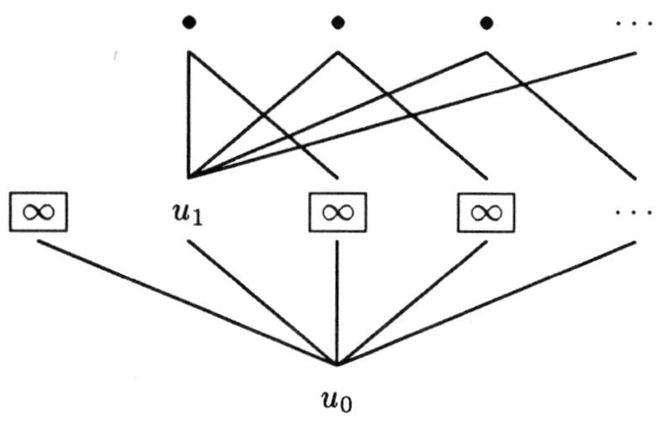

We list below the results obtained in [**HLW**] for birational extensions of polynomial rings:

1.4 PROPOSITION. [**HLW, Proposition 3.1**] *Let* $(R, \mathbf{m}_1, \ldots, \mathbf{m}_n)$ *be a semilocal domain of dimension one,* x *an indeterminate,* $A = R[x]$*,* $f \in A - \bigcup_{i=1}^{n} \mathbf{m}_i[x]$*, and let* B *be a finitely generated* A*-algebra strictly between* A *and* $A[1/f]$*. Then* j-Spec(B) *satisfies the following properties from Theorem 1.2:*

(i) (P0) holds if R *is countable.*

(ii) (P1)–(P3) hold without additional hypotheses.

(iii) There are only finitely many non-maximal height-one elements Q *of* j-Spec(B)*. Moreover, each of these height-one prime ideals contains a maximal ideal of* R*.*

(iv) If fA *has prime radical, then the number of these non-maximal height-one* j*-prime ideals* Q *is greater than the number* n *of maximal ideals of* R*, and the* Q*'s need not be pairwise comaximal.*

1.5 REMARK. For a prime ideal Q of B, recall that the j-radical of Q, j-rad(Q), equals $\bigcap\{M \text{ maximal in } B \mid M \supseteq Q\}$. We see that the essential thrust of Theorems 1.2 and 1.3 is that the prime spectrum of $R[x]$ is described by the j-spectrum and the specification of j-radicals. (For example, Property (P6) says that for each finite set of height-two maximals, there are infinitely many height-one prime ideals P for which j-rad(P) is the intersection of this finite set of maximals.) In what follows we consider the analogous situation for Spec(B).

2. The j-spectrum of B.

We begin by stating some elementary properties of j-primes.

2.1 REMARKS. (1) Clearly a nonmaximal j-prime must be the intersection of *infinitely* many maximal ideals.

(2) In a Noetherian domain of dimension 2, a height-one nonmaximal prime P contained in infinitely many maximal ideals is a j-prime, and it is the intersection of every infinite set of maximal ideals that contain it (since if $c \notin P$, then there exist only finitely many primes minimal over the ideal (c, P)).

(3) When R is a semilocal Noetherian domain of dimension one, the non-maximal height-one j-primes of $\operatorname{Spec}(R[x])$ are in one-to-one correspondence with the maximal ideals of R, via $P \mapsto P \cap R$.

(4) The spectra of birational extensions B of $R[x]$ may differ from $\operatorname{Spec}(R[x])$ in that we may have extra height-one j-primes of $\operatorname{Spec}(B)$ which contract to the same maximal ideal of R.

Suppose that $(R, \mathbf{m})$ is a local Noetherian domain of dimension one, K its field of fractions, $k = R/\mathbf{m}$, $f \in R[x] - \mathbf{m}[x]$, (f, g) is an $R[x]$-sequence, and

$$A = R[x] \subsetneq B = R[x][g/f] \subsetneq R[x, 1/f].$$

We will also consider B as $R[x, y]/(fy - g)$, where y is another indeterminate.

Before we proceed, we illustrate the construction, described in general below, of birational extensions with extra j-primes (that is, more than $R[x]$ has) by re-examining two examples constructed in [**HLW**]. In these examples, R is a discrete rank-one valuation domain with maximal ideal $\mathbf{m} = aR$.

EXAMPLE 1. A small birational extension with exactly two non-maximal j-primes, and they are comaximal.

Let $f = x^2 + a^3$ and $g = x$, so that $B = R[x][x/(x^2 + a^3)]$. Let $P_1 = aB[1/f] \cap B = (a, a^3/(x^2 + a^3))B$ and $P_2 = (a, x)B$. Then P_1 and P_2 are both height-one j-primes, and they are comaximal since

$$1 = \frac{x^2}{x^2 + a^3} + \frac{a^3}{x^2 + a^3} \in P_2 + P_1 \quad .$$

This becomes even clearer using the two-variable description of B:

$$\frac{B}{aB} \cong \frac{k[x,y]}{(x^2y-x)} = \frac{k[x,y]}{(x(xy-1))} .$$

Thus B/aB has two minimal primes, generated by the images of x and $xy-1$, obviously comaximal and corresponding to two height-one primes of B which contain aB. Note that $fR[x] = fK[x] \cap R[x]$ is a prime ideal, because a generates $\mathbf{m}$ and so no fractional root of a can be in K.

A partially ordered set diagram for j-Spec(B) in Example 1 is:

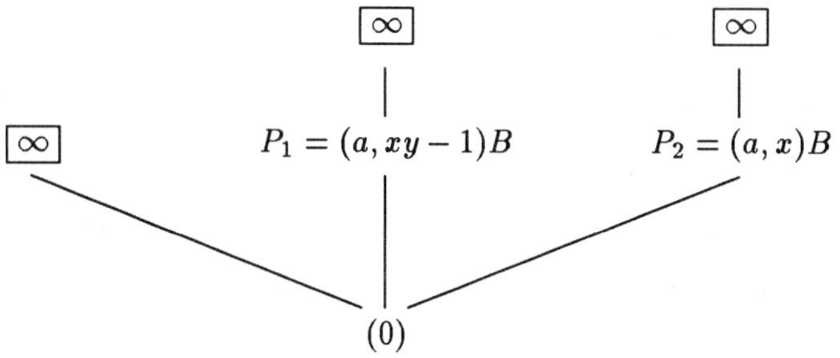

EXAMPLE 2. A small birational extension with exactly two non-maximal j-primes, and they are not comaximal.

Let $f = x$ and $g = a$, so that $B = A[a/x]$. Let $P_1 = (a/x)B = aB[1/f] \cap B$ and $P_2 = xB$. Then P_1 and P_2 are both height-one j-primes, and they are not comaximal since $(x, a/x)B$ contains both of them. In the other description of B,

$$\frac{B}{aB} \cong \frac{k[x,y]}{(xy)} .$$

Obviously the ideals generated by the images of x and y are not comaximal since they are both contained in the image of (x, y).

A partially ordered set diagram for j-Spec(B) in Example 2 is:

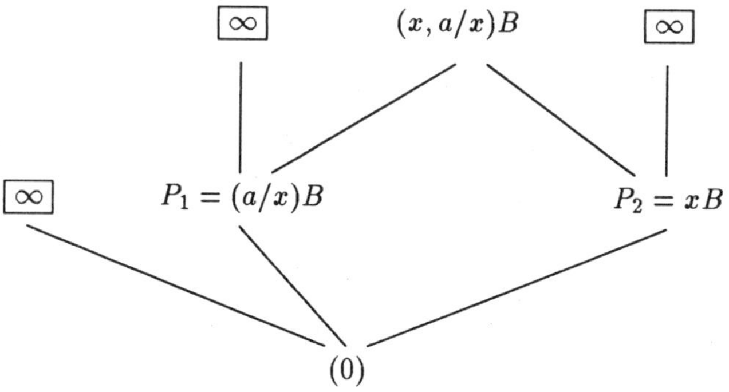

In both of these examples, only one of the nonmaximal height-one j-primes "survives" in $B[1/f]$, namely $P_1 = \mathbf{m}B[1/f] \cap B$.

2.2 Definition. The nonmaximal height-one j-primes P of B that *survive* in $B[1/f]$, that is, $PB[1/f] \subsetneq B[1/f]$, are called *survivors.* Those that do not survive, that is, $PB[1/f] = B[1/f]$, are called *transients.*

2.3 Remark. For any semilocal Noetherian domain R of dimension one, there exists exactly one survivor j-prime of B contracting to each maximal ideal of R, since the height-one j-primes of $B[1/f] = R[x, 1/f]$ are in one-to-one correspondence with the maximals of R. (Here we are using the fact that f was chosen outside $\bigcup_{i=1}^{n} \mathbf{m}_i R[x]$.)

2.4 A preview of (2.6) and (2.8):

(1) Let B be a a small birational extension of $R[x]$, for R semilocal. Given two height-one j-primes of B, there is at most one maximal ideal of B containing both the primes. Given a maximal ideal, there are at most two height-one j-primes contained in it. If a pair of height-one j-primes is contained in a maximal ideal, one of these will be a survivor, and the other will be a transient j-prime. Every pair of transient j-primes is comaximal. (Theorem 2.6 and Remark 2.9)

(2) For R local and $r > 1$ an integer, there exists a small birational extension B of $R[x]$ with exactly r height-one j-primes. Given r and $m < r$, it is possible to construct a birational extension with $r - 1$ transient j-primes, of which exactly m are co-maximal with the survivor j-prime. (Theorem 2.8)

2.5 Notation. Let $(R, \mathbf{m}, k)$ be a local domain of dimension one, (f, g) an $R[x]$-sequence, $f \notin \mathbf{m}[x]$, and consider the small birational extension $B =$

$R[x, g/f] \cong R[x,y]/(fy-g)$. We have $B/\mathbf{m}B \cong k[x,y]/(\bar{f}y-\bar{g})$ where $\bar{f}, \bar{g}$ are the images of f, g in $k[x]$.

In the polynomial ring $k[x,y]$ over the field k, write the factorization of $\bar{f}y - \bar{g}$ into a product of irreducibles in the form

$$\bar{f}y - \bar{g} = p(qy-s) = h_1 \cdot h_2 \cdots h_m \cdot p_1 \cdot p_2 \cdots p_n \cdot (qy-s),$$

where

(i) q and s are relatively prime elements of $k[x]$,

(ii) $h_1, h_2, \ldots, h_m, p_1, p_2, \ldots, p_n$ are powers of pairwise relatively prime irreducible elements of $k[x]$,

(iii) $h_1, h_2, \ldots, h_m$ divide a power of q, and

(iv) $p_1, p_2, \ldots, p_n$ are relatively prime to q.

Now let $H_1, \ldots, H_m, Q_1, \ldots, Q_n, P$, respectively, be the inverse images in B of the minimal primes in $B/\mathbf{m}B$ which are the radicals of the ideals generated by the images of $h_1(x), \ldots, h_m(x), p_1(x), \ldots, p_n(x), qy-s$, respectively. Then $H_1, \ldots, H_m, Q_1, \ldots, Q_n, P$ are height-one prime ideals of B.

2.6 THEOREM. *Let $(R, \mathbf{m}, k), B, f, g, m, n$ etc. be as in 2.5. Then*

(2.6.1) B has exactly $m+n+1$ nonmaximal nonzero j-primes, namely $H_1, \ldots, H_m$, $Q_1, \ldots, Q_n$, which are all transient, and $P = \mathbf{m}B[1/f] \cap B$, which is a survivor.

(2.6.2) The transient j-primes $H_1, \ldots, H_m$ are comaximal with P, but the transient j-primes $Q_1, \ldots, Q_n$ are not comaximal with P.

(2.6.3) Every pair of height-one j-primes of B is contained in at most one maximal ideal of B. If both of the height-one j-primes are transient, then they are comaximal.

In summary, every small birational extension B has j-$\mathrm{Spec}(B)$ like Diagram 2.6.4 below, for some choice of m and n.

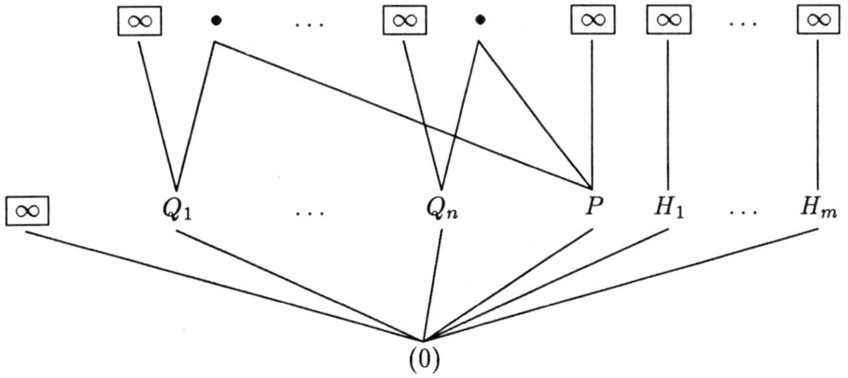

Diagram 2.6.4

PROOF. By 1.4, every non-maximal height-one j-prime of B contains $\mathbf{m}$ and therefore corresponds to a nonmaximal prime ideal of

$$B/\mathbf{m}B \cong k[x,y]/(\bar{f}y - \bar{g}) = S .$$

Since S is a Hilbert ring, every prime ideal of S is an intersection of maximal ideals. We conclude that the height-one j-primes of B are in one-to-one correspondence with the minimal primes of S; and hence with the primes of $k[x,y]$ associated to the polynomials $h_1(x), \ldots, h_m(x), p_1(x), \ldots, p_n(x)$ and $q(x)y - s(x)$. The prime ideal $(q(x)y - s(x))S$ corresponds to the survivor j-prime, namely P, of B since it survives in the localization $S[1/\bar{f}]$, and thus $P = \mathbf{m}B[1/f] \cap B$ by Remark 2.3. The prime radicals of the ideals (h_i) and (p_j) correspond to transient j-primes, since they do not survive in $S[1/\bar{f}]$.

(2.6.2) and (2.6.3): Since any two of $h_1, \ldots, h_m, p_1, \ldots, p_n$ are relatively prime in $k[x]$, any two transient j-primes of B are comaximal. To see which are comaximal with the survivor j-prime, we note that a maximal ideal containing the survivor j-prime corresponds to a maximal ideal in $k[x,y]$ containing $qy - s$, i.e., the set of all elements of $k[x,y]$ that vanish at a point (x_0, y_0) in the affine plane over the algebraic closure of k for which $q(x_0)y_0 - s(x_0) = 0$. Now if $q(x_0)y_0 - s(x_0) = 0$, we cannot have $h_i(x_0) = 0$, for then $q(x_0) = 0$ and hence $s(x_0) = 0$, contradicting the hypothesis that q, s are relatively prime. Thus, h_i is not in any maximal ideal containing $qy - s$, and the corresponding transient

j-prime in B is comaximal with the survivor j-prime. But if we choose x_0 so that $p_j(x_0) = 0$, then $q_j(x_0) \neq 0$, and by setting $y_0 = s(x_0)/q(x_0)$ we find a maximal ideal of $k[x, y]$ containing both p_j and $qy - s$; and hence the transient j-prime in B corresponding to p_j is not comaximal with the survivor j-prime. Finally, note that the maximal ideal of $k[x, y]$ containing p_j and $qy - s$ is the same no matter which root x_0 of p_j is used to find it; so the corresponding j-primes of B are contained in only one maximal ideal. □

2.7 DEFINITION. *Let m and n be nonnegative integers, not both zero. A countable partially ordered set is called j-birational of type (m, n) provided it is order-isomorphic to Diagram (2.6.4). (Here we interpret the ∞-boxes in the diagram to be countably infinite.)*

2.8 THEOREM. *Let $(R, \mathbf{m}, k)$ be a countable local domain of dimension one and let m and n be nonnegative integers, not both zero. Then there exists a small birational extension B of $R[x]$ so that j-$\operatorname{Spec}(B)$ is j-birational of type (m, n). (If R is not countable, then the form of j-$\operatorname{Spec}(B)$ will be like Diagram 2.6.4, but adjustments need to be made for uncountable cardinality.)*

Equivalently, for every countable partially ordered set U of the form of Diagram 2.6.4, there exists a small birational extension B of $R[x]$ with j-spectrum order-isomorphic to U.

PROOF. Choose, as in item 2.5(ii) above, $h_1, h_2, \dots, h_m, p_1, p_2, \dots, p_n$ to be pairwise relatively prime irreducible elements of $k[x]$.

Set $q = h_1 \cdot h_2 \cdot \ldots \cdot h_m$, and $p = h_1 \cdot h_2 \cdot \ldots \cdot h_m \cdot p_1 \cdot \ldots \cdot p_n$.

Then, as in 2.5(i), choose s in $k[x]$ to be relatively prime to q.

Then (iii) $h_1, h_2, \dots, h_m$ divide q, and (iv) $p_1, p_2, \dots, p_n$ are relatively prime to q.

Let $f(x)$ and $g(x)$ be preimages in $R[x]$ of $p(x)q(x)$ and $p(x)s(x)$, respectively. Then $f(x) \notin \mathbf{m}[x]$.

The one sticky point is whether the polynomials f, g can be adjusted to be an $R[x]$-sequence, to insure that $(fy - g)$ is prime. Since $R[x]$ is Cohen-Macaulay, it will suffice to find f, g with $\operatorname{ht}(f, g) = 2$. We have $f, g \neq 0 \mod \mathbf{m}$. There exist only finitely many primes in $R[x]$ minimal over (g). If $\operatorname{ht}(f, g) = 1$, then

$\text{ht}(f, g, a) = 2$, for some $a \in \mathbf{m}$. Now there are infinitely many elements of the form $f + a, f + a^2, \ldots$. Consider $I = (f + a^i, f + a^j)$, where $i < j$. Since $a^i(a^{j-i} - 1) \in I$ and $a^{j-i} - 1$ is a unit, $a^i, f \in I$, so $\text{ht}(I) = 2$. It follows that the height-one primes of $(f + a^i)$ are distinct for different i and so we can choose a^i to avoid the finite number of minimal primes of (g). Thus $(g, f + a^i)$ has height two.

Now, letting $B = R[x][g/f]$, we are in the situation of Theorem 2.6, and so j- Spec(B) is as desired.

REMARK 2.9. It is not hard to see that results extending Theorems 2.6 and 2.8 can be obtained for R semilocal, with more than one maximal ideal. There is some discussion of this in Section 4.

3. The exactly-less-than set of a singleton is infinite.

Now that the j-spectrum of B, for B a small birational extension of $R[x]$, has been characterized in Section 2, we direct our attention to the rest of the spectrum of B. As indicated in Theorem 1.3(2), the entire spectrum is determined by the exactly-less-than sets. Thus we need to consider the question: For T a finite set of height-two maximal elements of Spec(B), when is the exactly-less-than set infinite? Equivalently, when do there exist infinitely many height-one primes with j-radical equal to the intersection of the elements in T? If R is Henselian and T contains more than one element, then the exactly-less-than set of T is empty (Theorem 4.1). For singleton sets T, we prove in this section that the exactly-less-than set of T is infinite. (This is true both when R is Henselian and when it is not.)

3.1 THEOREM. *Suppose R is a semilocal domain of dimension one, x is an indeterminate, and B is a small birational extension of $R[x]$. Then for each height-two maximal ideal N of B, there exist infinitely many height-one primes P such that N is the only height-two maximal ideal containing P. That is, (P6) or (P6′) of 1.3 holds for singleton subsets T of the set of all height-two elements of Spec(B).*

We prove three lemmas in order to deduce the theorem. The first result is similar to Lemma 3 in [**rW1**], which is attributed to Wolmer Vasconcelos.

3.2 LEMMA. *Let A be a Noetherian ring of dimension one. Assume that all but at most finitely many of the height-one prime ideals P of A satisfy the two conditions: (i) P is the radical of a principal ideal, and (ii) A_P is a discrete (rank-one) valuation domain. Then every height-one prime ideal of A is the radical of a principal ideal.*

PROOF. Let $Q_1, Q_2, \ldots, Q_n$ be the height-one maximals for which either Q_i is not the radical of a principal ideal or A_{Q_i} is not a discrete valuation domain. Pick

$$x \in Q_1 - \ (Q_2 \cup \cdots \cup Q_n \cup \bigcup \{\text{height-zero primes of } A\}).$$

Then $\operatorname{rad}(xA) = Q_1 \cap P_1 \cap \cdots \cap P_m$, where $P_j = \operatorname{rad}(y_j A)$, for some $y_j \in A$ and A_{P_j} is a discrete valuation domain. Write $y_m = y$; since P_m does not consist of zerodivisors, y is not a zerodivisor. Since A_{P_m} is a discrete valuation domain, there are positive integers h, j such that $x^h A_{P_m} = y^j A_{P_m}$, so $z = x^h / y^j$ is a unit of A_{P_m}; thus, there exists an $s \in A - P_m$ such that $sz \in A$. Also $y^j z \in A$. Therefore $A = (s, y^j) = \{r \mid rz \in A\}$ and so $z \in A$. But since z is a unit in A_{P_m}, $z \notin P_m$. Hence $\operatorname{rad}(zA) = Q_1 \cap P_1 \cap \cdots \cap P_{m-1}$. Repeating this procedure, we get an element $t \in A$ such that $\operatorname{rad}(tR) = Q_1$. □

3.3 LEMMA. *Let A be a ring containing a field k. Suppose k^* is an algebraic closure of k, $A^* = A \otimes_k k^*$, and P is a prime ideal of A. If some prime ideal of A^* lying over P is the radical of a principal ideal, then P is the radical of a principal ideal of A.* (Cf. [**AEH, (2.13)**].)

PROOF. Let k' be the separable algebraic closure of k in k^*, and set $A' = A \otimes_k k'$, an integral extension ring of A that is a free A-module. Then A^* is purely inseparable over A'. Let P^* be a prime ideal of A^* lying over P and let $a^* \in A^*$ be such that $P^* = \operatorname{rad}(a^* A^*)$. Then $P' = P^* \cap A'$ is the radical of the principal ideal $(a^*)^{p^e} A'$, where p is the characteristic of k and e is chosen so that $(a^*)^{p^e} \in A'$.

Thus it suffices to prove the lemma for k' and A' in place of k^* and A^*.

Now let G be the Galois group of k'/k. Let $\{e_i\}$ be a vector space basis for A over k. For every $\varphi \in G$, extend φ to an A-automorphism φ' of A' by

$\varphi'(\Sigma r_i e_i) = \Sigma\varphi(r_i)e_i$, where the $r_i \in k'$. (This is clearly a well-defined k-vector space homomorphism, since $\{e_i\}$ is a k'-basis for A'.) Let $e_j e_k = \Sigma a_{jki} e_i$. Then

$$\begin{aligned}\varphi'((\Sigma r_j e_j)(\Sigma s_k e_k)) &= \varphi'(\sum_{j,k} r_j s_k e_j e_k)\\ &= \varphi'(\sum_{j,k} r_j s_k \sum_i a_{jki} e_i)\\ &= \varphi'(\sum_i(\sum_{j,k} a_{jki} r_j s_k) e_i)\\ &= \sum_i(\sum_{j,k} a_{jki}\varphi(r_j)\varphi(s_k) e_i)\\ &= \varphi'(\Sigma r_j e_j)\varphi'(\Sigma s_k e_k).\end{aligned}$$

Thus φ' is an A-automorphism of A'.

Let G' be the set of automorphisms of A'/A defined by extending the elements of G. An element $\Sigma r_i e_i$ of A' is fixed by every automorphism in G' if and only if each r_i is fixed by every automorphism in G. Thus G' is a group of automorphisms of A' having fixed ring A.

Now each conjugate $\varphi'(P')$ of P' is the radical of the principal ideal $\varphi'(aA')$ for some a in A'. Since a has only finitely many distinct images under elements of G', the set $\{\varphi'(P') : \varphi' \in G'\}$ is finite (even though the group G' may be infinite).

To complete the proof of Lemma 3.3, it suffices to prove the following Claim:

CLAIM. *P is the radical of the principal ideal bA, where b is the product of the distinct conjugates of a.*

PROOF OF CLAIM. Every minimal prime of bA' is of the form $\varphi'(P')$ for some φ' in G'. Thus we have

$$\operatorname{rad}(bA') = \bigcap\{\varphi'(P') : \varphi' \in G'\}.$$

Let Q be a prime in A that contains bA; then there is a prime Q' in A' lying over Q and hence containing bA'. It follows that Q' contains some $\varphi'(P')$, and so $P = \varphi'(P') \cap A \subseteq Q' \cap A = Q$. Therefore $\operatorname{rad}(bA) = P$. □

3.4 REMARK. For R an arbitrary commutative ring, to show that each height-one prime of a ring R is the radical of a principal ideal we may assume that R is reduced. This is because if $\mathbf{n}$ is the nilradical of R and $I/\mathbf{n} = \operatorname{rad}((x+\mathbf{n})R/\mathbf{n})$, then $I = \operatorname{rad}(xR)$, since every prime ideal of R contains $\mathbf{n}$. In fact, if I is a radical ideal of R such that $I/\mathbf{n}$ in $R/\mathbf{n}$ is the radical of an n-generated ideal in $R/\mathbf{n}$, then I is the radical of an n-generated ideal in R.

3.5 LEMMA. *Let k be a field and let x, y be indeterminates over k. Assume that $p, q, s \in k[x]$ with p, q nonzero and $(q, s)k[x] = k[x]$. Then every maximal ideal in the ring $S = k[x, y]/(p(qy - s))$ is the radical of a principal ideal.*

PROOF. By Lemma 3.3, we may assume k is algebraically closed by passing from S to $S \otimes_k k^*$, where k^* is an algebraic closure of k. Write the irreducible factorization of p in the form (from (2.5))

$$p = h_1 \cdot h_2 \cdots h_m \cdot p_1 \cdot p_2 \cdots p_n,$$

where $h_1, h_2, \dots, h_m, p_1, p_2, \dots, p_n$ are powers of pairwise relatively prime irreducible elements (i.e., nonassociate linear polynomials) of $k[x]$, $h_1, h_2, \dots, h_m$ divide a power of q, and $p_1, p_2, \dots, p_n$ are relatively prime to q. Finally, by Remark 3.4, we may assume the p_i's and h_i's are themselves linear and $p_i = x - c_i$, where $q(c_i) \neq 0$.

We have that

$$p(qy - s) = h_1 \cdot h_2 \cdots h_m \cdot p_1 \cdot p_2 \cdots p_n \cdot (qy - s).$$

Now for all points (a, b), if $h_i(a) = 0$, then $q(a) = 0$, which implies that $q(a)b - s(a) = -s(a) \neq 0$. Thus, $h_i, qy - s$ have no common points, i.e., the ideals they generate are comaximal. Therefore, by the Chinese Remainder Theorem,

$$S \cong k[x, y]/(h_1) \oplus \cdots \oplus k[x, y]/(h_m) \oplus S_1 \ ,$$
$$\text{where } S_1 = k[x, y]/(p_1 \cdots p_n)(qy - s) \ .$$

Note that if $s = 0$, then q must be a unit; consequently no h_i's occur and $S = S_1$.

Now the maximal ideals in the sum are of the form a maximal ideal in one summand and the unit ideal in the remaining summands, and the first summands, the $k[x, y]/(h_i)$, are all just isomorphic to $k[y]$. Thus once we show that every maximal ideal in S_1 is the radical of a principal ideal, the proof will be complete.

In fact, by Lemma 3.2, it suffices to show that each maximal ideal of S_1, except for the finite number corresponding to one of the points $(c_i, s(c_i)/q(c_i))$ at which the curve $qy = s$ meets the line $x = c_i$, is the radical of a principal ideal and the localization there is a discrete valuation domain. Let N denote a maximal ideal other than one of these intersection points; then $(S_1)_N$ is a discrete valuation domain, since N is a simple point of the variety V: $(qy - s)\prod_{i=1}^{n}(x - c_i)$. Suppose N corresponds to (a, b) on $qy = s$ but not on any $x = c_i$. Then the line $x = a$ meets V only at (a, b), so $N = \operatorname{rad}((x - a)S_1)$. Now suppose N corresponds to (c_i, b), where $b \neq s(c_i)/q(c_i)$, and set $h = (qy - s)\prod_{j \neq i}(x - c_j)$. Then the variety of $h(x, y) - h(c_i, b)$ meets V only at (c_i, b) (for, the curve $h(x, y) = h(c_i, b)$ meets the line $x = c_i$ only at the point (c_i, b), and for any point off the line $x = c_i$ at which $(qy - s)\prod_{j=1}^{n}(x - c_j)$ vanishes, h also vanishes, but $h(c_i, b) \neq 0$), so $N = \operatorname{rad}((h(x, y) - h(c_i, b))S_1)$. □

PROOF OF 3.1. Let $\mathbf{m}_1, \mathbf{m}_2, \ldots, \mathbf{m}_n$ be the maximal ideals of R. Then by the dimension inequality [**M, Theorem 15.5, p. 118**], $N \cap R = \mathbf{m}_i$, for some i. For convenience, let $\mathbf{m} = \mathbf{m}_i$. Now N is the preimage of a maximal ideal in

$$\frac{B}{\mathbf{m}B} = \frac{R[x, y]}{(\mathbf{m}, fy - g)} = \frac{k[x, y]}{(\bar{f}y - \bar{g})},$$

where $k = R/\mathbf{m}$ and overbars denote images mod $\mathbf{m}B$. By Lemma 3.5, the image of N in $B/\mathbf{m}B$ is the radical of a principal ideal $(\bar{h})$. Let h be a preimage in B of $\bar{h}$ and let P be a minimal prime of h contained in N. Then P is height-one and N is the only height-two prime containing P. Moreover, P does not contain $\mathbf{m}$ (because $(\mathbf{m}, h)B$ has height two). Thus $P \cap R = 0$. However, for all $0 \neq a \in \mathbf{m}$, $h + a$ is a preimage of $\bar{h}$ and $h + a \notin P$. (Otherwise, $a \in P$.) In this way we obtain infinitely many such P. □

4. Spectra of birational extensions.

4.1 THEOREM. *Let R be a Henselian local domain of dimension one, and let B be a small birational extension of $R[x]$. Then* Spec(B) *satisfies (P6′) of 1.3. If R is countable, then* Spec(B) *is uniquely determined by the partially ordered set j-*Spec(B).

PROOF. For the first part of (P6′), let T be a finite set of height-two maximals in B for which $\mathrm{L_e}(T) \neq \emptyset$, and take P in $\mathrm{L_e}(T)$. Then P is not a j-prime, so $P \cap R = 0$, and hence B/P is a domain of dimension one finitely generated over R. By the dimension formula [**M, Theorem 15.6, p. 118**], B/P is algebraic over R. Since R is a Henselian local domain of dimension one, the integral closure of R in every finite algebraic field extension is a DVR [**N, (43.12), (33.2)**]. It follows that every domain of dimension one algebraic over R is integral over R and has a unique maximal ideal. Therefore B/P is local, i.e., T is a singleton. The second part of (P6′) holds by Theorem 3.1. When R is countable, the uniqueness of Spec(B) follows from Theorem 1.3(2). □

We suspect that a similar result holds for non-Henselian semilocal domains, namely that the answer to the following question is "yes":

4.2 QUESTION. *Let R be a countable local non-Henselian domain of dimension one and let B be a small birational extension of $R[x]$. Suppose that j-*Spec(B) *is j-birational of type (m,n). Is* Spec(B) *uniquely determined? Equivalently, does (P6) always hold when R is not Henselian? (If R is not countable, cardinality adjustments are needed.)*

There is some evidence that this is true and we give some examples in Corollary 4.6 below, making use of results of Roger Wiegand:

4.3 PROPOSITION. [**rW1**][**rW2**] *The spectrum U =* Spec($\mathbb{Z}[x]$) *is order-isomorphic to* Spec($k[z,x]$), *where k is a field contained in the algebraic closure of a finite field and z and x are indeterminates over k. Furthermore, U is characterized among partially ordered sets by the following properties:*

(W1) U has a unique minimal element.

(W2) U has dimension 2.

(W3) For each element u of height one, U contains infinitely many elements $v > u$.

(W4) For each pair u, v of distinct elements of height one, U contains only finitely many elements t such that $t > u$ and $t > v$.

(W5) Let S be a finite set of height-one elements of U and T a finite subset of $\mathcal{M}(U)$. Then there exists a height-one element $w \in U$ such that $\mathrm{G}(s) \cap \mathrm{G}(w) \subseteq T \subseteq \mathrm{G}(w)$ *for all $s \in S$.*

4.4 PROPOSITION. *(1) If a partially ordered set U satisfies (W3) and (W4) of 4.3 and U contains infinitely many height-one elements, then for every finite set S of height-one elements of U, there exist infinitely many height-two elements not in* $\bigcup\{\mathrm{G}(s) \mid s \in S\}$.
(2) If a partially ordered set U satisfies (W3), (W4) and (W5) of 4.3, then U contains infinitely many height-one elements, and U satisfies:

(W) Let S be a finite set of height-one elements of U and T a finite subset of $\mathcal{M}(U)$. Then there exist infinitely many height-one elements $w \in U$ such that $\mathrm{G}(s) \cap \mathrm{G}(w) \subseteq T \subseteq \mathrm{G}(w)$, *for all $s \in S$.*

PROOF OF 4.4. For (1), let T be a finite set of height-two elements of U. Suppose that every height-two element of U is in the union

$$H = T \cup \bigcup\{G(s) \mid s \in S\} .$$

Choose a height-one element $u \notin S$. If $\mathrm{G}(u) \subseteq H$, then

$$\mathrm{G}(u) \subseteq (T \cap \mathrm{G}(u)) \cup \bigcup\{\mathrm{G}(s) \cap \mathrm{G}(u) \mid s \in S\} .$$

But (W4) implies that each $\mathrm{G}(s) \cap \mathrm{G}(u)$ is finite, so we have a contradiction to (W3) for u.

To see that (2) holds, note that (W5) and (W3) imply that U has infinitely many height-one elements. (The w in (W5) cannot be in S — or else some $\mathrm{G}(s) \cap \mathrm{G}(w)$ is infinite — so $S \cup \{w\}$ gives a new S to which to apply (W5), and so on.) With T and S as given, choose w_1 as in (W5), and set $S_1 = S \cup \{w_1\}$. Now by part (1), there exists a new height-two element $t_1 \notin T \cup \bigcup\{\mathrm{G}(s) \mid s \in S_1\}$. Let $T_1 = T \cup \{t_1\}$. By (W5), there exists an element w_2 with $\mathrm{G}(s) \cap \mathrm{G}(w_2) \subseteq T_1 \subseteq \mathrm{G}(w_2)$, for all $s \in S_1$. Note that $w_2 \neq w_1$ because $w_1 \not< t_1$ but $w_2 < t_1$. However we do have $T \subseteq T_1 \subseteq \mathrm{G}(w_2)$. Also, for each $s \in S$, since $s \not< t_1$, we have $\mathrm{G}(s) \cap \mathrm{G}(w_2) \subseteq T_1 \cap \mathrm{G}(s) \subseteq T$. Thus we have produced a second element which satisfies the condition relative to T. Continuing in this way, we can get infintely many elements. □

4.5 THEOREM. *Let R be a semilocal domain of dimension one, x an indeterminate over R, and suppose $B = R[x, g/f]$ is a small birational extension of $R[x]$*

(using the notation of 2.5). Suppose there exists a one-dimensional Noetherian domain $D \subset R$ such that R is a localization of D and such that $\operatorname{Spec}(D[x, g/f])$ *satisfies (W). Then* $\operatorname{Spec}(B)$ *satisfies (P6).*

PROOF. Let $\mathbf{m}_1, \mathbf{m}_2, ..., \mathbf{m}_n$ be the maximal ideals of R. If $P \in \operatorname{Spec}(B)$ is such that $G(P)$ is infinite, then $P \cap R = \mathbf{m}_i$, for some i. Using $\mathbf{m}$ for $\mathbf{m}_i$, we see that j-$\operatorname{Spec}(R_{\mathbf{m}}[x, g/f])$ has only finitely many nonmaximal height-one elements, by 1.4(iii) or (2.6.1).

To check that (P6) holds, let T be a finite nonempty set of height-two maximals of $\operatorname{Spec}(B)$, and let S be the (finite) set of all height-one nonmaximal j-primes of $\operatorname{Spec}(B)$. Let $C = D[x, g/f]$, and let Ψ be the localization map $\operatorname{Spec}(B) \to \operatorname{Spec}(C)$. Then $\Psi(T)$ is a finite set of height-two maximals of $\operatorname{Spec}(C)$ and $\Psi(S)$ is a finite set of height-one primes of $\operatorname{Spec}(C)$. By (W), there exist infinitely many elements $w \in \operatorname{Spec}(C)$ with $\mathrm{G}_C(s) \cap \mathrm{G}_C(w) \subseteq \Psi(T) \subseteq \mathrm{G}_C(w)$, for all $s \in \Psi(S)$. (Here $\mathrm{G}_C(u)$ means $\{v \in \operatorname{Spec}(C) \mid v > u\}$.) Now for each such w, since $\mathrm{G}_C(w) \cap \Psi(T) \neq \emptyset$, w is in the image of Ψ, i.e., $w = \Psi(w')$ for some $w' \in \operatorname{Spec}(B)$. It follows that $\mathrm{G}_B(w') \supseteq T$. (Clearly $\Psi(\mathrm{G}_B(w')) = \mathrm{G}_C(w) \cap \Psi(\operatorname{Spec}(B))$.) Suppose that $v \in \mathrm{G}_B(w')$. Since v is a height-two maximal of $\operatorname{Spec}(B)$, we have $v \in \bigcup\{\mathrm{G}_B(s) | s \in S\}$. But then $v \in \mathrm{G}_B(w') \cap \mathrm{G}_B(s)$ for some s in S, and so $\Psi(v) \in \mathrm{G}_C(w) \cap \mathrm{G}_C(\Psi(s)) \subseteq \Psi(T)$. Since Ψ is one-to-one, we must have $v \in T$. □

4.6 COROLLARY. *Suppose k is a field contained in the algebraic closure of a finite field. Let z and x be indeterminates over k, $\{\mathbf{p}_i \mid 1 \leq i \leq n\}$ a finite set of nonzero prime ideals of $k[z]$, $S = k[z] - \bigcup_{i=1}^n \mathbf{p}_i$, and $R = S^{-1}k[z]$. Let $B = R[x, g/f]$ be a small birational extension of $R[x]$. Then* $\operatorname{Spec}(B)$ *satisfies (P6) and so is uniquely determined by the type of j-*$\operatorname{Spec}(B)$.[2]

PROOF. By [**rW2, Theorem 2**], if a field k is contained in the algebraic closure of a finite field, then the spectrum of every domain of dimension two that is finitely generated as a k-algebra satisfies the axioms (W1)–(W5) and thus (W)

[2]We believe we can show, using Lemma 4 of [**rW1**], that, for an arbitrary ground field k and B as constructed here, $\operatorname{Spec}(B)$ satisfies property (P6). We hope to include this result in a future article.

by Proposition 4.4. In particular, for $D = k[z,x][g/f]$, Spec(D) satisfies (W). Thus by Theorem 4.5, (P6) holds. □

4.7 PROPOSITION. *Let R be a countable semilocal domain of dimension one with exactly t maximal ideals, and let x be an indeterminate over R.*

(1)If B is a small birational extension of $R[x]$, then j-Spec(B) is a union of t partially ordered sets U_i, $1 \le i \le t$, where each U_i is j-birational of type (m_i, n_i). The union is disjoint except that the minimal elements are identified. If (P6) or (P6′) holds, then Spec(B) *is uniquely determined.*

(2) Furthermore, for every countable partially ordered set U which is such a union, there exists a a small birational extension B of $R[x]$ with $U \cong j$-Spec(B).

(If R is uncountable, then j-Spec(B) is uncountable, but it has the same format as outlined in (1).)

PROOF. For each maximal ideal $\mathbf{m}_i = \mathbf{m}$, j-Spec($R_{\mathbf{m}}[x, g/f]$) is j-birational of type (m_i, n_i), so we may conclude that j-Spec($R[x, g/f]$) is a union of the desired sort. □

REFERENCES

[AEH] S. Abhyankar, P. Eakin, and W. Heinzer, *On the uniqueness of the coefficient ring in a polynomial ring*, J. Algebra **23(2)** (1972), 310-342.

[HW] W. Heinzer and S. Wiegand, *Prime ideals in two-dimensional polynomial rings*, Proc. Amer. Math. Soc. **107(3)** (1989), 577-586.

[HLW] W. Heinzer, D. Lantz, and S. Wiegand, *Projective lines over one-dimensional semilocal domains and spectra of birational extensions*, Algebraic Geometry and Applications (C. Bajaj, ed.), Collection of papers from Abhyankar's 60th Birthday conference, Springer Verlag, New York (to appear).

[M] H. Matsumura, *Commutative Ring Theory*, Cambridge Studies in Advanced Mathematics **8**, Cambridge University Press, 1989.

[N] M. Nagata, *Local Rings*, Interscience, New York, 1962.

[rW1] R. Wiegand, *Homeomorphisms of affine surfaces over a finite field*, J. London Math. Soc.(2) **18** (1978), 28-32.

[rW2] R. Wiegand, *The prime spectrum of a two-dimensional affine domain*, J. Pure Appl. Alg. **40** (1986), 209–214.

PURDUE UNIVERSITY, WEST LAFAYETTE, IN 47907-1395

COLGATE UNIVERSITY, HAMILTON, NY 13346-1398

UNIVERSITY OF NEBRASKA, LINCOLN, NE 68588-0323 AND PURDUE UNIVERSITY, WEST LAFAYETTE, IN 47907-1395

Contemporary Mathematics
Volume **159**, 1994

On the index of a homogeneous Gorenstein ring

JÜRGEN HERZOG

Introduction

In a minimal free resolution $F_{\bullet}$ of a Cohen-Macaulay module M of codimension t over a Gorenstein local ring $(R, \mathfrak{m})$ only the t-th syzygy module $\Omega^t(M)$ of M may possibly have a free summand. This can be easily seen by dualizing $F_{\bullet}$ into R and observing that $\operatorname{Ext}^i(M, R) = 0$ for $i \neq t$.

We may as well study $\Omega^t(M^\vee)$ where $M^\vee = \operatorname{Ext}^t(M, R)$. Since $(M^\vee)^\vee \cong M$, it doesn't really matter wether we replace M by $M^\vee$. However the advantage is that the number of free summands of $\Omega^t(M^\vee)$ may be interpreted as Auslander's δ-invariant of M. Of particular interest is the δ-invariant of $R/\mathfrak{m}^n$. Ding [**3**] shows in his thesis that $\delta(R/\mathfrak{m}^n) \leq 1$, and eventually becomes 1. The smallest n for which this happens is called the index of R. One of the main questions is how the index is related to other invariants of R. In [**4**] and [**5**] Ding gives partial answers to this question, and conjectures that it is the minimum of all integers n for which there exists a system of parameters $\mathbf{x}$ of R such that $\mathfrak{m}^n \subset (\mathbf{x})$. We answer his conjecture in the affirmative for homogeneous Gorenstein k-algebras. (Here one has to extend the above concepts in an obvious way to homogeneous k-algebras.)

All results in Section 1, except 1.5 can be found in Ding's thesis [**3**]. For the reader's convenience we present the basic properties of the δ-invariant which is not only defined for Cohen-Macaulay modules. For general modules it is defined via Cohen-Macaulay approximations as introduced by Auslander and Buchweitz [**2**].

Section 2 is devoted to the proof of Ding's conjecture for homogeneous Gorenstein k-algebras.

1991 *Mathematics Subject Classification.* Primary 13H10, secondary 13C05.
This paper is in final form and no version of it will be published elsewhere.
The author was partially supported by the DFG.

1. The δ-invariant

In this note $(R, \mathfrak{m}, k)$ will always be a Gorenstein local ring, and all R-modules are assumed to be finite. The minimal number of generators of an R-module M will be denoted by $\mu(M)$

Let M be an R-module. An exact sequence of R-modules

$$0 \to Y \xrightarrow{\psi} X \xrightarrow{\phi} M \to 0 \tag{1}$$

is a *Cohen-Macaulay approximation of M* if X is a maximal Cohen-Macaulay module, and Y a module of finite projective dimension. The Cohen-Macaulay approximation is *minimal* if the homomorphism ϕ is *right minimal*, i.e. if all endomorphisms $\alpha : X \to X$ with $\phi \circ \alpha = \phi$ are isomorphisms.

It has been shown by Auslander and Buchweitz [**2**] that Cohen-Macaulay approximations always exist. In [**1**] Auslander proves the existence of minimal Cohen-Macaulay approximations.

Given a maximal Cohen-Macaulay module X' and a homomorphism $\sigma : X' \to M$; then σ can be factored through ϕ. Indeed, the Cohen-Macaulay approximation of M gives rise to an exact sequence $\operatorname{Hom}_R(X', X) \to \operatorname{Hom}_R(X', M) \to \operatorname{Ext}^1_R(X', Y)$. Now since R is Gorenstein and X' is a maximal Cohen-Macaulay module one has $\operatorname{Ext}^1_R(X', R) = 0$; see for instance [**7**]. By induction on the projective dimension it follows then that $\operatorname{Ext}^1_R(X', Y) = 0$. Therefore, $\operatorname{Hom}_R(X', \phi)$ is surjective, and this proves the assertion.

From this factorization property of Cohen-Macaulay approximations it follows immediately that any two minimal Cohen-Macaulay approximations are isomorphic, and we denote by $0 \to Y_M \to X_M \to M \to 0$ or just by X_M a minimal Cohen-Macaulay approximation of M, and define the *δ-invariant of M* to be the f-rank(X_M) of X_M, i.e. the rank of a maximal free direct summand of X_M. Note that f-rank$(M) = \delta(M)$ for all maximal Cohen-Macaulay modules.

Minimal Cohen-Macaulay approximations can be characterized differently: the approximation (1) is minimal if and only if Y and X have no common free summand under ψ. Thus if we write X_M as $N \oplus F$, with F free and f-rank$(N) = 0$, then $\delta(M) = \mu(M/\phi(N))$. This observation leads to the following characterization of $\delta(M)$ which does not refer to the Cohen-Macaulay approximation of M.

LEMMA 1.1. *$\delta(M)$ is the minimum of the numbers* f-rank(X) *where the minimum is taken over all maximal Cohen-Macaulay modules X for which there exists an epimorphism $X \to M$ of rank $\delta'(M)$.*

PROOF. Denote this minimum by $\delta'(M)$. Then we have $\delta'(M) \leq \delta(M)$. In order to prove the other inequality we pick an epimorphism $\epsilon\ X \to M$ where X is a maximal Cohen-Macaulay module with f-rank$(X) = \delta'(M)$, and write $X = L \oplus F$ where f-rank$(L) = 0$ and F is free.

By the factorization property of Cohen-Macaulay approximations we obtain a commutative diagram

$$\begin{array}{ccccc} L & \xrightarrow{\epsilon} & M & \longrightarrow & 0 \\ {\scriptstyle\alpha}\downarrow & & \| & & \\ X_M & \xrightarrow{\phi} & M & \longrightarrow & 0 \end{array}$$

Write $X_M = N \oplus G$ with f-rank$(N) = 0$ and G free. Since f-rank$(L) = 0$, the projection of the submodule $\alpha(L) \subset X_M$ to G lies in $\mathfrak{m}G$. Therefore, $\delta'(M) =$ f-rank$(X) \geq \mu(M/\epsilon(L)) \geq \mu(M/\phi(N)) = \delta(M)$. □

COROLLARY 1.2. *Let N be an epimorphic image of M. Then $\delta(M) \geq \delta(N)$.*

COROLLARY 1.3. *For all R-modules we have $\delta(M) \leq \mu(M)$. Equality holds if the projective dimension of M is finite.*

PROOF. The module M is an epimorphic image of a free module of rank $\mu(M)$. Apply 1.2 to obtain the first assertion.

Assume $\operatorname{proj\,dim} M$ is finite, and let

$$0 \to F_p \to \ldots \to F_1 \xrightarrow{\partial_1} F_0 \to M \to 0$$

be a minimal free resolution of M. Then $0 \to \operatorname{Im}\partial_1 \to F_0 \to M \to 0$ is a minimal Cohen-Macaulay approximation of M. Thus $\mu(M) = \delta(M)$. □

The construction of a minimal Cohen-Macaulay approximation of a Cohen-Macaulay module M is rather simple. We set $M^\vee = \operatorname{Ext}_R^t(M, R)$ if M is a Cohen-Macaulay module of codimension t. Suppose M is such a module, and let

$$\cdots \to F_1 \xrightarrow{\partial_1} F_0 \to M^\vee \to 0$$

be a minimal free resolution of $M^\vee$. One has $\operatorname{Ext}_R^i(M^\vee, R) = 0$ for $i \neq t$, and $M \cong M^{\vee\vee} = \operatorname{Ext}_R^t(M^\vee, R)$; see for instance [**7**]. Therefore, if $\Omega^i(M)$ denotes the i-th syzygy module of M, we obtain, upon dualizing the resolution of $M^\vee$, the exact sequence

$$0 \to F_0^\vee \xrightarrow{\partial_1^\vee} F_1^\vee \xrightarrow{\partial_2^\vee} \ldots \xrightarrow{\partial_{t-1}^\vee} F_{t-1}^\vee \to \Omega^t(M^\vee)^\vee \to M \to 0.$$

Note that $\Omega^t(M^\vee)$ is a maximal Cohen-Macaulay module. Since for any Cohen-Macaulay module N, the module $N^\vee$ is again Cohen-Macaulay of same dimension, we conclude that $\Omega^t(M^\vee)^\vee$ is a maximal Cohen-Macaulay module, too. Now it is clear that $0 \to \operatorname{Coker}\partial_{t-1}^\vee \to \Omega^t(M^\vee)^\vee \to M \to 0$ is a Cohen-Macaulay approximation of M. It follows easily that it is minimal since the free resolution $F_\bullet$ of M is minimal.

COROLLARY 1.4. *Let M be a Cohen-Macaulay module of codimension t. Then $\delta(M) =$ f-rank$(\Omega^t(M^\vee)^\vee) =$ f-rank$(\Omega^t(M^\vee))$.*

We shall use this characterization of $\delta(M)$ in the proof of the next result.

PROPOSITION 1.5. *Let M be a Cohen-Macaulay module, and x an M-regular element. Then $\delta(M) = \delta(M/xM)$.*

PROOF. By 1.2 we have $\delta(M/xM) \leq \delta(M)$. In order to prove the converse inequality we assume that M is of codimension t. Then M/xM is Cohen-Macaulay of codimension $t+1$. Let $F_{\bullet}$ be a minimal free resolution of $M^{\vee}$, and let $G_{\bullet}$ be the mapping cone of $F_{\bullet} \xrightarrow{x} F_{\bullet}$. The complex $G_{\bullet}$ is a minimal free resolution of $M^{\vee}/xM^{\vee}$. Therefore the exact sequence of complexes $0 \to F_{\bullet} \to G_{\bullet} \to F_{\bullet}(-1) \to 0$ yields the exact sequence

$$0 \to \Omega^{t+1}(M^{\vee}) \to \Omega^{t+1}(M^{\vee}/xM^{\vee}) \to \Omega^{t}(M^{\vee}) \to 0.$$

Note that $(M/xM)^{\vee} \cong M^{\vee}/xM^{\vee}$; therefore

$$\delta(M/xM) = \text{f-rank}(\Omega^{t+1}(M^{\vee}/xM^{\vee})) \geq \text{f-rank}(\Omega^{t}(M^{\vee})) = \delta(M).$$

by 1.4. □

2. The index

Let M be an R-module, and set $\delta_n = \delta(M/\mathfrak{m}^n M)$, $n = 1, 2, \ldots$. According to 1.2 and 1.3 we have $\delta_1 \leq \delta_2 \leq \ldots \leq \delta_n \leq \ldots \leq \mu(M)$. Thus the sequence $(\delta_n)_{n \in \mathbb{N}}$ eventually stabilizes, and we define the *index* of M, denoted index(M), to be the least integer n for which $\delta_n = \delta_{n+1} = \ldots$. Of course we have $\lim \delta_n \leq \mu(M)$.

PROPOSITION 2.1. *Let M be a Cohen-Macaulay module, $\mathbf{x}$ a system of parameters* (sop) *of M, and $n \in \mathbb{N}$ such that $\mathfrak{m}^n M \subset (\mathbf{x})M$. Then $\delta_n = \delta(M)$. In particular,* index$(M) \leq n$ *and* $\lim \delta_n = \delta(M)$.

PROOF. We have $\delta(M) = \delta(M/(\mathbf{x})M) \leq \delta_n \leq \delta(M)$. The equality follows from 1.5, the inequalities from 1.2. □

Let N be a module of finite length. The *Loewy length* of N, denoted $\ell\ell(N)$, is the smallest $n \in \mathbb{N}$ for which $\mathfrak{m}^n N = 0$. Now 2.1 can be interpreted as follows: if M is Cohen-Macaulay, then

$$\text{index}(M) \leq \min\{\ell\ell(M/(\mathbf{x})M)\ \mathbf{x} \text{ is an sop of } M\}$$

Ding conjectures [**3**] that the index of R is indeed this minimum, and proves it in some cases, for instance for hypersurface rings; see [**5**].

We will prove this conjecture for homogeneous Gorenstein rings. Let k be a field. A graded k-algebra $R = \bigoplus_{i \geq 0} R_i$ is called *homogeneous* if $R_0 = k$ and $R = k[R_1]$. The homogeneous k-algebra has a unique graded maximal ideal, namely $\mathfrak{m} = \bigoplus_{i \geq 1} R_i$. All definitions we made and the results we obtained so far can be transferred accordingly to homogeneous Gorenstein algebras. In particular, if $R = \bigoplus_{i \geq 0} R_i$ is a homogeneous Gorenstein k-algebra with graded maximal ideal $\mathfrak{m}$, then $0 \leq \delta(R/\mathfrak{m}^n) \leq 1$, and the index of R is the smallest integer for which $\delta(R/\mathfrak{m}^n) = 1$.

Let $H_R(t) = \sum_{i\geq 0} \dim_k R_i t^i$ be the Hilbert series of R. Then $H_R(t) = Q(t)/(1-t)^d$, where $d = \dim R$. $Q(t)$ is a polynomial whose degree we denote by $s(R)$. Note that $s(R)$ equals the *reduction number of* R, that is, the smallest integer n for which there exists an sop $\mathbf{x}$ with $\mathfrak{m}^{n+1} = (\mathbf{x})\mathfrak{m}^n$.

The above minimum can now be interpreted as follows:

LEMMA 2.2. *Let R be a homogeneous Cohen-Macaulay k-algebra, where k is an infinite field. Then*

$$s(R) + 1 = \min\{\ell\ell(R/(\mathbf{x})R)\ \mathbf{x} \textit{ is a homogeneous sop of } R\}.$$

This minimum is attained for any sop which is generated by elements of degree 1.

PROOF. Let $\mathbf{x}$ be a homogeneous sop with $\deg x_i = a_i$ for $i = 1, \dots, d$, and set $\overline{R} = R/(\mathbf{x})R$. The k-algebra $\overline{R}$ is homogeneous with $\ell\ell(\overline{R}) = \max\{i\ \overline{R_i} \neq 0\} + 1$. On the other hand, $H_{\overline{R}}(t) = \prod_{i=1}^d (1 - t^{a_i}) H_R(t) = Q(t) \prod_{i=1}^d (1 + t + \dots t^{a_i - 1})$. Therefore $\ell\ell(\overline{R}) = \deg H_{\overline{R}}(t) + 1 = s(R) + 1 + \sum_{i=1}^d (a_i - 1)$. This formula shows that the minimum of $\ell\ell(\overline{R})$ is attained when all x_i have degree 1, in which case the minimum is indeed $s(R) + 1$. Note that sops of 1-forms exist if k is infinite. □

The following theorem combined with the preceding lemma yields Ding's conjecture for homogeneous Gorenstein algebras.

THEOREM 2.3. *Let k be an infinite field, and R a homogeneous Gorenstein k-algebra with graded maximal ideal* $\mathfrak{m}$. *Then* $\operatorname{index}(R) = s(R) + 1$.

PROOF. It remainsof M to show that $\operatorname{index}(R) \geq s(R) + 1$. Let $F_{\bullet}\ \dots \to F_2 \to F_1 \to F_0 = R \to 0$ be a minimal graded free resolution of $R/\mathfrak{m}^n$. Dualizing this resolution we obtain the exact sequence

$$0 \to R \to F_1^{\vee} \to F_2^{\vee} \dots \to F_{d-1}^{\vee} \to \Omega^d(R/\mathfrak{m}^n)^{\vee} \to (R/\mathfrak{m}^n)^{\vee} \to 0,$$

where $d = \dim R$. We will construct from this exact sequence a minimal graded free resolution of $(R/\mathfrak{m}^n)^{\vee}$, and then inspect the d-th syzygy module of $(R/\mathfrak{m}^n)^{\vee}$.

Let $G_{\bullet}$ be a minimal graded free resolution of $\Omega^d(R/\mathfrak{m}^n)^{\vee}$. The homomorphism $F_{d-1}^{\vee} \to \Omega^d(R/\mathfrak{m}^n)^{\vee}$ can be lifted to a homogeneous homomorphism $\phi_{\bullet}$ of complexes

$$\begin{array}{ccccccccccc}
 & & 0 & \longrightarrow & R & \longrightarrow & F_1^{\vee} & \longrightarrow & \dots & \longrightarrow & F_{d-1}^{\vee} & \longrightarrow & 0 \\
 & & {\scriptstyle\phi_d}\downarrow & & {\scriptstyle\phi_{d-1}}\downarrow & & {\scriptstyle\phi_{d-2}}\downarrow & & & & {\scriptstyle\phi_0}\downarrow & & \\
\dots & \longrightarrow & G_d & \longrightarrow & G_{d-1} & \longrightarrow & G_{d-2} & \longrightarrow & \dots & \longrightarrow & G_0 & \longrightarrow & 0.
\end{array}$$

The mapping cone $C(\phi_{\bullet})$ of $\phi_{\bullet}$ is a graded free resolution of $(R/\mathfrak{m}^n)^{\vee}$. Let $\partial_{\bullet}$ denote the differential of $C(\phi_{\bullet})$: Then $\partial_d : G_d \oplus R \to G_{d-1} \oplus F_1^{\vee}$ maps R into $\mathfrak{m}F_1^{\vee}$ and via ϕ_{d-1} into G_{d-1}. Of course $C(\phi_{\bullet})$ may not be minimal. This arises from the fact that $F_{d-1-i}^{\vee}$ and G_i may have common free summands under ϕ_i.

Canceling these summands yields a minimal graded free resolution of $(R/\mathfrak{m})^\vee$. It follows that $C(\phi_\bullet)$ is minimal in degree d if and only if $\operatorname{Im}\phi_{d-1} \subset \mathfrak{m}G_{d-1}$ in which case $\Omega^d((R/\mathfrak{m}^n)^\vee) \cong \operatorname{Ker}\partial_{d-1} \cong \Omega^d(M) \oplus R$ where $M = \Omega^d(R/\mathfrak{m}^n)^\vee$; otherwise $\Omega^d((R/\mathfrak{m}^n)^\vee) \cong \Omega^d(M)$. The module $\Omega^d(M)$ has no free summands since it is a syzygy module of a maximal Cohen-Macaulay module; see the discussion in the introduction. Therefore 1.4 implies that $\delta(R/\mathfrak{m}^n) = 1$ if and only if $\operatorname{Im}\phi_{d-1} \subset \mathfrak{m}G_{d-1}$.

From now on let us assume that $\delta(R/\mathfrak{m}^n) = 1$. Then we have $\operatorname{Im}\phi_{d-1} \subset \mathfrak{m}G_{d-1}$. We will show in a moment that $\phi_{d-1} \neq 0$. Since R has its generator in degree zero, this will imply that there is a non-zero element of degree zero in $\mathfrak{m}G_{d-1}$. On the other hand, we will show that G_{d-1} is generated by elements of degree at least $s(R) - n$. Therefore, $0 > s(R) - n$, as desired.

To see that $\phi_{d-1} \neq 0$, note that according to the above discussions the exact sequence

$$0 \longrightarrow \operatorname{Im}\partial_d^\vee \longrightarrow \Omega^d(M)^\vee \oplus R \xrightarrow{f} R/\mathfrak{m}^n \longrightarrow 0$$

is the minimal Cohen-Macaulay approximation of $R/\mathfrak{m}^n$, where the restriction of f to $\Omega^d(M)^\vee$ is the map induced by $\phi_{d-1}^\vee$. Thus, if $\phi_{d-1} = 0$, then f restricted to $\Omega^d(M)^\vee$ is zero, contradicting the right-minimality of f.

It remains to prove that the generators of G_{d-1} have degree at least $s(R) - n$. For $i = 0, \dots, d-2$ let D_i be the common free summand under ϕ_i, and write $F_{d-i-1}^\vee = D_i \oplus E_{i-1}$ and $G_i = D_i \oplus L_i$. Then, cancelling the D_i, we obtain a minimal graded free resolution of $(R/\mathfrak{m}^n)^\vee$ which is of the form

$$\dots \longrightarrow G_{d-1} \oplus E_{d-1} \longrightarrow L_{d-2} \oplus E_{d-2} \longrightarrow \dots \longrightarrow L_1 \oplus E_1 \longrightarrow L_0 \longrightarrow 0.$$

We choose a maximal regular sequence $\mathbf{x}$ of elements of degree n in $\mathfrak{m}^n$, and set $R/(\mathbf{x}) = \bar{R}$. Then

$$(R/\mathfrak{m}^n)^\vee = \operatorname{Ext}_R^d(R/\mathfrak{m}^n, R) \cong \operatorname{Hom}_{\bar{R}}(R/\mathfrak{m}^n, \bar{R})(dn).$$

Since $\bar{R}$ is Gorenstein, it is self injective and its socle is generated by one element, say σ. Let $a_1, \dots, a_k$ be a k-basis of the socle $\mathfrak{m}^{n-1}/\mathfrak{m}^n$ of $R/\mathfrak{m}^n$. By the self injectivity of R, for each $i = 1, \dots, k$ there exists a homogeneous element $\phi_{a_i} \in \operatorname{Hom}_{\bar{R}}(R/\mathfrak{m}^n, \bar{R})$ which maps a_i to σ and the other basis elements to zero. These ϕ_{a_i} generate $\operatorname{Hom}_{\bar{R}}(R/\mathfrak{m}^n, \bar{R})$. Thus we see that $\operatorname{Hom}_{\bar{R}}(R/\mathfrak{m}^n, \bar{R})$ is minimally generated by elements which are all of degree $\deg\sigma - (n-1)$. The degree of σ is the degree of $H_{\bar{R}}(t)$ (which is a polynomial). We have

$$H_{\bar{R}}(t) = H_R(t)(1-t^n)^d = \frac{P(t)}{(1-t)^d}(1-t^n)^d = P(t)(1+\dots+t^{n-1})^d.$$

Thus $\deg H_{\bar{R}}(t) = \deg P(t) + d(n-1) = s(R) + d(n-1)$, and so $\operatorname{Hom}_{\bar{R}}(R/\mathfrak{m}^n, \bar{R})$ is minimally generated by elements of degree $s(R) + d(n-1) - (n-1) = s(R) + (d-1)(n-1)$ and $(R/\mathfrak{m}^n)^\vee$ by elements of degree $s(R) + (d-1)(n-1) - dn = s(R) - d - n + 1$, and this is also the degree of the generators of H_0. Since the above resolution of $(R/\mathfrak{m}^n)^\vee$ is minimal the degrees of the generators of $G_i \oplus H_i$ are at least $s(R)-d-n+1+i$. For $i=d-1$ we obtain the desired assertion. □

In his thesis [8] (see also [9]) Levin proved the following conjecture of Kaplansky: Suppose $(R, \mathfrak{m})$ is a Noetherian local ring, M a finite R-module such that $\mathfrak{m}M \neq 0$ and $\mathfrak{m}M$ has finite projective dimension. Then R is regular. In particular, if $\operatorname{depth} R = t$, and $\Omega^t(R/\mathfrak{m}^n)$ is free for some n, then R is regular.

With the above methods we are able to prove the following

COROLLARY 2.4. *Let k be a field, and R a homogeneous Gorenstein k-algebra of dimension d. If for some n, $\Omega^d(R/\mathfrak{m}^n)$ has a free summand, then R is regular.*

PROOF. Assume $\Omega^d(R/\mathfrak{m}^n)$ has a free summand, and write $\Omega^d(R/\mathfrak{m}^n) = U \oplus R(-a)$. Dualizing the minimal graded free resolution $F_{\bullet}$ of $R/\mathfrak{m}^n$, we obtain the exact sequence

$$\dots \to F_{d-2}^\vee \to F_{d-1}^\vee \xrightarrow{\alpha} U^\vee \oplus R(a) \xrightarrow{\epsilon} (R/\mathfrak{m}^n)^\vee \to 0.$$

The homomorphism α composed with the projection $U^\vee \oplus R(a) \to R(a)$ maps into $\mathfrak{m}R(a)$. Therefore $R(a)$ is mapped under ϵ onto a minimal generator of $(R/\mathfrak{m}^n)^\vee$. We have seen in the proof of 2.3 that these generators have degree $s(R)-d-n+1$. Thus $a = d+n-s(R)-1$ is the degree of a minimal generator of $R(-a)$. Inspecting the resolution $F_{\bullet}$, we see that the degree of a minimal generator of $R(-a)$ is at least $n+d-1$, and so $d+n-s(R)-1 \geq n+d-1$. That is, $s(R)=0$ since $s(R) \geq 0$, always. However, $s(R)=0$ if and only if R is regular. □

Open questions and comments. If M is not Cohen-Macaulay, is it still true that $\lim \delta(M/\mathfrak{m}^n M) = \delta(M)$, and if M is Cohen-Macaulay (and even graded) what meaning has $\operatorname{index}(M)$?

If $(R, \mathfrak{m})$ is a d-dimensional local Gorenstein ring, and I an $\mathfrak{m}$-primary ideal. What can be said about $\delta(R/I)$? Here is a partial result: we know already that $0 \leq \delta(R/I) \leq 1$, and $\delta(R/I) = 1$ if $I \subset (\mathbf{x})$ for some sop $\mathbf{x}$. On the other hand, if I properly contains an sop $\mathbf{x}$ which is part of a minimal system of generators of $\mathfrak{m}$, then $\delta(R/I) = 0$. Indeed, let $\bar{R} = R/(\mathbf{x})$. The socle of $\bar{R}$ is generated by one element, say σ. In view of 1.2 it suffices to show that $\delta(\bar{R}/(\sigma)) = 0$. Note that $(R/(\sigma))^\vee$ is isomorphic to the maximal ideal $\bar{\mathfrak{m}}$ of $\bar{R}$. Let $X_{\bullet}$ be the Tate resolution of k (over R) and $K_{\bullet}$ the Koszul complex for the sequence $\mathbf{x}$. Then $K_{\bullet}$ is a minimal free R-resolution of $\bar{R}$, and $K_{\bullet}$ maps split injective into $X_{\bullet}$. It follows that

$$\dots \to X_{d+1} \to X_d/K_d \to \dots \to X_1/K_1 \to 0$$

is minimal free R-resolution of $\bar{\mathfrak{m}}$. Therefore, $\Omega_R^d(\bar{\mathfrak{m}}) \cong \Omega_R^{d+1}(k)$, and hence it has no free summands.

In the meantime Ding [**6**] has generalized Theorem 2.3 to the case where $(R, \mathfrak{m})$ is a Gorenstein local ring whose associated graded ring $\mathrm{gr}_{\mathfrak{m}}(R)$ is Cohen-Macaulay. For such rings the generalized Loewy length of R equals the reduction number minus 1. His proof, which uses induction on the dimension of R, is completely different from the one presented here. In a similar manner one can generalize 2.4 to local rings (not necessarily Gorenstein) whose associated graded ring is Cohen-Macaulay.

REFERENCES

1. M. Auslander, Minimal Cohen-Macaulay approximations, in preparation.
2. M. Auslander and R. O. Buchweitz, The homological theory of maximal Cohen-Macaulay approximations, Soc. Math. de France, Mem. No.38(1989). 5-37.
3. S. Ding, Cohen-Macaulay approximations over a Gorenstein local ring, Thesis, Brandeis University 1990.
4. S. Ding, Cohen-Macaulay approximations and multiplicity, J. Algebra 153 (1953).
5. S. Ding, A note on the index of Cohen-Macaulay local rings, to appear in Comm. Algebra.
6. S. Ding, The associated graded ring and the index of a Gorenstein local ring, to appear in Proc. Amer. Math. Soc.
7. J. Herzog and E. Kunz, Der kanonische Modul eines Cohen-Macaulay rings, LNM 238, Springer, 1971.
8. G. Levin, Homology of local rings, Thesis, University of Chicago 1965.
9. G. Levin and W. V. Vasconcelos, Homological dimensions and Cohen-Macaulay rings, Pacific Journal of Mathematics, 25(2) (1968), 315-323.

FB 6 MATHEMATIK, UNIVERSITÄT-GESAMTHOCHSCHULE ESSEN, UNIVERSITÄTSTR. 3, 4300 ESSEN, GERMANY

Contemporary Mathematics
Volume **159**, 1994

Solid Closure

MELVIN HOCHSTER

In memory of my father, Lothar Hochster,
April 20, 1906 – February 17, 1991

CONTENTS

1. Introduction
2. Solid modules and solid algebras
3. Formally solid modules and algebras
4. Generic forcing algebras
5. Solid closure
6. Minimal solid algebras
7. S-regular rings
8. Comparison with tight closure in characteristic p
9. A formal power series criterion
10. Shadow homology
11. Big Cohen-Macaulay algebras and tight closure in equal characteristic zero
12. The case of dimension two
13. Regular rings revisited
14. Questions

1991 *Mathematics Subject Classification.* Primary 13E05, 13B99, 13A35, 13H05, 13H10, 13D25.

The author was supported in part by a grant from the National Science Foundation.

This paper is in final form and no version of it will be submitted for publication elsewhere.

1. Introduction

Throughout this paper, unless otherwise specified, all rings are assumed to be commutative, associative, with identity, and all modules are assumed to be unital. In **[HH4]**, the author, jointly with Craig Huneke, introduced the notion of tight closure for submodules of finitely generated modules over certain Noetherian rings. The definition is first made in characteristic p using the action of the Frobenius endomorphism. A notion for finitely generated algebras over a field of characteristic zero is then obtained by reduction to characteristic p. No satisfactory notion in mixed characteristic has yet been proposed. Tight closure theory has produced a host of new results and improvements of old results. We refer to the introductions of [**HH1-4, 6, 8-10**] for more detail, and to these papers as well as [**Ab1-2, AHH, FeW, HH11, Hu1-2, Gla, Sm1-3, Vel**] and [**Wil**] for the full development of tight closure theory. The applications include invariant theory (for background, cf. [**B**], [**Hr1-2**], [**Ke**]), the Briançon-Skoda theorem (for background, cf. [**BrS, LS, LT, Sk**]) and the local homological conjectures (for background, cf. [**Du, EvG1-3, Ho1-3, 5-7, PS1-2, Ro1-5, S**]).

The existence of a "sufficiently good" parallel theory in mixed characteristic would settle many long standing conjectures.

Our objective in this paper is to introduce a new closure operation, solid closure, defined *a priori* in a characteristic-free manner. The author originally hoped that this theory might play, in mixed characteristic, a role analogous to that of tight closure theory in equal characteristic. Solid closure does turn out to agree, in characteristic p, with the notion of tight closure if the ring is well-behaved (for example, if the ring is finitely generated over an excellent local ring, or if $R^p \subseteq R$ is module-finite (cf. [**Ku2**]), or, more generally, if the ring has a completely stable weak test element in the sense of §6 of [**HH4**]; see §8). However, an example of Paul Roberts [**Ro6**] proves that, in equal characteristic zero, solid closure is "too big." Cf. (7.22-4). Roberts' example shows, for example, that in the ring $K[[x, y, z]]$ or $K[x, y, z]$, where K is a field of characteristic zero and x, y, z are formal indeterminates, the element $x^2y^2z^2$ is in the solid closure of the ideal (x^3, y^3, z^3). Thus, ideals of regular rings of dimension three need not be solidly closed in equal characteristic zero, which is quite different from the situation in positive characteristic.

However, it is still not clear whether every ideal is solidly closed in a regular ring of mixed characteristic. Oddly, if (V, pV) is a discrete valuation ring of mixed characteristic p then $x^2y^2z^2$ is not in the solid closure of (x^3, y^3, z^3) in $V[[x, y, z]]$ (but we do not know whether the ideal is solidly closed), while (x^3, y^3, z^3) is solidly closed in $\mathbb{Z}[x, y, z]$. We do not know whether (p^2, x^2, y^2) is solidly closed in $V[[x, y]]$ (nor in $V[x, y]$). Roberts' result is discouraging, but in some ways it makes the study of solid closure even more intriguing. The reader is referred to §7 and §13 for further discussion of the issues raised here. We note one more point: as a consequence of these remarks, solid closure does not commute with localization, because the solid closure of (x^3, y^3, z^3) in $\mathbb{Q}[x, y, z]$ is not the expansion of the solid closure of (x^3, y^3, z^3) in $\mathbb{Z}[x, y, z]$.

Regardless of whether every ideal turns out to be solidly closed in regular rings of mixed characteristic (that would imply the direct summand conjecture in general), solid closure has a number of aspects that make it worthwhile to study. It gives a novel perspective on tight closure in characteristic p that leads to new results: e.g., Theorem (5.9), viewed as a result about tight closure (this is done explicitly in Corollary (8.8)), is new. Although ideals of regular rings are not solidly closed in general, the solid closure of an ideal is contained in the integral closure and is usually much smaller, even in equal characteristic zero and in mixed characteristic. For example, if V is the ring of p-adic integers, $\mathbf{x} = x_1, \dots, x_n$, and $R = V[[\mathbf{x}]]$, every ideal of R containing p is solidly closed, while the integral closure of the ideal $(p, x_1^k, \dots, x_n^k)$ is $pR + (\mathbf{x})^k R$. Moreover, the solid closure point of view leads to a characterization of tight closure in complete local domains of characteristic p in terms of contracted expansions from a suitable balanced big Cohen-Macaulay algebra: see Theorem (11.1).

Insofar as possible, this manuscript is self-contained. We shall not need to make much use of the theory of tight closure except in §8 and §11, where we compare the two theories, and in those sections we can easily spell out just what is needed.

We next give a brief description of the notion of solid closure for the case of ideals. First note that when we refer to a "local ring (R, m, K)" we mean a Noetherian ring R with a unique maximal ideal m and residue field $K = R/m$.

(1.1) DEFINITIONS. *If R is a domain, we shall say that an R-module M is solid if $\mathrm{Hom}_R(M, R) \neq 0$. We shall say that an R-algebra S is solid if it is solid as an R-module.*

(1.2) DEFINITION. *Let $I \subseteq R$ be an ideal of a Noetherian ring R and let $x \in R$. If R is a complete local domain we say that x is in the solid closure $I^\bigstar$ of I if there exists a solid R-algebra S such that $x \in IS$. (Note: the notation $I^\blacksquare$ was used for the solid closure of I in an earlier version of this manuscript.) More generally, $x \in I^\bigstar$ if for every complete local domain B arising as the quotient by a minimal prime of the completion of R_m for some maximal ideal m of R, the image of x in B is in $(IB)^\bigstar$.*

There is a similar notion for submodules of a finitely generated R-module: see (5.1).

(1.3) REMARKS. A module-finite extension S of a Noetherian domain R is always a solid R-algebra. However, in more general situations, even when S and R are both finitely generated over a base field K, it is a subtle and difficult problem to decide whether S is a solid R-algebra for specific choices of R and S. The problem is that the elements of $\mathrm{Hom}_R(S, R)$ correspond to the solutions of an infinite system of linear equations over R (although only countably infinite in the main case, where S is finitely generated as an R-algebra).

We view the condition that S be solid as an R-algebra as a nondegeneracy condition. The reason for the choice of the term is that if R is normal and S is a solid R-algebra then S does not contain any elements of the fraction field of R that are not in R: see Proposition (2.9).

The manuscript is structured as follows. In §2 we develop the basic properties of solid modules and algebras. In §3 we do likewise for formally solid modules and algebras over a Noetherian ring: this notion is convenient in studying solid closure, since the latter is defined in terms of passage to various complete local domains associated with R. In §4 we introduce the notion of a generic forcing algebra for a triple (M, N, u) where $N \subseteq M$ are modules and $u \in M$; these map to the other algebras S such that $1 \otimes u$ is "forced" into $\mathrm{Im}\,(S \otimes N \to S \otimes M)$. This enables us to characterize when u is in the solid closure ${N^\bigstar}_M$ of N in M as follows: $u \in {N^\bigstar}_M$ iff the triple (M, N, u) has a formally solid generic forcing algebra. These ideas and a number of others are explored in §5, where the basic properties of solid closure are proved. Many of the results are parallel to those of [**HH4**] for tight closure, but the proofs are different. In some instances we have referred to [**HH4**] for "isolated" arguments that can be read without reference to other parts of [**HH4**].

In §6 we study minimal solid algebras (they have no proper solid quotients), and exhibit better behavior than other solid algebras in a number of useful ways. In §7 we study rings in which every ideal is solidly closed, producing a theory parallel to the theory of weakly F-regular rings initiated in [**HH4**]. In §8 we show that solid closure agrees with tight closure for sufficiently good rings of characteristic p, including algebras essentially of finite type over an excellent local ring. In §11 we show more: over a complete local domain R, an element is in the tight closure of an ideal if and only if it is in its contracted expansion from a big Cohen-Macaulay algebra, and a big Cohen-Macaulay algebra is solid. This result also enables us to see that a certain equal characteristic zero notion of tight closure is contained in the solid closure. Moreover, the arguments give a new proof of the existence of big Cohen-Macaulay algebras in characteristic p which is simpler than the argument of [**HH7**]. However, it does not yield the weakly functorial behavior which is a very important consequence of the results of [**HH7**]: see [**HH12**].

In §9 a rather elementary criterion for when an element is in a solid closure over a complete local domain is developed in terms of the behavior of multiples of certain formal power series.

In §10 we introduce a theory of shadow homology, which is parallel to the theory of phantom homology initiated in [**HH4**] and pursued in [**Ab1**], [**HH8**], and [**AHH**]. We are able to obtain an equicharacteristic analogue of the phantom acyclicity criteria developed in [**HH4**] and [**HH9**], but the proof makes use of the existence of big Cohen-Macaulay algebras in equal characteristic, which one has either from [**HH7**] or from the arguments of §11 here (see also [**HH5**] and [**Hu2**]). Note that the result of [**HH7**] asserts that the integral closure of a complete (or excellent) local domain of characteristic p in an algebraic closure of its fraction field is a big Cohen-Macaulay algebra for the ring, which implies the existence of big Cohen-Macaulay algebras for all local rings of equal characteristic.

In §12, we consider solid closure over rings of dimension two. We prove, in particular, that over a complete local domain of dimension two, an element is in the solid closure of an ideal if and only if it is in the contracted expansion of the ideal from a big Cohen-Macaulay algebra. The point is that in dimension two any solid algebra can be mapped to a big Cohen-Macaulay algebra. But this is false in dimension three in equal characteristic zero.

In §13 we make some further remarks concerning the behavior of solid closure in regular rings. We discuss what is known about which ideals are, or are not, solidly closed. Connections with the Briançon-Skoda theorem are discussed.

The title of the final section is self-explanatory.

To maximize its accessibility the theory of solid closure is presented here as independently as possible of tight closure theory. Nonetheless, it is, most definitely, an offshoot of tight closure theory, and it would not exist without the enormous contributions of Craig Huneke to the development of tight closure.

2. Solid modules and solid algebras

The definitions of solid module and solid algebra were given in (1.1). Although we are mainly interested in the case where R is a Noetherian ring, for the moment we shall not impose any finiteness conditions. The following results give some basic properties of solid modules and algebras. Although most of these results are elementary, they are very important, particularly Corollaries (2.3) and (2.4) and Theorem (2.11). While R is often Noetherian in the applications, it is rare for a solid R-module M to be finitely generated as an R-module: the main case is where M is a finitely generated R-algebra.

(2.1) PROPOSITION. *Let R be a domain.*

(a) *If M, N are solid R-modules (or algebras) then $M \otimes_R N$ is a solid R-module (or algebra). (The same then applies to a nonempty finite family of solid R-modules or solid R-algebras.)*

(b) *If $M \twoheadrightarrow N$ is a surjection of R-modules and N is a solid R-module then M is a solid R-module.*

(c) *If S is an R-algebra and some S-module M is a solid R-module then S is a solid R-algebra.*

(d) *If S is a solid R-algebra then there exists an R-module homomorphism $\alpha: S \to R$ such that $\alpha(1) \neq 0$.*

(e) *If M is a solid R-module and S is an extension domain of R then $S \otimes_R M$ is a solid S-module (in particular, we may take S to be any localization of R). In fact, if $\alpha : M \to R$ has image $J \neq 0$, then $id_S \otimes_R \alpha : S \otimes_R M \to S$ has image $JS \neq 0$.*

(f) *Let T be a solid R-algebra. Then every R-algebra S that has an R-algebra homomorphism to T is a solid R-algebra. In particular, T and a polynomial ring (in an arbitrary number of variables over T) are solid or not alike.*

(g) *If M is a solid R-module (or R-algebra) and N is the submodule (or ideal) of M consisting of all elements that are killed by an element of $R - \{0\}$, then M/N is a solid R-module (or algebra). In fact, any R-homomorphism $M \to R$ kills N and so factors $M \twoheadrightarrow M/N \to R$.*

(h) *If M is a solid R-module (or algebra) and $\{I_\lambda\}_\lambda$ is a family of ideals of R such that $\bigcap_\lambda I_\lambda = (0)$, then $M/(\bigcap_\lambda I_\lambda M)$ is a solid R-module (or algebra). In fact, any R-homomorphism $M \to R$ kills $\bigcap_\lambda I_\lambda M$ and so has a factorization $M \twoheadrightarrow M/(\bigcap_\lambda I_\lambda M) \to R$.*

(i) *A finitely presented R-module is solid if and only if it is faithful. In particular, a finitely generated module over a Noetherian domain R is solid if and only if it is faithful.*

(j) *A direct sum of R-modules is solid if and only if at least one of the summands is solid.*

(k) *If M is an R-module such that $N \otimes_R M$ is solid for some choice of R-module N then M is solid.*

(l) *If M is has a finite filtration $M = M_h \supseteq M_{h-1} \supseteq \cdots \supseteq M_0$ and M is a solid R-module then at least one of the factors M_{i+1}/M_i, for $0 \le i \le h-1$, is solid.*

(m) *Let S be a solid R-algebra. If J is an ideal of S such that $J^h = 0$, then S/J is a solid R-algebra. In particular, if S is Noetherian then S_{red} is a solid R-algebra.*

PROOF. (a) If $\alpha: M \to R$ and $\beta: N \to R$ are nonzero maps with images I, J (nonzero ideals of R) then there is a bilinear map $M \times N \to R$ sending (u, v) to $\alpha(u)\beta(v)$, and the corresponding map $\gamma: M \otimes_R N \to R$ has nonzero image IJ.

(b) It is clear that if $N \to R$ is nonzero then the composite map $M \twoheadrightarrow N \to R$ is nonzero (it has the same image).

(c) Fix $u \in M$ that has nonzero image under an R-homomorphism $\alpha: M \to R$ and define $\beta: S \to R$ by $\beta(s) = \alpha(su)$.

(d) Fix a nonzero R-homomorphism $\beta: S \to R$ and suppose that $\beta(s) \neq 0$ for some fixed element $s \in S$. Then $\gamma: S \to R$ defined by $\gamma(t) = \beta(st)$ has the required property.

(e) This is obvious.

(f) The first statement follows from (c), taking $M = T$ (or from (d), choosing $\alpha: T \to R$ with $\alpha(1) \neq 0$ and composing with the structural R-homomorphism $S \to T$). The second statement follows because each of T and the polynomial ring over T has a T-algebra map to the other (one is the obvious inclusion; for the other, kill the variables), and these T-algebra maps are necessarily R-algebra maps as well.

(g) This is obvious.

(h) This follows from the observation that, since $I_\lambda M$ will map into I_λ for all λ, $\bigcap_\lambda I_\lambda M$ will map into $\bigcap_\lambda I_\lambda = (0)$.

(i) It is clear from (g) that a solid module must be faithful. Now suppose that M is finitely presented and faithful. Let $W = R - \{0\}$. Then $W^{-1}M \neq 0$, and so can be mapped onto the field $F = W^{-1}R$. Since M is finitely presented, $W^{-1}\mathrm{Hom}_R(M, R) \cong \mathrm{Hom}_F(W^{-1}M, F)$, and so $\mathrm{Hom}_R(M, R) \neq 0$.

(j) This is obvious.

(k) Map a (necessarily nonzero) free R-module G onto N. Then $G \otimes_R M$ maps onto $N \otimes_R M$ and so is solid by (b). Since $G \otimes_R M$ is a direct sum of copies of M, M must be solid by (j).

(l) If one has that $0 \to N \to M \to M/N \to 0$ is exact then a map $M \to R$ is either nonzero on N, so that N is solid, or factors through M/N, so that M/N is solid. This handles the case $h = 2$ (the case $h = 1$ is trivial), while the general case follows easily by induction on h.

(m) S has a finite filtration $S \supseteq J \supseteq J^2 \supseteq \cdots J^i \supseteq \cdots \supseteq J^h = (0)$. Hence, by part (l), J^i/J^{i+1} is solid for some i. Since this is an (S/J)-module, S/J is solid by part (c). □

(2.2) PROPOSITION. *Let $R \subseteq S$ be domains. Let M be an S-module viewed also as an R-module via restriction of scalars.*

(a) *If S is embeddable, as an R-module, in a product of copies of R (this is true, for example, if S is embeddable in a free R-module) and M is solid as an S-module then M is solid as an R-module.*

(b) *If $\mathrm{Hom}_R(S, R)$ is embeddable, as an S-module, in a product of copies of S (this is true, for example, if $\mathrm{Hom}_R(S, R)$ is embeddable in a free S-module), and M is solid as an R-module, then M is solid as an S-module.*

PROOF. (a) If $S \subseteq P$, where P is a product of copies of R, and $\alpha\colon M \to S$ is nonzero, then some element of $\alpha(M) \subseteq P$ has a nonzero entry in one coordinate: let $P \to R$ be the product projection corresponding to that coordinate. Then the composite map $M \to S \hookrightarrow P \to R$ is R-linear and nonzero.

(b) Suppose that $\alpha\colon M \to R$ is nonzero and R-linear. Define

$$\beta\colon M \to \mathrm{Hom}_R(S, R)$$

to be the S-linear map whose value on $u \in M$ is the homomorphism θ_u defined by $\theta_u(s) = \alpha(su)$. Note that β is nonzero, for if $\alpha(u) \neq 0$ then $\theta_u(1) = \alpha(u) \neq 0$. Suppose that $\mathrm{Hom}_R(S, R) \subseteq Q$, a product of copies of S. Then $\beta(M) \subseteq Q$, and we may choose a product projection $Q \to S$ that is nonzero on some element of $\beta(M)$ (since $\beta(M) \neq 0$). The composite map

$$M \xrightarrow{\beta} \mathrm{Hom}_R(S, R) \hookrightarrow Q \to S$$

yields a nonzero S-linear homomorphism $M \to S$, as required. □

The following immediate corollary is very important in the applications:

(2.3) Corollary (independence of ring for module-finite extensions). *Let $R \subseteq S$ be a module-finite extension of domains, where R is Noetherian. Let M be an S-module. Then M is solid as an S-module if and only if M is solid as an R-module.*

Proof. Since S (respectively, $\mathrm{Hom}_R(S, R)$) is finitely generated and torsion-free as an R-module (respectively, S-module) it is embeddable in a free R-module (respectively, S-module). Thus, both parts of Proposition (2.2) apply. □

This also yields:

(2.4) Corollary (local cohomology criterion). *Let (R, m, K) be a complete local domain of Krull dimension d. An R-module M is solid if and only if $H^d_m(M) \neq 0$.*

Proof. We can represent R as a module-finite extension of a complete regular local ring (A, q, K). Then M is solid as an R-module if and only if it is solid as an A-module. Moreover, since qR is primary to m, $H^d_q(M) = H^d_m(M)$. Thus, there is no loss of generality in assuming that R is regular. In this case $E = H^d_m(R)$ is an injective hull for the residue field of R and $\mathrm{Hom}_R(\_, E)$ is faithfully exact. Then $H^d_m(M) \cong M \otimes_R H^d_m(R) = M \otimes_R E$ is nonzero if and only if $\mathrm{Hom}_R(M \otimes_R E, E) \neq 0$. By the adjointness of $\otimes_R$ and Hom_R the latter may be identified with $\mathrm{Hom}_R(M, \mathrm{Hom}_R(E, E)) \cong \mathrm{Hom}_R(M, R)$ (since R is complete, we have that $\mathrm{Hom}_R(E, E) \cong R$, by Matlis duality). Thus, $H^d_m(M) \neq 0$ if and only if its Matlis dual, $\mathrm{Hom}_R(M, R)$, is not zero. □

(2.5) Remarks on local cohomology. We use **[GrHa]** as a general reference for local cohomology.

(a) Let m be an ideal of a Noetherian ring R and let $\mathbf{x} = x_1, \ldots, x_d$ be a sequence of elements of R such that m and $I = (\mathbf{x})R$ have the same radical. Then the functors $H^j_I(\_)$ and $H^j_m(\_)$ may be identified. (E.g., if (R, m, K) is local we may choose $\mathbf{x}$ to be a system of parameters for R.) Let $x = x_1 \cdots x_d$ and for each i, $1 \leq i \leq d$, let $y_i = x_1 \cdots x_{i-1} x_{i+1} \cdots x_d$, so that $x_i y_i = x$ for $1 \leq i \leq d$. Then $H^d_m(R) \cong H^d_I(R)$ may be identified with $R_x/(\sum_i \mathrm{Im}\, R_{y_i})$ and if M is any R-module we have that $H^d_m(M) \cong H^d_m(R) \otimes_R M \cong M_x/(\sum_i \mathrm{Im}\, M_{y_i})$. (We have written "Im" for precision: the maps $M_{y_i} \to M_x$ may not be injective if one or more of the x_i is a nonzerodivisor.)

(b) With notation as in (a), let $\mathbf{x}^t$ denote $x_1^t, \ldots, x_d^t$. Note that, alternatively, we may view $H^d_m(R)$ as $\varinjlim_t R/(\mathbf{x}^t)$ where the maps between consecutive terms in the direct limit system are induced by multiplication by $x = x_1 \cdots x_d$ on the copies of R in the numerators, or as $(R_{x_1}/\mathrm{Im}\, R) \otimes_R \cdots \otimes_R (R_{x_d}/\mathrm{Im}\, R)$. From this last characterization it is easy to see that if m has the same radical as $(x_1, \ldots, x_d)R$, n has the same radical as $(z_1, \ldots, z_e)R$ and q has the same radical as $m + n$, then $H^{d+e}_q(R) \cong H^d_m(R) \otimes_R H^e_n(R)$.

(c) As a consequence of the discussion in (a), if R is complete local ring with system of parameters $\mathbf{x}$, then M is solid if and only if $M_x \neq \sum_i \mathrm{Im}\, M_{y_i}$.

(d) If S is an R-algebra then S_x and $\sum_i \operatorname{Im} S_{y_i} \subseteq S_x$ are both S-modules, and S_x is generated as an S-module by the elements $1/x^t$. Let z_i denote the image of y_i in S_x. Then $H^d_m(S) \neq 0$ if and only if for some t (equivalently, for all large t) $1/x^t \notin \bigcup_N \big(\sum_i S \cdot (1/z_i)^N\big)$ in S_x. This holds if and only if there exists t such that (equivalently, for all sufficiently large t) $x^{N-t} \notin (x_1^N, \ldots, x_d^N)S$ for all $N \in \mathbb{N}$ (if the x's are zerodivisors one can still multiply by a sufficiently high power of x to get a valid equation with denominators cleared). Said slightly differently:

(2.6) OBSERVATION. *Let R be a Noetherian ring, let $x_1, \ldots, x_d \in R$, let m be an ideal of R having the same radical as $(x_1, \ldots, x_d)R$, and let S be an R-algebra. (E.g., we may have that (R, m, K) is a local ring of dimension d and that $x_1, \ldots, x_d$ is a system of parameters.) Then $H^d_m(S) \neq 0$ if and only if there exists an integer t such that whenever $(x_1 \cdots x_d)^k \in (x_1^N, \ldots, x_d^N)S$ for $N, k \in \mathbb{N}$ then $N \leq k + t$.* $\square$

(2.7) REMARK. Let $(R, m, K) \to (S, n, L)$ be a local homomorphism of local rings and let $\mathbf{x} = x_1, \ldots, x_d$ be a system of parameters for R. Note that the images of the x's in S are part of a system of parameters for S if and only if killing them drops the dimension of S by $d = \dim R$. Since the radicals of $(\mathbf{x})S$ and mS are the same, an equivalent condition is that (#) $\dim S = \dim R + \dim S/mS$, and this condition is evidently independent of the system of parameters. It suffices if ht $mS \geq \dim R$ (in which case ht $mS = \dim R$, since mS is generated up to radicals by $\dim R$ elements), and the condition that ht $mS = \dim R$ is equivalent to (#) if S is equidimensional and catenary (for then ht $J + \dim S/J = \dim S$ for every ideal J of S).

Corollary (2.4) and the above discussion imply:

(2.8) COROLLARY. *Let (R, m, K) be a complete local domain, let (S, n, L) be a local ring, and let $(R, m, K) \to (S, n, L)$ be a local homomorphism such that*

$$\dim S = \dim R + \dim S/mS.$$

Let M be any S-module such that $H^{\dim S}_n(M) \neq 0$; if S is a complete local domain this simply means that M is solid as an S-module. (The condition $H^{\dim S}_n(M) \neq 0$ holds for any finitely generated S-module with $\dim M = \dim S$.) Then M is a solid R-module.

PROOF. Let $d = \dim R$. Choose a system of parameters $x_1, \ldots, x_d$ for R. The hypothesis implies that $x_1, \ldots, x_d$ can be extended to a system of parameters $x_1, \ldots, x_d, z_1, \ldots, z_e$ for S, where $\dim S = d + e$. (For finitely generated modules over a local ring, the highest nonvanishing local cohomology module with support in the maximal ideal occurs at the dimension of the module.) In any case, assume that $H^{d+e}_n(M) \neq 0$. This module may be identified with $H^{d+e}_n(S) \otimes_S M$. Let I be the ideal generated by the x's in R and let J be the ideal generated by the z's in S. By (2.5c) we have that $H^{d+e}_n(S) \cong H^d_{IS}(S) \otimes_S H^e_J(S)$ and so $H^{d+e}_n(M) \cong \big(H^d_{IS}(S) \otimes_S H^e_J(S)\big) \otimes_S M \cong \big(H^d_I(R) \otimes_R S\big) \otimes_S H^e_J(S) \otimes_S M \cong$

$H^d_I(R) \otimes_R (S \otimes_S M \otimes_S H^e_J(S) \cong H^d_m(R) \otimes_R (M \otimes_S H^e_J(S))$, and this equals $(H^d_m(R) \otimes_R M) \otimes_S H^e_J(S) \cong H^d_m(M) \otimes_S H^e_J(S)$. Since this module is not zero, we must have that $H^d_m(M) \neq 0$, and then Corollary (2.4) shows that M is solid over R. □

The following result (that solid extensions tend not to adjoin fractions) is part of the reason for the use of the word "solid."

(2.9) PROPOSITION. *Let R be a Noetherian domain, let $a, b \in R$ with $b \neq 0$ and let S be a solid R-algebra. Suppose that $a \in bS$. Then the element a/b of the fraction field of R is integral over R. Hence, if R is normal, $a/b \in R$.*

In consequence, if S is a domain solid over the Noetherian domain R and $\theta \in S$ is algebraic over R, then θ is integral over R.

PROOF. We begin with the proof of the statement in the first paragraph. We may assume that $a \neq 0$. Since the normalization of R in its fraction field F is an intersection of discrete valuation rings contained in F, if a/b is not integral over R we may choose a discrete valuation ring V with $R \subseteq V \subseteq F$ and $a/b \notin V$. We may replace R, S by $V, V \otimes_R S$ by Proposition (2.1e). Thus, we may assume without loss of generality that $R = V$ is a discrete valuation ring and that $a/b \notin V$. By Proposition (2.1g) we may replace S by a quotient that is torsion-free over V. Then $c = b/a$ is in the maximal ideal of V, and we have that $a \in bS = acS$. Since a is not a zerodivisor in S, it follows that $1 \in cS$, i.e., that c is a unit in S, and so $S = c^n S$ for every positive integer n. But then, by Proposition (2.1h), $S = \bigcap_n c^n S$ must map to zero in V, since $\bigcap_n c^n V = (0)$. This is a contradiction.

Now consider the situation in the second paragraph. Since θ is algebraic over R it satisfies an equation $\sum_{i=0}^n b_i \theta^i = 0$ for some positive integer n with the $b_i \in R$ and $b = b_n \neq 0$. It follows that $b\theta$ is integral over R and so $R[b\theta]$ is module-finite over R. Then S is also solid over $R[b\theta]$ by Theorem 2.3. Since $b\theta \in bS$ it follows from the statement in the first paragraph that the fraction $b\theta/b = \theta$ is integral over $R[b\theta]$, and since this ring is module-finite over R, θ is integral over R. □

(2.10) DISCUSSION. In Proposition (2.1e) it was noted that, quite trivially, if M is a solid R-module and S is an extension domain of R then $S \otimes_R M$ is a solid S-module. Somewhat surprisingly, when R is a Noetherian domain this remains true without the hypothesis that the map of domains $R \to S$ be injective. Before giving the proof, we note the following:

(2.11) LEMMA. *Let J be a nonzero ideal of a Noetherian domain R and let $P_1, \dots, P_k$ be height one prime ideals of R.*

(a) *If each of the rings R_P is a discrete valuation ring for $P \in \{P_1, \dots, P_n\}$ then there is an R-homomorphism $\theta: J \to R$ such that θ takes on a value not in any of the P_i, i.e., $\theta(J) \not\subseteq \bigcup_i P_i$.*
(b) *There is a module-finite extension R' of R within its fraction field and an R'-linear map $JR' \to R'$ that takes on a value outside any of the primes of*

R' that lie over one of the P_i: in fact, it takes on a value in R outside any of the P_i.

PROOF. Let $W = \bigcup_i P_i$.

(a) $W^{-1}R$ is a semilocal Dedekind domain and, hence, a PID. It follows that there is an isomorphism $\phi\colon W^{-1}J \to W^{-1}R$ as modules over $T = W^{-1}R$. Since $\mathrm{Hom}_T(W^{-1}J, T) \cong W^{-1}\mathrm{Hom}_R(J, R)$, ϕ has the form $w^{-1}\theta$ for some $w \in W$ and $\theta \in \mathrm{Hom}_R(J, R)$. Since $\mathrm{Im}\,\phi$ contains $1 \in W^{-1}R$, we have that θ takes on the value $w \in W$.

(b) The normalization of a one-dimensional semilocal Noetherian domain D is also a one-dimensional semilocal Noetherian domain, although it need not, in general, be module-finite over D. (Cf. [**N**], Theorem 33.2, which guarantees that the integral closure is Noetherian. The fact that there are only finitely many maximal ideals follows easily from the fact that there only finitely many maximal ideals in D.) Such a normalization is therefore a PID. Let S be the normalization of R. Then $W^{-1}S$ is the normalization of $W^{-1}R$, which is a one-dimensional semilocal ring. Thus, $W^{-1}S$ is a PID, and we can choose a $W^{-1}S$-linear map $\phi\colon JW^{-1}S \to W^{-1}S$ that is an isomorphism. Suppose that u/w' maps to 1 under this homomorphism, where $u \in JS$ and $w' \in W$. Then u maps to w'. Choose a finite set of generators j_ν of J as an R-module and choose $w \in W$ such that the elements $\phi(j_\nu)$ are all of the form s_ν/w with every $s_\nu \in S$. Choose a module-finite extension R' of R contained in S such that $u \in JR'$ and such that the s_ν are in R'. The restriction of $w\phi$ to JR' has image contained in R' (if we have several elements $a_\nu \in R'$ then $w\phi$, which is $W^{-1}S$-linear, maps $\sum_\nu a_\nu j_\nu$ to $\sum_\nu a_\nu w\phi(j_\nu) = \sum_\nu a_\nu s_\nu$, which is in R'.) Thus, the restriction of $w\phi$ to JR' yields an R'-linear map of JR' to R' whose value on u is $ww' \in W$. □

We are now ready to prove a very important result:

(2.12) THEOREM (PERSISTENCE OF SOLIDITY). *Let R be a Noetherian domain and let M be a solid R-module. Then for any homomorphism $R \to S$, where S is a domain, $S \otimes_R M$ is a solid S-module.*

PROOF. Let $P = \mathrm{Ker}\,(R \to S)$ and let $P = P_h \supset \cdots \supset P_0 = (0)$ be a saturated chain of primes in R descending from P. The map $R \to S$ factors $R \to R/P_1 \to \cdots \to R/P_h \hookrightarrow S$. By the associativity of $\otimes$, in order to prove the result for the composite map it suffices to prove it for each map in the chain. The problem of proving that $S \otimes_R M$ is solid over S thereby reduces to proving the result in two cases: one is the case of an inclusion, and the other is the case where the domain S is obtained from the domain R by killing a height one prime. Since we have already done the case of an inclusion, we may assume that $S = R/P$, where P is a height one prime of S.

Fix a nonzero homomorphism $\alpha\colon M \to R$ with image $J \neq (0)$. By Lemma (2.11) we can choose a module-finite extension domain R' of R within the fraction field of R and an R'-linear map $\theta\colon JR' \to R'$ that takes on a value in $R - P$. Let Q be a prime ideal of R' lying over P. Then $M' = R' \otimes_R M$ is solid over R', by

Proposition (2.1e), and, in fact, $\mathrm{id}_{R'} \otimes_R \alpha$ gives an R'-linear map α' of $R' \otimes_R M$ to R' with image JR'. It follows that the composite map $\beta = \theta\alpha'$ from M' to R' takes on a value in $R - P$: this value will be an element of $R' - Q$. Thus, if we tensor with $S' = R'/Q$ we obtain a nonzero (R'/Q)-module homomorphism $M'/QM' \to R'/Q$, so that M'/QM' is solid over R'/Q.

Now, R'/Q is a module-finite extension of R/P, and so it follows from Corollary (2.3) that M'/QM' is solid over R/P. But then

$$M'/QM' \cong (R'/Q) \otimes_R M \cong (R'/Q) \otimes_{R/P} \big((R/P) \otimes_R M\big)$$

is solid over R/P, and it follows from Proposition (2.1k) that $(R/P) \otimes_R M$ is solid over R/P, as required. □

3. Formally solid modules and algebras

We begin with two definitions.

(3.1) DEFINITION. *Let m be a maximal ideal of R. We shall refer to the R-algebra obtained by completing the local ring R with respect to its maximal ideal and then killing a minimal prime as a complete local domain of R.*

(3.2) DEFINITION. *Let R be a Noetherian ring. We shall say that an R-module M is formally solid if for every complete local domain B of R, $B \otimes_R M$ is solid over B.*

From this definition we have at once:

(3.3) PROPOSITION. *Let R be a Noetherian ring and let M be an R-module.*
(a) *M is formally solid if and only if for every maximal ideal m of R, M_m is formally solid over R_m.*
(b) *If R is local, M is formally solid over R if and only if $\widehat{R} \otimes_R M$ is formally solid over $\widehat{R}$.*
(c) *If R is complete local, M is formally solid if and only if for every minimal prime $\mathfrak{p}$ of R, $(R/\mathfrak{p}) \otimes_R M$ is formally solid over $R/\mathfrak{p}$.*
(d) *If R is complete local, M is formally solid over R if and only if M is solid over R.* □

We also note:

(3.4) PROPOSITION. *Let R be a Noetherian ring.*
(a) *If M, N are formally solid R-modules (or algebras) then $M \otimes_R N$ is a formally solid R-module (or algebra). (The same then applies to a nonempty finite family of solid R-modules or solid R-algebras.)*
(b) *If $M \twoheadrightarrow N$ is a surjection of R-modules and N is a formally solid R-module then M is a formally solid R-module.*
(c) *If S is an R-algebra and some S-module M is a formally solid R-module then S is a formally solid R-algebra.*

(d) *If an R-algebra has S has an R-algebra map to a formally solid R-algebra T, then S is formally solid. In particular, T and a polynomial ring (in an arbitrary number of variables) over T are formally solid or not alike.*

(e) *M is formally solid over R if and only if $M_{red} = R_{red} \otimes_R M$ is formally solid over R_{red} (if and only if M_{red} is formally solid over R).*

(f) *A finitely generated R-module M is formally solid if and only if $M_{red} = R_{red} \otimes_R M$ is faithful over R_{red}.*

(g) *If M is an R-module such that $N \otimes_R M$ is formally solid for some choice of R-module N then M is formally solid.*

PROOF. (a) This follows from the fact that

$$B \otimes_R (M \otimes_R N) \cong (B \otimes_R M) \otimes_B (B \otimes_R N)$$

and part (a) of Proposition (2.1).

(b) This follows from the right exactness of $\otimes$ and part (b) of Proposition (2.1).

(c) Let B be a complete local domain of R and apply part (c) of Proposition (2.1) to $B \otimes_R M$ viewed as a module over $B \otimes_R S$.

(d) This is immediate from part (c).

(e) This is immediate from the fact that the complete local domains of R are the same as the complete local domains of R_{red}, viewed as R-algebras.

(f) By part (e), we may assume that R is reduced. Note that M is faithful if and only if R can be embedded in a finite direct sum of copies of M. Thus, faithfulness is preserved by flat base change, and is preserved when we localize and complete. Hence, if M is faithful, it remains faithful when we pass to the completed localization of R at a maximal ideal. If $\mathfrak{p}$ is a minimal prime of R, then $(R/\mathfrak{p}) \otimes_R M$ is still supported at $\mathfrak{p}$, and so $(R/\mathfrak{p}) \otimes_R M$ is faithful over $R/\mathfrak{p}$, and hence, solid over $R/\mathfrak{p}$. This shows that if M_{red} is faithful over R_{red}, then M is formally solid.

Now suppose that R is reduced but that M is not faithful. Then M is killed by some element x that is not nilpotent. Localize R at a maximal ideal m containing $\bigcup_t \mathrm{Ann}_R x^t$. Then the image of x is still not nilpotent in R_m, and kills M_m. Likewise, the image of x in the completion C of R is not nilpotent, and so we may kill a minimal prime $\mathfrak{p}$ of C to obtain a complete local domain B of R such that the image of x is nonzero in B.

(g) For every complete local domain B of R,

$$B \otimes_R (N \otimes_R M) \cong (B \otimes_R N) \otimes_B (B \otimes_R M)$$

is solid over B, and so $B \otimes_R M$ is solid over $B$by Proposition (2.1k). □

In (3.7) we shall prove that the property of being formally solid is preserved by arbitrary Noetherian base change. We first note:

(3.5) LEMMA. *Let M be a formally solid module over a complete local ring R and let S be a Noetherian R-algebra. Then $S \otimes_R M$ is formally solid over S.*

PROOF. It suffices to prove that $C \otimes_R M$ is solid over C after a further base change to a complete local domain C. Therefore, we may assume without loss of generality that $S = C$ is a complete local domain. Choose a minimal prime $\mathfrak{p}$ of R contained in $\mathrm{Ker}\,(R \to C)$. Then $(R/\mathfrak{p}) \otimes_R M$ is solid over $R/\mathfrak{p}$ by the hypothesis on M, and we may apply Theorem (2.12) to the base change $R/\mathfrak{p} \to C$. □

We shall also need:

(3.6) LEMMA. *Let $(R, m, K) \to (S, n, L)$ be a local homomorphism of local rings and suppose that at least one of the following conditions holds:*
(a) *R, S are complete local domains and $ht\, mS \geq ht\, m$ ($= \dim R$).*
(b) *R, S are complete local rings and for every minimal prime $\mathfrak{p}$ of R there is a prime ideal $\mathfrak{q}$ of S lying over $\mathfrak{p}$ such that $ht\, m(S/\mathfrak{q}) \geq ht\,(m/\mathfrak{p})$ ($= \dim R/\mathfrak{p}$).*
(c) *S is faithfully flat over R.*

Then every S-module W that is formally solid as an S-module is formally solid as an R-module.

Moreover, if M is an R-module such that $S \otimes_R M$ is formally solid as S-module then M is formally solid as an R-module.

PROOF. The second conclusion follows from the first conclusion by Proposition (3.4g).

(a) By (2.8) (see also Remark (2.7)) and the height condition on S, we have that W is solid over R.

(b) Fix a minimal prime $\mathfrak{p}$ of R and fix a prime ideal $\mathfrak{q}$ of S such that $\mathfrak{q}$ lies over $\mathfrak{p}$ and ht $m(S/\mathfrak{q}) \geq$ ht $m/\mathfrak{p}$. Since W is formally solid over S, by (3.3e) it is solid over S. By Theorem (2.12), $(S/\mathfrak{q}) \otimes_S W \cong S/\mathfrak{q} \otimes_{R/\mathfrak{p}} (R/\mathfrak{p} \otimes_R W)$ is solid over $S/\mathfrak{q}$, and so part (a) applied to $R/\mathfrak{p} \to S/\mathfrak{q}$ shows that $R/\mathfrak{p} \otimes_R W$ is solid over $R/\mathfrak{p}$. Since this holds for every minimal prime $\mathfrak{p}$ of R, we see that (b) is implied by (a) (which is, of course, a special case of (b)).

(c) The map $\widehat{R} \to \widehat{S}$ is also faithfully flat and, in consequence, satisfies the condition in (b). To see this, note that if $\mathfrak{p}$ is a minimal prime of R then $S/\mathfrak{p}S$ is faithfully flat over $R/\mathfrak{p}$, and we can choose a minimal prime $\mathfrak{q}$ of $\mathfrak{p}S$ (which will be a minimal prime of S, since $S_\mathfrak{q}$ is nilpotent modulo $\mathfrak{p}S_\mathfrak{q}$ and $\mathfrak{p}R_\mathfrak{p}$ is nilpotent) such that $\dim S/\mathfrak{q} = \dim S/\mathfrak{p}S$. Now, $\dim S/\mathfrak{p}S = \dim R/\mathfrak{p} + \dim\big((S/\mathfrak{p}S)/m(S/\mathfrak{p}S)\big)$ ($= \dim R/\mathfrak{p} + \dim S/mS$), by the flatness of $S/\mathfrak{p}S$ over $R/\mathfrak{p}$ (cf. **[Mat]** (13.B) Theorem 19 (2)). Thus,

$$\dim S/\mathfrak{q} = \dim R/\mathfrak{p} + \dim\,(S/mS) \geq \dim R/\mathfrak{p} + \dim\big((S/\mathfrak{q})/m(S/\mathfrak{q})\big),$$

so that the dimension of $S/\mathfrak{q}$ falls by at least $\dim R/\mathfrak{p}$ when one kills $m(S/\mathfrak{q})$, and this shows that ht $m(S/\mathfrak{q}) \geq \dim R/\mathfrak{p}$, as required.

Now, the fact the W is formally solid over S implies that $\widehat{S} \otimes_R W \cong \widehat{S} \otimes_{\widehat{R}} (\widehat{R} \otimes_R W)$ is formally solid over $\widehat{S}$, and so, by part (b), over $\widehat{R}$. This implies that $\widehat{R} \otimes_R W$ is formally solid over $\widehat{R}$, and so W is formally solid over R, as required. □

Part (c) of this result is globalized in Theorem (3.8) below.

Notice that in part (b) there may be a suitable choice of $\mathfrak{q}$ although there is no *minimal* prime of S that gives a suitable choice of $\mathfrak{q}$. For example, let $R = K[[x, y]]$ with K a field and let $S = R[[u, v]]/(xv - yu)$. Then $R \subseteq S$ and has the unique minimal prime $\mathfrak{p} = (0)$, and ht $mS = 1$, so that we may not choose $\mathfrak{q} = (0)$ (the only minimal prime of S), but we may take $\mathfrak{q} = (u, v)S$.

We can now prove the following result of basic importance:

(3.7) THEOREM (PERSISTENCE OF FORMAL SOLIDITY). *If M is a formally solid module over a Noetherian ring R and S is a Noetherian R-algebra then $S \otimes_R M$ is formally solid over S.*

PROOF. It suffices to prove this when S is replaced by a complete local domain to which it maps. We may therefore assume that S is a complete local domain. Let P be the contraction of the maximal ideal of S to R, let m be a maximal ideal of R containing P, and let T be the completion of the ring R_m. Let Q be a prime ideal of T lying over P. Then the map $R \to S$ factors

$$R \to R_m \to R_P \to (R_P)^\frown \to S$$

while the faithfully flat map $R_P \to T_Q$ induces a faithfully flat map $(R_P)^\frown \to (T_Q)^\frown$, and the map $R \to (R_P)^\frown \to (T_Q)^\frown$ also factors $R \to R_m \to T \to (T_Q)^\frown$. The fact that M is formally solid over R is preserved when we make a base change from R to T by Proposition (3.3a,b), and since T is complete we may then make a further base change to $(T_Q)^\frown$ by Lemma (3.5). By Lemma (3.6), since $(T_Q)^\frown$ is faithfully flat over $(R_P)^\frown$, we have that $(R_P)^\frown \otimes_R M$ is formally solid over $(R_P)^\frown$. But we may then make a base change from $(R_P)^\frown$ to S by a second application of Lemma (3.5), since $(R_P)^\frown$ is complete. □

(3.8) THEOREM. *If the Noetherian ring S is faithfully flat over a Noetherian ring R and W is a formally solid S-module then W is a formally solid R-module.*

Hence, if M is an R-module such that $S \otimes_R M$ is formally solid over S, then M is formally solid over R.

PROOF. As in the proof of (3.6), the second conclusion follows from the first.

Let m be a maximal ideal of R and let Q be a prime ideal of S lying over m. Then $R_m \to S_Q$ is faithfully flat. By Theorem (3.7), W_Q is formally solid over S_Q. By Theorem (3.6), W_Q is formally solid over R_m. Since $W_Q \cong S_Q \otimes_{R_m} W_m$ it follows from Proposition (3.4g) that W_m is formally solid over R as well. Since this holds for every maximal ideal m of R, the result follows from Proposition (3.3a). □

We also note:

(3.9) PROPOSITION. *If R is Noetherian, a faithfully flat R-module M is formally solid.*

PROOF. The hypothesis is preserved when we pass from R to a complete local domain of R. Thus, we may assume that (R, m, K) is a complete local domain with $\dim R = d$. But then $H^d_m(M) \cong H^d_m(R) \otimes_R M \neq 0$. □

4. Generic forcing algebras

(4.1) DEFINITION. *Let R be a Noetherian ring, let $N \subseteq M$ be finitely generated R-modules, and let $u \in M$. We shall say that an R-algebra S is a forcing algebra for the triple (M, N, u) if the image of u in $S \otimes_R M$ is in $Im(S \otimes_R N \to S \otimes_R M)$.*

It is trivial to verify that:

(4.2) OBSERVATION. *The R-algebra S is a forcing algebra for (M, N, u) if and only if it is a forcing algebra for the triple $(M/N, 0, \overline{u})$, where $\overline{u}$ denotes the image of u in M/N. If T is a forcing algebra for (M, N, u) and there is a homomorphism $T \to S$ then S is also a forcing algebra for (M, N, u).* □

(4.3) DISCUSSION AND DEFINITION. Let $R^k \xrightarrow{A} R^h \to M/N \to 0$ be a finite presentation of M/N, where $A = (r_{ij})$ is an $h \times k$ matrix over R that gives the map and if we choose a vector $(r_1, \dots, r_h) \in R^h$ that represents $\overline{u}$ in $R^h/\mathrm{Im}\, R^k \cong M/N$, then S will be a forcing algebra for the triple (M, N, u) if and only if the image of $\overline{u}$ in $S \otimes_R (M/N)$ is 0, which is equivalent to the statement that the image of $(r_1, \dots, r_h)$ is zero in $S \otimes_R (R^h/\mathrm{Im}\, R^k)$. If we write $(r_1, \dots, r_h)$ instead as a column vector ρ this is equivalent to the existence of a column vector σ in S^k such that $A\sigma$ is the image of ρ.

We can therefore construct a forcing algebra for (M, N, u) by adjoining indeterminates $X_1, \dots, X_h$ to R and then killing the ideal generated by the entries of the matrix $AX - \rho$ in the polynomial ring $R[X_1, \dots, X_h]$, where X is the column vector whose entries are the X_i. We shall refer to this algebra as the *generic forcing algebra* for (M, N, u) for the data A, ρ. We shall also say that it is a *generic forcing algebra* for (M, N, u). Let T be the generic forcing algebra for the data A, ρ. Given any other forcing algebra S for (M, N, u), we can map T to S by sending the X_i to the entries of the vector σ whose existence was observed above.

Thus, if one fixes a generic forcing algebra for (M, N, u), all other forcing algebras may be viewed as, simply, the algebras to which it maps. This explains the terminology.

If $I = (a_1, \dots, a_k)R$ and $u = r \in R$ then there is a generic forcing algebra for (R, I, u) of the particularly simple form

$$R[X_1, \dots, X_h]/(\sum_{i=1}^{k} a_i X_i - r).$$

Here, we are using the data $A = (\, a_1 \quad \dots \quad a_k \,)$ and $\rho = (\, r \,)$.

With this terminology, we observe the following:

(4.4) PROPOSITION. *Let R be a Noetherian ring. Let $N \subseteq M$ be finitely generated R-modules and let $u \in M$. Let T be a generic forcing algebra for (M, N, u) for the data A, ρ.*

(a) *If S is any R-algebra, let A', ρ' denote the images of A, ρ under the homomorphism $R \to S$ applied to every entry. Then $S \otimes_R T$ is a generic forcing algebra for the triple $(S \otimes_R M, S \otimes_R N, 1 \otimes u)$ for the data A', ρ'.*

(b) *T maps as an R-algebra to every other forcing algebra for (M, N, u).* □

(4.5) DISCUSSION. It is natural to ask how the generic forcing algebras for (M, N, u) over R vary with choice of data A, ρ. Roughly speaking, they are uniquely determined as R-algebras up to "adjunction of indeterminates": see Proposition (4.6) below for a precise statement. We make the following observations:

(a) If one changes only ρ the algebra does not change, up to R-isomorphism. For ρ will change by adding a vector of the form $A\beta$, where β has entries in R, and the matrix $AX - (\rho + A\beta)$ becomes $AX' - \rho$ if we let $X' = X - \beta$, i.e., if we make a translation of coordinates over R.

(b) Suppose that we change the presentation of M/N by using the same generators for M but a possibly different set of generators for the relations. We may understand the effect on the generic forcing algebra by comparing each of the algebras with the algebra obtained by using the union of the two sets of relations. This enables us to consider only what happens when one enlarges the set of relations by adjoining some additional "redundant" relations. This is equivalent to giving the matrix A some additional columns, each of which is a linear combination of the columns already present.

For simplicity, we may consider the case where there is just one column, which we may suppose has the form $A\delta$ for some column vector δ over R. Then there is one new variable, X_{h+1}, and if X is the column vector whose entries are $X_1, \dots, X_h$, then to define the new generic forcing algebra we must kill the entries of $AX + A\delta X_{h+1} - \rho = AX' - \rho$ if we let $X' = X + X_{h+1}\delta$. We see that by using coordinates X', X_{h+1} (this is a linear change of coordinates) we have changed the R-isomorphism class of the generic forcing algebra by adjoining an indeterminate.

Thus, if S, T are two generic forcing algebras for (M, N, u) that arise from data that differs only by altering the relations chosen in the presentation of M/N, then there are indeterminates $\mathbf{Y} = Y_1, \dots, Y_\mu$ and $\mathbf{Z} = Z_1, \dots, Z_\nu$ such that $S[\mathbf{Y}] \cong T[\mathbf{Z}]$ over R.

(c) Suppose that we change the generators of M. By comparing what happens for each of two sets of generators with what happens when their union is used, we see that it suffices to understand what happens when a set of generators is enlarged, and it is enough to consider the case where we insert a single new redundant generator. Indeed, we may assume that we have chosen generators $u_1, \dots, u_h$ for M and that the new generator is $-(a_1u_1 + \cdots + a_hu_h)$. Then we get a new presentation in which the matrix, in block form, is $\begin{pmatrix} A & \alpha \\ 0 & 1 \end{pmatrix}$, where α is a column vector whose entries are the a's, the 0 is a row vector of length k, and the element 1 represents a size one block with the single entry 1. We may then take the replacement of ρ to be $\begin{pmatrix} \rho \\ 0 \end{pmatrix}$. If we write our vector of indeterminates as $X' = \begin{pmatrix} X \\ X_{h+1} \end{pmatrix}$, where X retains its former meaning, then

$$\begin{pmatrix} A & \alpha \\ 0 & 1 \end{pmatrix} \begin{pmatrix} X \\ X_{h+1} \end{pmatrix} - \begin{pmatrix} \rho \\ 0 \end{pmatrix} = \begin{pmatrix} AX + \alpha X_{h+1} - \rho \\ X_{h+1} \end{pmatrix}.$$

Killing the entries of this matrix produces the same generic forcing algebra as before, up to R-isomorphism.

We can summarize the content of this discussion as follows:

(4.6) PROPOSITION. *If R is a Noetherian ring, $N \subseteq M$ are finitely generated R-modules, $u \in M$, and S, T are two generic forcing algebras for (M, N, u) for possibly different data, then there are finite sets of indeterminates $\mathbf{Y}$, $\mathbf{Z}$ such that $S[\mathbf{Y}] = T[\mathbf{Z}]$ as R-algebras.* □

5. Solid closure

We first want to extend Definition (1.2) to the case of submodules of finitely generated modules.

(5.1) DEFINITION. *Let R be a Noetherian ring, let $N \subseteq M$ be finitely generated R-modules, and let $u \in M$. If R is a complete local domain we say that u is in the solid closure $N^\bigstar$ (or ${N^\bigstar}_M$) of N in M over R if there is a solid R-algebra S such that the image $1 \otimes u$ of u in $S \otimes_R M$ is in $Im(S \otimes_R N \to S \otimes_R M)$. In other words, $u \in N^\bigstar$ if and only if (M, N, u) has a forcing algebra S that is solid as an R-algebra.*

In the general case, we say that x is in the solid closure $N^\bigstar$ (or ${N^\bigstar}_M$) of N in M over R if for every complete local domain B of R, the image of x in $B \otimes_R M$ is in the solid closure of $Im(B \otimes_R N \to B \otimes_R M)$ in $B \otimes_R M$ over B. In other words, every complete local domain B of R has a solid B-algebra S such that the image of x in $S \otimes_R M$ is in $Im(S \otimes_R N \to S \otimes_R M)$.

(5.2) DISCUSSION AND NOTATIONS. Let $\overline{x}$ denote the image of x in M/N. For any given R-algebra S, the image of x in $S \otimes_R M$ is in Im $(S \otimes_R N \to S \otimes_R M)$ if and only if the image of $\overline{x}$ in $S \otimes_R (M/N)$ is 0. It follows that x is in ${N^\bigstar}_M$ if and only if $\overline{x}$ is in ${0^\bigstar}_{M/N}$. We may also map a finitely generated free module G onto M, let H be the inverse image of N in G and let y be an element of G that maps to x under the surjection $G \twoheadrightarrow M$. Then $x \in {N^\bigstar}_M$ if and only if $y \in {H^\bigstar}_G$.

It follows that issues about solid closures of submodules may be reduced either to the case where the submodule is zero, or to the case where the ambient module is free.

When G is free we denote by HS the module $\mathrm{Im}\,(S \otimes_R H \to S \otimes_R G)$. This is analogous to the notation used when $G = R$ and H is an ideal of R. When the base ring, say R, is understood we shall often use the notation M_S for $S \otimes_R M$, and the notation $\langle N_S \rangle$ to denote $\mathrm{Im}\,(N_S \to M_S)$. Of course, $\langle N_S \rangle$ depends not only on N and S but also on the map $N \to M$, the base ring R, and the R-algebra structure of S. Nonetheless, this notation is convenient, and it will only be used when the ambiguity is not a problem.

Solid closure may be described alternatively in terms of formal solidity for generic forcing algebras.

(5.3) PROPOSITION. *Let R be a Noetherian ring, let $N \subseteq M$ be finitely generated R-modules, and let $u \in M$. Then the following conditions are equivalent:*
(a) $u \in N^{\bigstar}{}_M$.
(b) *For every complete local domain B of R, there is a solid forcing algebra for the triple $(B \otimes_R M, B \otimes_R N, 1 \otimes u)$ over B.*
(c) *For every complete local domain B of R, there is a finitely generated solid forcing algebra for $(B \otimes_R M, B \otimes_R N, 1 \otimes u)$ over B.*
(d) *For every complete local domain B of R, some (equivalently, every) generic forcing algebra for $(B \otimes_R M, B \otimes_R N, 1 \otimes u)$ over B is solid.*
(e) *Some forcing algebra for (M, N, u) over R is formally solid.*
(f) *Some (equivalently, every) generic forcing algebra for (M, N, u) over R is formally solid.*

PROOF. (a) and (b) are equivalent by virtue of the definition of solid closure. The weak form of (d) ("some") implies (c) and (c) evidently implies (b). On the other hand, (b) implies the strong form of (d) ("every"), since if T is a forcing algebra over B then for every generic forcing algebra S, S maps as a B-algebra to T and we may apply Proposition (2.1f). This shows that (a), (b), (c) and both forms of (d) are equivalent. Assume these conditions. Let S_R denote a generic forcing algebra for (M, N, u) over R. Then for every complete local domain B of R, $S_B = B \otimes_R S_R$ is a generic forcing algebra for $(B \otimes_R M, B \otimes_R N, 1 \otimes u)$ over B. It is then clear that if some choice of S_R is formally solid then the weak form of (d) holds, while if the strong form of (d) holds then every choice of S_R must be formally solid. Thus, both the strong and weak forms of (f) are equivalent to (d). Since (f) $\Rightarrow$ (e) $\Rightarrow$ (d) in its weak form (similarly), all these conditions are equivalent. □

In the next two propositions we explore the basic properties of solid closure without using the notion of a generic forcing algebra nor the notion of formal solidity. There is little cost in doing so. However, these notions provide a very easy proof for Theorem (5.6), given the stability theorem for formally solid modules that we have already established.

(5.4) PROPOSITION. *Let R be a Noetherian ring, let M be a finitely generated R-module, let N, N', and P denote submodules of M, and let I denote an ideal of R. If there is no subscript on a solid closure, the subscript is understood to be M.*

(a) *$N^{\bigstar}{}_M$ is a submodule of M containing N.*

(b) *If $P \subseteq N \subseteq M$ then $P^{\bigstar}{}_M \subseteq N^{\bigstar}{}_M$ and $P^{\bigstar}{}_N \subseteq P^{\bigstar}{}_M$.*

(c) *If $N_1 \subseteq M_1, \dots, N_h \subseteq M_h$ are finitely many inclusions of finitely generated modules over R then for every complete local domain B of R there is a solid B-algebra S such that the image of $N_i{}^{\bigstar}{}_{M_i}$ in $S \otimes_R M_i = (M_i)_S$ is contained $\langle (N_i)_S \rangle$ for $1 \le i \le h$.*

(d) $(N^{\bigstar})^{\bigstar} = N^{\bigstar}$.

(e) $(N \cap N')^{\bigstar} \subseteq N^{\bigstar} \cap N'^{\bigstar}$.

(f) $(N + N')^{\bigstar} = (N^{\bigstar} + N'^{\bigstar})^{\bigstar}$.

(g) $(IN)^{\bigstar} = (I^{\bigstar} N^{\bigstar})^{\bigstar}$.

(h) *An arbitrary intersection of solidly closed submodules of M is solidly closed.*

(i) *The solid closure of the ideal* (0) *in R is the nilradical J of R. The solid closure $N^{\bigstar}$ of N in M contains JM.*

(j) *Let $R_{red} = R/J$ with J as in* (i) *just above and let $P_{red} = R_{red} \otimes_R P$ for every R-module P. Then the solid closure of N in M over R is the same as the inverse image in M of the solid closure of $\mathrm{Im}(N_{red} \to M_{red})$ in M_{red}, and the latter may be calculated either considering these modules over R_{red} or considering them over R.*

(k) *An element $u \in M$ is in $N^{\bigstar}{}_M$ if and only if for every minimal prime $\mathfrak{p}$ of R, the image of u in $M/\mathfrak{p}M$ is in the solid closure of $\mathrm{Im}(N/\mathfrak{p}N \to M/\mathfrak{p}M)$ in $M/\mathfrak{p}M$, calculated over $R/\mathfrak{p}R$.*

(l) *$(N :_M I)^{\bigstar}{}_M \subseteq N^{\bigstar}{}_M :_M I$ and $(N :_R P)^{\bigstar}{}_R \subseteq N^{\bigstar}{}_M :_R P$. Also, $N^{\bigstar}{}_M :_M I$ and $N^{\bigstar}{}_M :_R P$ are solidly closed in M and R, respectively.*

(m) *If $M_1, \dots, M_h$ are finitely generated R-modules, $N_i \subseteq M_i$ for $1 \le i \le h$, and P_i denotes the solid closure of N_i in M_i, then the solid closure of $N_1 \oplus \cdots \oplus N_h$ in $M_1 \oplus \cdots \oplus M_h$ is $P_1 \oplus \cdots \oplus P_h$.*

PROOF. (a) Suppose that $u_1, u_2 \in N^{\bigstar}$ and that r is an element of R. We must show that ru_1 and $u_1 + u_2$ are in $N^{\bigstar}$. Let B be a complete local domain of R. We must show that there is a solid B-algebra S such that the images of ru_1 and $u_1 + u_2$ in M_S are in $\langle N_S \rangle$. There exist solid B-algebras S_1, S_2 such that the image of u_i is in $\langle N_{S_i} \rangle$ for $i = 1, 2$. This remains true when S_i is mapped further. Since both S_1 and S_2 map as B-algebras to $S = S_1 \otimes_B S_2$, it suffices to observe that $S_1 \otimes_B S_2$ is also a solid B-algebra, by Proposition (2.1a). (It is obvious that $N^{\bigstar} \supseteq N$.)

(b) This is immediate from the definition.

(c) Choose finitely many generators u_{ij} of N_i for every i. Given B, for every i, j there is a solid B-algebra S_{ij} such that the image of u_{ij} in $S_{ij} \otimes_R M$ is in $\langle (N_i)_{S_{ij}} \rangle$. Let S be the tensor product over B of the finitely many solid B-algebras S_{ij}, which is solid by Proposition (2.1a).

(d) By part (a), $N^\bigstar \subseteq (N^\bigstar)^\bigstar$, and so it suffices to show that if $u \in (N^\bigstar)^\bigstar$ then $u \in N^\bigstar$. Let B be any complete local domain of R. We must show that there is a solid B-algebra S such that the image of u in M_S is in $\langle N_S \rangle$. By part (c) we know that we can choose a solid B-algebra S_1 such that the image of $N^\bigstar$ in M_{S_1} is contained in $\langle N_{S_1} \rangle$. We also know that we can choose a solid B-algebra S_2 such that image of u in M_{S_2} is in $\langle N_{S_2} \rangle$. It follows easily that with $S = S_1 \otimes_B S_2$ the image of u is in contained in $\langle N_S \rangle$ (the fact that S_1 maps to S shows that this image contains the image of $N^\bigstar$, and the fact that S_2 maps to S then shows that it contains u).

(e) This is a consequence of the first part of (b).

(f), (g) By (a), $N + N' \subseteq N^\bigstar + N'^\bigstar$ (respectively, $IN \subseteq I^\bigstar N^\bigstar$), so that $\subseteq$ follows from (b). Let $P = N + N'$ (respectively, IN) and let $Q = N^\bigstar + N'^\bigstar$ (respectively, $I^\bigstar N^\bigstar$). Let B be a complete local domain of R. We must show that there is a solid B-algebra S such that the image of $Q^\bigstar$ in M_S is contained in $\langle P_S \rangle$. But by part (c) we can choose S such that the images of $Q^\bigstar$, $N^\bigstar$, and $N'^\bigstar$ (respectively, $Q^\bigstar$, $N^\bigstar$, and $I^\bigstar$) are contained respectively in $\langle Q_S \rangle$, $\langle N_S \rangle$, and $\langle N'_S \rangle$ (respectively, in $\langle Q_S \rangle$, $\langle N_S \rangle$, and IS). It follows that the image of $Q^\bigstar$ is contained in the sum of the images of N_S and N'_S (respectively, the product of IS with the image of N_S), and this is the same as the image of P_S.

(h) This follows from the first part of (b).

(i) Since any nilpotent of R is killed by a map to a complete local domain of R, it is clear that $(0)^\bigstar \supseteq J$. But if f is an element of R that is not nilpotent then there is a maximal ideal m of R that contains $\operatorname{Ker}(R \to R_f)$, and the image of f in R_m is also not nilpotent. Since R_m injects into its completion C, the image of f in C is not nilpotent, and so there is a minimal prime ideal $\mathfrak{p}$ of C that does not contain f. It follows that the image of f in $B = C/\mathfrak{p}$ (a complete local domain of R) is not zero, and so $f \notin (0)^\bigstar$, as required.

The second statement follows from the fact that the image of JM is zero in M_B for any complete local domain B of R.

(j) The complete local domains B of R are obviously in bijective correspondence with the complete local domains of R_{red}, and the nilradical of R is killed in mapping to any complete local domain B of R. For each solid algebra S over a complete local domain B of R,

$$S \otimes_R (M/N) \cong S \otimes_{R_{red}} (R_{red} \otimes_R (M/N)) \cong S \otimes_{R_{red}} (M_{red}/\operatorname{Im} N_{red}).$$

The result is immediate from these remarks.

(k) Let $\mathfrak{p}_1, \ldots, \mathfrak{p}_h$ be the minimal primes of R. For every localization R_m of R, the minimal primes of R_m are given by the expansions of those $\mathfrak{p}_i$ contained in m. Every minimal prime $\mathfrak{q}$ of the completion B of R_m lies over a unique minimal prime of $\mathfrak{p}R_m$ of R_m (where $\mathfrak{p}$ is a unique minimal prime of R) since $R_\mathfrak{p} \to B_\mathfrak{q}$ will be faithfully flat. Conversely, if $\mathfrak{p}$ is a minimal prime of R contained in m and $\mathfrak{q}$ is a minimal prime of $\mathfrak{p}B$ then $\mathfrak{q}$ is a minimal prime of B (again consider the map $R_\mathfrak{p} \to B_\mathfrak{q}$; the elements of $\mathfrak{p}R$ are nilpotent and $\mathfrak{q}B_\mathfrak{q}$ is nilpotent modulo $\mathfrak{p}B_\mathfrak{q}$).

It follows that every complete local domain B of R corresponds to a complete local domain of some $R/\mathfrak{p}$ for a (unique) choice of minimal prime $\mathfrak{p}$ of R, and every complete local domain of any $R/\mathfrak{p}$ is a complete local domain of R as well. The stated result is then immediate.

(l) The second statement yields the first, since $N^\bigstar{:}_M I$ (respectively, $N^\bigstar{:}_R P$) is solidly closed and contains $N{:}_M I$ (resepctively, $N{:}_R P$). Let u be an element of $(N^\bigstar{:}_M I)^\bigstar$ (respectively, $(N^\bigstar{:}_R P)^\bigstar$). Let B be a complete local domain of R and let S be a solid B-algebra such that u is in $\langle (N^\bigstar{:}_M I)_S\rangle$ (respectively, such that u is in $\langle (N^\bigstar{:}_R P)_S\rangle$). It follows that the image of Iu (respectively, uP) is in $\langle (N^\bigstar)_S\rangle$. After replacing S by a possibly larger solid B-algebra, we may assume that this image in turn is contained in $\langle N_S\rangle$. Thus, for every complete local domain B of R there is a solid B-algebra S such that the image of Iu (respectively, uP) is contained in $\langle N_S\rangle$. It follows that Iu (respectively, uP) $\subseteq N^\bigstar$, and so $u \in N^\bigstar{:}_M I$ (respectively, $N^\bigstar{:}_R P$).

(m) It is an immediate consequence of the definition that if

$$u = (u_1, \dots , u_h) \in (N_1 \oplus \cdots \oplus N_h)^\bigstar$$

then u_i is in $N_i^\bigstar$ for all i. The fact that if $u \in \bigoplus_i N_i^\bigstar$ then $u \in (\bigoplus_i N_i)^\bigstar$ is also immediate if one uses part (c) and observes that for every complete local domain B of R, there is a single solid B-algebra S such that $u_i \in \operatorname{Im}(S\otimes_R N_i \to S\otimes_R M_i)$ for $1 \le i \le h$. $\square$

(5.5) PROPOSITION. *Let $N \subseteq M$ be finitely generated modules over a Noetherian ring R. Let $u \in M$.*

(a) *u is in the solid closure of N in M over R if and only if for every maximal ideal m of R, $u/1$ is in the solid closure of N_m in M_m over R_m.*

(b) *If R is local, then u is in the solid closure of N in M over R if and only if its image in $\widehat{M}$ is in the solid closure of $\widehat{N}$ over $\widehat{R}$.*

(c) *Suppose that u is multiplied into N by a power of a maximal ideal m of R. Then u is in the solid closure of N in M over R if and only if $u/1$ is in the solid closure of N_m in M_m over R_m.*

(d) *If M/N is supported only at one maximal ideal m of R (so that the natural map $M/N \to (M/N)_m \cong M_m/N_m$ is an isomorphism), then the solid closure of N_m in M_m over R_m is the expansion of the solid closure of N in M over R.*

(e) *Suppose that $R = \prod_{i=1}^h R_i$ is a finite product of rings. (Then every R-module M has a canonical decomposition $M \cong \prod_{i=1}^h M_i$ where M_i is an R_i-module, and if $N \subseteq M$ the decomposition $N \cong \prod_{i=1}^h N_i$ is such that for all i, $N_i \subseteq M_i$.) In this situation an element $u = (u_1, \dots , u_h)$ of a finitely generated R-module $M = \prod_{i=1}^h M_i$ is in the solid closure of $N = \prod_{i=1}^h N_i \subseteq M$ over R if and only if for all i, $1 \le i \le h$, u_i is in the solid closure of N_i in M_i over R_i.*

PROOF. (a) Every complete local domain of R_m is a complete local domain of R, and every complete local domain of R is a complete local domain of R_m for some m.

(b) The complete local domains of $\widehat{R}$ are the same as the complete local domains of R.

(c) For any maximal ideal $q \neq m$, the image of u in M_q is in the image of N_q, and so the statement follows at once from part (a).

(d) By Discussion (3.3) we may assume that $N = 0$, and so $M = M/N$ is supported only at m and $M \cong M_m$. The result is then immediate from part (c).

(e) This is immediate from the fact (a) that one may test for membership in a solid closure locally on the maximal ideals of R. □

(5.6) THEOREM (PERSISTENCE OF SOLID CLOSURE). *Let R be a Noetherian ring. Let $N \subseteq M$ be finitely generated R-modules and suppose that $u \in M$ is in ${N^\bigstar}_M$. Then for every ring homomorphism $h: R \to S$, where S is Noetherian, we have that the image $1 \otimes u$ of u in M_S is in the solid closure of $\langle N_S \rangle$ in M_S over S.*

In particular, if $u \in R$ is in ${I^\bigstar}_R$ for an ideal I of R then $h(u) \in {(IS)^\bigstar}_S$ over S.

PROOF. The statement of the second paragraph is immediate from the assertion in the first paragraph. To prove the former, let T denote a generic forcing algebra for (M, N, u) over R. By the equivalence of (a) and (f) in Proposition (5.3), T is formally solid over R. By the persistence theorem for formally solid modules, Theorem (3.7), $S \otimes_R T$ is formally solid over S, and it is a generic forcing algebra for $(S \otimes_R M, S \otimes_R N, 1 \otimes u)$ over S, by Proposition (4.4a). Thus, a second application of the equivalence of (a) and (f) in Proposition (5.3) yields the desired conclusion. □

A very simple but extremely useful consequence of the above is:

(5.7) COROLLARY. *Let $R \to S$ be any homomorphism of Noetherian rings and let J be a solidly closed ideal of S. Then the contraction I of J to R is solidly closed in R.*

PROOF. Let $u \in {I^\bigstar}_R$. Then Theorem (5.6) implies that the image of u in S is in ${(IS)^\bigstar}_S$ and, hence, in ${J^\bigstar}_S$, since I maps into J. But ${J^\bigstar}_S = J$, so that u is in the contraction of J to R. □

(5.8) COROLLARY. *Let R be a Noetherian ring, let $N \subseteq M$ be finitely generated R-modules, and let $u \in M$. Then $u \in {N^\bigstar}_M$ if and only if for every homomorphism of R to a complete local domain B, the image of u in M_B is in the solid closure over B of $\langle N_B \rangle$ in M_B.*

PROOF. The "if" part is clear from the definition of solid closure, while the "only if" part is immediate from Theorem (5.6). □

(5.9) THEOREM. *Let $R \to S$ be a ring homomorphism of Noetherian rings, let $N \subseteq M$ be finitely generated R-modules, and let $u \in M$. Suppose that the* *image of u in M_S is in the solid closure of $\langle N_S \rangle$ in M_S over S. Suppose also that at least one of the following three conditions holds:*

(a) *S is faithfully flat over R.*

(b) *$R \to S$ is a local homomorphism of complete local rings and for every minimal prime $\mathfrak{p}$ of R there is a prime ideal $\mathfrak{q}$ of S lying over $\mathfrak{p}$ such that $ht\, m(S/\mathfrak{q}) \geq ht\, m/\mathfrak{p}$.*

(c) *For every maximal ideal m of R and minimal prime $\mathfrak{p}$ of $(R_m)^\frown$ there is a prime ideal Q of S lying over m and a prime ideal $\mathfrak{q}$ of $(S_Q)^\frown$ lying over $\mathfrak{p}$ such that $ht\, m\big((S_Q)^\frown/\mathfrak{q}\big) \geq dim\,\big((R_m)^\frown/\mathfrak{p}\big)$.*

Then u is in the solid closure of N in M over R.

PROOF. By the equivalence of (a) and (f) in Proposition (5.3), we know that there is a formally solid S-algebra T that is a forcing algebra for $(S \otimes_R M, S \otimes_R N, 1 \otimes u)$, which trivially implies that T is a forcing algebra for (M, N, u) over R. By applying Theorem (3.8) in case (a) and Lemma (3.6b) in case (b), we see that T is formally solid over R.

To prove (c), note that to show that $u \in N^\bigstar{}_M$ it suffices to prove this for $1 \otimes u$ in M_B with $\langle N_B \rangle$ replacing N for every ring B of the form $(R_m)^\frown/\mathfrak{p}$. By Theorem (5.6), if $C = (S_Q)^\frown/\mathfrak{q}$ then $1 \otimes u$ is in the solid closure of $\mathrm{Im}\,(C \otimes_S N)$ in $C \otimes_S M$ over C. Since $B \to C$ satisfies the hypothesis of (b), the result is immediate from part (b). □

Proposition (2.9) coupled with Theorem (5.6) yields:

(5.10) THEOREM. *Let R be a Noetherian ring and let I be an ideal of R. Then $I^\bigstar$ is contained in the integral closure $\overline{I}$ of I.*

Moreover, if I is a principal ideal then $I^\bigstar = \overline{I}$.

PROOF. Let $u \in I^\bigstar$ and suppose that $u \notin \overline{I}$. Then there is a ring homomorphism from R to a discrete valuation ring V such that the image a of u in V is not in IV. By Theorem (5.6), a is in $(IV)^\bigstar{}_V$ over V. We therefore have a counterexample in the discrete valuation ring V. Since IV is principal, say $IV = bV$ with $b \in V$, and since V is normal, the fact that $a \in (bV)^\bigstar$ implies, by Proposition (2.9), that $a/b \in V$, i.e. that $a \in bV$, a contradiction. This completes the proof of the statement in the first paragraph.

To prove the statement in the second paragraph we may assume that $I = bR$. It will suffice to show that if $a \in \overline{I}$ then a is in $I^\bigstar$; we have already established the other inclusion in general. To accomplish this it suffices to consider what happens after we replace R by a complete local domain of R. The expansion of I is still generated by the image of b and the image of a will still be in $\overline{I}$. Thus, we may assume without loss of generality that R is a complete local domain. If $b = 0$ then $(0) = I = \overline{I} = I^\bigstar$ and we are done. If $b \neq 0$ then an equation of integral dependence for a on bR shows that a/b is integral over R. But then $S = R[a/b]$ is a solid R-algebra, and $a \in bS$, so that $a \in I^\bigstar$, as required. □

The following result is a strong parallel, for solid closure, of Proposition (8.18) of [**HH4**], which is a corresponding result for tight closure. It shows, in particular, that over a normal Noetherian ring the solid closure of a torsion-free module embedded in a projective module is independent of the embedding.

(5.11) PROPOSITION. *Let R be a reduced Noetherian ring and let M, N, F, G be finitely generated R-modules.*

(a) *If M/N is torsion-free, then N is solidly closed in M. More generally, $N^{\bigstar}{}_M$ may be identified with a submodule of $N' = Ker\big(M \to (R^{\circ})^{-1}(M/N)\big)$, where R° denotes the set of elements of R not in any minimal prime of R. If N is torsion-free, then $N' \subseteq (R^{\circ})^{-1}N$.*

(b) *If $N \subseteq G \subseteq F$, where G is projective and F is any module, then $N^{\bigstar}{}_F \cap G = N^{\bigstar}{}_G$. Hence, if G is solidly closed in F, then $N^{\bigstar}{}_F = N^{\bigstar}{}_G$.*

(c) *If R is normal, and $G \subseteq F$ with G projective and F torsion-free, then G is solidly closed. If an arbitrary module N has embeddings in two possibly distinct finitely generated projective modules F and G, then $N^{\bigstar}{}_F \cong N^{\bigstar}{}_G$ canonically.*

PROOF. (a) The argument is identical with that used for part (a) of Proposition (8.18) of [**HH4**], and is omitted.

(b) Suppose that there is an element x in $N^{\bigstar}{}_F$ but not in $N^{\bigstar}{}_G$. Then this remains true after replacing R by a suitable localization and killing a submodule of F maximal with respect to being disjoint from the image of G. We may therefore suppose that (R, m, K) is local, that G is free, and that F is an essential extension of G. It follows that F is torsion-free, and consequently we may replace F by an essential extension that is free. Thus, we may assume that $\beta \colon G \to F$ is a map of free modules with matrix B, and the injectivity of β implies that $b = \det(B)$ is a nonzerodivisor in R. This remains true when we complete, which does not affect any relevant issue. Finally, we may replace the complete ring R by its quotient by a minimal prime $\mathfrak{p}$: we replace G, F by their tensor products with $R/\mathfrak{p}$ and N by the image of $(R/\mathfrak{p}) \otimes_R N$ in $(R/\mathfrak{p}) \otimes_R F$ (or its image in $(R/\mathfrak{p}) \otimes_R G$; since $b \notin \mathfrak{p}$, the induced map $(R/\mathfrak{p}) \otimes_R G \to (R/\mathfrak{p}) \otimes_R F$ remains injective). By the definition of solid closure, $\mathfrak{p}$ may be chosen so that the image of x is not in the solid closure of (the new) N in (the new) G. As remarked above, the image of b will be nonzero, and so the induced map $G \xrightarrow{B} F$ will still be injective. Thus, there is no loss of generality in assuming that R is a complete local domain and that $G \to F$ is an injection of free modules of equal rank. Choose data A, ρ for the triple (G, N, x). Then BA, $B\rho$ give data for the triple (F, N, x). To complete the argument, it will suffice to show that if the generic forcing algebra T for the data BA, $B\rho$ for the triple (F, N, x) is solid over R, then so is the generic forcing algebra S for the data A, ρ for the triple (G, N, x) over R.

But T is obtained from $R[X]$ by killing the entries of the matrix $BAX - B\rho = B(AX - \rho)$, while S is obtained from $R[X]$ by killing the entries of the matrix $AX - \rho$. Thus, there is an obvious surjection $T \twoheadrightarrow S$. We claim that the kernel of this map is an R-torsion module. The point is that we may multiply $B(AX - \rho)$

by the classical adjoint $\operatorname{adj} B$ of B, and this produces $b(AX - \rho)$, which shows that b kills $\operatorname{Ker}(T \to S)$. By Proposition (2.1g), T is solid over R if and only if S is solid over R.

(c) The argument is the same as for part (c) of Proposition (8.18) of [**HH4**] (the argument requires knowing that over one-dimensional regular rings, every submodule of every finitely generated module is solidly closed: this is proved in (7.13) and (7.17) below). □

6. Minimal solid algebras

Many of the problems that we shall encounter concerning the properties of solid algebras can be reduced to the case of what we shall call *minimal* solid algebras. The minimal solid algebras, which are defined just below, are better behaved than ordinary solid algebras in a number of ways. We shall illustrate this by showing that the minimal solid algebras finitely generated over a complete one-dimensional local domain are simply the module-finite extension domains. (This cannot, however, be true in higher dimensions: see Example (6.6).)

(6.1) DEFINITION AND DISCUSSION. Let R be a domain. We shall say that a solid R-algebra S is a *minimal* solid R-algebra if S has no proper homomorphic image that is a solid R-algebra. Thus, every solid Noetherian R-algebra maps onto a minimal solid R-algebra. Likewise, if R is Noetherian, then every finitely generated solid R-algebra maps onto a minimal finitely generated solid R-algebra. The solid Noetherian R-algebras can be described as the Noetherian R-algebras that map onto a minimal solid R-algebra. When a solid R-algebra S is not Noetherian, it is not clear whether the family of ideals I such that S/I is solid will have a maximal element.

Of major interest is the case where R is Noetherian and S is finitely generated over R.

(6.2) PROPOSITION. *Let R be a domain and let S be a minimal solid R-algebra.*

(a) *If M is a solid S-module then M is a solid R-module.*

(b) *If S is Noetherian then S is a domain. (Thus, S is an extension domain of R.)*

(c) *If $S \subseteq T$ is a ring extension such that every nonzero element of T has a nonzero T-multiple in S, (e.g., if T is a domain and the extension of fraction fields $S_{(0)} \subseteq T_{(0)}$ is algebraic), and T is a solid R-algebra then T is a minimal solid R-algebra.*

(d) *If S is a domain then every extension domain T that is finitely presented as an S-module is a minimal solid R-algebra.*

(e) *If R is a Noetherian domain and R' is a module-finite extension domain of R then an R'-algebra S is minimal solid over R' if and only if it is minimal solid over R.*

PROOF. (a) Choose a nonzero S-module map $M \to S$ and call the image I, so that I is a nonzero ideal of S. There exists a nonzero R-module map $S \to R$. If this map kills I then it factors $S \twoheadrightarrow S/I \to R$, which implies that S/I is a solid R-algebra, a contradiction. If it does not kill I then the composite map $M \to S \to R$ is nonzero.

(b) S has a finite filtration as an S-module with factors that are of the form S/P, where P is a varying prime ideal of S. By Proposition (2.1), part (l), at least one of these factors is a solid R-module. Since no proper homomorphic image of S is a solid R-module, P must be (0), so that S is a domain.

(c) Suppose that T has a proper homomorphic image T/J, $J \neq 0$, that admits a nonzero R-module map to R. By Proposition (2.1d) we may assume that this map is nonzero on the identity element of T/J. When we restrict the map to S, we get a nonzero R-module map $S \to R$ that kills $J \cap S$ and so induces a nonzero R-module map $S/(J \cap S) \to R$. The hypothesis implies that $J \cap S \neq (0)$, which contradicts the minimality of S.

(d) T is solid over S by Proposition (2.1i) and so is solid over R by (a) above. It is then minimal by (c) above.

(e) By Theorem (2.3), S is solid over R if and only if it is solid over R'. But it is also true that each given proper homomorphic image S/J of S is solid over R if and only if it is solid over R', by the same result. □

We next want to characterize the minimal solid finitely generated algebras over a one-dimensional local ring. We first observe:

(6.3). *Proposition. Let (R, m, K) be a complete one-dimensional local domain, let $x \in m - \{0\}$, and let M be a torsion-free R-module. Then M is solid if and only if $M \neq xM$.*

Thus, if M is a torsion-free R-algebra , M is solid if and only if the image of x in R is not a unit.

Proof. The single element x is a system of parameters for R. By the local cohomology criterion (2.4), a necessary and sufficient condition for M to be solid is that $M_x/\mathrm{Im}\, M$ not be zero. If M is torsion-free the condition that u/x^h be in the image of M in M_x for all $u \in M$ and all h is that $M = x^h M$ for all h. This conditon when $h = 1$ is that $M = xM$, which in turn implies that $M = x^h M$ for all h. □

(6.4) Theorem. *Let (R, m, K) be a complete local domain of dimension one. A finitely generated R-algebra S is a minimal solid R-algebra if and only if S is a module-finite extension domain of R.*

Proof. S is a domain, and x is not invertible in S. It follows that we may choose a maximal ideal $\mathfrak{q}$ of S containing x. If S has dimension two or more then it has infinitely many primes of height one. Not all of these can contain x, since those that contain x will be minimal primes of xS. If P is a height one prime not containing x such that $P \subseteq \mathfrak{q}$ then S/P is torsion-free as an R-module (since $x \notin P$) and x is not invertible in S/P. Thus, S/P is also solid, contradicting the minimality of S. It follows that S has a maximal ideal containing x (and, hence, m) of height at most one. Let τ be the transcendence degree of the fraction field of S over R and let δ be the transcendence degree of $S/\mathfrak{q} = S_\mathfrak{q}/\mathfrak{q}S_\mathfrak{q}$ over $K = R/m$. Since q is maximal, $\delta = 0$. By [Mat] (14.C) Theorem 23 (the dimension formula), we have that ht $\mathfrak{q} =$ ht $P + t - \delta$, so that ht $\mathfrak{q} = 1 + \tau$. Since ht $\mathfrak{q} \leq 1$, we must have that ht $\mathfrak{q} = 1$ and $\tau = 0$. Thus, S is algebraic over R. By Theorem (2.9), since S is solid over R, S must be integral over R and, hence, module-finite over R. □

(6.5) COROLLARY. *Let R be a complete local domain of dimension one. Then a finitely generated R-algebra S is solid if and only if S has a homomorphic image that is a module-finite extension of R.*

PROOF. If S is solid it can be mapped onto a minimal solid R-algebra, and the result is then immediate from Theorem (6.4). On the other hand, an algebra that can be mapped onto a module-finite extension of R is solid by Proposition (2.1), parts (i) and (f). □

(6.6) EXAMPLE. It cannot be true in the equal characteristic case that a minimal solid finitely generated algebra over a complete local domain of dimension 3 is necessarily module-finite over the domain, even when the domain is regular. To see this, choose T to be a complete normal domain of dimension 3 containing a field of characteristic zero such that T is not Cohen-Macaulay. Such domains exist even when the coefficient field is algebraically closed. Represent T as module-finite over a complete regular local ring R (necessarily a formal power series ring in three variables over a field). Let x, y, z be the variables in R. Then, since T is not Cohen-Macaulay there is a relation $uz = vx + wy$ with $u, v, w \in T$ and $u \notin (x, y)T$. By Theorem (10.11) below, u is in the solid closure of $(x, y)T$ in T, and so the generic forcing algebra $T[Z, Z']/(u - xZ - yZ')$ is a finitely generated solid T-algebra. Thus, it can be mapped onto a minimal solid T-algebra S, which will also be a minimal solid R-algebra. We claim that S cannot be module-finite over T (which is equivalent to being module-finite over R). Suppose that it were module-finite. Let d be the degree of the extension of fraction fields, $[G : F]$, with $G = S_{(0)}$ and $F = T_{(0)}$. Then, since T is normal, $(1/d)\,\mathrm{Trace}_{G/F}$ gives a T-module retraction of S to T. Since $u \in (x, y)S$, it would follow (applying the retraction) that $u \in (x, y)T$, a contradiction. □

7. S-regular rings

In parallel with tight closure theory, we define a Noetherian ring R to be *weakly S-regular* if every ideal of R is solidly closed, and *S-regular* if $W^{-1}R$ is weakly S-regular for every multiplicative system W in R. (Solid closure does not commute with localization; we do not know whether weakly S-regular implies S-regular.) In characteristic p, and in dimension at most two, regular rings are S-regular. We shall see later that for well-behaved rings of characteristic p, the notion of weak S-regularity coincides with that of weak F-regularity. Cf. Corollary (8.9). On the other hand, by a result of Paul Roberts [Ro6] (cf. (7.22) and (7.23)), an S-regular ring of equal characteristic zero is forced to have a rather small dimension: see Corollary (7.24) below. The situation for regular local rings of mixed characteristic remains mysterious.

Quite generally, S-regular rings are normal, and a weakly S-regular ring containing a field is Cohen-Macaulay. We do not know whether an S-regular ring is Cohen-Macaulay in mixed characteristic.

We first observe:

(7.1) PROPOSITION. *Let R be a Noetherian ring.*

(a) *R is weakly S-regular if and only if every ideal of R primary to a maximal ideal is solidly closed.*

(b) *R is weakly S-regular if and only if R_m is weakly S-regular for every maximal ideal m of R.*

(c) *If (R, m, K) is local, then R is weakly S-regular if and only if $\widehat{R}$ is weakly S-regular.*

(d) *If (R, m, K) is local, a sufficient condition for it to be S-regular is that there exist a sequence of solidly closed irreducible m-primary ideals $\{I_t\}_t$ cofinal with the powers of m. In fact, if such a sequence exists, then for all finitely generated R-modules $N \subseteq M$, N is solidly closed in M.*

PROOF. (a) This is a consequence of the fact that every proper ideal of a Noetherian ring is an intersection of a (usually infinite) family of ideals primary to maximal ideals.

(b) The ideals q primary to m in R are in bijective with correspondence with the ideals primary to mR_m in R_m via expansion and contraction, and q is solidly closed in R if and only if qR_m is solidly closed in R_m, by Proposition (5.5d).

(c) The ideals q primary to m in R are in bijective correspondence with the ideals primary to $m\widehat{R} = \widehat{m}$ in $\widehat{R}$ via expansion and contraction, and q is solidly closed in R if and only if $q\widehat{R} = \widehat{q}$ is solidly closed in $\widehat{R}$, by Proposition (5.5b).

(d) Since $N = \bigcap_t (N + m^t M)$ it suffices to do the case where M/N has finite length. By Discussion (5.2) we may assume that $N = 0$ and that M has finite length. Choose I so that it kills M and view M as a module over R/I_t, which is self-injective (i.e., 0-dimensional Gorenstein), since I is m-primary and irreducible. Then R/I_t is the only indecomposable injective module over R/I_t, and the injective hull of M over R/I_t will be a finite direct sum of copies of R/I_t. Thus, $M \subseteq (R/_t)^h$ for some nonnegative integer h. Since I_t is solidly closed in R, 0 is solidly closed in R/I_t, and so 0 is solidly closed in $(R/I_t)^h$ by Proposition (5.4m). But then 0 is solidly closed in M by the second part of Proposition (5.4b). □

(7.3) DEFINITION. *We shall say that elements $x_1, \dots, x_d$ in a Noetherian ring R are parameters if for every prime ideal $P \supseteq (x_1, \dots, x_d)R$, the images $x_1/1, \dots, x_d/1$ of these elements in R form part of a system of parameters for R_P. In particular, an element x not in the union of the minimal primes of R is called a parameter.*

(7.4) DEFINITION. *We shall say that a Noetherian ring R is S-rational if every ideal generated by parameters is solidly closed.*

(7.5) REMARKS. The terminology is parallel to that used for tight closure: a Noetherian ring R for which tight closure is defined such that every ideal generated by parameters is tightly closed is called *F-rational*. See [**HH9**] and [**FeW**] for further discussion. The reason for this terminology is that, in equal characteristic 0, the F-rational rings may coincide with the rings R such that $\operatorname{Spec} R$ has rational singularities. Cf. [**Sm1,3**].

(7.6) PROPOSITION. *Let R be a Noetherian ring. Suppose either that:*
(a) *(0) is solidly closed (this is equivalent to the statement that R is reduced) and that every principal ideal generated by a parameter is solidly closed or*
(b) *every ideal generated by a single parameter is solidly closed and the zero-dimensional connected components of $\operatorname{Spec} R$, if any, are reduced.*
Then R is normal.

PROOF. Since (b) $\Rightarrow$ (a) it will suffice to prove that R is normal assuming (b). Suppose that $R = R_1 \times \cdots \times R_h$, where each $\operatorname{Spec} R_i$ is connected. The parameters in R correspond to h-tuples $(x_1, \ldots, x_h)$ such that every x_j is a parameter in R_j. If x_i is a parameter in R_i then

$$y = (1, \ldots, 1, x_i, 1, \ldots, 1)$$

is a parameter in R. From the fact that yR is solidly closed in R, it follows that x_iR_i is solidly closed in R_i, by Proposition (5.5e). Thus, the condition that every parameter generate a solidly closed ideal is inherited by the rings occurring as factors in the product decomposition of R, and so we may assume without loss of generality that $\operatorname{Spec} R$ is connected.

If $\dim R = 0$, then, by hypothesis, R is reduced and so R is a field. If $\dim R \geq 1$ then every principal ideal generated by a parameter in R is integrally closed, by Theorem (5.10). But this implies that R is normal: see Lemma (5.9) of [**HH4**]. □

(7.7) COROLLARY. *If R is S-regular (or S-rational) then R is normal.*

PROOF. If R is S-rational then the condition in Proposition (7.6a) holds. □

In order to show that the property of being S-rational passes to local rings at maximal ideals we first note:

(7.8) LEMMA. *Let m be a maximal ideal of Noetherian ring R and let I be an ideal of R_m generated by a system of parameters of length d (thus, $\dim R_m = d$). Then there is an ideal J of R generated by d parameters such that $JR_m = I$.*

PROOF. We use induction on d. If $d = 0$ then $I = (0)$ in R_m and we may take $J = (0) \subseteq R$. Assume that $d \geq 1$. We claim that there is an element $x \in R$, not in any minimal prime of R, such that $x/1$ is part of a minimal set of generators for I. To see this, let I_0 and I_1 denote the contractions of I and

mI to R and let $\mathfrak{p}_1, \dots, \mathfrak{p}_h$ denote the minimal primes of R. We wish to choose x in $I_0 - (I_1 \cup \bigcup_\nu \mathfrak{p}_\nu)$. Since all but at most one of the ideals I_1, $\mathfrak{p}_\nu$ are prime, we can do this unless $I_0 \subseteq I_1$ or $I \subseteq \mathfrak{p} = \mathfrak{p}_\nu$ for some minimal prime of $\mathfrak{p}$ of R. The former is impossible because I, mI are equal to the expansions of their contractions to R, which would imply $I \subseteq mI$ (and so $I = (0)$) if $I_0 \subseteq I_1$. On the other hand, I_0 is the contraction of an R_m-primary ideal of R_m and so is primary to m. If $I_0 \subseteq \mathfrak{p}$ then $m \subseteq \mathfrak{p}$ and so $m = \mathfrak{p}$ and $\dim R_m = 0$.

Thus, we can choose x as specified. We may now apply the induction hypothesis to the ring R/xR, the maximal ideal m/xR, and the ideal IR_m/xR_m. Since $x/1$ is part of a minimal set of generators for IR_m, it is part of a system of parameters generating I, and the ideal IR_m/xR_m will be generated by $d-1$ parameters. By the induction hypothesis we can choose parameters $\overline{y}_1, \dots, \overline{y}_{d-1}$ in R/xR (where the $y_i \in R$ and the bar indicates images modulo xR) such that the images of the $\overline{y}_i$ in R_m/xR_m generate IR_m/xR_m. The fact that x is not in any minimal prime of R then implies that $x, y_1, \dots, y_{d-1}$ are parameters in R, and these elements evidently satisfy the required condition. □

(7.9) PROPOSITION. *Let R be a Noetherian ring.*

(a) *R is S-rational if and only if R_m is S-rational for every maximal ideal m of R.*

(b) *If (R, m, K) is local, then R is S-rational if and only if every ideal generated by a (full) system of parameters is solidly closed.*

(c) *If (R, m, K) is local, then R is S-rational if and only if $\widehat{R}$ is S-rational.*

(d) *If (R, m, K) is Cohen-Macaulay local and $x_1, \dots, x_d$ is one system of parameters, then R is S-rational if and only if $(x_1^t, \dots, x_d^t)$ is solidly closed for infinitely many values of t.*

(e) *If R is Gorenstein, then R is weakly S-regular if and only if it is S-rational.*

PROOF. (b) Let $x_1, \dots, x_k$ be part of a system of parameters and extend it to a full system of parameters $x_1, \dots, x_d$ for some $d \geq k$. Then

$$(x_1, \dots, x_k)R = \bigcap_t (x_1, \dots, x_k, x_{k+1}^t, \dots, x_d^t)R,$$

and each of the ideals in the intersection is generated by a full system of parameters and, hence, solidly closed.

(a) To prove "if" assume that every R_m is S-rational and let I be an ideal of R generated by parameters. Suppose that $u \in I^{\bigstar}{}_R - I$. Then this is preserved upon localization at some maximal ideal m of R, and we must have $I \subseteq m$. But then IR_m is generated by part of a system of parameters in R_m, a contradiction.

To prove "only if", note that by part (b) it suffices to show that if I is generated by a system of parameters in R_m then I is solidly closed. If not, we can choose $u \in R$ such that $u/1 \in I^{\bigstar} - I$. Choose J, an ideal generated by parameters in R, such that $JR_m = I$. Then $J = J^{\bigstar}$, since R is S-rational. Let J' be the contraction of JR_m to R, which is the same as the set of elements multiplied into J by an element of $R - m$. Thus, $J' = J:_R w$ for a single element $w \in R - m$. Then

J' is an m-primary ideal of R and is solidly closed by part (1) of Proposition (5.4). Since $u/1 \notin I = JR_m$, we have that $u \notin J'$. But then the fact that $u \notin J'^{\bigstar} = J'$ is preserved by localization at m, by Proposition (5.5c), and so u is not in the solid closure of $J'R_m = I$, a contradiction.

(c) By part (b), we need only consider ideals I generated by full systems of parameters in R (and $\widehat{R}$), and there is a bijection between such ideals in R and such ideals in $\widehat{R}$ given by expansion (or completion) and contraction. The issue of whether such an ideal is solidly closed is unaffected by completion, since $R/I \cong \widehat{R}/I\widehat{R}$ and we may apply Proposition (5.5b).

(d) If $\mathbf{y} = y_1, \dots, y_d$ is any full system of parameters then the local cohomology module $H^d_m(R) = H^d_{(\mathbf{y})R}(R)$ is the direct limit of modules $R/(y_1^N, \dots, y_d^N)R$, and the maps in the system are injective. Thus, $R/(\mathbf{y})R$ embeds in $H^d_m(R)$, which is also the increasing union of submodules isomorphic with $R/(x_1^t, \dots, x_d^t)R$. It follows that $R/(\mathbf{y})R$ embeds in $R/(x_1^t, \dots, x_d^t)R$ for any sufficiently large value of t. If $(x_1^t, \dots, x_d^t)R$ is solidly closed in R, then 0 is solidly closed in $R/(x_1^t, \dots, x_d^t)R$, which implies that 0 is solidly closed in $R/(y_1, \dots, y_d)R$, and so $(y_1, \dots, y_d)R$ is solidly closed in R.

(e) Since both properties are local on the maximal ideals of R it suffices to prove the case where (R, m, K) is local of dimension d. Let $x_1, \dots, x_d$ be a system of parameters for R and let $I = (x_1^t, \dots, x_d^t)R$. Then the sequence $\{I_t\}_t$ consists of m-primary irreducible ideals and is cofinal with the powers of m. By the F-rationality of R, the ideals I are solidly closed, and by Proposition (7.1d) this implies that R is weakly F-regular. □

(7.10) REMARK. In §10 we shall see that every S-rational ring containing a field is Cohen-Macaulay. We do not know whether this is true in mixed characteristic.

(7.11) COROLLARY. *Every local ring of a weakly S-regular ring or of an S-rational ring is analytically normal.*

PROOF. The properties of weak S-regularity and S-rationality are preserved by localization at a maximal ideal and by completion. Thus, the result is clear for maximal ideals. But if P is a prime ideal of R, m is a maximal ideal of R containing P and Q is a prime ideal of the completion T of R that lies over Q, then $(R_P)^\frown \to (T_Q)^\frown$ is faithfully flat. Since T is normal and complete, T_Q is excellent and normal, and so $(T_Q)^\frown$ is normal. It follows that $(R_P)^\frown$ is normal as well. □

The normality of weakly S-regular rings enables us to prove a converse to Proposition (7.1d).

(7.12) PROPOSITION. *A local ring (R, m, K) is weakly S-regular if and only if there exists a sequence of solidly closed irreducible m-primary ideals $\{I_t\}_t$ cofinal with the powers of m.*

PROOF. The sufficiency of the condition was established in Proposition (7.1d).

On the other hand, if (R, m, K) is weakly S-regular then it is normal and so *approximately Gorenstein* in the sense of [**Ho4**] (see also Discussion (8.6) on p.75 of [**HH4**] for a summary), which means precisely that there exists a sequence of m-primary irreducible ideals $\{I_t\}_t$ cofinal with the powers of m. These ideals will all be solidly closed, since R is S-regular. □

(7.13) THEOREM. *If R is weakly S-regular and $N \subseteq M$ are finitely generated R-modules then N is solidly closed in M.*

PROOF. Suppose, to the contrary, that $u \in N^{\bigstar}{}_M - N$. Then this will be preserved upon localization at a suitable maximal ideal m of R. Hence, there is no loss of generality in assuming that R is local. But then the result follows from Proposition (7.12) and Proposition (7.1d). □

(7.14) DISCUSSION. We recall that a map of R-modules $N \to M$ is called *pure* if for every R-module Q, $Q \otimes_R N \to Q \otimes_R M$ is injective. In particular, $N \to M$ itself must be an injection. This is a weakening of the condition that the map embed N as a direct summand of M over R. This weaker condition has the advantage that it is stable under taking direct limits. See §6 of [**HR1**], §5 (a) of [**HR2**], and [**Ho4**]. If M/N is finitely presented over R then $N \hookrightarrow M$ is pure if and only if it splits (see, for example, Corollary 5.2 on p. 142 of [**HR2**]). When R is Noetherian, an injection $N \hookrightarrow M$ is pure if and only if $N \hookrightarrow M_0$ splits for every submodule M_0 of M containing the image of N such that $M/\mathrm{Im}\, N$ is finitely generated. If R is a complete local ring and $R \to M$ is pure, then it splits without any finiteness condition on M: see the second paragraph of the proof of Corollary (6.24), p. 59, of [**HH4**].

The purity of $R \hookrightarrow M$ implies that for every ideal I of R, $IM \cap R = I$, since applying $(R/I) \otimes_R \_$ yields an injection.

(7.15) PROPOSITION. *Suppose that $A \subseteq R$ are Noetherian rings and that the injection $A \hookrightarrow R$ is pure, or even that every ideal of A is contracted from R. Then if R is weakly S-regular, so is A. In particular, if A is a direct summand of R as an A-module, and R is weakly S-regular then so is A.*

PROOF. If $I \subseteq A$, then I is the contraction of $IR \subseteq R$, and so I is solidly closed in A by Corollary (5.7). □

Although the following result is easy, it is rather important:

(7.16) PROPOSITION. *Let R be a Noetherian ring.*

(a) *If R is a Noetherian ring and S is a module-finite extension (or, more generally, a formally solid extension), then $IS \cap R \subseteq I^{\bigstar}$ for every ideal I of R.*

(b) *If R is weakly S-regular, then R is pure in every formally solid extension algebra.*

(c) *If R is weakly S-regular, then R is a direct summand, as a module over itself, of every module-finite extension ring.*

(d) *If R is weakly S-regular and complete local, then R is a direct summand, as a module over itself, of every solid R-algebra.*

PROOF. Part (a) is obvious, and (b) follows from (a) together with the fact that weakly S-regular rings are normal, hence, approximately Gorenstein, and so the contractedness of ideals implies purity. Part (c) is immediate from (b) and the fact that pure implies split when the cokernel is finitely presented. Part (d) is immediate from (b) and the fact that when a complete local ring is a pure submodule of a module the map splits: cf. the second paragraph of the proof of (6.24) in [**HH4**]. □

We next note:

(7.17) PROPOSITION. *A Noetherian ring of dimension at most one is weakly S-regular if and only if it is regular, i.e., if and only if it is a finite product of Dedekind domains, in which case it is S-regular.*

PROOF. Since a weakly S-regular ring is normal, and a normal Noetherian ring of dimension at most one is regular, "only if" is clear. Thus, we only need to show that a regular Noetherian ring R of dimension at most one is S-regular. It will suffice to show that every local ring of R is weakly S-regular. But the local rings of R are discrete valuation rings or fields. In such a ring every ideal is principal. Since the solid closure of a principal ideal is equal to the integral closure and since every principal ideal in a normal ring is integrally closed, it follows that in a field or discrete valuation ring every ideal is solidly closed. □

We shall see later that not every regular local ring is S-regular: this is false in equal characteristic zero in dimension bigger than or equal to 3, by a result of Paul Roberts [**Ro6**]. But it is true in characteristic p and in dimension at most two that regular rings are S-regular, and it is an open question for regular local rings of mixed characteristic. To explore this question further, we first observe:

(7.18) PROPOSITION. *Suppose that (R, m, K) is an analytically irreducible local Gorenstein domain of dimension d with system of parameters $x_1, \dots, x_d$. Let $I_t = (x_1^{t+1}, \dots, x_d^{t+1})R$, for $t \geq 0$, let $I = I_0$, let $x = x_1 \cdots x_d$, and let u denote an element of R whose image in R/I generates the socle.*

Let $Y_1, \dots, Y_d$ denote indeterminates over R and let $G_t = x^t u - \sum_{i=1}^d x_i^{t+1} Y_i$. Then R is weakly S-regular if and only if for all $t \geq 0$ (equivalently, for infinitely many values of $t \geq 0$) $H_m^d(R[Y]/(G_t)) = 0$.

Note here that $H_m^d(R[Y]/(G_t)) = H_I^d(R[Y]/(G_t))$.

PROOF. If R is weakly S-regular then, evidently, all the ideals I_t are solidly closed. On the other hand, these ideals form a sequence of m-primary ideals cofinal with the powers of the maximal ideal, and so R is weakly S-regular if and only if infinitely many of the I_t are solidly closed, by Proposition (7.12). Since the image of $x^t u$ generates the socle in R/I_t for every $t \geq 1$, any ideal strictly larger than I_t must contain $x^t u$. Thus, I_t is solidly closed if and only if $x^t u$ is not in its solid closure, i.e., if and only if a generic forcing algebra for $(R, I_t, x^t u)$ is not formally solid. $R[Y]/(G_t)$ is evidently a generic forcing algebra for $(R, I_t, x^t u)$. Thus, the condition for it not to be formally solid is that $\widehat{R} \otimes_R R[Y]/(G_t)$ not be solid. By Corollary (2.4), this is equivalent to the conditon that $H^d_m(\widehat{R} \otimes_R R[Y]/(G_t)) = 0$, and this module may be identified with $\widehat{R} \otimes_R H^d_m(R[Y]/(G_t)) \cong H^d_m(R[Y]/(G_t))$, since every element of $H^d_m(N)$ is killed by a power of m for every R-module N. □

In particular, the criterion of this result is valid when R is a regular local ring and the x's are a system of parameters and, in that case, the x's may be chosen to be a minimal set of generators of the maximal ideal of R, in which case we may take $u = 1$. In this very important special case we therefore have:

(7.19) COROLLARY. *Let (R, m, K) be a regular local ring of dimension d and suppose that $x_1, \dots, x_d$ is a regular system of parameters, i.e., that $m = (x_1, \dots, x_d)R$. Let $Y_1, \dots, Y_d$ denote indeterminates over R and let G_t denote the polynomial $x_1^t \cdots x_d^t - \sum_{i=1}^d x_i^{t+1} Y_i$. Then R is weakly S-regular if and only if for all $t \geq 0$ (equivalently, for infinitely many values of $t \geq 0$) $H^d_m(R[Y]/(G_t)) = 0$.* □

(7.20) THEOREM. *Let R be a regular Noetherian ring.*
(a) *If R has positive prime characteristic p then R is S-regular.*
(b) *If $\dim R \leq 2$ then R is S-regular.*

PROOF. In both cases it suffices to show that the local rings of R are weakly S-regular, and these may be replaced by their completions. Thus, we need only show that a complete regular local ring satisfying one of the two hypotheses is weakly S-regular.

(a) Let $x_1, \dots, x_d$ be a minimal set of generators for the maximal ideal of R. We use the notation of Proposition (7.18) and Corollary (7.19) here. We may take $u = 1$. Consider $R[Y]/(G_t)$. In this ring S we have that $x^t \in I_{t+1}S$. Taking q^{th} powers where $q = p^e$ yields that $x^{qt} \in I_{qt+q}S$. Since $(qt + q) - qt = q$ is unbounded, Observation (2.6) shows that $H^d_I(S) = 0$.

(b) Since we have already handled the case where $\dim R \leq 1$ we may assume that R is a complete regular local domain of dimension 2, with regular system of parameters x_1, x_2, and take $u = 1$. Let $T = \mathbb{Z}[X_1, X_2, Y_1, Y_2]/(G_t)$ where G_t denotes the polynomial $(X_1X_2)^t - \sum_{i=1}^2 Y_i X_i^{t+1}$. We can map T to $S = R[Y]/(G_t)$ by sending X_i to x_i and Y_i to Y_i for $i = 1, 2$. Then $H^2_I(S) \cong S \otimes_T H^2_{(X_1,X_2)T}(T)$.

But $H^2_{(X_1,X_2)T}(T)$ is shown to vanish in §6 of **[Ho7]**, pp. 545-547. The groups $H_{2,t,c}$ studied there are the graded pieces (with respect to a certain $\mathbb{Z}^2$-grading: $c \in \mathbb{Z}^2$ here) of $H^2_{(X_1,X_2)T}(T)$. See Corollary (6.11) of **[Ho7]**. □

(7.21) REMARKS. The result of part (a) is implicit in **[Ho1]** as well as in **[Ho7]**, while the result of part (b) is implicit in **[Ho7]**.

(7.22) DISCUSSION: CERTAIN MULTIGRADED LOCAL COHOMOLOGY MODULES. Let d and t be positive integers, let $X_1, \dots, X_d$, $Y_1, \dots, Y_d$ be indeterminates over $\mathbb{Z}$, and let

$$G = (X_1 \cdots X_d)^t - \sum_{i+1}^{d} Y_i X_i^{t+1}.$$

In [**Ho7**], pp.544-552, there is a study of the dth local cohomology modules $H_I^d(R_{d,t})$ of the rings

$$R_{d,t} = \mathbb{Z}[X_1, \ldots, X_d, Y_1, \ldots, Y_d]/(G_{d,t}),$$

with support in $I = (X_1, \ldots, X_d)R$. More specifically, a $\mathbb{Z}$-grading is introduced such that the degree of X_i is the ith row of a size d identity matrix and the degree of Y has -1 in the ith spot and t's elsewhere. The cth graded piece $H_{d,t,c}$ is calculated, in a certain sense, as an abelian group. These groups are shown to be divisible. When $d = 2$ they are also shown to be finitely generated, and, hence, 0. The question is raised as to whether these groups must always vanish. This has recently been answered negatively by Paul Roberts.

(7.23) DISCUSSION: ROBERTS' CALCULATION. We continue the notation of (7.22). In these terms, Roberts' main result in [Ro6] is that if $d = 3$ then $\mathbb{Q} \otimes_{\mathbb{Z}} H_{3,2,c} \neq 0$ when $c = (-2,-2,-2)$. This is the same as the degree $(-2,-2,-2)$ graded piece of $H_I^3(\mathbb{Q} \otimes_{\mathbb{Z}} R_{3,2})$ with $I = (x_1, x_2, x_3)$. In particular, $H_I^3(\mathbb{Q} \otimes_{\mathbb{Z}} R_{3,2}) \neq 0$. As an immediate consequence we have:

(7.24) COROLLARY. *Let f, g, h be any three elements in a Noetherian ring S of equal characteristic zero. Then $f^2g^2h^2$ is in $(f^3, g^3, h^3)^\bigstar$. In particular, if x_1, x_2, x_3 are part of a system of parameters for an equicharacteristic zero regular local ring of dimension three (or any equicharacteristic zero local ring of dimension at least three) then $x_1^2x_2^2x_3^2$ is in $(x_1^3, x_2^3, x_3^3)^\bigstar$.*

In consequence, a weakly S-regular ring (or an S-rational ring) containing the rationals has dimension at most two.

PROOF. To prove the first statement note that it suffices to do so after passing to a complete local domain of R. Thus, we may assume that R is a complete local domain. If any of f, g, or h is a unit the conclusion is clear. Thus, we may assume that all three are in the maximal ideal of the ring. Then there is a local homomorphism of $A = \mathbb{Q}[[X_1, X_2, X_3]]$ to R carrying X_1, X_2, X_3 to f, g, h respectively, and by the persistence of solid closure, Theorem (5.6), it suffices to prove the result when the ring is $\mathbb{Q}[[X_1, X_2, X_3]]$ and f, g, h are X_1, X_2, X_3. The relevant generic forcing algebra is then $B = A[Y_1, Y_2, Y_3]/(G)$ where

$$G = X_1^2X_2^2X_3^2 - \sum_{i=1}^{3} Y_i X_i^3,$$

and so it will suffice to prove that $H_J^3(B) \neq 0$ with $J = (X_1, X_2, X_3)B$. But $H_J^3(B)$ is $\varinjlim_t B/(X_1^t, X_2^t, X_3^t)B$ (where the maps between consecutive modules are induced by multiplication by $X_1X_2X_3$ acting on the copies of B in the numerators). Let $R = Q[X_1, X_2, X_3, Y_1, Y_2, Y_3]/(G)$. Then $R/(X_1^t, X_2^t, X_3^t)R \cong B/(X_1^t, X_2^t, X_3^t)B$, and it follows that $H_J^3(B) \cong H_I^3(R)$ with $I = (X_1, X_2, X_3)R$. But this R is precisely the ring $\mathbb{Q} \otimes_{\mathbb{Z}} R_{3,2}$ discussed in (7.23), and the result is now immediate from Paul Roberts' result in [Ro6].

The final statement follows at once: if one had an S-regular ring of dimension at least three containing $\mathbb{Q}$, it would have a local ring (at some maximal ideal) of dimension at least three, and that local ring would still be S-regular, by Proposition (7.1b). Choose x_1, x_2, x_3 in the local ring so that they are part of a system of parameters. Then $x_1^2x_2^2x_3^2 \in (x_1^3, x_2^3, x_3^3)^\star = (x_1^3, x_2^3, x_3^3)$, and this contradicts the monomial conjecture, which is known in equal characteristic. (The proof for the S-rational case is the same, using Proposition (7.9a).) □

(7.25) REMARKS. Despite this result, if $R = \mathbb{Z}[\mathbf{x}]$ or $S = Z[[\mathbf{x}]]$ with $\mathbf{x} = x_1, \ldots, x_n$, any ideal I generated by monomials in the x's is solidly closed either in R or in S. In fact, over R, N is solidly closed in M, where M is a finitely generated R-module, if M/N is torsion-free over $\mathbb{Z}$: see Theorem (13.1b). The result for $I \subseteq S$ follows from Theorem (13.1a), since $I = \bigcap\{I + pS : p \text{ prime in } \mathbb{Z}\}$. E.g., (x_1^3, x_2^3, x_3^3) is solidly closed in $\mathbb{Z}[x_1, x_2, x_3]$ and in $\mathbb{Z}[[x_1, x_2, x_3]]$.

The reader is referred to §13 for further discussion.

8. Comparison with tight closure in characteristic p

Our main objective in this section is to show that in many good cases in characteristic p, including the case of algebras essentially of finite type over an excellent local ring, the notion of tight closure defined in [**HH4**] coincides with the notion of solid closure defined here. The key point is that both can be tested after passing to a complete local domain of R. One of the main results is Theorem (8.6), which shows that the tight closure is always contained in the solid closure and that the two notions agree when the ring has a completely stable weak test element (this includes the case where the ring is essentially of finite type over an excellent local ring).

In §11 we shall prove that for a complete local domain of characteristic p, an element is in the tight closure of a submodule of a finitely generated module M if and only if it is in the expansion after tensoring with a big Cohen-Macaulay algebra. Since big Cohen-Macaulay algebras over a complete local domain are always solid, this is a refinement of the result discussed in the preceding paragraph. In §11 we shall also use this result on big Cohen-Macaulay algebras to show that the solid closure contains the tight closure (defined via reduction to characteristic p) for rings containing $\mathbb{Q}$, insofar as tight closure can be defined.

In the theory of tight closure one needs a technical device, namely, the theory of completely stable test elements (described in Definition (8.3c) below), to control behavior as one passes to the completion of a local ring.

The definitions, discussion, and results given in (8.1) through (8.5) below sketch rapidly all that we need concerning the theory of tight closure from [**HH4**], [**HH3**], and [**HH9**].

(8.1) Discussion and definitions. We recall that if R is a ring of positive prime characteristic p, and F^e, for $e \in \mathbb{N}$, denotes the eth iteration of the Frobenius endomorphism of R (so that $F^e(r) = r^{p^e}$), we can define a right exact functor $\mathbf{F}^e$ (or, more precisely, $\mathbf{F}^e_R$) from R-modules to R-modules that preserves finite generation by applying the functor $S \otimes_R \_$, where S denotes R viewed as an R-algebra via the structural homomorphism F^e. Quite generally, for any R-algebra S, $S \otimes_R \_$ is a right exact functor from R-modules to S-modules that preserves finite generation, whose value on R is S, and such that images of R-free modules are S-free. If g denotes the structural homomorphism $R \to S$, then this functor sends $M = \operatorname{Coker}(R^h \xrightarrow{A} R^k)$, where A is given by the matrix (r_{ij}), to the cokernel of the map of free S-modules given by the matrix $g(A) = \big(g(r_{ij})\big)$.

In the present instance, rather confusingly, $S = R$, so that the functor sends R-modules to R-modules. Note that $\mathbf{F}^e(\operatorname{Coker}(r_{ij})) \cong \operatorname{Coker}(r_{ij}^q)$, where $q = p^e$.

The functors $\mathbf{F}^e$ are known as the *Peskine-Szpiro* or *Frobenius* functors. In the general situation of a base change from R-modules to S-modules, where S is an R-algebra , there is a map $M \to S \otimes_R M$ as R-modules. We shall write the image of $u \in M$ in $\mathbf{F}^e(M)$ as u^{p^e}. Because $S = R$ in our case, $S \otimes_R M$ has two R-module structures. We shall always work with the one obtained by viewing it as an S-module and then "remembering" that $S = R$. With this convention, the map $u \mapsto u^{p^e}$ is linear in the sense that $(ru)^{p^e} = r^{p^e} u^{p^e}$. When $N \subseteq M$ we shall denote the image of $\mathbf{F}^e(N)$ in $\mathbf{F}^e(M)$ as $N^{[p^e]}$. This notation is somewhat imprecise, since one needs to know what M and the embedding $N \subseteq M$ are. When $M = R$ and $N = I$ is an ideal of R, $I^{[p^e]}$ denotes the ideal of R generated by the $(p^e)^{th}$ powers of the elements of I.

(8.2) Notation. We shall denote by R° the multiplicative system consisting of all elements of R not in any minimal prime ideal of R. If R is a domain, $R^{\circ} = R - \{0\}$.

(8.3) Definitions. *Let R be a Noetherian ring of positive prime characteristic p.*

(a) *If $N \subseteq M$ are finitely generated R-modules, we say that $u \in M$ is in the tight closure $N^*{}_M$ (or simply N^* if M is clear from the context) of N in M over R if there exists an element $c \in R^{\circ}$ such that $cu^{p^e} \in N^{[p^e]}$ in $\mathbf{F}^e(M)$ for all sufficient large nonnegative integers e.*

(b) *An element $c \in R^{\circ}$ is called a q'-weak test element, where q' is a power of p, if for every pair of finitely generated R-modules $N \subseteq M$ and every element $u \in M$, u is in $N^*{}_M$ if and only if $cu^{p^e} \in N^{[p^e]}$ in $\mathbf{F}^e(M)$ for all $p^e \geq q'$. If $q' = 1$, then c is called a test element.*

(c) *An element $c \in R^{\circ}$ is called a completely stable q'-weak test element if its image in the completion of every local ring of R is a q'-weak test element for that local ring. If $q' = 1$ then c is called a completely stable test element.*

(8.4) Remarks. Tight closure and the notion of test element are introduced in [**HH4**], but see also [**HH1-3**] and [**Hu1**]. The theory is further pursued in [**HH6**] and [**HH8-11**]. The following result on the existence of completely stable test elements, which follows from Theorems (5.10) and (6.20) of [**HH9**] will suffice for our purpose here:

(8.5) THEOREM. *Let R be a Noetherian ring of characteristic p and suppose either*

(i) *that R is an algebra essentially of finite type over an excellent local ring or*

(ii) *that R is module finite over $R^p = F(R)$, i.e., that the Frobenius endomorphism is a finite morphism.*

Suppose that $c \in R_{red}$ is such that $(R_{red})_c$ is regular (such elements c always exist). Then c has a power that is a completely stable q'-weak test element for some q'. If R is reduced then c has a power that is a test element.

Thus, a ring satisfying (i) *or* (ii) *has a completely stable weak test element.* □

(8.6) THEOREM. *Let $N \subseteq M$ be finitely generated modules over a Noetherian ring R of characteristic p.*

(a) ${N^*}_M \subseteq {N^{\bigstar}}_M$.

(b) *If R has a completely stable weak test element, then ${N^*}_M = {N^{\bigstar}}_M$.*

PROOF. Suppose that $u \in {N^*}_M$. This is preserved upon completion and also upon killing a minimal prime of the completion. Thus, (a) will follow if one can show tight closure and solid closure agree for complete local domains. (b) will also follow from the complete local domain case if one can show that if a given element $u \in M$ is not in ${N^*}_M$, then one can choose a complete local domain B of R such that u_B is not in ${N_B}^*{}_{M_B}$ over B. But the fact that one has a completely stable q'-weak test element c shows that $cu^{p^e} \notin N^{[p^e]}$ in $\mathbf{F}^e(M)$ for some $p^e \geq q'$, and this continues to be the case after localizing at a suitable maximal ideal and then after completion at that maximal ideal. We can then preserve that the image of u is not in the tight closure after killing a suitable minimal prime of the complete local ring so obtained by (6.25) of [**HH4**] generalized to the case of modules (the argument is valid without essential change).

Thus, both parts of the theorem follow if we can establish that for a complete local domain (R, m, K), one has ${N^*}_M = {N^{\bigstar}}_M$. The fact that ${N^{\bigstar}}_M \subseteq {N^*}_M$ is implicit in Theorem (5.22) of [**HH10**]. Let $u \in M$ and suppose that there is a solid R-algebra S such that $1 \otimes u \in \langle N_S \rangle$. We may choose an R-module map $\theta: S \to R$ such that $\theta(1) = c \in R - \{0\}$. If we apply $\mathbf{F}^e_S$ we find that $1 \otimes u$ is in

$$\operatorname{Im}\left(S \otimes_R \mathbf{F}^e_R(N) \to S \otimes_R \mathbf{F}^e_R(M)\right) = \operatorname{Im}\left(S \otimes_R N^{[p^e]} \to S \otimes_R \mathbf{F}^e_R(M)\right).$$

For every R-module W the map $\theta: S \to R$ induces a map $S \otimes_R W \to W$ sending $s \otimes w$ to $\theta(s)w$. If $W_0 \subseteq W$ the diagram:

$$\begin{array}{ccc} S \otimes_R W_0 & \longrightarrow & S \otimes_R W \\ \downarrow & & \downarrow \\ W_0 & \longrightarrow & W \end{array}$$

commutes. Applying this with $W = \mathbf{F}^e(M)$ and $W_0 = N^{[p^e]}$ we find that the image of $1 \otimes u^{p^e}$ is cu^{p^e} and is in $N^{[p^e]}$. Since this holds for all e, we have that $u \in N^*{}_M$.

Now suppose that R is a complete local domain of dimension d and that $u \in N^*{}_M$ for some $u \in M$. Let S be the generic forcing algebra for (M, N, u) for, say, the data A, ρ, so that $S = R[X]/J$, where J is generated by the entries of $AX - \rho$. To complete the argument, it will suffice to show that $H^d_m(S) \neq 0$.

Let $x_1, \dots, x_d$ be a system of parameters for R. Let $x = x_1 \cdots x_d$. We shall prove that $x^t \notin (x_1^{t+1}, \dots, x_d^{t+1})S$ for all t. By Observation (2.6) this condition is sufficient to imply that $H^d_m(S) \neq 0$.

Suppose to the contrary that $x^t \in (x_1^{t+1}, \dots, x_d^{t+1})S$ for some fixed t. Then we can lift this to $R[X]$: we have that

$$(\#) \qquad x^t = \sum_{i=1}^{d} x_i^{t+1} F_i(X) + G, \qquad \text{where } G \in J.$$

We shall obtain a contradiction. To this end, let b be a positive integer that is an upper bound for all the numbers $\deg F_i$.

The fact that u is in the tight closure of N in M implies that ρ is in the tight closure of the column space of A. We therefore have that for all $e \gg 0$, $c\rho^{p^e} = F^e(A)y_e$, where $F^e(A)$ denotes the matrix obtained from A by raising every entry to the p^e power and where y_e is a vector over R. Taking $(p^e)^{th}$ roots, if we let $c_e = c^{1/p^e}$, then $c_e\rho = Az_e$ for every $e \gg 0$, where z_e is a vector over R^{1/p^e} depending on e. Consider the equations (#) over the domain $(R^{1/p^e})_{c_e}$, and substitute $X = c_e^{-1}z_e$. This substitution makes the entries of $AX - \rho$ vanish, and since these generate J, it makes G vanish. This yields

$$x^t = \sum_{i=1}^{d} X_i^{t+1} F_i(c_e^{-1} z_e).$$

If we multiply both sides by c_e^b we clear denominators on the right, and we obtain

$$c^{b/p^e} x^t \in (x_1^{t+1}, \dots, x_d^{t+1})R^{1/p^e}$$

for all $e \gg 0$. Raising both sides to the $(p^e)^{th}$ power yields

$$c^b (x^t)^{p^e} \in \left((x_1^{t+1}, \dots, x_d^{t+1})R\right)^{[p^e]}$$

for all $e \gg 0$. Let K be a coefficient field for R and let $A = K[[x_1, \dots, x_d]]$, a regular ring over which R is module-finite. Replace c^b by a nonzero multiple c' in A. Then for all $e \gg 0$ we have that

$$c'(x^t)^{p^e} \in \left((x_1^{t+1})^{p^e}, \dots, (x_n^{t+1})^{p^e}\right)R \cap A,$$

and by the main result of [**Ho1**] (or by Theorem (7.20a) and Proposition (7.16c) here, which recover that result) this is

$$\left((x_1^{t+1})^{p^e}, \dots, (x_n^{t+1})^{p^e}\right)A.$$

(One may also use the main result of §7 of **[HH4]** concerning operations on ideals generated by monomials in parameters to get a contradiction here.) This shows that x^t is in the tight closure of the ideal $(x_1^{t+1}, \ldots, x_d^{t+1})A$, a contradiction, since every ideal of a regular ring of characteristic p is tightly closed by Theorem (4.4) of **[HH4]**. □

(8.7) REMARK. If one defines the *formal tight closure* of a submodule N of a finitely generated R-module M over a Noetherian ring R of positive prime characteristic to consist of all elements $u \in M$ such that for every complete local domain B of R, the image $1 \otimes u$ of u in $B \otimes_R M$ is in the tight closure, over B, of $\langle N_S \rangle$, then it is readily apparent from (8.6) that the formal tight closure of N in M is the same as $N^{\bigstar}{}_M$.

(8.8) COROLLARY. *Let $R \to S$ be a homomorphism of Noetherian rings of characteristic p such that R has a completely stable weak test element. Suppose also that*

(*) *for every maximal ideal m of R and minimal prime $\mathfrak{p}$ of $(R_m)^\frown$ there is a prime ideal Q of S lying over m and a prime ideal $\mathfrak{q}$ of $(S_Q)^\frown$ lying over $\mathfrak{p}$ such that ht $m\big((S_Q)^\frown/\mathfrak{q}\big) \geq$ dim $\big((R_m)^\frown/\mathfrak{p}\big)$.*

(In particular, this condition holds whenever S is faithfully flat over R.)

Suppose as well that the image of u in $S \otimes_R M$ is in the tight closure of the $\langle N_S \rangle$ in $S \otimes_R M$ over S. Then u is in the tight closure of N in M over R.

PROOF. Since the image of u is in the tight closure $\langle N_S \rangle$ in $S \otimes_R M$, by Theorem (8.6a) it is in the solid closure. We may then apply (5.9c) to conclude that $u \in N^{\bigstar}{}_M$ over R, and by Theorem (8.6b) it follows that $u \in N^*{}_M$ over R. □

The result above is a very great improvement upon Theorem (5.31) of **[HH10]**, in two ways. First, the height condition (*) above is very much weaker than the condition that $R \to S$ be "formally height preserving" in the sense of **[HH10]**. The latter condition, in essence, restricts Q to be a minimal prime of mS, so that m expands to an ideal primary to the maximal ideal of $(S_Q)^\frown$, and then requires that $\dim (S_Q)^\frown/\mathfrak{p}(S_Q)^\frown \geq \dim (R_m)^\frown/\mathfrak{p}(R_m)^\frown$. (Cf. (5.30a) of **[HH10]**.) To simplify notation, let $B = (R_m)^\frown/\mathfrak{p}(R_m)^\frown$ and let $C = (S_Q)^\frown/\mathfrak{p}(S_Q)^\frown$. Thus, $B \to C$ is a local homomorphism of complete local rings such that the maximal ideal of B expands to the maximal ideal of C. The condition that $\dim C \geq \dim B$ implies that $\dim C = \dim B$ and that $B \to C$ is injective (otherwise, the image of B has dimension d smaller than that of B, and the maximal ideal of B will be the radical of an ideal generated by d elements). Thus, while no prime $\mathfrak{q}$ of C is mentioned in the definition, any minimal prime $\mathfrak{q}$ of C such that $\dim C/\mathfrak{q} = \dim C$ must lie over (0) in B. It follows that $R \to S$ is formally height preserving if and only if for every maximal ideal m of R and minimal prime $\mathfrak{p}$ of $(R_m)^\frown$, there are a minimal prime Q of mS and a minimal prime $\mathfrak{q}$ of $(S_Q)^\frown$ lying over $\mathfrak{p}$ in $(R_m)^\frown$ such that ht $m(S_Q)^\frown/\mathfrak{q} \geq \dim (R_m)^\frown/\mathfrak{p}$. (This is equivalent to $\dim (S_Q)^\frown/\mathfrak{q} = \dim (R_m)^\frown/\mathfrak{p}$.)

It is now clear that condition (*) is much weaker than "formally height preserving". (Consider the example given after the proof of Lemma (3.6): mS is a prime of height one in S.)

The second point is that, even in situations where $R \to S$ is formally height preserving (even, in fact, when $R \to S$ is faithfully flat), Corollary (8.8) above has an advantage, because one does not need to worry about the technical hypotheses (i) or (ii) that are needed for Theorem (5.31) of [**HH10**].

We also have the following immediately from Theorem (8.6):

(8.9) COROLLARY. *Let R be a Noetherian ring of characteristic p.*

(a) *If R is weakly S-regular (respectively, S-regular, respectively, S-rational) then R is weakly F-regular (respectively, F-regular, respectively, F-rational).*

(b) *If R has a completely stable weak test element then R is weakly F-regular (respectively, F-regular, respectively, F-rational) if and only if R is weakly S-regular (respectively, S-regular, respectively, S-rational).* □

(8.10) COROLLARY. *Let R be a Noetherian ring of characteristic p such that either:*

(i) *R is weakly F-regular and has a completely stable weak test element or*

(ii) *The completion of every local ring of R at a maximal ideal is weakly F-regular.*

(In particular, (ii) *is satisfied if R is regular.) Let S be a formally solid R-algebra. Then $R \to S$ is pure as a map of R-modules. In particular, if R is a regular domain and S is a solid R-algebra, then $R \to S$ is pure as a map of R-modules.*

PROOF. Since (i) implies (ii), it suffices to prove the result when (ii) holds. $R \to S$ is pure if and only if $R_m \to S_m$ is pure for every maximal ideal m of R, and the issue of whether $R_m \to S_m$ is pure is unaffected by applying $(R_m)^\frown \otimes_{R_m} \_$. It follows that we may assume that R is complete local, weakly F-regular, and that $R \to S$ is solid. But then R is S-regular and normal. Since R is normal, it is approximately Gorenstein in the sense of [**Ho4**] (cf. also Discussion (8.6) on p. 75 of [**HH4**] and the proof of (7.12) here), and so to prove that $R \to S$ is pure it suffices to prove that every ideal of R is contracted from S. But this is immediate from the definition of solid closure and the fact that every ideal of R is solidly closed. □

(8.11) REMARK. The result of P. Roberts [Ro6] shows that, in equal characteristic zero, if R is regular and $R \to S$ is solid it is not the case that R is necessarily pure in S. For example, let K be a field of characteristic zero, let $R = K[[x_1, x_2, x_3]]$, and let $S = R[y_1, y_2, y_3]/(g)$ where $g = x_1^2x_2^2x_3^2 - \sum_{i=1}^3 y_i x_i^3$. By Roberts' result $R \to S$ is solid, but $(x_1^3, x_2^3, x_3^3)S \cap R$ contains $x_1^2x_2^2x_3^2$, so that $(x_1^3, x_2^3, x_3^3)R$ is not contracted from S.

We do not know what the situation is if, for example, R is a regular local ring of mixed characteristic.

9. A formal power series criterion

Our objective is to give a remarkably elementary characterization of when an element of a complete local domain is in the solid closure of an ideal I. The main result is Theorem (9.3) below. The idea can be generalized to the case of modules: see Remark (9.4c). We need some preliminary definitions and discussion and a lemma.

(9.1) DEFINITIONS AND DISCUSSION. Let $\mathbf{X} = X_1, \ldots, X_n$ be indeterminates over a ring R and let $\mathbf{y} = y_1, \ldots, y_n$ be analytic indeterminates over R. Then there is an R-module isomorphism between the R-module homomorphisms of $R[\mathbf{X}]$ to R and the formal power series ring $R[[\mathbf{y}]]$ that sends the homomorphism $\phi\colon R[\mathbf{X}] \to R$ to the power series $\sum_\nu \phi(\mathbf{X}^\nu)\mathbf{y}^\nu$, where $\nu = (\nu_1, \ldots, \nu_n)$ is a multi-index running through $\mathbb{N}^n$, $\mathbf{X}^\nu$ denotes $X_1^{\nu_1} \cdots X_n^{\nu_n}$, and similarly for $\mathbf{y}^\nu$.

Call a power series $\sum_\nu r_\nu \mathbf{y}^\nu$ *special* if for every ν with all entries positive, $r_\nu = 0$. A special power series may alternatively be described as one which is a finite sum of elements from the power series rings over R in the proper subsets of the variables $y_1, \ldots, y_n$.

Thus, a special power series is simply one that does not "honestly involve" all the variables in any one of its nonzero monomial terms. The special power series are the obvious R-module complement in $R[[\mathbf{y}]]$ for the principal ideal $(y_1 \cdots y_n)$.

(9.2) LEMMA. *Let notation be as in (9.1). Let $r_1, \ldots, r_n, r \in R$, and let*

$$f = r - \sum_{i+1}^{n} r_i X_i.$$

Let $z = y_1 \cdots y_n$ and let $z_i = \prod_{j \neq i} y_j$, so that $z = y_i z_i$ for every i. Then an R-module homomorphism $\phi\colon R[\mathbf{X}] \to R$ kills the ideal $fR[\mathbf{X}]$ (and so corresponds to an R-module homomorphism of $S = R[\mathbf{X}]/fR[\mathbf{X}]$ to R) if and only if the corresponding power series g has the property that $(rz - \sum_{i+1}^{n} r_i z_i)g$ is special in the sense of Definition (9.1).

PROOF. Let $e_1, \ldots, e_n$ be the standard free basis for $\mathbb{Z}^n$. The ideal

$$(r - \sum_{i+1}^{n} r_i X_i)R[\mathbf{X}]$$

is spanned as an R-module by the elements

$$X^\nu(r - \sum_{i+1}^{n} r_i X_i) = rX^\nu - \sum_{i+1}^{n} r_i X^{\nu+e_i}$$

as ν runs through $\mathbb{N}^n$. Thus, the power series $g = \sum_\nu a_\nu y^\nu$ represents a map that factors through S if and only if

$$ra_\nu = \sum_i r_i a_{\nu+e_i}$$

for all $\nu \in \mathbb{N}^n$. This condition can be described in terms of the power series g as follows: $rg \equiv \sum r_i y_i^{-1} g$ modulo terms on the right with a negative exponent. Note that the terms with a negative exponent will actually involve only one y_i to a negative power, and that negative power will be -1. Let T_i denote the ring of formal power series over R in the variables y_j for $j \neq i$. Then g satisfies the condition cited above if and only if we have

$$rg = \sum_i r_i y_i^{-1} g + \sum_i y_i^{-1} h_i$$

with $h_i \in T_i$. Multiplying through by $z = y_1 \cdots y_n$ yields $(rz - \sum_i r_i z_i)g \in \sum_i z_i T_i$. Thus, if g corresponds to a map $S \to R$ then $(rz - \sum_i r_i z_i)g$ is special. On the other hand, if $(rz - \sum_i r_i z_i)g$ is special, so that it is in $\sum_i T_i$, it is clear that the value must actually be in $\sum_i z_i T_i$: since z and the z_i's are all in the ideal generated by the z_i's, every monomial term in the expansion of $(rz - \sum_i r_i z_i)g$ must be divisible by at least one of the z_i (i.e., $(z_1, \ldots, z_n)R[[y]] \cap \sum_i T_i = \sum_i z_i T_i$). □

(9.3) THEOREM. *Suppose that (R, m, K) is a complete local domain, let $I = (r_1, \ldots, r_n)R$ be an ideal, and let $r \in R$. Let $\mathbf{y} = y_1, \ldots, y_n$ be analytic indeterminates over R. Let $z = y_1 \cdots y_n$ and for $1 \leq i \leq n$ let $z_i = \prod_{j \neq i} y_j$. If $r \neq 0$ or if $I \neq (0)$ then $r \in I^\star$ if and only if $rz - \sum_i r_i z_i$ has a nonzero multiple in $R[[\mathbf{y}]]$ that is special in the sense of (9.1).*

PROOF. Let $S = R[X_1, \ldots, X_n]/(r - \sum_{i=1}^n r_i X_i)$, which is a generic forcing algebra for (R, I, r). The result follows at once from Lemma (9.2), since S is solid (i.e., has a nonzero R-module homomorphism to R) if and only if the power series g corresponding to the induced map $R[X] \twoheadrightarrow S \to R$ is nonzero and $(rz - \sum_i r_i z_i)g$ is special. Since $R[[\mathbf{y}]]$ is a domain and the elements r, r_i are not all zero, g is nonzero if and only if $(rz - \sum_i r_i z_i)g$ is nonzero. □

(9.4) REMARKS. (a) Evidently, one can use the criterion of Theorem 9.3 to define solid closure for ideals in complete local domains. This definition has the advantage of being very elementary, in a certain sense: one does not even have to know what an R-module homomorphism is to understand it. However, virtually every other property of solid closure becomes more obscure from this point of view. It is not even immediately clear that $I \subseteq I^\star$. (One can see this directly as follows: if, in fact, $r = \sum_i a_i r_i$ then

$$rz - \sum_i r_i z_i = \sum_i a_i r_i z - \sum_i r_i z_i = \sum_i (1 - a_i y_i) r_i z_i,$$

and multiplication by $\prod_i (1 - a_i y_i)^{-1}$ produces

$$\sum_i \Big(\prod_{j \neq i} (1 - a_j y_j)^{-1}\Big) r_i z_i.$$

The ith term in the sum does not involve y_i, and so we have produced the required special multiple.)

(b) Let notation be as in the proof of Theorem (9.3), so that S is a generic forcing algebra for (R, I, r). The proof of (9.2) shows that $\mathrm{Hom}_R(S, R)$ is simply the R-module of power series $g \in R[[y]]$ such that $(rz - \sum_i r_i z_i)g$ is special.

When $\dim R = d$ and (R, m, K) is Gorenstein (as well as being a complete local domain), and, in particular, when R is regular, $\mathrm{Hom}_R(S, R)$ is the same as the dual, into an injective hull of K over R, of $H_m^d(S) \cong S \otimes_R H_m^d(R)$ (the injective hull of K over R may be identified with $H_m^d(R)$). Thus, in this complete Gorenstein local domain case, we have identified the dual of $H_m^d(S)$ with the R-module of power series g such that $(rz - \sum_i r_i z_i)g$ is special.

(c) We may also characterize tight closure for modules in similar terms. Let $A = (r_{ij})$ (an $h \times n$ matrix over R) and ρ (an $h \times 1$ column vector over R) be data for a generic forcing algebra for a triple (M, N, u). The question of when u is in the solid closure of N in M is equivalent to the question of when the column vector ρ is in the solid closure of the column space of A in R^h. Let r_i be the ith entry of ρ. The generic forcing algebra S may be written as $R[X_1, \dots, X_n]/(f_1, \dots, f_h)$ where f_i is the polynomial $\sum_j r_{ij}X_j - r_i$. The R-module maps from S to R may be viewed as the maps from $R[\mathbf{X}]$ to R that kill the ideal of $R[\mathbf{X}]$ generated by the f_i. Since this ideal is the sum (as an abelian group) of the principal ideals generated by the f_i, it is necessary and sufficient that the map $R[\mathbf{X}] \to R$ kill all of the ideals $f_iR[\mathbf{X}]$. As before, the map $R[\mathbf{X}] \to R$ is represented by a formal power series $g \in R[[\mathbf{y}]]$, and the condition on g is that for all i, $1 \le i \le k$, $(r_i z - \sum_j r_{ij} z_j)g$ be special. Thus, u is in the solid closure of N in M if and only if there is a nonzero power series g such that all of the products $(r_i z - \sum_j r_{ij} z_j)g$ are special.

10. Shadow homology

We introduce the notion of shadow homology, which is parallel to the notion of phantom homology in tight closure theory (cf. [**HH4**], §9 and [**HH8**]). We show that certain of the results of [**HH4**] and [**HH8**] concerning phantom homology have analogues for shadow homology. For example, the conditions developed for the phantom acyclicity of a finite free complex in Theorem (9.8) of [**HH4**] and Theorem 3.22 of [**HH8**] are essentially valid for shadow acyclicity if the ring contains a field. However, the proof of this fact in the equal characteristic zero case uses the existence of big Cohen-Macaulay algebras!

We begin with a definition.

(10.1) DEFINITION. *Let $M_\bullet$ denote a complex of modules*

$$\cdots \to M_{i+1} \xrightarrow{d} M_i \xrightarrow{e} M_{i-1} \to \cdots$$

over a Noetherian ring R. We shall say that that an element $\overline{u} \in H_i(M_\bullet)$ is a shadow element of the homology group if for some (equivalently, every) element of $\mathrm{Ker}\, e$ that represents it is in the solid closure of $\mathrm{Im}\, d$ in M_i. We shall say that $M_\bullet$ has shadow homology at the ith spot if every element of $H_i(M_\bullet)$ is a shadow *element. We shall say that a left complex (so that M_i vanishes for $i < 0$) is shadow acyclic if $M_\bullet$ has shadow homology at the ith spot for all $i \geq 1$.*

The following result is trivial but of great usefulness:

(10.2) PROPOSITION. *Let $R \to S$ be a homomorphism of Noetherian rings and let $M_\bullet$ be a complex of finitely generated R-modules. Then the induced map $H_i(M_\bullet) \to H_i(S \otimes_R M_\bullet)$ sends shadow elements to shadow elements (now working over S).*

In particular, if $M_\bullet$ has shadow homology at the ith spot and S is weakly S-regular then the map $H_i(M_\bullet) \to H_i(S \otimes_R M_\bullet)$ is zero.

PROOF. Suppose that $u \in Z = \mathrm{Ker}\,(M_i \to M_{i-1})$ represents a shadow element in the homology, so that it is in the solid closure of $B = \mathrm{Im}\,(M_{i+1} \to M_i)$. Then by Theorem (5.6) on the persistence of solid closure, $1 \otimes u$ will be in the solid closure of the image of $S \otimes_R B$ in $S \otimes_R M_i$, and this is the same as the image of $S \otimes_R M_{i+1}$ in $S \otimes_R M_i$.

The second statement is then immediate. □

To pursue the theory further we shall make use of the the existence of big Cohen-Macaulay algebras in the equal characteristic case.

(10.3) DEFINITION. *If (R, m, K) is a local ring we shall say that an R-module M is a big Cohen-Macaulay module if there is some system of parameters $x_1, \ldots, x_d$ for R that is a regular sequence on M, by which we mean not only that x_i is a nonzerodivisor on $M/(x_1, \ldots, x_{i-1})M$ for $1 \leq i \leq d-1$, but also that $(x_1, \ldots, x_d)M \neq M$. If every system of parameters is a regular sequence on M, then R is called a balanced big Cohen-Macaulay module (cf.* **[Sh]***).*

An R-algebra S is called a (balanced) big Cohen-Macaulay algebra if it is a (balanced) big Cohen-Macaulay R-module.

The existence of big Cohen-Macaulay modules for equicharacteristic local rings was established in [**Ho2**]. Quite recently, the existence of big Cohen-Macaulay algebras in the equicharacteristic case was proved as well, in [**HH7**] (see also [**HH5**], [**Hu2**] and [**HH12**]):

(10.4) THEOREM. *If (R, m, K) is an equicharacteristic local ring, then there is a ring homomorphism $R \to S$ such that $mS \neq S$ and such that every system of parameters for R is a regular sequence on S. Thus, S is a balanced big Cohen-Macaulay algebra for R. Moreover, S may be chosen to be quasilocal (and $R \to S$ local).*

If R is a complete (or excellent) local domain of positive prime characteristic p, one may choose S to be the integral closure R^+ of R in an algebraic closure of its fraction field. (Note that $R \to R^+$ is local if R is complete). □

The connection with solid closure comes from the following observation:

(10.5) PROPOSITION. *Let (R, m, K) be a local ring of dimension d. Let M be a big Cohen-Macaulay module (or algebra) for R.*

(a) *$\widehat{R} \otimes_R M$ is a big Cohen-Macaulay module (or algebra) for $\widehat{R}$.*
(b) *$(\widehat{R}/\mathfrak{p}) \otimes_R M$ is solid over $\widehat{R}/\mathfrak{p}$ for at least one minimal prime $\mathfrak{p}$ of $\widehat{R}$ such that $\dim \widehat{R}/\mathfrak{p} = \dim \widehat{R}$ $(= \dim R)$.*
(c) *If $\widehat{R}$ has only one minimal prime, then M is formally solid.*

PROOF. (a) Since $\widehat{R}$ is a faithfully flat extension of R and a system of parameters for R is a system of parameters for $\widehat{R}$, $\widehat{R} \otimes_R M$ is a big Cohen-Macaulay module for $\widehat{R}$.

(b) We may replace R, M by $\widehat{R}$, $\widehat{R} \otimes_R M$. Thus, we may assume that R is complete. Let $\mathbf{x} = x_1, \ldots, x_d$ be a system of parameters for R that is a regular sequence on M. Then $H^d_m(M) \cong H^d_{(\mathbf{x})}(M)$ is nonzero, since the maps

$$M/(x_1^t, \ldots, x_d^t)M \to M/(x_1^{t+1}, \ldots, x_d^{t+1})M$$

in the direct limit system which may be used to compute $H^d_{(\mathbf{x})}(M)$ (these maps are induced by multiplication by $x_1 \cdots x_d$ on the numerators) are all injective. Then R has a finite filtration

$$R = J_0 \supseteq J_1 \supseteq \cdots \supseteq J_h = (0)$$

in which each factor J_i/J_{i+1} occurring is of the form R/P_i for some prime ideal P_i of R, and there is a corresponding filtration for M, namely

$$M = J_0M \supseteq J_1M \supseteq \cdots \supseteq J_hM = (0).$$

It follows that $H^d_m(J_iM/J_{i+1}M) \neq 0$ for at least one choice of i. Since the surjection $J_i \otimes_R M \twoheadrightarrow J_iM$ sends the image of $J_{i+1} \otimes_R M$ to $J_{i+1}M$, it follows that there is a surjection of $(J_i/J_{i+1}) \otimes_R M \cong (R/P_i) \otimes_R M$ onto $J_{i+1}M/J_iM$. Thus, there is a prime ideal P_i of R such that $H^d_m\big((R/P_i) \otimes_R M\big) \neq 0$. But since $(R/P_i) \otimes_R M$ may be viewed as a module over R/P_i for the purpose of computing local cohomology (replacing m by $m(R/P_i)$), this cannot happen unless $\dim R/P = d$, i.e., unless $P_i = \mathfrak{p}$ is a minimal prime of R such that $\dim R/\mathfrak{p} = d$.

(c) This is immediate from (b). □

Thus:

(10.6) COROLLARY. *Let R be a complete local domain. Then an R-algebra that has an R-algebra homomorphism to a big Cohen-Macaulay algebra for R is solid.* □

It is reasonable to ask, whether, conversely, when R is a complete local domain, every solid R-algebra can be mapped to a big Cohen-Macaulay algebra for R. This is true if $\dim R \leq 2$: see §12. But if $\dim R \geq 3$ it is false, in general, when R contains a field of characteristic 0, as the examples just below show. We do not know what the situation is when R has characteristic p or is of mixed characteristic.

(10.7) EXAMPLES. (a) Let $G = SL(n, K)$, where K is a field of characteristic zero, and let $X = (x_{ij})$ denote an $n \times (n+1)$ matrix of indeterminates over K. Let

$S = K[x_{ij}]_{ij}$ be the polynomial ring in the entries of X, and let G act K-linearly on R by sending the entries of X to the entries of αX for all $\alpha \in G$. Then the ring of invariants $R = S^G$ is well known (cf. [**We**]) to be generated by the $n+1$ size n minors $\Delta_1, \ldots, \Delta_{n+1}$ of X, and these are algebraically independent. To be precise, we shall use Δ_j to denote the product of $(-1)^j$ with the determinant of the $n \times n$ matrix obtained by deleting the jth column of X. Every row of X gives a relation on the Δ_j's, namely

$$(\#) \qquad \sum_{j=1}^{n+1} x_{ij}\Delta_j = 0.$$

The Reynolds operator is a K-linear retraction ρ that sends the hth graded piece $[S]_h$ of S to the hth graded piece $[R]_h$ of R. (It kills the sum of irreducible G-submodules of $[S]_h$ on which G acts non-trivially.) Moreover, ρ yields a degree preserving R-module retraction $S \to R$. We refer to [**We**] and [**Mum**] for details. Let $\widehat{R}$, $\widehat{S}$ denote the completions of R and S with respect to the ideals $(\Delta_j)_j R$ and $(x_{ij})_{ij} S$, respectively. Evidently, ρ also induces a retraction $\widehat{S} \to \widehat{R}$ as $\widehat{R}$-modules, so that $\widehat{S}$ is a solid $\widehat{R}$-algebra.

We want to observe that S cannot, however, be embedded in an R-algebra such that $\Delta_1, \ldots, \Delta_{n+1}$ is a regular sequence if $n \geq 2$. If T were such an algebra then we would have from the relations (#) given above that each x_{ij} is in the ideal of T generated by the Δ_ν for $\nu \neq j$ (since x_{ij} multiplies Δ_j into that ideal). But then $(x_{ij})_{ij}T \subseteq (\Delta_j)_j T$. On the other hand, each Δ_j is in the nth power of the ideal $(x_{ij})_{ij}R$ even over R, so that $(\Delta_j)_j T \subseteq \big((x_{ij})_{ij}R\big)^n$. But this shows that $(\Delta_j)_j T \subseteq \big((\Delta_j)_j T\big)^n$, which is impossible if the Δ_j form a regular sequence on T.

(b) In the example considered in Remark (8.11), S cannot be mapped to a big Cohen-Macaulay R-algebra T, since $x_1^2x_2^2x_3^2$ will still be in the ideal (x_1^3, x_2^3, x_3^3) in T, and this cannot happen if x_1, x_2, x_3 is a regular sequence on T.

(10.8) FINITE PROJECTIVE COMPLEXES AND FORMAL MINHEIGHT. Suppose that $\alpha: G \to G'$ is any map of finitely generated projective modules. Let $I_1(\alpha)$ denote the ideal that is the image of the map

$$G \otimes_R \operatorname{Hom}_R(G', R) \to R$$

sending $g \otimes f$ to $f(\alpha(g))$, and let $I_t(\alpha) = I_1(\wedge^t \alpha)$ for $t \geq 0$ with the convention $I_0(\alpha) = R$. Let rank α denote the largest integer t such that $I_t(\alpha) \neq 0$ (this is determinantal rank: when G, G' are free, $I_t(\alpha)$ is the ideal generated by the size t minors of a matrix for α).

Let $G_\bullet$ denote a finite complex of finitely generated projective modules over the nonzero Noetherian ring R, say

$$0 \to G_d \to \cdots \to G_i \xrightarrow{\alpha_i} G_{i-1} \to \cdots \to G_0 \to 0,$$

and assume for simplicity that each G_i is locally free of constant rank b_i (the ranks are automatically constant if Spec R is connected). Let

$$r_i = \sum_{j=i}^{d} (-1)^{j-i} b_i \ .$$

Let δ be a function from the ideals of R to $\mathbb{N} \cup \{\infty\}$, such as depth, height, or minheight. We say that the complex $G_\bullet$ *satisfies the standard conditions on rank and* δ if for all i, $1 \leq i \leq d$, rank $\alpha_i = r_i$ and $\delta\big(I_{r_i}(\alpha_i)\big) \geq i$. Thus, a complex satisfies the standard conditions on rank and depth if and only if it is acyclic ([**BE**]). Moreover, by the results of [**HH4**] §9, [**HH8**] §3, and, especially, Proposition (5.4) of [**AHH**], if R is a homomorphic image of a Cohen-Macaulay ring and of characteristic p, then $G_\bullet$ and all its images under the iterates of the Frobenius functor have phantom homology in positive degree (i.e., $G_\bullet$ is *stably phantom acyclic*) if and only if $R_{red} \otimes_R G_\bullet$ satisfies the standard conditions on rank and minheight. (The *minheight* mnht I of an ideal I is defined as

$$\min\{\text{ht } I(R/\mathfrak{p}) : \mathfrak{p} \text{ is a minimal prime of } R\}.$$

See [**HH8**], §2.) Thus, one would hope to have an analogous result for solid closure, and this is the case in equal characteristic. We do not know whether the analogous result in mixed characteristic holds. The equal characteristic result is remarkably simple to prove, given the existence of big Cohen-Macaulay algebras. To avoid issues arising from pathology of the ring, we shall work with formal minheight of the ideals: we define the *formal minheight* of I as

$$\inf\{\text{mnht } I(R_m)^\frown : m \text{ is a maximal ideal of } R\}.$$

Evidently, it suffices to consider $m \supseteq I$. If R is universally catenary then minheight does not change upon completion, and the formal minheight of I is the same as the minheight of I. We refer the reader to §2 of [**HH8**].

(10.9) THEOREM. *Let R be a Noetherian ring of equal characteristic and let $G_\bullet$ be a finite complex finitely generated projective modules of constant rank. Suppose that $R_{red} \otimes_R G_\bullet$ satisfies the standard conditions on rank and formal minheight. Then G has shadow homology in positive degree.*

PROOF. If we tensor with a complete local domain of R the standard conditions on rank and formal minheight continue to hold (note that the nilpotents are automatically killed; the ranks cannot go up, and do not decrease because all the ideals of minors have formal minheight at least one). Hence, it suffices to consider the case where R is a complete local domain, so that minheight agrees with height. Let S denote a balanced big Cohen-Macaulay algebra for R. By Appendix B of [**Nor**] (cf. also Theorem (1.2.3) of [**Ab1**]), $S \otimes_R G_\bullet$ is acyclic. It follows that, for $i \geq 1$, every cycle in G_i is in the solid closure of the boundaries, since it is in the expansion of the boundaries to $S \otimes_R G_i$, and S is a solid R-algebra by (10.5). $\square$

(10.10) REMARK. The conclusion of (10.9) is valid in mixed characteristic if $\dim R \leq 2$, or, more generally, whenever every complete local domain of R has a big Cohen-Macaulay algebra (in dimension two, one may use the normalization).

Similarly, we have:

(10.11) THEOREM. *Let R be a Noetherian ring of equal characteristic and let $x_1, \dots, x_n$ be elements of R such that the formal minheight of $I = (x_1, \dots, x_n)R$ is at least n. Let $J = (x_1, \dots, x_{n-1})R$. Then $J{:}_R x_n R \subseteq J^\bigstar$.*

PROOF. It suffices to see this once we expand I, J to a complete local domain of R. The result is clear if any element of J becomes a unit or if x_n becomes a unit. Thus, we may assume that R is a complete local domain and $x_1, \dots, x_n$ are in the maximal ideal. The desired conclusion now follows by expanding to a big Cohen-Macaulay algebra S for R, as in (10.9). □

(10.12) REMARK. Suppose that a finite complex $G_\bullet$ of finitely generated projective R-modules is such that $R_{red} \otimes_R G_\bullet$ satisfies the standard conditions on rank and formal minheight. (We may work with minheight if the ring is universally catenary.) Then this is preserved when one tensors with an R-algebra S provided that the map $R \to S$ preserves formal minheight, i.e., provided that for every ideal I of R, the formal minheight of IS is at least as big as the formal minheight of I. For example, if R is a complete local domain and the standard conditions on rank and height hold over R, then they hold after applying $S \otimes_R \_$ for any complete local domain S module-finite over R.

(10.13) REMARK. One would really like to have much stronger results than (10.11) comparable to the results on iterated operations for tight closure obtained in §7 of [**HH4**]. For example, suppose for simplicity that $x_1, \dots, x_n$ is a system of parameters in a complete local domain R. One would like to have that for integers $t \geq 1$,

$$(\dagger) \qquad (x_1, \dots, x_{n-1}, x_n^t)^\bigstar{:}_R x_n \subseteq (x_1, \dots, x_{n-1}, x_n^{t-1})^\bigstar$$

While this can be proved in characteristic p using tight closure results, it is false in equal characteristic zero, and we do not know whether it holds in mixed characteristic, even if the ring is regular. In fact, if the ring is regular, the iterated use of instances of (†) can be used to show that

$$(x_1^t, \dots, x_n^t)^\bigstar{:}_R (x_1 \cdots x_n)^{t-1} \subseteq (x_1, \dots, x_n)^\bigstar \neq R,$$

for all $t \geq 1$, and then it follows that that $(x_1^t, \dots, x_n^t)$ is solidly closed for all t. This in turn implies that regular rings are S-regular. Because of the example of Roberts [**Ro6**], we know that regular rings of dimension three or more are not S-regular in equal characteristic zero, and this shows that (†) fails in general in dimension three in equal characteristic zero. We do not know what happens in complete local domains of mixed characteristic.

11. Big Cohen-Macaulay algebras and tight closure in equal characteristic zero

Let $N \subseteq M$ be finitely generated modules over a locally excellent domain R of characteristic p. It is an open question whether

(‡) an element $u \in M$ is in $N^*{}_M$ if and only if it is in the expansion of N to $R^+ \otimes_R M$, where R^+ is the integral closure of R in an algebraic closure of its fraction field.

Some evidence for this can be found in [**Sm1, Sm2**], where it is shown to be true for ideals of R generated by parameters. (More generally, it is established in [**Ab3**] that, given the result of [**Sm2**], then (‡) holds whenever M/N has finite phantom projective dimension.)

It should be noted that the question can be reduced to studying complete local domains. For a complete local domain R of characteristic p, R^+ is a balanced big Cohen-Macaulay module. The following result can therefore be viewed as a weakened version of (‡). However, the result is of considerable interest in its own right: for example, it gives a new proof that balanced big Cohen-Macaulay algebras exist in characteristic p. However, this proof does not yield the existence of big Cohen-Macaulay algebras in the weakly functorial sense discussed in detail in [**HH12**], which is very important for certain applications.

(11.1) THEOREM. *Let R be a complete local domain (or analytically irreducible excellent local domain) of characteristic p, and let $N \subseteq M$ be finitely generated R-modules. Let $u \in M$. Then $u \in N^\bigstar{}_M$ $(= N^*{}_M)$ if and only if there exists a balanced big Cohen-Macaulay algebra S over R such that $1 \otimes u \in \langle N_S \rangle$.*

PROOF. The "if" direction is clear, since a big Cohen-Macaulay algebra over an analytically irreducible domain is formally solid: see (10.5). To prove "only if" we begin with the generic forcing algebra T over R for the triple (M, N, u) and perform successive algebra modifications with respect to relations on various systems of parameters for R (the precise meaning is given below). If no finite sequence of such modifications

$$T = T_0 \to T_1 \to \cdots \to T_r$$

is such that

$$(\#) \qquad 1 = \sum_{j=1}^{n} y_j \theta_j$$

with the θ_j in T_r and the y's a set of generators for the maximal ideal of R, then a certain direct limit of such modifications will serve as a balanced big Cohen-Macaulay module for R. The details of this kind of argument are discussed at length in [**HH12**] (see §§(3.1)–(3.7) of [**HH12**], especially (3.6) and (3.7)), and here we shall only give the proof of the key point, that a finite sequence of modifications cannot lead to the relation (#). Here, when we say that T_{i+1} is an *algebra modification* of T_i with respect to a relation on parameters for R we mean that for that value of i there exists part of a system of parameters

$$x_1^{(i)}, \dots, x_{k_i}^{(i)}, x_{k_i+1}^{(i)}$$

for R and a relation

$$x_{k_i+1}^{(i)} s^{(i)} = \sum_{t=1}^{k_i} x_t^{(i)} s_t^{(i)}$$

with $s^{(i)}, s_t^{(i)} \in T_i$ such that

$$T_{i+1} = T_i[z_1^{(i)}, \dots, z_{k_i}^{(i)}]/(s^{(i)} - \sum_{t=1}^{k_i} x_t^{(i)} z_t^{(i)})$$

with the $z_t^{(i)}$ indeterminates over T_i.

We can assume without loss of generality that M is free here, and that N is the column space of a matrix $\alpha = (a_{ij})$ which is, say, ν by μ. Then we may take T to be the algebra $R[z_1^{(0)}, \dots, z_\mu^{(0)}]/\mathfrak{A}$ where $\mathfrak{A}$ is the ideal generated by the entries of the column matrix $\alpha Z^{(0)} - u$: here, $Z^{(0)}$ is the $\mu \times 1$ column whose entries are the $z_j^{(0)}$ and $u \in M$ has been written as a $\nu \times 1$ column.

For $0 \le i \le r$ we define a certain finite subset $\Sigma_i \subseteq T_i$ and an integer $b(j) \in \mathbb{N}$ by reverse induction on i in the following manner:

(1) Σ_r is the set whose elements are the θ's occurring in the relation (#)

$$1 = \sum_{j=1}^{n} y_j \theta_j$$

in T_r displayed in the first paragraph of this proof.

(2) If $0 < j < r$ and $\Sigma_i \subseteq T_i$ has been defined for $i \ge j$ while $b(i)$ has been defined for $r \ge i > j$ define Σ_{j-1} and $b(j)$ as follows. Express each element of Σ_j as a polynomial over T_{j-1} in the standard generators (the images of the $z_t^{(j)}$) for T_j over T_{j-1}. Let $b(j)$ be the greatest degree of any of these polynomials. Let Σ_{j-1} consist of the coefficients of these polynomials and the elements $s^{(j-1)}, s_t^{(j-1)}$ occurring in the relation used in the construction of T_j from T_{j-1}.

(3) Express the elements of Σ_0 as polynomials in the standard generators (the images of the $z_t^{(0)}$) of T_0 over R, and let $b(0)$ be the greatest degree of any of these polynomials.

Thus, (1) gets the Σ's started. Repeated application of (2) enables one to construct $b(r)$ and Σ_{r-1}, $b(r-1)$ and Σ_{r-2}, ..., $b(1)$ and Σ_0. Finally, (3) specifies $b(0)$.

We next define two "intertwined" sequences of integers

$$\beta(0), \dots, \beta(r) \in \mathbb{N} \text{ and } B(0), \dots, B(r) \in \mathbb{N}$$

from the $b(i)$ by the following recursive rules:

(i) $\beta(0) = 1$ and $B(0) = b(0)$.

(ii) For $0 < i \le r$, $\beta(i+1) = B(i) + 1$ and $B(i+1) = b(i+1)\beta(i+1) + B(i)$.

We let $B = B(r)$.

Next, choose a test element c for tight closure in R: this is possible by Theorem (6.1a) of [**HH9**]. Then for any sequence of elements $x_1, \dots, x_{k+1}$ of R that are part of a system of parameters, if $w \in (x_1, \dots, x_k) :_R x_{k+1}$ then $cw^q \in (x_1^q, \dots, x_k^q)R$ for all $q = p^e$, since tight closure "captures" colon ideals for parameters: see, for example, Theorem (4.7) of [**HH4**] or the much more extended discussion in §7 of [**HH4**]. Of course, since $u \in N^*{}_M$ we also have that for all q, $cu^q \in N^{[q]}$. Taking q^{th} roots we can assert instead that the following two conditions hold:

(i) For all q, $c^{1/q}u \in NR^{1/q} \subseteq R^{1/q} \otimes_R M$ and

(ii) For all q, if $x_1, \dots, x_{k+1}$ is part of a system of parameters for R then

$$w \in (x_1, \dots, x_k)R :_R x_{k+1} \text{ implies that } c^{1/q}w \in (x_1, \dots, x_k)R^{1/q}.$$

Now let $R^\infty = \bigcup_q R^{1/q}$. We note the following fact:

(ii°) For all q, if $x_1, \dots, x_{k+1}$ is part of a system of parameters for R then

$$w \in (x_1, \dots, x_k)R^\infty :_{R^\infty} x_{k+1} \text{ implies that } c^{1/q}w \in (x_1, \dots, x_k)R^\infty.$$

To see why, suppose that $wx_{k+1} = \sum_{i=1}^k x_i w_i$ with $w, w_1, \dots, w_k \in R^\infty$. The point is that we may choose a power of p, say Q, so large that $Q \ge q$ and such that all the elements w^Q, w_i^Q are in R. Taking Q^{th} powers we see that $w^Q \in (x_1^Q, \dots, x_k^Q)R :_R x_{k+1}^Q$. It follows that $cw^Q \in (x_1^Q, \dots, x_k^Q)R$ from (ii). Since $Q/q \ge 1$ we have as well that $c^{Q/q}w^Q \in (x_1^Q, \dots, x_k^Q)R$. Taking Q^{th} roots yields that $c^{1/q}w \in (x_1, \dots, x_k)R^\infty$.

The next point is this: we shall show that for every integer $q = p^e$ there are R-algebra homomorphisms $\psi_i : T_i \to R_c^\infty$, $0 \le i \le r$, such that the images of the generators of T_i over T_{i-1} (or over R if $i = 0$) are contained in the cyclic R^∞-submodule of R_c^∞ generated by $c^{-\beta(i)/q}$, while the image of $\Sigma_i \subseteq T_i$ is contained in the cyclic R^∞-submodule of R_c^∞ spanned by $c^{-B(i)/q}$. Moreover, each ψ_{i+1} extends ψ_i, $0 \le i < r$.

Notice that we defined the integers $\beta(i)$, $B(i)$, and, in particular, the integer $B = B(r)$ before introducing q, so that, evidently, these integers are independent of q.

To define $\psi_0 : T \to R_c^\infty$ we note that the fact that $c^{1/q}u \in NR^{1/q}$ yields a solution of the matrix equation $\alpha Z_0 = u$ for the z_t^0 in $c^{-1/q}R^{1/q}$. We let ψ_0 be the unique R-algebra map sending z_t^0 to the element of $c^{-1/q}R^{1/q}$ given by this solution. Since the elements of Σ_0 are polynomials of degree $b(0)$ in the z_t^0, they map into $c^{-b(0)/q}R^{1/q}$.

Now suppose that we have constructed $\psi_i : T_i \to R_c^\infty$ such that the generators map into $c^{-\beta(i)/q}R^\infty$ and Σ_i maps into $c^{-B(i)/q}R^\infty$, where $0 \le i < r$. We want to extend ψ_i to

$$(\#\#) \qquad T_{i+1} = T_i[z_1^{(i)}, \ldots, z_{k_i}^{(i)}]/(s^{(i)} - \sum_{t=1}^{k_i} x(i)_t z_t^{(i)}),$$

where there is a relation

$$x_{k_i+1}^{(i)} s^{(i)} = \sum_{t=1}^{k_i} x_t^{(i)} s_t^{(i)}.$$

In the sequel we sometimes omit the superscripts $^{(i)}$ and write $k = k_i$. By the construction of the Σ's, the elements $s^{(i)}$, $s_t^{(i)}$ are in Σ_i. It follows that we obtain a relation

$$x_{k+1}\psi_i(s) = \sum_{t=1}^{k} x_t \psi_i(s_t),$$

where the elements $\psi_i(s)$, $\psi_i(s_t)$ are all in $c^{-B(i)/q}R^\infty$. Let $\psi_i(s) = \sigma C^{-B(i)/q}$ and $\psi_i(s_t) = \sigma_t c^{-B(i)/q}$, with σ, σ_t in R^∞. Then

$$x_{k+1}\sigma = \sum_{t=1}^{k} x_t \sigma_t$$

in R^∞, and so by condition (ii°) above we have that

$$c^{1/q}\sigma = \sum_{t=1}^{k} x_t \tau_t,$$

where the elements $\tau_t \in R^\infty$. This yields an equation

$$\sigma = \sum_{t=1}^{k} x_t(\tau_t c^{-1/q})$$

and hence

$$\begin{aligned}\psi_i(s) &= \sigma c^{-B(i)/q} \\ &= \sum_{t=1}^{k} x_t(\tau_t c^{-B(i)/q})(c^{-1/q}) \\ &= \sum_{t=1}^{k} x_t(\tau_t c^{-\beta(i+1)/q}).\end{aligned}$$

By the formula (##) defining T_{i+1} we can extend ψ_i from T_i to T_{i+1} by sending each $z_t^{(i)}$ to $\tau_t c^{-\beta(i+1)/q}$. The generators are mapping into $c^{-\beta(i+1)/q}R^\infty$, as required. Now each element of Σ_{i+1} can be written as a polynomial of degree at most $b(i+1)$ in these generators with coefficients in Σ_i, and the value of such a polynomial will lie in

$$(c^{-\beta(i+1)/q})^{b(i+1)} c^{-B(i)/q} R^\infty = c^{-B(i+1)/q} R^\infty,$$

as required.

We are now ready for the dénouement. We apply ψ_r to the equation (#) $1 = \sum_{j=1}^{n} y_j\theta_j$ to obtain the equation

$$1 = \sum_{j=1}^{n} y_j\psi_r(\theta_j)$$

holding in R_c^∞. The value of $\psi_r(\theta_j)$ lies in $c^{-B/q}R^\infty$, where $B = B(r)$. Multiplying by $c^{B/q}$ we obtain that $c^{B/q} \in mR^\infty$ for every $q = p^e$. Taking q^{th} powers, we have that $c^B \in m^{[q]}R^\infty \cap R$ for every $q = p^e$. Since $IR^\infty \cap R \subseteq I^*$ for any ideal I ($u \in IR^\infty \cap R$ implies that $1 \cdot u^{qq'} \in I^{[qq']}$ for all q'), we find that $c^B \in (m^{[q]})^*$ for all q. We emphasize again that $B = B(r)$ is independent of q. Since c is a test element, we have that

$$c \cdot c^B = c^{B+1} \in \bigcap_q m^{[q]} \subseteq \bigcap_q m^q = (0),$$

a contradiction. □

(11.2) Remark. We may apply (11.1) in instances where we know that the element u is in N^* or even N. For example, we can take $M = R$, $N = R$, and $u = 0$. The result of (11.1) then shows that R has a balanced big Cohen-Macaulay algebra. This is a new proof of the existence of big Cohen-Macaulay algebras in characteristic p, independent of the results of [**HH7**].

(11.3) Comparison of solid closure with tight closure in equal characteristic zero. We next want to show that that if $N \subseteq M$ are finitely generated modules over a Noetherian ring R containing $\mathbb{Q}$ then $N^*{}_M \subseteq N^{\bigstar}{}_M$ insofar as * is defined. To this end we want to use the "biggest" version of $N^*{}_M$ considered in [**HH11**]. The appropriate notation is N^{*EQ}: we refer the reader to [**HH11**] for details, but we give a brief description below.

There is no loss of generality in assuming that M is free and that N is the span of the columns of a $\nu \times \mu$ matrix $\alpha = (a_{ij})$ over R. The following then gives the "biggest" version of tight closure: $u \in N^{*EQ}{}_M$ if for every map of R to a complete local domain S there exists a finitely generated $\mathbb{Z}$-algebra B, a prime ideal Q of B, and a map of B_Q to S such that:

(1) B has a $\nu \times \mu$ matrix α_B that maps to the image of α over S, so that if we define $M_B = B^\nu$ and N_B as the column space of α_B in M_B, then $S \otimes_B (M_B/N_B) \cong S \otimes_R (M/N)$. Moreover, M_B has an element u_B that maps to $1 \otimes u$ in $S \otimes_R M$.

(2) B/Q is smooth over $\mathbb{Z}$: this may be achieved by localizing B at one element not in Q and $\mathbb{Z}$ at one nonzero integer h. (It then follows that for every prime p not dividing h, $B/(Q + pB)$ is regular of characteristic p.) For p not dividing h we denote by W_p the multiplicative system in B consisting of elements not in any minimal prime of the radical ideal $Q + pB$, and we let $B\langle p\rangle$ denote $W_p^{-1}B/pB$. We write $M_{B\langle p\rangle}$ for $B\langle p\rangle \otimes_B M_B$ and $N_{B\langle p\rangle}$ for the image of $B\langle p\rangle \otimes_B N_B$ in $M_{B\langle p\rangle}$.

(3) For all but finitely many primes p of $\mathbb{Z}$ with p not dividing h, we have that $1 \otimes u \in {N_{B\langle p\rangle}}^*{}_{M_{B\langle p\rangle}}$ over $B\langle p\rangle$.

(11.4) THEOREM. *Let R be any Noetherian ring containing $\mathbb{Q}$. Let $N \subseteq M$ be finitely generated R-modules. Then $N^{*EQ}{}_M \subseteq N^{\bigstar}{}_M$. Moreover, if R is a complete local domain and $u \in N^{*EQ}{}_M$, then there is a balanced big Cohen-Macaulay R-algebra $\mathcal{T}$ such that $1 \otimes u$ is in the expansion of N to $\mathcal{T} \otimes_R M$.*

PROOF. Let $u \in N^{*EQ}{}_M$ and let S be a complete local domain of R. It suffices to show the result after replacing R, M, N, u by S, $S \otimes_R M$, $\langle N_S \rangle$, $1 \otimes u$. Thus, we may assume that R is a complete local domain. We have $B \to B_Q \to R$ as in the definition of ^{*EQ}. The contraction of the maximal ideal of R to B is a prime Q_0 contained in Q. It is easy to see that we can preserve (1), (2), (3) working with B_{Q_0} instead of B_Q. Thus, we may assume that $B_Q \to R$ is a local homomorphism. We may replace B_Q by its image in R without affecting any relevant issues. We may therefore also assume that the map $B_Q \to R$ is injective. We shall complete the argument by proving the final assertion: since a big Cohen-Macaulay algebra over a complete local domain is solid, this will suffice.

We note that if we factor $B_Q \to B'_{Q'} \to R$ (where the maps are local and B' is a finitely generated $\mathbb{Z}$-algebra) then the image $u_{B'}$ of u_B continues to be in ${N_{B'}}^{*EQ}{}_{M'_B}$, where $M_{B'} = B' \otimes_B M$ and $N_{B'}$ is $\mathrm{Im}\,(B' \otimes_B N_B \to M_{B'})$.

Our objective is to show that u is forced into the expansion of N to a big Cohen-Macaulay algebra. As in the proof of (11.1), if not, a generic forcing algebra $T = T^{(0)}$ for (M, N, u) will have a sequence of algebra modifications with respect to relations on segments of various systems of parameters,

$$T^{(0)} \to \cdots \to T^{(r)},$$

such that $1 = \sum_{j=1}^{n} y_i \theta_i$, where the y_i generate the maximal ideal of R and the $\theta_i \in T^{(r)}$.

Because R is the direct limit of local rings $B'_{Q'}$ where B' is a finitely generated $\mathbb{Z}$-subalgebra of R and Q' is the contraction to B' of the maximal ideal of R, we have that the map $B_Q \to R$ factors $B_Q \to B'_{Q'} \to R$ in such a way that all the systems of parameters involved, the y's, all coefficients from R needed in the construction of the sequence $T^{(i)}$, all coefficients from R needed to describe the θ's in terms of the standard generators of the T's, etc. are actually in B'. Moreover, it is shown in [**HH11**], Theorem (3.5.1), using the main result of [**ArR**], that $B'_{Q'} \to R$ factors $B'_{Q'} \to B''_{Q''} \to R$ where B'' is a domain of the same sort in which a specified set of sequences of parameters (i.e., each sequence is part of a system of parameters) retain the property of being sequences of parameters. (Here, the dimension of B'' may be much larger than that of R.) We do not assume that the map $B'' \to R$ is necessarily injective. We change notation and write B_Q for $B''_{Q''}$. Note that by localizing B at one element not in Q we may further assume that any given finite set of elements of B_Q is in B.

Thus, we may assume that we have a sequence of algebras

$$T_B = T_B^{(0)} \to T_B^{(1)} \to \cdots \to T_B^{(r)}$$

over B such that

(1) T_B is a generic forcing algebra for (M_B, N_B, u_B).
(2) For $0 \le i < r$, $T_B^{(i+1)}$ is an algebra modification of $T_B^{(i)}$ with respect to a relation on *parameters* for Q (by which we mean that the "parameters" can be extended to a sequence of ht Q elements that generate an ideal whose radical is Q).
(3) $T^{(i)} \cong R \otimes_B T_B^{(i)}$.
(4) $1 = \sum_{i=1}^n y_i\theta_i$ in $T_B^{(r)}$.

Notice that the map $B \to R$ will factor $B' \to R''$ where B' is the normalization of B and R'' is a suitable module-finite extension of R. Thus, there is no loss of generality in assuming that B is normal: these replacements do not affect any relevant issues. It follows that for all but finitely many primes p of $\mathbb{Z}$, B/pB is normal (although it may not be a domain): see, for example, [**HH11**], Proposition (2.3.17).

We now get a contradiction by passing to characteristic p for suitable p: we only require that

(1) B/pB be normal,
(2) the image of each sequence of parameters used in defining the sequence of algebra modifications $T^{(i)}$ remain a sequence of parameters, and
(3) $u_{B\langle p\rangle}$ be in the tight closure of $N_{B\langle p\rangle}$ in $M_{B\langle p\rangle}$.

We consider B/pB localized at a minimal prime P of $Q(B/pB)$: call this ring C. Notice that C is a localization of $B\langle p\rangle$, so that u_C is in the tight closure of N_C in M_C where $M_C = C \otimes_{B\langle p\rangle} M_{B\langle p\rangle}$, and u_C, N_C indicate the images of $u_{B\langle p\rangle}$, $C \otimes_{B\langle p\rangle} N_{B\langle p\rangle}$ respectively in M_C. Then $T_C = C \otimes_B T_B$ is a generic forcing algebra for (M_C, N_C, u_C) and the algebras $T_C^{(i)} = C \otimes_B T_B^{(i)}$ form a sequence of modifications of exactly the sort considered in (11.1). The image of the relation $1 = \sum_{i=1}^n y_i\theta_i$ then contradicts (11.1), since C is normal and, hence, analytically irreducible. □

(11.5) Corollary (Briançon-Skoda theorem for solid closure). *If I is an ideal of an equicharacteristic ring R generated by at most d elements and $k \in \mathbb{N}$, then $\overline{I^{d+k}} \subseteq (I^{k+1})^\bigstar$.*

Proof. The issues are unaffected by passing to a complete local domain. In characteristic p the result follows Theorem (5.4) of [**HH4**]. In equal characteristic zero the result follows from the fact that $(I^{k+1})^{*EQ} \subseteq (I^{k+1})^\bigstar$ just established and the results of [**HH11**]. □

(11.6) Remark. Likewise, whenever a result for tight closure in equal characteristic zero shows that it is "big enough" to contain a certain ideal or module, the same will hold for solid closure. On the other hand, we generally do not understand the situation in mixed characteristic.

12. The case of dimension two

In this section we make a detailed exploration of solid algebras and solid closure in dimension two. The main result here is that an algebra over a two-dimensional complete local domain is solid if and only if it cam be mapped to a big Cohen-Macaulay algebra: see Theorem (12.5). We first need:

(12.1) DEFINITION. *Let R be a ring, let $x, y \in R$ and let M be an R-module. Let $I = (x, y)R$. We denote by $\Theta = \Theta(x, y; M)$ the submodule*

$$\{u \in M_{xy} : \textit{for some } h \in \mathbb{N}, I^h u \in Im M\} \subseteq M_{xy}.$$

Thus, $\Theta/(Im M) \subseteq M_{xy}/(Im M)$ is $H_I^0(M_{xy}/(Im M))$.

Note that given an arbitrary R-linear map $M \to N$ there is an induced map $M_{xy} \to N_{xy}$, and that this map carries $\Theta(x, y; M)$ to $\Theta(x, y; N)$. Thus, $\Theta(x, y; \_)$ is a covariant functor from R-modules to R-modules. Notice also that if we have a homomorphism $R \to R'$ carrying x, y to x', y' and M is an R'-module, then

$$\Theta(x', y'; {}_{R'}M) \cong \Theta(x, y; {}_RM),$$

where the subscript on the left of M indicates over which ring we are considering it as a module.

If $M = R$ is an integral domain with fraction field L and $\mathfrak{A}$ is an ideal of R, then Nagata defines the $\mathfrak{A}$-transform $S(\mathfrak{A}; R)$ as

$$\{f \in L : \text{for some } h \in \mathbb{N}, \mathfrak{A}^h f \subseteq R\}.$$

In case $\mathfrak{A} = I = (x, y)R$ this agrees with $\Theta(x, y; R)$. Nagata makes a deep study of the $\mathfrak{A}$-transform in Chapter V of **[N1]**, pp. 41-60, including examples where it is not finitely generated as an R-algebra, even though R is an affine domain over a field.

(12.2) DISCUSSION. We want to make some observations about regular sequences x, y of length two on an R-module M. Let $I = (x, y)$.

(a) If $xu = yv$ then $u \in yM$ and $v \in xM$. If $x^s u = y^t v$, where s, t are positive integers, then $u \in y^t M$ and $v \in x^s M$. (That $v \in xM$ is part of the definition of regular sequence. But if $v = xv'$ then $x(u - yv') = 0$ and since x is not a zerodivisor on M, $u = yv'$. The second statement follows because x^s, y^t is also a regular sequence on M.)

(b) If x, y is a regular sequence on M then y, x is a regular sequence on M if and only if y is not a zerodivisor on M. (The other condition that is needed is proved in (a)).

(c) If x, y is a regular sequence on M, then x, y and y, x are both regular sequences on the I-adic completion $\widehat{M}$ $(= \varprojlim_t M/I^tM)$ of M. (Let $\{m_n\}_n$ be a Cauchy sequence in M representing an element μ of the completion. If $y\mu = 0$ then ym_n is eventually in $(x^t, y^{t+1})M$ for any given t, say $ym_n = x^t u_n + y^{t+1} v_n$. Then $y(m_n - y^t v_n) = x^t u_n$ shows that $m_n - y^t v_n$ is in $x^t M$ and so, for all t, m_n is eventually in $(x^t, y^t)M$. The argument for x is the same. Now suppose that $\{m_n\}_n$ represents $\mu \in \widehat{M}$ and $\{p_n\}_n$ represents $\pi \in \widehat{M}$. Suppose that $x\mu = y\pi$. For all t, $xm_n - yp_n$ is in $(x^{t+1}, y^{t+1})M$ for all sufficiently large n, so that we may choose $n(t)$ arbitrarily large such that

$$xm_{n(t)} - yp_{n(t)} = x^{t+1} u_{n(t)} + y^{t+1} v_{n(t)},$$

and then

$$x(m_{n(t)} - x^t u_{n(t)}) = y(p_{n(t)} + y^t v_{n(t)}).$$

Thus we may choose an increasing sequence $n(t)$ such that the preceding equation holds, and $\{m_{n(t)} - x^t u_{n(t)}\}_t$, $\{p_{n(t)} - y^t v_{n(t)}\}_t$ are Cauchy sequences indexed by t with limits μ, π as before. Using these to replace the original sequences, we see that there is no loss of generality in assuming that $xm_n = yp_n$ for all n. It follows that $p_n = xz_n$ for all n (notice that z_n is uniquely determined). For all t, $xz_n - xz_{n+1} \in (x^{t+1}, y^t)M$ for large n, say

$$x(z_n - z_{n+1}) = x^{t+1} g_n + y^t h_n,$$

so that $x(z_n - z_{n+1} - x^t g_n) = y^t h_n$, and then $z_n - z_{n+1} - x^t g_n \in y^t M$, which shows that for all t and for all sufficiently large n, $z_n - z_{n+1} \in (x^t, y^t)M$. Thus, $\{z_n\}_n$ is a Cauchy sequence, and it follows that $x\mu = y\pi$ implies that $\pi \in x\widehat{M}$. Since we have already seen that x, y are both nonzerodivisors on $\widehat{M}$, x, y is a permutable regular sequence on $\widehat{M}$).

(d) There is an obvious map $M \to \widehat{M}$ whose kernel is $\bigcap_t I^t M$.

We also note:

(12.3) COROLLARY. *Let (R, m) be a local ring of dimension two, let x, y be a system of parameters, and let M be an R-module such that x, y is a regular sequence on M. Then every system of parameters for R is a regular sequence on M.*

PROOF. Given systems of parameters x, y and u, v, we can choose z so that x, z and u, z are both systems of parameters. The chain of systems of parameters

$$x,y;\ x,z;\ z,x;\ z,u;\ u,z;\ u,v$$

shows that it suffices to consider the case where we permute the two parameters, handled in (12.2c) (note that every pair of elements determines the same topology on M), and the case where the two systems of parameters have their first elements in common, say x, y and x, z. But this case is clear, since, modulo xR, y has a power that is divisible by z. □

With these preliminaries out of the way, we next observe:

(12.4) LEMMA. *Let R be a ring, let $x, y \in R$, and let M be an R-module. Let $\Theta = \Theta(x, y; M)$. Let $I = (x, y)R$. Then:*

(a) *x, y are nonzerodivisors on Θ and x, y is a possibly improper regular sequence on Θ.*

(b) *If M is an R-algebra, then Θ is an R-subalgebra of M_{xy}.*

(c) *If N is any R-module such that x, y are nonzerodivisors on N and x, y is a possibly improper regular sequence on N, then every R-module map $\phi: M \to N$ factors uniquely $\phi: M \to \Theta \to N$. Moreover:*

  (1) *If, in addition, N is I-adically separated and there exists a nonzero R-module map $\phi: M \to N$, then x, y is a regular sequence on Θ. (It suffices that $\operatorname{Im}\phi$ not be contained in $\bigcap_t I^t N$.)*

  (2) *If, instead, we assume in addition that M, N are R-algebras and the map $M \to N$ is an R-algebra homomorphism, then $\Theta \to N$ is also an R-algebra homomorphism.*

(d) *The following conditions are equivalent:*

  (i) *M can be mapped R-homomorphically to an R-module N on which x, y form a regular sequence in such a way that the image of M is not contained in $\bigcap_t I^t N$.*

  (ii) *There is a nonzero R-linear map from M to an I-adically separated R-module N on which x, y form a regular sequence.*

  (iii) *There is a nonzero R-linear map from M to an I-adically complete and separated R-module N on which x, y form a regular sequence.*

  (iv) *$\operatorname{Im}(M \to \Theta)$ is not contained in $\bigcap_t I^t \Theta$.*

(e) *If M is an R-algebra, then M has an R-algebra homomorphism to an R-algebra S such that x, y is a regular sequence on S if and only if $(x, y)\Theta \neq \Theta$, i.e., if and only if $1 \notin (x, y)\Theta$. In this case, Θ itself is an R-algebra on which x, y is a regular sequence.*

PROOF. The map ϕ kills the kernel of the map $M \to M_{xy}$. There is no loss of generality in replacing M by its quotient by this kernel, and we henceforth assume that x, y are nonzerodivisors on M, so that $M \subseteq M_{xy}$. This does not change Θ.

(a) It is clear that x, y are nonzerodivisors on $\Theta \subseteq M_{xy}$. If $x\theta = y\theta'$ with $\theta, \theta' \in \Theta$ then $\tau = \theta/y = \theta'/x$ is multiplied into M by a power of I, since both $x\tau$ and $y\tau$ are multiplied into M by I^s for $s \gg 0$, and then $I^{s+1}\tau \subseteq M$. Thus, $\tau \in \Theta$, which shows that $\theta' \in x\Theta$.

(b) If a power of I multiplies each of θ, θ' into the image of M, then the sum of the two exponents will yield a power of I that multiplies $\theta + \theta'$ (respectively, $\theta\theta'$) into the image of M.

(c) The map $\phi: M \to N$ induces a map $\psi : M_{xy} \to N_{xy}$. The key point is that the image of Θ is contained in N. For suppose $\theta \in \Theta$. Then for large t, $x^t\theta = u$ and $y^t\theta = v$ are in M. Then

$$x^t\psi(\theta) = \psi(x^t\theta) = \psi(u) = \phi(u) = u' \in N$$

and, similarly, $y^t\psi(\theta) = v' \in N$. But then $y^tu' = x^tv'$ shows that u' is a multiple of x^t in N, say $u' = x^tu''$, and then $x^t\psi(\theta) = x^tu''$ shows that $\psi(\theta) = u'' \in N$. The uniqueness is obvious, since xy is not a zerodivisor on N.

To prove (1), note that if $I\Theta = \Theta$ then $I^t\Theta = \Theta$ for all $t \in N$, and then the same will be true for the image of Θ in N. Since N is I-adically separated, this implies that the image of Θ is 0, a contradiction, since the image of M is not zero. (The parenthetical comment follows from the fact that we may replace N by $\widehat{N}$, and the induced map $\Theta \to \widehat{N}$ will be nonzero provided that $\operatorname{Im}\Theta$ is not contained in $\bigcap_t I^tN$.)

(2) is a consequence of the fact that the map is constructed as the restriction of the induced may $\psi: M_{xy} \to N_{xy}$, and ψ is evidently an algebra homomorphism.

(d) (iii) $\Rightarrow$ (ii) $\Rightarrow$ (i) is obvious. Given (i), we replace N by $\widehat{N}$: since $\operatorname{Ker}(N \to \widehat{N}) = \bigcap_t I^tN$, we obtain a complete module as required. Thus, (i), (ii), and (iii) are equivalent. Part (c) (1) shows that (ii) $\Rightarrow$ (iv). To complete the proof, it will suffice to show that (iv) $\Rightarrow$ (iii). But given (iv), we have that the induced map $M \to \widehat{\Theta}$ is not zero, and so $\widehat{\Theta} \neq 0$. It is then automatic that $\widehat{\Theta} \neq I\widehat{\Theta}$, and x, y is a regular sequence on $\widehat{\Theta}$ by part (a) and (12.2c).

(e) If $1 \notin (x,y)\Theta$ then it is clear from part (a) that x, y is a regular sequence on Θ, since $(x,y)\Theta \neq \Theta$. On the other hand, from part (c) (2) the R-algebra map $M \to S$ factors $M \to \Theta \to S$, where these are maps of R-algebras. If $(x,y)S \neq S$ then $1 \notin (x,y)S$, from which it is clear that $1 \notin (x,y)\Theta$. □

(12.5) THEOREM. *Let (R, m, K) be a complete local domain of dimension at most two with system of parameters x, y. Then an R-algebra S is solid if and only if it has an R-algebra homomorphism $S \to T$ such that x, y are nonzerodivisors on T and x, y is a regular sequence on T, so that T is a big Cohen-Macaulay algebra for R. We may choose $T = \Theta(x, y; R' \otimes_R S)$ where R' may be chosen to be either the normalization of R or the S_2-ification of R (the latter may be identified with $\Theta(x, y; R)$ and is a subring of the normalization of R). We may also choose $T = \Theta(x, y; S)$.*

PROOF. We have that $R' \otimes_R S$ is solid over R' and admits an R'-linear map ϕ to R' such that the image of 1 is not zero. It then follows from Lemma (12.4) that x, y are nonzerodivisors on $T = \Theta(x, y; R' \otimes_R S)$ and that x, y is a regular sequence on T, since x, y is a regular sequence on R', since R' is $(x,y)R'$-adically separated, and since the map ϕ is not zero.

Since $R \to S \to \Theta(x, y; R' \otimes_R S) = T$ factors $R \to S \to T' \to T$ with $T' = \Theta(x, y; S)$, we must also have that $1 \notin (x,y)T'$, and so T' is a big Cohen-Macaulay algebra over R to which S maps as well. □

(12.6) COROLLARY. *Let (R, m, K) be a complete local domain of dimension two with system of parameters $x, y \in m$, let $N \subseteq M$ be finitely generated R-modules, let $u \in M$, and let S be a generic forcing algebra for the triple (M, N, u). Then u is in the solid closure of N if and only if with $\Theta = \Theta(x, y; S)$ (defined in (12.1)) we have that $(x,y)\Theta \neq \Theta$, i.e., $1 \notin (x,y)\Theta$.* □

We use this criterion in an example:

(12.7) EXAMPLE. Let $R = K[[X,Y,Z]]/(g) = K[[x,y,z]]$, where K is a field of characteristic zero and $g = X^3 + Y^3 + Z^3$, and let $I = (x,y)R$. Then $I^{*EQ} = I^{\bigstar}$. First note that $z^2 \in I^{*eq} \subseteq I^{*EQ}$: see **[HH11]**, Example (2.2.4). Since we know that $I^{*EQ} \subseteq I^{\bigstar}$, if we show that $z \notin I^{\bigstar}$ then it will follow that $I^{*eq} = I^{\bigstar} = (x,y,z^2)$ (and also that $I^{*EQ} = (x,y,z^2)$), for any ideal strictly larger than (x,y,z^2) must contain z. But $z \in I^{\bigstar}$ if and only if R can be mapped to an algebra S on which x, y is a regular sequence and $z \in IS$.

Thus, it will suffice to show that if $x^3 + y^3 + z^3 = 0$ in S, with 3 invertible in S (actually, we know that $S \supseteq \mathbb{Q}$ here) and $z = ux + vy$, then x, y cannot be a regular sequence in S (we have changed notation, using the same letters x, y, z for the images of the original x, y, z in S). But then $x^3 + y^3 + (ux+vy)^3 = 0$ which can be rewritten as:

$$(u^3+1)x^3 + (3u^2v)(x^2y) + (3uv^2)(xy^2) + (v^3+1)y^3 = 0.$$

If x, y is a regular sequence on S then the relations on the elements x^3, x^2y, xy^2, y^3 are spanned by the quadruples

$$(y,-x,0,0),\ (0,y,-x,0) \quad \text{and} \quad (0,0,y,-x).$$

This implies that

$$(u^3+1, 3u^2v, 3uv^2, v^3+1) = a(y,-x,0,0) + b(0,y,-x,0) + c(0,0,y,-x)$$

with $a,b,c \in S$, yielding

(1)	$u^3 + 1 = ay$	or	(1°)	$u^3 = -1 + ay$
(2)	$3u^2v = -ax + by$			
(3)	$3uv^2 = -bx + cy$			
(4)	$v^3 + 1 = -cx$	or	(4°)	$v^3 = -1 - cx$

Multiplying (1°) and (4°) together, we obtain that $u^3v^3 \in 1 + (x,y)S$ while equation (2) (or (3)) shows that $3u^3v^3 \in (x,y)S$, and, since 3 is invertible, $u^3v^3 \in (x,y)S$. Thus, $1 \in (x,y)S$, a contradiction. □

(12.8) REMARK. We do not know, in general, whether, over a complete local domain of dimension two and equal characteristic zero, the big equational tight closure, ^{*EQ}, agrees with the solid closure, ${}^{\bigstar}$.

13. Regular rings revisited

We have already noted (cf. (7.24)) that, as a consequence of a local cohomology calculation of P. Roberts, in the regular rings $K[x,y,z]$ and $K[[x,y,z]]$, where K is a field of characteristic zero, we have that $x^2y^2z^2 \in (x^3,y^3,z^3)^{\bigstar}$. Nonetheless, there are several positive results on when ideals in regular rings are solidly closed that yield useful information.

(13.1) THEOREM. *Let R be a regular Noetherian ring. Let $I \subseteq R$ be an ideal and let $N \subseteq M$ be finitely generated R-modules.*

(a) *Suppose that p is a prime integer and that R/pR is regular. Then every ideal of R containing p is solidly closed in R. More generally, if M/N is killed by p, then N is solidly closed in M.*

(b) *Suppose that R is a finitely generated $\mathbb{Z}$-algebra of characteristic zero. Then every ideal I of R such that R/I is torsion-free over $\mathbb{Z}$ is solidly closed. More generally, if M/N is torsion-free over $\mathbb{Z}$, then N is solidly closed in M.*

PROOF. (a) It suffices to prove the final statement. We may assume without loss of generality that $N = 0$. Suppose that $u \in M$ is in the solid closure of 0 in M. Then this remains true when we take images in $S \otimes_R M$, where $S = R/pR$. Since S is regular of characteristic p and $S \otimes_R M = M$, we have that u is in $0^{\star}{}_M$ over S, which is 0, and so $u = 0$ as required.

(b) By part (a), it will suffice to prove that $N = \bigcap_p (N + pM)$ as p runs through the positive prime integers such that R/pR is regular, since for every such p, $N + pM$ is solidly closed in M. Since R is regular, $\mathbb{Q} \otimes_{\mathbb{Z}} R$ is smooth over $\mathbb{Q}$, and so for some $b \in \mathbb{Z} - \{0\}$, R_b is smooth over $\mathbb{Z}_b$. It follows that for all positive prime integers p not dividing b that R/pR is smooth over $\mathbb{Z}/p\mathbb{Z}$, and, hence, regular. Thus, the set of p such that R/pR is regular is infinite. The result now follows from (13.2), given immediately below, applied to M/N. □

(13.2) THEOREM. *Let M be a finitely generated module over R, where R is a finitely generated $\mathbb{Z}$-algebra, and suppose that M is torsion-free as a $\mathbb{Z}$-module. Then $\bigcap_{p \in P} pM = 0$ for any infinite set P of prime integers.*

PROOF. By the lemma of generic freeness (see [**HR1**], Lemma (8.1), p. 146), we may choose a nonzero integer $a \in \mathbb{Z}$ such that M_a is free over $\mathbb{Z}_a$. Since M is torsion-free over $\mathbb{Z}$, $M \subseteq M_a$. Let Q be the set of primes in P not dividing a: it is still infinite. We have that $\bigcap_{p \in P} pM \subseteq \bigcap_{p \in Q} pM_a$, which is zero simply because M_a is Z_a-free and $\bigcap_{p \in Q} pZ_a = 0$. □

(13.3) COROLLARY. *Let R be a regular domain finitely generated over $\mathbb{Z}$, with $\mathbb{Z} \subseteq R$, and suppose that S is a solid R-algebra. Then $\mathbb{Q} \otimes_{\mathbb{Z}} R \to \mathbb{Q} \otimes_{\mathbb{Z}} S$ is pure.*

PROOF. Suppose that $T \subseteq U$ where T is a normal Noetherian domain and U is an extension ring. We claim that $T \hookrightarrow U$ is pure if and only if every ideal of T is contracted from U (of course, "only if" is trivial). To see this, note that the issue is local on the maximal ideals m of T, and so it suffices that, for every such m, T_m be pure in U_m. Since T_m is normal, it is approximately Gorenstein by the results of [**Ho4**], and it follows from the results of that paper that T_m is pure in U_m if every ideal $\mathfrak{B}$ of T_m primary to mT_m is contracted from U_m, and $\mathfrak{B}$ will be the expansion of its contraction $\mathfrak{A}$ to T, which is primary to m. But if $t/f \in \mathfrak{B}U_m = \mathfrak{A}U_m$ with $t \in T - \mathfrak{A}$ and $f \in R - m$, then gt is in the contraction of $\mathfrak{A}U$ to T for some $g \in T - m$, and so if $\mathfrak{A}$ is contracted from U it follows that $gt \in \mathfrak{A}$. Since $\mathfrak{A}$ is m-primary, it follows that $t \in \mathfrak{A}$, a contradiction. This establishes the claim.

We now apply this with $T = \mathbb{Q} \otimes_{\mathbb{Z}} R$ and $U = \mathbb{Q} \otimes_{\mathbb{Z}} S$.

Thus, it suffices to show that every ideal of $\mathbb{Q} \otimes_{\mathbb{Z}} R$ is contracted from $\mathbb{Q} \otimes_{\mathbb{Z}} S$. Such an ideal J will have the form $I(\mathbb{Q} \otimes_{\mathbb{Z}} R)$, where I is its contraction to R. Since R/I embeds into $\mathbb{Q} \otimes_{\mathbb{Z}} (R/I)$, it follows that R/I is torsion-free over $\mathbb{Z}$. Thus, by (13.1b), I is solidly closed. Since $R \to S$ is solid, it is injective; moreover, since I is solidly closed, I is contracted from S. But if $u \in J(\mathbb{Q} \otimes_{\mathbb{Z}} S) \cap (\mathbb{Q} \otimes_{\mathbb{Z}} R)$ then a multiple of u by some nonzero integer h will be an element of R, and, multiplying further by a nonzero integer k, we have that $khu \in IS \cap R = I$. Since kh is a unit of $\mathbb{Q} \otimes_{\mathbb{Z}} R$, we have that $u \in J$. □

(13.4) REMARK. In the case where S is module-finite over R this follows rather trivially from a trace argument, but it does not appear that a trace argument can be used to recover the general case of (13.3).

(13.5) PROPOSITION. *Let R be any Noetherian ring that does not contain $\mathbb{Q}$, and let $X_1, \ldots, X_n$ be formal indeterminates over R. Let $\mathcal{M}_1, \ldots, \mathcal{M}_h$, and $\mathcal{M}$ be monomials in the X_i in $S = R[[X_1, \ldots, X_n]]$, and let $I = (\mathcal{M}_1, \ldots, \mathcal{M}_h)S$. Then $\mathcal{M} \in I^\star$ in S if and only if $\mathcal{M} \in I$.*

PROOF. Since $\mathbb{Q}$ is not contained in R, for some positive prime integer p, p has no inverse in R. Then pR is contained in a maximal ideal m of R, where $K = R/m$ has characteristic p. Given a counterexample, it remains a counterexample when we pass to $K[[X_1, \ldots, X_n]]$, a contradiction, since this is a regular ring of characteristic p. □

(13.6) EXAMPLE. In the ring $\mathbb{Z}[X_1, X_2, X_3]$ the ideal (X_1^3, X_2^3, X_3^3) is solidly closed, by (13.1b). If V is a discrete valuation ring of mixed characteristic p, then in $V[[X_1, X_2, X_3]$ we do not know whether (X_1^3, X_2^3, X_3^3) is solidly closed, but we know at least that $X_1^2X_2^2X_3^2$ is not in the solid closure, by (13.5). Both these examples show that solid closure does not commute with localization, since $X_1^2X_2^2X_3^2$ is in the solid closure after we localize at $\mathbb{Z} - \{0\}$ (respectively, $V - \{0\}$). Cf. (7.24).

(13.7) REMARKS. Let (V, pV) be a discrete valuation ring of mixed characteristic p such that p generates the maximal ideal and let $R = V[[X_1, \ldots, X_n]]$. Note that an ideal generated by monomials in p, $X_1, \ldots, X_n$ is solidly closed provided that p is in the ideal (by (13.1a)), and that we know something about the behavior of the solid closure of an ideal generated by monomials in the X's only (i.e., with no positive powers of p occurring) from (13.5). However, we do not know whether, when $n = 3$, there is a positive integer h such that $p^h X_1^2 X_2^2 X_3^2$ is in $(X_1^3, X_2^3, X_3^3)^\star$.

Some weak results about the solid closures of monomial ideals can be obtained by mapping to rings of lower dimension. To illustrate the technique, we prove:

(13.8) THEOREM. *Let x, y, z be elements generating an ideal of height three in a Noetherian ring R. Assume either*

(i) *that* $2 \cdot 1_R$ *is not in any height three minimal prime of* $(x, y, z)R$ *(this is automatic if* $2 \cdot 1_R$ *is invertible in* R*) or*

(ii) *that* R_{red} *has characteristic* 2.

Then $xyz \notin (x^2, y^2, z^2)^\bigstar$.

PROOF. If we have a counterexample we can localize at a suitable minimal prime of (x, y, z) (not containing $2 \cdot 1_R$ in case (i)) complete, and kill a minimal prime. By the persistence of solid closure, we still have a counterexample. Thus, there is no loss of generality in assuming that x, y, z is part of a system of parameters in a complete local domain R, and that either (i) $2 = 2 \cdot 1_R$ is a unit or (ii) R has characteristic 2. In the second case the result is immediate from Theorem (7.15a) of [**HH4**] and the fact that solid closure agrees with tight closure in this case, and we henceforth assume that we are in case (i), where 2 is a unit of R. Now map R to

$$S = R[[U, V]]/\big(x - (U^2 + V^2), y - (U^2 - V^2), z - UV\big).$$

Since R has dimension 3, $R[[U, V]]$ has dimension 5, and it is clear that U, V, x, y, z is a system of parameters, whence

$$U, V, x - (U^2 + V^2), y - (U^2 - V^2), z - UV$$

is a system of parameters for $R[[U, V]]$. It follows that the images u, v of U, V in S are a system of parameters for the two-dimensional complete local ring S. In S, x, y, z map to $u^2 + v^2$, $u^2 - v^2$, and uv. Thus, it will suffice to show that if S is a complete local ring of dimension two in which 2 is a unit and u, v is a system of parameters, then with $x = u^2 + v^2$, $y = u^2 - v^2$, $z = uv$ we have that $xyz \notin (x^2, y^2, z^2)^\bigstar$ in S. As before, we may kill a suitable minimal prime. Thus, we may assume that S is a two-dimensional complete local domain. But then if $xyz \in (x^2, y^2, z^2)^\bigstar$ we can find a big Cohen-Macaulay algebra T over S such that u, v is a regular sequence on T and $xyz \in (x^2, y^2, z^2)T$, by (12.5). This says that $(u^4 - v^4)uv \in (u^4 + v^4, u^2v^2)T$. Moreover, we may complete T with respect to the $(u, v)T$-adic topology (equivalently, with respect to the mT-adic topology, where m is the maximal ideal of S, since each of $(u, v)S$ and m has a power contained in the other). It follows (cf. Corollary (12.3)) that every system of parameters for S is a regular sequence on T. Now, since uv, $u^4 + v^4$ is also a system of parameters for S it is a regular sequence on T, and so

$$u^4 - v^4 \in \big(u^4 + v^4, (uv)^2\big) :_T uv = (u^4 + v^4, uv)$$

which yields

$$u^4 - v^4 = a(u^4 + v^4) + buv$$

or

$$(a - 1)u^4 + buv + (a + 1)v^4 = 0.$$

It follows that

$$(a - 1, b, a + 1) = c(v, -u^3, 0) + d(0, -v^3, u),$$

i.e., $a = 1 + cv$, $b = -cu^3 - dv^3$, $a = -1 + du$. Subtracting the third equation from the first shows that $2 \in (u, v)T$ is not a unit of T, and, hence, not a unit of R, a contradiction. □

(13.9) REMARK. It is hopeless to try to show, *by mapping to lower dimensional regular rings*, that if f, g, h are a system of parameters in a local ring of mixed characteristic and of dimension 3 then $f^2g^2h^2 \notin (f^3, g^3, h^3)^\bigstar$. The key point is that, by the Briançon-Skoda theorem, one has that $f^2g^2h^2 \in (f^3, g^3, h^3)$ for any three elements f, g, h of a regular ring of dimension at most two. A detailed explanation is given in (6.8) of [**Ho8**].

14. Questions

In this concluding section we discuss a number of questions about the behavior of solid closure which, so far as the author knows, are open.

(14.1) QUESTION. *Is every ideal of a regular local ring of mixed characteristic solidly closed?*

Roberts' calculation, discussed in (7.23) and (7.24), is discouraging, but the question seems to remain open in mixed characteristic. An affirmative answer would show that a regular local ring of mixed characteristic is pure in any solid algebra over it, and, in particular, is a direct summand of every module-finite extension algebra.

It would be interesting to know, given an ideal I generated by monomials in a regular system of parameters for a regular local ring R, precisely when another such monomial is in the solid closure of I. We understand the situation in characteristic p, where every ideal is solidly closed in a regular ring, and Roberts' calculation shows that the answer may be complicated in the equal characteristic case. Mixed characteristic remains mysterious. But we may ask:

(14.2) QUESTION. *If it is not true that every ideal of a mixed characteristic regular local ring is solidly closed, can one find a variant closure operation for which this is true and that is sufficiently well-behaved to yield a proof of the direct summand conjecture?*

Suppose we denote this enigmatic closure for an ideal $I \subseteq R$ by $I^\diamond$. If we know that when S is a module-finite extension domain of a complete local domain R then $IS \cap R \subseteq I^\diamond$ and that $I^\diamond = I$ if the ring R is regular, then the operation is "sufficiently well-behaved."

(14.3) QUESTION. *If a complete local domain (R, m) has residual characteristic p, is it true that an R-algebra S is solid if and only if S can be mapped to a balanced big Cohen-Macaulay algebra T for R?*

This is true in dimension two without restriction on the characteristic, by Theorem (12.5), and false in equal characteristic zero in dimension greater than or equal to three, by Examples (10.7a,b). The author feels that it is much more likely to be true in characteristic p than in mixed characteristic.

The following is a closely related question:

(14.4) QUESTION. *Let (R, m) be a complete local domain of residual characteristic p and let S be a solid R-algebra. Is every algebra modification of S with respect to a system of parameters for R again solid over R?*

See the first paragraph of the proof of Theorem (11.1) for a discussion of algebra modifications. Briefly, if $\mathbf{x} = x_1, \ldots, x_n$ is a system of parameters for a complete local domain R, and S is an R-algebra, by an *algebra modification* (*of type* k, where $0 \leq k < n$) with respect to $\mathbf{x}$ of S over R we mean an S-algebra T of the form

$$R[z_1, \ldots, z_k]/(s_{k+1} - \sum_{j=1}^{k} x_j z_j),$$

where $s_{k+1} \in S$ is such that there exist $s_1, \ldots, s_k \in S$ satisfying $x_{k+1}s_{k+1} = \sum_{j=1}^{k} x_j s_j$ in S.

In fact, (14.3) and (14.4) can easily be shown to be equivalent. Assume (14.3). Then S can be mapped to a balanced big Cohen-Macaulay algebra T for R, and it is easy to see that any algebra modification S' of S can also be mapped to T. Since T is solid, it follows that S' is solid. On the other hand, assume that (14.4) holds. Then every finite sequence of algebra modifications of S (the systems of parameters and types may vary) yields a solid R-algebra T, and for such an algebra we have that m is contracted from T, i.e., $1 \notin mT$. But then, as in §3 of [**HH12**], we may construct a balanced big Cohen-Macaulay algebra for R to which S maps as a direct limit of algebras obtained by finite sequences of algebra modifications of S.

We note that (14.3) reduces to the case where S is finitely generated over R. (It suffices to check the "only if" part. But if S cannot be mapped to a balanced big Cohen-Macaulay algebra for R then it has a finite sequence of algebra modifications, $S = T_0 \to \cdots \to T_h$ such that $1 \in mT_h$. It is easy to see that some finitely generated subalgebra S_0 of S will also have such a sequence of modifications, and S_0 is also solid over R if S is.)

Any finitely generated solid algebra over R maps onto a minimal solid algebra (cf. (6.1)), also finitely generated over R. If we view R as a finite module over a complete regular local ring A, the minimal solid algebras over R are also minimal solid algebras over A (cf. (6.2e)). We are therefore led to the following question:

(14.5) QUESTION. *What are the minimal solid algebras over a complete regular local domain, especially in characteristic p and in mixed characteristic?*

Of course, the question is answered in dimension one by Theorem (6.4).

Finally, we ask:

(14.6) QUESTION. *Is the Briançon-Skoda theorem for "solid closure" valid in arbitrary Noetherian rings? That is, given an ideal I generated by n elements in an arbitrary Noetherian ring of mixed characteristic p, is $\overline{I^n} \subseteq I^\bigstar$? (One may also ask whether various versions of the Briançon-Skoda theorem hold, e.g., is $\overline{I^{n+k}} \subseteq (I^{k+1})^\bigstar$ for every nonnegative integer k.)*

To prove this for a given ring R, it suffices to prove it for the complete local domains of R. The result is known in equal characteristic, by Corollary (11.5). Thus, one may reduce to the case of a complete local domain of mixed characteristic.

The above questions give just a small taste of what remains to be done in the study of solid algebras and solid closure.

REFERENCES

[Ab1] Aberbach, I., *Finite phantom projective dimension*, Amer. J. Math., to appear.

[Ab2] ———, *Test elements in excellent rings with an application to the uniform Artin-Rees property*, Proc. Amer. Math. Soc., to appear.

[Ab3] ———, *Tight closure in F-rational rings*, preprint.

[AHH] Aberbach, I., M. Hochster, and C. Huneke, *Localization of tight closure and modules of finite phantom projective dimension*, J. Reine Angew. Math. (Crelle's Journal) **434** (1993), 67-114.

[ArR] Artin, M. and C. Rotthaus, *A structure theorem for power series rings*, in Algebraic Geometry and Commutative Algebra: in honor of Masayoshi Nagata, Vol. I, Kinokuniya, Tokyo, 1988, 35-44.

[B] Boutot, J.-F., *Singularités rationelles et quotients par les groupes réductifs*, Invent. Math. **88** (1987), 65-68.

[BrS] Briancon, J. and H. Skoda, *Sur la clôture intégrale d'un idéal de germes de fonctions holomorphes en un point de C^n*, C. *R.* Acad. Sci. Paris Sér. A **278** (1974), 949-951.

[BuE] Buchsbaum, D. and D. Eisenbud, *What makes a complex exact*, J. of Algebra **25** (1973), 259-268.

[Du] Dutta, S. P., *On the canonical element conjecture*, Trans. Amer. Math. Soc. **299** (1987), 803-811.

[EvG1] Evans, E.G. and Griffith P., *The syzygy problem*, Annals of Math. **114** (1981), 323–333.

[EvG2] ———, *Syzygies*, London Math. Soc. Lecture Note Series **106**, Cambridge University Press, Cambridge, 1985.

[EvG3] ———, *The syzygy problem*, in Commutative Algebra, Math. Sci. Research Inst. Publ. **15**, Springer-Verlag, New York · Berlin · Heidelberg, 1989, pp. 213-225.

[EGA] Grothendieck, A. (rédigés avec la collaboration de J. Dieudonné), *Éléments de géométrie algébrique IV: Étude local des schémas et des morphismes de schèmas (Seconde partie)*, I. H. E. S. Publ. Math. (Paris) N° **24** (1965), 1-231.

[FeW] Fedder, R. and K. Watanabe, *A characterization of F-regularity in terms of F-purity*, in Commutative Algebra, Math. Sci. Research Inst. Publ. **15**, Springer-Verlag, New York · Berlin · Heidelberg, 1989, pp. 227–245.

[Gla] Glassbrenner, D. J., *Invariant rings of group actions, determinantal rings, and tight closure*, Thesis, University of Michigan, 1992.

[GrHa] Grothendieck, A. (notes by R. Hartshorne), *Local Cohomology*, . Lecture Notes in Math. No. **41**, Springer-Verlag, New York · Berlin · Heidelberg, 1967.

[Ho1] Hochster, M., *Contracted ideals from integral extensions of regular rings*, Nagoya Math. J. **51** (1973), 25–43.

[Ho2] ———, *Topics in the homological theory of modules over commutative rings*, C.B.M.S. Regional Conf. Ser. in Math. No. **24**, A.M.S., Providence, *R.*I., 1975.

[Ho3] ———, *Big Cohen-Macaulay modules and algebras and embeddability in rings of Witt vectors*, in Proceedings of the Queen's University Commutative Algebra Conference, Queen's Papers in Pure and Applied Math. **42**, 1975, pp. 106–195.

[Ho4] ———, *Cyclic purity versus purity in excellent Noetherian rings*, Trans. Amer. Math. Soc. **231** (1977), 463–488.

[Ho5] ———, *Some applications of the Frobenius in characteristic* 0, Bull. Amer. Math. Soc. **84** (1978), 886–912.

[Ho6] ———, *Cohen-Macaulay rings and modules*, Proc. of the International Congress of Mathematicians, Helsinki, Finland, Vol. I, Academia Scientarium Fennica, 1980, pp. 291–298.

[Ho7] ———, *Canonical elements in local cohomology modules and the direct summand conjecture*, J. of Algebra **84** (1983), 503–553.

[Ho8] ———, *Tight closure in equal characteristic, big Cohen-Macaulay algebras, and solid closure*, in Proc. of the Summer Research Conference on Commutative Algebra (held at Mt. Holyoke College, July, 1992), to appear (in these proceedings).

[HoE] Hochster, M. and J. A. Eagon, *Cohen-Macaulay rings, invariant theory, and the generic perfection of determinantal loci*, Am. J. Math. **93** (1971), 1020–1058.

[HH1] Hochster, M. and C. Huneke, *Tightly closed ideals*, Bull. Amer. Math. Soc. **18** (1988), 45–48.

[HH2] ———, *Tight closure*, in Commutative Algebra, Math. Sci. Research Inst. Publ. **15**, Springer-Verlag, New York · Berlin · Heidelberg, 1989, pp. 305–324.

[HH3] ———, *Tight closure and strong F-regularity*, Mémoires de la Société Mathématique de France, numéro **38** (1989), 119–133.

[HH4] ———, *Tight closure, invariant theory, and the Briançon-Skoda theorem*, J. Amer. Math. Soc. **3** (1990), 31–116.

[HH5] ———, *Absolute integral closures are big Cohen-Macaulay algebras in characteristic p*, Bull. Amer. Math. Soc. (New Series) **24** (1991), 137–143.

[HH6] ———, *Tight closure and elements of small order in integral extensions*, J. of Pure and Appl. Algebra **71** (1991), 233–247.

[HH7] ———, *Infinite integral extensions and big Cohen-Macaulay algebras*, Annals of Math. **135** (1992), 53–89.

[HH8] ———, *Phantom homology*, Memoirs Amer. Math. Soc. Vol. **103**, No. **490** (1993), 1-91.

[HH9] ———, *F-regularity, test elements, and smooth base change*, preprint.

[HH10] ———, *Tight closures of parameter ideals and splitting in module-finite extensions*, preprint.

[HH11] ———, *Tight closure in equal characteristic zero*, in preparation.

[HH12] ———, *Applications of the existence of big Cohen-Macaulay algebras*, preprint.

[HR1] Hochster, M. and J.L. Roberts, *Rings of invariants of reductive groups acting on regular rings are Cohen-Macaulay*, Advances in Math. **13** (1974), 115–175.

[HR2] ———, *The purity of the Frobenius and local cohomology*, Advances in Math. **21** (1976), 117–172.

[Hu1] Huneke, C., *An algebraist commuting in Berkeley*, Mathematical Intelligencer **11** (1989), 40–52.

[Hu2] ———, *Absolute integral closures and big Cohen-Macaulay algebras*, in Proc. of the 1990 Intern. Congress of Mathematicians, Kyoto 1990, Vol. I, Math. Soc. of Japan, Springer-Verlag, New York · Berlin · Heidelberg, 1991, pp. 339–349.

[Hu3] ———, *Uniform bounds in Noetherian rings*, Invent. Math. **107** (1992), 203–223.

[Ku1] Kunz, E., *Characterizations of regular local rings of characteristic p*, Am. J. Math. **91** (1969), 772-784.

[Ku2] ———, *On Noetherian rings of characteristic p*, Am. J. Math. **98** (1976), 999–1013.

[LS] Lipman J. and A. Sathaye, *Jacobian ideals and a theorem of Briançon-Skoda*, Michigan Math. J. **28** (1981), 199–222.

[LT] Lipman, J. and B. Teissier, *Pseudo-rational local rings and a theorem of Briançon-Skoda about integral closures of ideals*, Michigan Math. J. **28** (1981), 97–116.

[Mat] Matsumura, H., *Commutative Algebra*, Benjamin, 1970.

[Mum] Mumford, D., *Geometric Invariant Theory*, Springer-Verlag, New York, 1965.

[N1] Nagata, M., *Lectures on the Fourteenth Problem of Hilbert*, Tata Institute of Fundamental Research, Bombay, 1965.

[N2] ———, *Local Rings*, Interscience, New York, 1972.

[Nor] Northcott, D. G., *Finite Free Resolutions*, Cambridge Tracts in Math. **71**, Cambridge University Press, Cambridge, 1976.

[PS1] Peskine, C. and L. Szpiro, *Dimension projective finie et cohomologie locale*, I.H.E.S. Publ. Math. **42** (Paris) (1973), 323-395.

[PS2] ———, *Syzygies et multiplicités*, C. R. Acad. Sci. Paris Sér. A **278** (1974), 1421-1424.

[Ro1] Roberts, P., *Two applications of dualizing complexes over local rings*, Ann. Sci. Ec. Norm. Sup. **9** (1976), 103–106.

[Ro2] ———, *Cohen-Macaulay complexes and an analytic proof of the new intersection conjecture*, J. of Algebra **66** (1980), 225–230.

[Ro3] ———, *The vanishing of intersection multiplicities of perfect complexes*, Bull. Amer. Math. Soc. **13** (1985), 127–130.

[Ro4] ———, *Le théorème d'intersection*, C. *R.* Acad. Sc. Paris Sér. I **304** (1987), 177–180.

[Ro5] ———, *Intersection theorems*, in Commutative Algebra, Math. Sci. Research Inst. Publ. **15**, Springer-Verlag, New York · Berlin · Heidelberg, 1989, pp. 417–436.

[Ro6] ———, *A computation of local cohomology*, in Proc. of the Summer Research Conference on Commutative Algebra (held at Mt. Holyoke College, July, 1992), to appear (in these proceedings).

[S] Serre, J.-P., *Algèbre Locale · Multiplicités*, Springer-Verlag Lecture Notes in Math. **11**, Springer-Verlag, New York · Berlin · Heidelberg, 1965.

[Sh] Sharp, R. Y., *Cohen-Macaulay properties for balanced big Cohen-Macaulay modules*, Math. Proc. Cambridge Philos. Soc. **90** (1981), 229-238.

[Sk] Skoda, H., *Applications des techniques L^2 a la théorie des idéaux d'une algèbre de fonctions holomorphes avec poids*, Ann. Scient. Ec. Norm. Sup. 4ème série t. **5** (1972), 545–579.

[Sm1] Smith, K. E., *Tight closure of parameter ideals and F-rationality*, Thesis, University of Michigan, 1993.

[Sm2] ———, *Tight closure of parameter ideals*, Invent. Math., to appear.

[Sm3] ———, *F-rational rings have rational singularities*, preprint.

[Vel] Velez, J., *Openness of the F-rational locus, smooth base change, and Koh's conjecture*, Thesis, University of Michigan, 1993.

[W1] Watanabe, K.-I., *Study of F-purity in dimension two*, in Algebraic Geometry and Commutative Algebra in honor of Masayoshi Nagata, Vol. II, Kinokuniya, Tokyo, 1988, pp. 791–800.

[W2] ———, *F-regular and F-pure normal graded rings*, J. of Pure and Applied Algebra **71** (1991), 341–350.

[We] Weyl, H., *The Classical Groups*, Princeton Univ. Press, Princeton, New Jersey, 1946.

[Wil] Williams, L., *Uniform stability of kernels of Koszul cohomology indexed by the Frobenius endomorphism*, Thesis, University of Michigan, 1992; and preprint.

Department of Mathematics, University of Michigan, Ann Arbor, MI 48109-1003, USA

Contemporary Mathematics
Volume **159**, 1994

Tight Closure in Equal Characteristic, Big Cohen-Macaulay Algebras, and Solid Closure

MELVIN HOCHSTER

ABSTRACT. We first discuss joint work of Craig Huneke and the author, giving an overview of the status of tight closure theory both in characteristic p and in equal characteristic 0, including recently discovered interconnections with the existence of big Cohen-Macaulay algebras, especially their existence in a weakly functorial sense. For example, either tight closure or the functorial existence of big Cohen-Macaulay algebras can be used to prove that direct summands of regular rings are Cohen-Macaulay in equal characteristic. Later we present and explore a new notion, solid closure, defined *a priori* in all characteristics, even mixed characteristic, but agreeing with tight closure in positive characteristic.

Prologue

Let A be a complete local domain. We can define an apparently frivolous closure operation on ideals of A as follows. If $a_1, \dots, a_n \in A$ let $y_1, \dots, y_n$ be new formal variables over A and define $a \in A$ to be in the *closure* of the ideal $(a_1, \dots, a_n)A$ if the power series

$$(\#) \qquad ay_1 \cdots y_n - \sum_{i=1}^{n} a_i y_1 \cdots y_{i-1} y_{i+1} \cdots y_n$$

has a nonzero multiple in $A[[y_1, \dots, y_n]]$ in which none of the monomials occurring involves all of the y's. This condition is independent of the choice of

---

1991 *Mathematics Subject Classification.* Primary 13E05, 13B99, 13A35, 13H05, 13H10, 13D25.

The author was supported in part by a grant from the National Science Foundation.

This paper is in final form and no version of it will be submitted for publication elsewhere.

generators of the ideal. It is true that if $a = \sum_{i=1}^{n} r_i a_i$ is in the ideal, then it satisfies this condition: the difference (#) can be rewritten as

$$\sum_{i=1}^{n}(r_i y_i - 1)a_i y_1 \cdots y_{i-1} y_{i+1} \cdots y_n$$

and multiplication by $\prod_i (r_i y_i - 1)^{-1}$ will have the desired effect. It is an incredibly subtle problem to determine whether an element of A is in the closure of an ideal in this sense. But is it important? Yes!!! Let V be the p-adic integers, where p is a positive prime integer, and let $A = V[x_2, x_3]$. Let $x_1 = p$. Then it is not even known whether $x_1^2 x_2^2 x_3^2$ is in the closure of the ideal (x_1^3, x_2^3, x_3^3). On the other hand, if $A = K[[x_1, x_2, x_3]]$ with K a field, it is known that $x_1^2 x_2^2 x_3^2$ is not in the closure of (x_1^3, x_2^3, x_3^3) if K has positive characteristic p, but, by a very recent result of P. Roberts **[Ro6]**, it is in the closure of the ideal generated by the cubes when K has characteristic zero.

In the course of this paper I will explain how one is led to ask this apparently innocuous, somewhat strange, but, in fact, quite difficult and important question. There is a concluding discussion in (6.8).

Problem for thought. Let f, g, h be three elements in a regular ring of dimension two. Is it the case that $(fgh)^2$ must be in (f^3, g^3, h^3)? What if the ring is $\mathbb{C}[[x, y]]$? $\mathbb{C}[x, y]$?

Must fgh be in (f^2, g^2, h^2) in these rings?

1. Introduction

Except where otherwise indicated, all of the results on tight closure in equal characteristic and the existence of big Cohen-Macaulay algebras discussed here are the fruit of a more than six year collaboration of the author with Craig Huneke. The reader is referred to **[HH1-12]**, **[AHH]**, **[Hu1,3]** for the detailed account, as well as to **[Ab1-3]**, **[FeW]**, **[Fe]**, **[Gla]**, **[Hu2]**, **[Sm1-3]**, **[Sw1-2]**, **[Vel]**, **[W2]** and **[Wil]** for additional information and applications. The results on solid algebras and solid closure discussed in §6 were developed by the author in **[Ho8]** as part of a rather speculative assault on the problem of defining an analogue of tight closure in mixed characteristic. The notion of solid closure does agree with the notion of tight closure in characteristic p in good cases, and does make sense quite generally. It has many of the right properties to be a good analogue of tight closure. By a recent result of Paul Roberts **[Ro6]** it does not have the right properties in equal characteristic zero. The question of whether it has the right properties in mixed characteristic remains open, although Roberts' result is discouraging. In particular, if it could be proved that every ideal of a regular local ring of mixed characteristic is solidly closed, the direct summand conjecture, in a greatly strengthened form, would follow. This appears to be a difficult question. What we know at the moment is that ideals of regular rings are solidly closed in positive characteristic, not necessarily so in equal characteristic zero, despite which, the issue remains open in mixed characteristic.

We begin by summarizing the properties of tight closure in the equal characteristic case, without even giving the definition. Later, we shall backtrack and give a definition. But before we do that we explain briefly how a number of applications are obtained from the properties that we state.

This approach will enable us to see quickly the merits and defects of the theory of solid closure when it is developed.

Before discussing the theory of solid closure we shall spend some time explaining how the investigation of tight closure led to the discovery of the existence of big Cohen-Macaulay algebras, and how the two theories intertwine.

2. Basic Properties of Tight Closure and Applications

For simplicity, we assume that given rings of characteristic p are localizations of finitely generated algebras over an excellent local ring (more generally, we may admit rings that have a faithfully flat extension such that the Frobenius endomorphism is finite). We assume that rings containing $\mathbb{Q}$ are locally excellent. These hypotheses are often unnecessary but avoid many technicalities. Let us call a Noetherian ring R *admissible* if it is equal characteristic and has the appropriate properties specified above.

In [**HH12**] a notion of tight closure is developed for the class of Noetherian K-algebras, where K is a given field of equal characteristic zero. This type of tight closure is denoted ^{*K} in [**HH12**].

We shall focus here on the notion that corresponds to the case $K = \mathbb{Q}$, which is also called *equational tight closure* in [**HH12**] and in is denoted there either by ${}^{*\mathbb{Q}}$ or ^{*eq}. However, here we shall often simply denote it *.

Tight closure is an operation on submodules of finitely generated modules defined over any admissible base ring. We shall denote the tight closure of N in M by N^* or ${N^*}_M$. If $N = I$ is an ideal of R then M is understood to be R unless otherwise specified: this is a very important case. If $N \subseteq M$ are R-modules and $R \to S$ is a homomorphism we shall write M_S for $S \otimes_R M$ and $\langle N_S \rangle$ for the image of N_S in M_S. This is an abuse of notation since $\langle N_S \rangle$ depends on $N \to M$ and not just on N. In the case when M is free, forming $\langle N_S \rangle$ is easily seen to be parallel to expanding an ideal of R to an ideal of S (which is the case $M = R$).

By a *complete local domain* of R we mean a ring obtained by localizing R at a prime ideal, completing, and then killing a minimal prime. We define the *minheight* of $I \subseteq R$ as $\min\{\operatorname{ht} I(R/P) : P \text{ is a minimal prime of } R\}$. We write $\overline{J}$ for the integral closure of the ideal J.

(2.1) THEOREM (PROPERTIES OF TIGHT CLOSURE). *Let R, S be admissible rings, $R \to S$ a homomorphism, and let $N \subseteq M \subseteq Q$ be finitely generated R-modules. Tight closures are taken in M unless otherwise specified. Let u denote an element of M, and I an ideal of R.*

(a) *$u \in {N^*}_M$ if and only if $u + N \in {0^*}_{M/N}$.*

(b) *${N^*}_M \subseteq {N^*}_Q$; ${N^*}_Q \subseteq {M^*}_Q$; $N^{**} = N^*$.*

(c) *(Persistence of *.) If $u \in N$ then $1 \otimes u \in \langle N_S \rangle^*{}_{M_S}$.*
(d) *$u \in N^*$ iff for every complete local domain B of R, $1 \otimes u \in \langle N_B \rangle^*{}_{M_B}$.*
(e) *If R is a regular ring, $N^* = N$ for all $N \subseteq M$.*
(f) *(Generalized Briançon-Skoda theorem) If I has at most n generators then for all $k \in \mathbb{N}$, $\overline{(I^{n+k})} \subseteq (I^{k+1})^*$.*
(g) *(Capturing colon ideals) If $I = (x_1, \dots, x_n)R$ is such that minheight$I \geq n$, and $I_{n-1} = (x_1, \dots, x_{n-1})R$, then $I_{n-1} :_R x_n R \subseteq I^*_{n-1}$.*
(Better: $I^*_{n-1} :_R x_n R = I^*_{n-1}$.)
(h) *(Phantom acyclicity criterion) Suppose that R is reduced, that $G_\bullet$ is a finite free complex of length n over R with rank $G_i = b_i$, and let $r_i = \sum_{j \geq i} (-1)^{j-i} b_j$. If $G_\bullet$ is such that the rank of the ith map is r_i for $1 \leq i \leq n$ and such that the minheight of the ideal generated by the r_i size minors of a matrix of the map $G_i \to G_{i-1}$ is at least i, then for all $i \geq 1$ the cycles in G_i are in the tight closure of the boundaries. (We say that $G_\bullet$ is phantom acyclic in this case.)*
(i) *If S is a module-finite extension of R then $u \in N^*{}_M$ over R if and only if $u \in \langle N_S \rangle^*{}_{M_S}$ over S. Moreover, the contraction of IS to R is contained in I.*

PROOF. The reader is referred to [**HH12**] for the characteristic zero assertions, especially §(3.2), §(3.6) and, for (h), Chapter 4. For characteristic p see [**HH4**], §5 for (f), §8 for (a), (b), (e), [**HH9**], §6 for (c), [**HH8**], [**AHH**] for (h), [**HH4**] §7 for (g) and (5.23) of [**HH10**] for (i). Also, §§(1.4) - (1.6) in [**HH12**] give a summary of characteristic p theory. □

(Two comments on part (h). First, it is easy to generalize the result to complexes consisting of projective modules that are locally free of constant rank. Second, in characteristic p, if $\mathbf{F}^e$ is the eth iterate of the Frobenius functor then $\mathbf{F}^e(G_\bullet)$ is phantom acyclic for all e if and only if the specified conditions on the ranks and heights given above in (h) hold. See [**HH8**].)

Assuming these properties for tight closure one gets many striking results. Let us call a ring *weakly F-regular* if every ideal is tightly closed. Let us call a ring *F-rational* if every ideal I generated by parameters (i.e., elements that are part of a system of parameters in every local ring at a prime containing I) is tightly closed. It is not difficult to show that if R is weakly F-regular then every submodule of every finitely generated module is tightly closed. (In characteristic zero, one gets a notion for Noetherian K-algebras relative to every field K. Here, we are working with the notion for $K = \mathbb{Q}$.) R is called *F-regular* if all of its local rings are weakly F-regular. F-rational, weakly F-regular, and F-regular are known to be equivalent conditions if the ring is Gorenstein. Cf. [**HH9**], §4.

By the *regular closure* N^{reg} of $N \subseteq M$ we mean the set of all $u \in M$ such that for every map of R to a regular ring S, $u_S \in \langle N_S \rangle$ (in M_S).

(2.2) COROLLARIES. *Let R be an admissible ring of equal characteristic. Let $N \subseteq M$ be finitely generated R-modules and let I be an ideal of R.*

(a) $N^* \subseteq N^{reg}$ *(this is known to be strict: cf.* [HH8], §5*).*
(b) $I^* \subseteq I^{reg} \subseteq \overline{I}$ *(the latter since integral closure is tested by mapping to DVR's).*
(c) *Weakly F-regular rings are normal and Cohen-Macaulay. In fact, F-rational rings are normal and Cohen-Macaulay.*
(d) *A direct summand (or a pure subring) of a weakly F-regular ring is weakly F-regular.*
(e) *A direct summand (or pure subring) of a regular ring is Cohen-Macaulay.*
(f) *A weakly F-regular ring is a direct summand of every module-finite extension.*

PROOF. (a) is immediate from (2.1c,e). The first inclusion in (b) follows from (a), and the second from the fact that $u \in \overline{I}$ iff $u \in IV$ for every homomorphism from R to a discrete valuation ring V. It follows from (b) and (2.1f) in the case $n = 1, k = 0$ that $I^* = \overline{I}$ if I is principal, and from this it follows easily that weakly F-regular rings are normal. The Cohen-Macaulay property is then immediate from (2.1g). (d) is immediate from the persistence of tight closure, and (e) follows at once from (c) and (d).

Part (f) follows from (i) and the results of **[Ho4]**: see §5 of **[HH10]** for details. (The converse is true for locally excellent Gorenstein rings of characteristic p: see §6 of **[HH10]**. On the other, hand every normal ring containing $\mathbb{Q}$ is a direct summand of every module-finite extension, so that the converse is quite false in equal characteristic zero.) Of course, (f) may be viewed as a generalization of the fact that equicharacteristic regular rings are direct summands of all their module-finite extensions. □

Tight closure gives some rather important results in a more general form. See **[BrS]**, **[LS]**, **[LT]**, and **[Sk]** for background for the Briançon-Skoda theorem. The result that pure subrings of regular rings are Cohen-Macaulay has aroused a great deal of interest: see **[Bou]**, **[HR1-2]**. The original interest arose from studying the question, is the ring of invariants of a linearly reductive linear algebraic group over a field K acting K-rationally on a regular K-algebra Cohen-Macaulay.

The term "F-rational" is used because there is a strong connection between this notion and the property of having rational singularities. Let us say that an affine K-algebra R has *$\mathcal{P}$-type*, where $\mathcal{P}$ is a certain property (e.g., weakly F-regular, F-rational, F-regular) if it arises as $K \otimes_A R_A$, where A is a finitely generated $\mathbb{Z}$-subalgebra of K, R is a finitely generated A-algebra, and for all maximal ideals μ of A, with $\kappa = A/\mu$, $R_\kappa = \kappa \otimes_A R_A$ has property $\mathcal{P}$. K. E. Smith has shown **[Sm1,3]** that affine K-algebras of F-rational type have rational singularities. The converse is open (but true in the graded case in dimension two: cf. **[Fe]**.) I. Aberbach has shown **[Ab3]** that an admissible ring of characteristic p is F-rational if and only if whenever M/N has finite projective dimension then ${N^*}_M = N$.

One also gets:

(2.3) THEOREM (VANISHING THEOREM FOR MAPS OF TOR). *Let A be a regular equicharacteristic domain, let R be module-finite extension domain of A and let $R \to S$ be any map to a ring that is regular (or admissible and weakly F-regular). Then for any finitely generated A-module M and every $i \geq 1$; the map $Tor_i^A(M, R) \to Tor_i^A(M, S)$ is zero.*

PROOF. Take a finite free resolution $G_\bullet$ of M over A. One sees easily that, since $G_\bullet$ satisfies the criterion of **[BuE]**, $R \otimes_A G_\bullet$ satisfies the hypothesis of (2.1h), and so has phantom homology ($A \to R$ does not preserve depth but does preserve height). By the persistence of tight closure the image z of a cycle in $S \otimes_A G_\bullet$ is still in the tight closure of the boundaries. But since S is weakly F-regular, z is now a boundary. □

This extremely powerful vanishing theorem becomes a very interesting conjecture in mixed characteristic. The case where R is local and S is the residue field of R is equivalent to the direct summand conjecture. The case where R is a direct summand of S easily yields another proof that direct summands of regular rings are Cohen-Macaulay. See §4 of **[HH11]** for a discussion. (Cf. **[PS1-2]**, **[Ho1-3, 5-7]**, **[Ro1-5]**, **[Du]** and **[EvG]** for background on these and other related questions.)

3. What is Tight Closure in Characteristic p?

We next give the definition of tight closure in characteristic p. If R is a ring, let R^o be the set of elements of R not in any minimal prime. If I is an ideal of R, we say that an element $u \in R$ is in the tight closure I^* of I if there exists $c \in R^o$ such that $cu^{p^e} \in I^{[p^e]}$ for all sufficiently large $e \in \mathbb{N}$. Here, $I^{[p^e]}$ is generated by all the elements i^{p^e} for $i \in I$. If we replace R by a free module M and I by a submodule N, we can make the same definition: the $(p^e)^{th}$ power of an element of M may be defined by taking $(p^e)^{th}$ powers in each coordinate. In the general case, one may map a finitely generated free module onto M and work with it and inverse images for the submodule N and the element u in the free module. The notion one gets is independent of how one chooses a free basis and of how one maps a free module onto M. There is a coordinate-free treatment making use of the Peskine-Szpiro or Frobenius functors of **[PS1]**.

It turns out that one may test tight closure modulo minimal primes of R. In the reduced case, one can choose the multiplier c so that $cu^{p^e} \in N^{[p^e]}$ for all $e \in \mathbb{N}$ (even $e = 0$). In the admissible case one can test tight closure in all the complete local domains of R, i.e., in all the rings obtained by localizing, completing, and killing a minimal prime.

Since this is the case it is of considerable interest to have alternative characterizations of tight closure when R is a complete local domain (several of those stated below are, in fact, valid under much weaker hypotheses). We give some of these characterizations in Theorem (3.1): we shall discuss additional characterizations later. For simplicity we only treat the case of the tight closure of an ideal of R: in all instances, there is an analogous result for modules.

For a reduced ring R of characteristic p we shall write R^∞ for $\bigcup_{e\in\mathbb{N}} R^{1/p^e}$. For a domain R, by R^+ , the *absolute integral closure* of R (cf. [**Ar2**]), we mean the integral closure of R in an algebraic closure of its fraction field.

(3.1) THEOREM. *Let (R, m, K) be a complete local domain of characteristic p. Let I be an ideal of R and let $u \in R$. Then each of the following conditions, sometimes in the presence of a supplementary hypothesis, is equivalent to the condition that u be in I^*.*

(a) *Fix a discrete $\mathbb{Z}$-valuation nonnegative on R and positive on m and extend it to a valuation of R^+ to $\mathbb{Q}$. Then there exist elements $\epsilon \in R^+ - \{0\}$ of arbitrarily small order (i.e., value under the valuation) such that $\epsilon u \in IR^+$!*

(b) *Assume that I is m-primary. With $J = I + uR$, we have that*

$$\lim_{e\to\infty} \frac{\ell(R/J^{[p^e]})}{\ell(R/I^{[p^e]})} = 1 \,!$$

(Here, "ℓ" indicates length.)

(c) *(K. E. Smith [**Sm1,2**].) Assume that I is generated by part of a system of parameters. Then $u \in IR^+$!*

(d) *There exists a big Cohen-Macaulay algebra S for R such that $u \in IS$!*

(e) *The element u is in the closure of I in the formal power series sense described in the Prologue!*

PROOF AND REMARKS. By and large we just supply references, but it is easy to see that $u \in I^*$ implies (a), for if we have $c \neq 0$ such that $cu^{p^e} \in I^{[p^e]}$ for all $e \gg 0$, then taking $(p^e)^{th}$ roots yields that $(c^{1/p^e})u \in IR^{1/p^e} \subseteq IR^\infty \subseteq IR^+$, and as $e \to +\infty$, $\operatorname{ord}(c^{1/p^e}) = (1/p^e)\operatorname{ord} c \to 0$. The converse is one of the main results of [**HH6**].

Part (b) is a consequence of Theorem (8.17) of [**HH4**] and the results of [**Mo**]: see the discussion below. We want to comment here that this result really does characterize tight closure in complete local domains of characteristic p, since an element is in the tight closure of $I \subseteq m$ if and only if it is in the tight closure of all m-primary ideals containing I.

Part (c) was proved for ideals generated by at most three parameters in [**HH11**] and in general in [**Sm1,2**] by a different method. Whether tight closure in locally excellent domains of characteristic p is, in general, simply the contracted expansion from R^+ is a tantalizing question. It does reduce to the case of complete local domains. (One knows that the contracted expansion from R^+ is contained in the tight closure by (2.1i).) This question is open even in dimension 2.

Part (d) and the peculiar fact (e) are results that are obtained in [**Ho8**] from the perspective of solid closure: we shall discuss them again in §5 and §6 (cf. (5.6) and (6.7)). □

Let us write $q = p^e$. When I is primary to m, $\ell(R/I^{[q]})$ as a function of e is called the *Hilbert-Kunz function* (cf. [**Ku**], [**Mo**], [**HaMo**]). (Again, "ℓ" denotes length.) By a result of [**Mo**], if $\dim R = d$ there is a positive real number C_I

such that for all $q = p^e$, $\ell(R/I^{[q]}) = C_I q^d + O(q^{d-1})$. If $\dim R = 1$ then C_I is a positive integer. In general, C_I is conjectured to be rational, but this is an open question even in dimension two! By Theorem (8.17) of [**HH4**], when $I \subseteq J$ are m-primary we have that $J \subseteq I^*$ iff

$$\lim_{e \to \infty} \frac{\ell(J^{[q]}/I^{[q]})}{q^d} = 0,$$

which, by Monsky's result, is equivalent to the condition that $C_I = C_J$. This in turn is equivalent to condition (b) above.

The behavior of the Hilbert-Kunz functions is quite surprising. For example, if $R = (\mathbb{Z}/5\mathbb{Z})[[x_1, x_2, x_3, x_4]]/(G)$ where $G = \sum_{i=1}^4 x_i^4$, then

$$\ell(R/m^{[5^e]}) = \frac{168}{61}(5^e)^3 - \frac{107}{61}(3^e).$$

Cf. [**HaMo**]. See also [**Ch**].

We want to comment briefly on the theory of strongly F-regular rings and the theory of test elements in characteristic p. A reduced Noetherian ring R of characteristic p is called *strongly F-regular* if R is F-finite (i.e., R is module-finite over the image $F(R) = R^p$ of the Frobenius endomorphism) and for every $d \in R^o$ there exists $q = p^e$ such that (equivalently, for all sufficiently large $q = p^e$) the inclusion of the cyclic R-module $R \cdot d^{1/q} \hookrightarrow R^{1/q}$ splits as a map of R-modules. F-finite Gorenstein rings are strongly F-regular if and only if they are weakly F-regular. It is an open question whether an F-finite weakly F-regular ring must be strongly F-regular: cf. [Wil], where this is established in dimension less than or equal to three.

J. Velez has developed an analogous theory for F-rational rings in characteristic p. See [Vel].

An element $c \in R^o$ is called a *test element* if whenever $u \in N^*{}_M$ for some pair of finitely generated modules $N \subseteq M$ (one may assume that M is free), $cu^{p^e} \in N^{[p^e]}$ for all $e \in \mathbb{N}$. Thus, c may be used in all tight closure tests. A crucial point in the development of the theory is that if R is F-finite, reduced, and $c \in R^o$ is such that R_c is strongly F-regular (in particular, if R_c is regular) then c has a power that is a test element in R, in every local ring of R, and in the completion of every local ring of R. A test element satisfying this stronger condition is called *completely stable*. See §3 of [**HH3**] and §5 of [**HH9**]. In consequence, one can show that if R is admissible and reduced then every element $c \in R^o$ such that R_c is regular has a power that is a completely stable test element.

It is natural to study ideals generated by R-sequences that consist of test elements. The issue arises in trying to prove the results of §6 of [**HH10**]. It was in this context that it was first realized that:

(3.2) THEOREM. *Let I be an ideal of a local domain R such that I is generated by the elements of a regular sequence consisting of test elements. Then $IR^+ \cap R = I^*$.*

(There is now a much better result: see (3.1c) and [**Sm1,2**].) The proof depended on an Equational Lemma that was eventually published in [**HH7**], where the circle of ideas that had arisen in [**HH10**] was used to prove the existence of big Cohen-Macaulay algebras. We shall return to this subject in §5, but we first explain how to define tight closure in equal characteristic 0.

4. What Is Tight Closure in Equal Characteristic Zero?

We begin by discussing a family of characteristic p examples. Suppose that we let $R = K[X,Y,Z]/(X^3+Y^3+Z^3) = K[x,y,z]$, where K is a field of positive characteristic p. We assume that $p \neq 3$. We want to observe that z^2 is in $(x,y)^*$ regardless of the value of p. In fact, $x(z^2)^q \in (x,y)^{[q]} = (x^q, y^q)$ for all q, because if $2q = 3k+r$ where $r \in \{1,2\}$ we can rewrite xz^{2q} as $\pm x(x^3+y^3)^k z^r$ and it turns out that $xx^{3i}y^{3(k-i)} \in (x^q, y^q)$ for all i, $0 \le i \le k$, since $(3i+1)+(3k-3i) = 3k+1 \ge 2q-1$, so that one of the exponents is at least q. This is an interesting example, because it turns out that z is not in the tight closure of (x,y) but it is in the regular closure (and, of course, in the integral closure). See §5 of [**HH8**].

We shall adopt the following conventions: If R_A is an A-algebra, and M_A is an A-module, u_A is in R_A or M_A, and $A \to B$ is a homomorphism, we write R_B, M_B, u_B for $B \otimes_A R_A$, $B \otimes_A M_A$ and $1 \otimes u_A$ (in R_B or M_B, as the case may be), respectively. If $N_A \subseteq M_A$ we write $\langle N_B \rangle$ for $\operatorname{Im}(N_B \to M_B)$.

Now consider the $\mathbb{Z}$-algebra $R_{\mathbb{Z}} = \mathbb{Z}[X,Y,Z]/(X^3+Y^3+Z^3) = \mathbb{Z}[x,y,z]$. If we imagine tight closure to be a geometric notion, then $z^2/1 \in R_{\mathbb{Q}}$ ought to be in the tight closure of $(x,y)R_{\mathbb{Q}}$ because it is true that the image of z^2 in $R_{\mathbb{Z}/p\mathbb{Z}}$ is in the tight closure of $(x,y)R_{\mathbb{Z}/p\mathbb{Z}}$ for almost all the closed fibers ("almost all" means on a Zariski dense open subset of Max Spec $\mathbb{Z}$ here): what happens at the generic fiber should be governed by what happens on a Zariski dense open subset of the closed fibers.

This is exactly the viewpoint we take to define K-tight closure over a field K of characteristic zero. Suppose that we are given a domain A contained in K and finitely generated over $\mathbb{Z}$, a ring R_A finitely generated over A, a pair of modules $N_A \subseteq M_A$, and an element $u_A \in M_A$. If $u_\kappa \in {N_\kappa}^*{}_{M_\kappa}$ for almost all maximal ideals μ of A ($\kappa = A/\mu$ will be a finite field), we decree that $u_K \in {N_K}^*{}_{M_K}$ over R_K . We require this not only for R_K: we also require that the image of u_K in $S \otimes_{R_K} M_K$ be in the tight closure, over S, of the image of $S \otimes_{R_K} N_K$ in $S \otimes_{R_K} M_K$ for all Noetherian K-algebras S to which R_K maps.

We note that the results of [**Ar1**], [**ArR**], and [**Rot**] play an important role in establishing that this notion has the right properties. (When R is not necessarily admissible, we define u_R to be in ${N_R}^*{}_{M_R}$ if u_B is in $\langle N_B \rangle^*_{M_B}$ in the sense above for all complete local domains B of R.)

The theory we have been using here is ${}^{*\mathbb{Q}}$. We do not know whether ^{*K} and ^{*L} really are different when $K \subseteq L$: quite generally, $N^{*K}{}_M \subseteq N^{*L}{}_M$. There is yet another notion, ^{*EQ}, which we may describe briefly as follows: let R be a local ring of a finitely generated $\mathbb{Q}$-algebra, and write R as C_P where C is a finitely generated $\mathbb{Z}$-algebra and P is a prime disjoint from $\mathbb{Z} - \{0\}$. Let $N_C \subseteq M_C$ be finitely generated C-modules and $u_C \in M_C$. Let $B(p)$ be the localization of C/pC at the multiplicative system of nonzerodivisors in C/pC on $C/(P + pC)$. Then $u_R \in {N_R^{*EQ}}_{M_R}$ if for almost all p, $u_{B(p)} \in {M_{B(p)}^{*}}_{N_{B(p)}}$.

More generally, if S is admissible, we define $u \in {N^{*EQ}}_M$ if for every prime Q of S, $(S_Q, M_Q, N_Q, u/1)$ may be obtained as $S \otimes_R (R, M_R, N_R, u_R)$ for some homomorphism $R \to S$, where R is as above and $u_R \in {N_R^{*EQ}}_{M_R}$ in the sense defined in the preceding paragraph.

It turns out that if S is a Noetherian K-algebra, where K is a field of characteristic 0, then $N^{\mathbb{Q}} \subseteq N^{*K} \subseteq N^{*EQ}$ when $N \subseteq M$ are finitely generated S-modules. We do not know whether $N^{*\mathbb{Q}} = N^{*EQ}$.

5. Big Cohen-Macaulay Algebras

We recall that M is a *balanced big Cohen-Macaulay module* for a local ring (R, m) if every system of parameters for R is a regular sequence on M. (The term *balanced* is used in [**Sh**].) This implies that $mM \neq M$. Henceforth we omit the term "balanced": by a *big Cohen-Macaulay module* we always mean a balanced big Cohen-Macaulay module. It may be the case that a big Cohen-Macaulay module M is also an R-algebra, in which case we shall say that M is a *big Cohen-Macaulay algebra* for R.

As noted in the last paragraph of §3, the investigation of tight closures of parameter ideals eventually lead to the following theorem, which is the main result of [**HH7**]. (Note: R^+ is defined in the discussion preceding (3.1).)

(5.1) THEOREM. *Let R be an excellent local domain of characteristic p. Then R^+ is a big Cohen-Macaulay algebra for R.*

This is false in equal characteristic zero if the Krull dimension of R is at least three!

The applications of (5.1) are numerous. One gets a new equicharacteristic proof of the Faltings connectedness theorem (cf. (6.8) of [**HH7**]). The result is used in [**Ab1**] to show that if a finite free complex $G_\bullet$ over a reduced admissible local ring of characteristic p satisfies the conditions of (2.1h) and is minimal (i.e., the entries of the matrices of the maps are in the maximal ideal) then the ranks of the free modules occurring depend only on what $H_0(G_\bullet)$ is (the complex is referred to as a *finite phantom free resolution* of $H_0(G_\bullet)$). Note that the isomorphism class of such a complex is not uniquely determined by $H_0(G_\bullet)$. See [**Ab1**]. (5.1) also implies that locally excellent rings of characteristic p that are direct summands of all their module-finite extensions are Cohen-Macaulay (this may actually characterize F-regularity: it does for locally excellent Gorenstein rings, by Theorem (6.7) of [**HH10**]): see §7 of [**HH7**]. Moreover, a graded version of (5.1) can be used to bound the numbers of generators of certain homogeneous primes (cf. §8 of [**HH11**]) and to give some rather odd constraints on pairs of systems of parameters (cf. §9 of [**HH11**]).

But the point that we really want to focus on here is that (5.1) gives a weakly functorial method of obtaining big Cohen-Macaulay algebras: if we have any map of domains $R \to S$, one can extend it (not uniquely) to a map $R^+ \to S^+$. To see this, factor the map as the composition of a surjection $R \twoheadrightarrow \operatorname{Im} R$ and an inclusion $\operatorname{Im} R \hookrightarrow S$. It suffices to solve the problem for each map separately. In the case of an inclusion, $R \hookrightarrow S$, note that an algebraic closure of the fraction field of S will contain an algebraic closure of the fraction field of R, which gives the result at once. In the case of a surjection, we may assume that $S = R/P$ with P prime. Since R^+ is integral over R it has a prime ideal Q lying over P. Then $S = R/P$ injects into R^+/Q, and R^+/Q is an integral extension of S such that every monic polynomial factors into linear factors. This implies that R^+/Q may be identified with S^+.

Call a local homomorphism of local rings $R \to S$ *permissible* if the induced homomorphism $h : \hat{R} \to \hat{S}$ has the property that for every minimal prime $\mathfrak{q}$ of $\hat{S}$ such that $\dim \hat{S}/\mathfrak{q} = \dim \hat{S}$, there is a minimal prime $\mathfrak{p}$ of $\hat{R}$ such that $\mathfrak{p} \subseteq h^{-1}(\mathfrak{q})$ and $\dim \hat{R}/\mathfrak{p} = \dim \hat{R}$. One of the main results of [**HH11**] uses the weak functoriality of R^+, Theorem (5.1), and a rather non-routine application of Artin approximation [**Ar1**] to show the following:

(5.2) THEOREM. *One may assign to every equicharacteristic local ring R a big Cohen-Macaulay algebra $\mathcal{B}(R)$ in such a way that whenever $R \to S$ is a permissible local homomorphism it extends to a homomorphism $\mathcal{B}(R) \to \mathcal{B}(S)$.*

This yields at once a proof that direct summands (or pure subrings) of regular rings are Cohen-Macaulay. Recall that $R \to S$ is *pure* if $R \otimes_R M \to S \otimes_R M$ is injective for every R-module M. This implies that for every ideal I of R, the contraction of IS to R is I.

(5.3) COROLLARY. *If S is an equicharacteristic regular ring and R is pure in S, then R is Cohen-Macaulay.*

SKETCH OF PROOF. One can reduce easily to the case where $R \to S$ is local. Since S is regular, $\mathcal{B}(S)$ is faithfully flat over S, which implies that S is pure in $\mathcal{B}(S)$. Then R is pure in $\mathcal{B}(S)$. Since we have $R \to \mathcal{B}(R) \to \mathcal{B}(S)$ this implies that R is pure in $\mathcal{B}(R)$. Since a system of parameters for R is a regular sequence in $\mathcal{B}(R)$, it is a regular sequence in R. □

Similarly:

(5.4) THEOREM. *If R is a domain module-finite over an equicharacteristic regular ring A and $R \to S$ is a homomorphism to a regular ring, then for every finitely generated A-module M the map $Tor_i^A(M, R) \to Tor_i^A(M, S)$ is zero for all $i \geq 1$.*

SKETCH OF PROOF. (Cf. Theorem (4.1) of [**HH11**].) One reduces to the case where the rings are local. Let τ indicate $\operatorname{Tor}_i^A(M, \_)$. Then we have a factorization $\tau(R) \to \tau(S) \to \tau(\mathcal{B}(S))$. Since S is pure in $\mathcal{B}(S)$ (as before) it suffices to show that the composite is zero. But the map $\tau(R) \to \tau(\mathcal{B}(S))$

also factors $\tau(R) \to \tau(\mathcal{B}(R)) \to \tau(\mathcal{B}(S))$. Since $\mathcal{B}(R)$ is a big Cohen-Macaulay module for R, it is also a big Cohen-Macaulay module for A, and so flat over A. It follows that $\tau(\mathcal{B}(R)) = 0$. □

This result does not really depend on the fact that S is regular. It suffices to know that after localizing S at a maximal ideal, the ring one obtains is pure in a big Cohen-Macaulay algebra to which a big Cohen-Macaulay algebra for R can be mapped.

To make this idea precise, we define a 0th *level big Cohen-Macaulay algebra* for an equicharacteristic local ring S to be the same as a big Cohen-Macaulay algebra. We define an $(n+1)^{th}$ *level big Cohen-Macaulay algebra* C for S to be one such that every permissible local homomorphism $R \to S$ extends to a map from some n^{th} level big Cohen-Macaulay algebra for R to C. We define C to be an ωth *level big Cohen-Macaulay algebra* for S if it is an nth level big Cohen-Macaulay algebra for S for all $n \in \mathbb{N}$. We define an equicharacteristic Noetherian ring S to be *weakly* CM^n*-regular*, where $n \in \mathbb{N} \cup \omega$, if each of its local rings at a maximal ideal is pure in an nth level big Cohen-Macaulay algebra. If $n = 0$ this just means that S is Cohen-Macaulay. For $n \geq 1$ this notion is not transparent but we do have:

(5.5) THEOREM. *The following results hold for equicharacteristic Noetherian rings:*

(a) *Theorem (5.4) remains valid if S, instead of being assumed regular, is instead assumed to be pure in a first or higher level big Cohen-Macaulay algebra for S.*
(b) *A weakly CM^n-regular ring for $n \geq 1$ is normal (and Cohen-Macaulay).*
(c) *An equicharacteristic local ring R is weakly CM^n-regular if and only if it is pure in $\mathcal{B}(R)$.*
(d) *If R is a locally excellent ring of characteristic p and R is weakly F-regular then R is CM^ω-regular.*
(e) *If R is an affine K-algebra in which every ideal is tightly closed in the sense of $*^K$, then R is weakly CM^ω-regular.*

COMMENTS ON THE PROOF AND REFERENCES. For (a), the proof of Theorem (5.4) goes through without essential change. See Theorem (4.12) of [**HH11**]. Part (b) is Theorem (4.9d) of [**HH11**]. Part (c) is Proposition (5.5.e) of [**HH11**]. Parts (d) and (e) are parts (g) and (h) of Theorem (4.9) of [**HH11**]. □

FURTHER REMARKS. We want to emphasize here that once one realizes that there are two similar vanishing theorems for maps of Tor, (5.5a) here and Theorem (2.3), one in which S is assumed to be admissible and weakly F-regular, the other in which S is assumed to be CM^n-regular for some $n \geq 1$, it is obvious that there ought to be some connection between the two properties. What is more, one starts to suspect that tight closure may be a contracted expansion from some sort of big Cohen-Macaulay algebra! Parts (d) and (e) above provide some evidence in this direction. Further evidence is provided by the following result:

(5.6) THEOREM. *Let R be an excellent equicharacteristic local ring. Let $N \subseteq M$ be finitely generated R-modules and let $u \in M$.*

(a) *Suppose that $\hat{R}$ is equidimensional and $\mathcal{B} = \mathcal{B}(R)$. If $u_{\mathcal{B}} \in \langle N_{\mathcal{B}} \rangle$ then $u \in N^*{}_M$.*

(b) *If R has characteristic p and $\hat{R}$ is a domain, then $u \in N^*$ if and only if there exists a big Cohen-Macaulay algebra B for R such that $u_B \in \langle N_B \rangle$.*

(c) *If R is a complete local domain of characteristic 0 and $u \in N^{*EQ}{}_M$ (in particular, if $u \in N^*{}_M$) then there exists a big Cohen-Macaulay algebra B for R such that $u_B \in \langle N_B \rangle$. (See also Theorem (6.6).)*

PROOF. Part (a) is part of Theorem (5.12) of [**HH11**]. Part (b), which has already been stated in a special case in Theorem (3.1d), is Theorem (11.1) of [**Ho8**]: it is proved from the perspective of solid closure. Part (c) is the main part of Theorem (11.4) of [**Ho8**]. All three arguments are lengthy and difficult. □

It is a challenging problem to refine these results connecting various kinds of tight closure with contracted expansions from big Cohen-Macaulay algebras. Notice that this circle of ideas offers the possibility of characterizing tight closure in equal characteristic zero in a way that does not make any reference to reduction to characteristic p.

Here is a tantalizing example.

(5.7) EXAMPLE. Let $R = K[[X,Y,Z]]/(X^3+Y^3+Z^3) = K[[x,y,z]]$, where K is a field of characteristic different from 3. We have seen that in all cases, $z^2 \in (x,y)^*$. What we want to show here is that, as one might expect, in all cases, if B is any higher level Cohen-Macaulay algebra for R ("higher" meaning level at least 1) then $z^2 \in (x,y)B$. (5.6a) then implies, as we already know, that $z^2 \in (x,y)^*$. Thus, we are establishing the converse of (5.6a) in an example.

The trick is to take new variables s, t and form the subring

$$R_0 = K[[xs, ys, zs, xt, yt, zt]] \subseteq R[[s,t]].$$

There is a local K-homomorphism of R_0 to R that sends the formal generators xs, ys, zs, xt, yt, zt to x, y, z, x, y, z respectively. In R_0, xs, yt, $xt+ys$ is part of a system of parameters: call these elements u, v, w. One has the relation $(zs)(zt)(w) = (zt)(zt)u + (zs)(zs)v$. It follows that in any big Cohen-Macaulay algebra B_0 for R_0, $(zs)(zt) \in (xs,yt)B_0$. But some such B_0 maps to the first level or higher big Cohen-Macaulay algebra B for R, and so $z^2 \in (x,y)B$, as claimed. □

We conclude this section with the following closely related result, which does not appear elsewhere.

(5.8) THEOREM. *Let (R,m,K) be a local ring, let I be an ideal of R, and let $u \in R$. Suppose that there exists an integral extension T of R such that $u \in IT$. Then for any first or higher level big Cohen-Macaulay algebra S for R, $u \in IS$.*

In particular, if R is a domain, then if $u \in IR^+$ it follows that $u \in IS$ for every higher level big Cohen-Macaulay algebra S for R.

PROOF. Let Λ denote either a field contained in R or else the localization of $\mathbb{Z}$ at the prime $p\mathbb{Z}$, where p is the residual characteristic of R. In either case, there is an obvious local homomorphism $\Lambda \to R$. Let $a_1, \ldots, a_n$ denote generators for I, and let $x_1, \ldots, x_n$ be indeterminates over Λ. Write $u = \sum_{i=1}^n a_i\theta_i$, where every θ_i is an element of T (hence, integral over R). For every i choose an equation of integral dependence

$$g_i(z_i) = z_i^{\delta(i)} + \sum_{j=0}^{\delta(i)-1} r_{ij} z_i^j = 0$$

for θ_i over R, where the $r_{ij} \in R$. Let v_{ij} be new indeterminates, and let

$$G_i(z_i) = z_i^{\delta(i)} + \sum_{j=0}^{\delta(i)-1} v_{ij} z_i^j$$

in the polynomial ring $W = \Lambda[x_i, z_i, v_{ij}, t]$, where t is yet another indeterminate. There is a homomorphism $W \to T$ extending the map $\Lambda \to R$, and which sends the x_i to the a_i, the v_{ij} to the r_{ij}, the z_i to the θ_i, and t to 0. This homomorphism kills all the G_i, and so induces a homomorphism of the quotient $V = W/(G_i : i)$ to T. Let $y = \sum_{i=1}^n z_i x_i$. Note that y maps to u under $W \to T$. The subring

$$U = \Lambda[x_i, y, v_{ij}, t, z_i t]$$

of V therefore maps to R, and, consequently, so does $U_{\mathcal{N}}$, where $\mathcal{N}$ is the contraction of the maximal ideal of R to U.

Let $B = \Lambda[x_i, y_{ij}, t]$, which is regular, let $\mathcal{M}$ denote the contraction of the maximal ideal of R to B, which contains the x_i and t, and let $C = B_{\mathcal{M}}$, which is a regular local ring. Then V is module-finite over B, and the monomials $z_1^{\nu_1} \cdots z_n^{\nu_n}$ in the z's such that $\nu_i < \delta(i)$ for all i are a free basis for V over B. Since $B \subseteq U \subseteq V$, the ring U is module-finite and torsion-free over B. It follows that $U_{\mathcal{N}}$ is a localization of $U_{\mathcal{M}}$ (which is semilocal) at an ideal lying over $\mathcal{M}C$. Here, $U_{\mathcal{M}}$ is module-finite and torsion-free over the regular local ring C. The ideals of $U_{\mathcal{M}}$ lying over $\mathcal{M}C$ are precisely the maximal ideals of $U_{\mathcal{M}}$, and since C is normal and $U_{\mathcal{M}}$ is torsion-free over C, we have going-down as well as going-up holding for the inclusion $C \hookrightarrow U_{\mathcal{M}}$. It follows that all the maximal ideals of $U_{\mathcal{M}}$ have the same height, equal to the dimension of C. Moreover, it follows that the image of any system of parameters for C is a system of parameters in each of the local rings of $U_{\mathcal{M}}$ at a maximal ideal, and, hence, in particular, in $U_{\mathcal{N}}$. Since $x_1, \ldots, x_n, t$ is a regular sequence of B contained in $\mathcal{M}$, it is part of a system of parameters for C and, hence, its image in $U_{\mathcal{N}}$ is part of a system of parameters for $U_{\mathcal{N}}$.

Now, the completion of $U_{\mathcal{M}}$ with respect to $\mathcal{M}C$ will be the product of the completions of its local rings at its maximal ideals, one of which will be $(U_{\mathcal{N}})\widehat{\ }$.

Since $U_{\mathcal{M}}$ is torsion-free as a C-module, so is its $(\mathcal{M}C)$-adic completion $\mathcal{U}$ as a $\hat{C}$-module, and it follows that every nonzero $\hat{C}$-submodule of $\mathcal{U}$ has the same dimension as $\hat{C}$. This shows that $(U_{\mathcal{N}})^{\wedge}$ is equidimensional (and has no embedded primes). It follows that $U_{\mathcal{N}} \to R$ is permissible, and we have already seen that $x_1, \ldots, x_n, t$ is part of a system of parameters in $U_{\mathcal{N}}$.

Since S is a first level big Cohen-Macaulay algebra for R, there exists a big Cohen-Macaulay algebra $\mathcal{B}$ for $U_{\mathcal{N}}$ such that $U_{\mathcal{N}} \to R \to S$ also has the factorization $U_{\mathcal{N}} \to \mathcal{B} \to S$. In $U_{\mathcal{N}}$ we have the relation

$$yt = \sum_{i=1}^{n} x_i(z_i t),$$

by the definition of $y = \sum_{i=1}^{n} x_i z_i$. But then, since $x_1, \ldots, x_n, t$ is part of a system of parameters in $U_{\mathcal{N}}$, we must have that $y = \sum_{i=1}^{n} \beta_i x_i$ in $\mathcal{B}$. When we then map further to S, we obtain that the image of u in S is in $(a_1, \ldots, a_n)S = IS$, as required, since, under the map $U_{\mathcal{N}} \to R$, y maps to u and x_i to a_i for all i. □

(5.9) REMARK. There is a corresponding result for modules instead of ideals. One can reduce to looking at submodules of finitely generated free modules, where the statement and proof are entirely similar.

(5.10) CONCLUDING DISCUSSION. For simplicity, we assume that (R, m, K) is a complete local domain of equal characteristic and consider only the case of ideals. Let I be an ideal of R and let $u \in R$.

In characteristic p, we know that $u \in I^*$ if and only u is in the expansion of I to a big Cohen-Macaulay algebra for R. However, it need not be in the expansion of I to every big Cohen-Macaulay algebra: R itself can be Cohen-Macaulay without having the property that every ideal is tightly closed. Theorem (5.8) suggests the possibility that $u \in I$ if and only if it is in the expansion of I to every first or higher level big Cohen-Macaulay algebra for R. This will be the case if and only if it turns out that in characteristic p, $I^* = IR^+ \cap R$. Indeed, since R^+ is such an algebra, this statement forces I^* to be contained in $IR^+ \cap R$. On the other hand, Theorem (5.8) shows every element of $IR^+ \cap R$ is forced into the expansion of I to every first or higher level big Cohen-Macaulay algebra.

It is possible that in equal characteristic zero, I^{*eq} is simply what is in the contracted expansion of I from every first or higher level big Cohen-Macaulay algebra. Theorem (5.12) of **[HH11]** implies that any element that is in such a contracted expansion for all first or higher level big Cohen-Macaulay algebras is in I^{*eq}. (It actually shows this for a specific ω level big Cohen-Macaulay algebra. This is also stated as Theorem (5.6a) here.) Moreover, Theorem (11.4) of **[Ho8]** shows, at least, that every element of I^{*eq} (and even of I^{*EQ}) is in the expansion of I to some big Cohen-Macaulay algebra: what we don't know is whether requiring the algebra to be first or higher level automatically ensures that it is "big enough." (This is stated as Theorem (5.6c) here.) Example (5.7) simply presents one instance in which we can verify that this is, indeed, the case.

6. Solid Closure

In this section we shall discuss a closure operation defined in a characteristic-free manner that coincides with tight closure for admissible rings of characteristic p.

(6.1) SOLID MODULES AND ALGEBRAS: SOME BASIC PROPERTIES. If R is a domain we call an R-module M *solid* if $\mathrm{Hom}_R(M, R) \neq 0$. We call an R-algebra S *solid* if it is solid as an R-module. In the case of a solid algebra it is always possible to choose a nonzero R-module map $h : S \to R$ such that $h(1) \neq 0$ ([**Ho8**], Proposition (2.1d)). One reason for the name "solid" is that solid algebras tend not to adjoin fractions: if S is a solid R-algebra, $a, b \in R$, $b \neq 0$, and $a \in bS$, then a/b is integral over R; if θ is an element of a solid R-algebra and θ is algebraic over R, then θ is integral over R ([**Ho8**], Proposition (2.9)).

Moreover, if $R \to S$ is any homomorphism of domains with R Noetherian and M is a solid R-module then $S \otimes_R M$ is a solid S-module ([**Ho8**], Theorem (2.12)).

Also, it turns out that a module S over a complete local domain (R, m) of dimension d is solid if and only if the highest local cohomology $H^d_m(S) \neq 0$ (cf. [**Ho8**], Corollary (2.4)).

In consequence, a big Cohen-Macaulay algebra over a complete local domain is always solid. On the other hand, so is any module-finite extension algebra.

(6.2) SOLID CLOSURE. Let R be a Noetherian ring, let $N \subseteq M$ be finitely generated R-modules, and let $u \in M$. We shall say that $u \in N^\bigstar$ (or $N^\bigstar{}_M$ if greater precision is needed), the *solid closure* of N in M, if for every complete local domain B of R, there exists a solid B-algebra S such that $u \in \langle N_S \rangle$. It suffices to consider complete local domains arising from completions at maximal ideals of R. (Note: the notation $N^\blacksquare$ was used for solid closure in a previous version of this manuscript.)

(6.3) A CONNECTION WITH TIGHT CLOSURE. It is easy to see that if R is a Noetherian domain of characteristic p, I is an ideal, and $u \in IS$ for a solid R-algebra S, where $u \in R$, then $u \in I^*$. For we may choose an R-module map $h : S \to R$ such that $h(1) = c \neq 0$, and the fact that $u \in IS$ implies that $u^q \in I^{[q]}S$ for all $q = p^e$ (we may apply Frobenius). Applying h to both sides yields that $cu^q \in I^{[q]}$ for all q. For complete local domains, the converse is true: see Theorem (8.6) of [**Ho8**]. From this one can deduce:

(6.4) THEOREM. *If R is an admissible ring of characteristic p, $N \subseteq M$ are finitely generated R-modules, and $u \in M$, then $u \in N^*$ if and only if $u \in N^\bigstar$.*

Does the new notion coincide with one of the equal characteristic zero tight closure notions already discussed? The answer was not known until recently, when a calculation of Paul Roberts [**Ro6**] revealed that in the rings $K[[x_1, x_2, x_3]]$ and $K[x_1, x_2, x_3]$, where K is a field of equal characteristic zero, one has that

$x_1^2x_2^2x_3^2 \in (x_1^3, x_2^3, x_3^3)^\bigstar$. This means that in equal characteristic zero, solid closure is "too big." See also (7.22-5) in [**Ho8**].

As for mixed characteristic, the verdict is not yet in concerning whether solid closure gives the "right" notion or, at least, a sufficiently good notion, although Roberts' example is discouraging. Note, however that (x_1^3, x_2^3, x_3^3) (and every other ideal generated by monomials in a polynomial ring over $\mathbb{Z}$) is solidly closed. Cf. (13.1), (13.3) and (13.5-7) of [**Ho8**].

Even if it should turn out that solid closure is too big in mixed characteristic, it is still much smaller than integral closure, and so a result that asserts that a certain ideal is in the solid closure of another is much sharper than the corresponding result for integral closure. It remains possible that in any regular local ring of mixed characteristic, every ideal is solidly closed. This is true in dimension two, but not known in dimension three or higher. (As already mentioned, it is true for regular rings of characteristic p and false for regular rings of equal characteristic zero of dimension 3.) Settling the issue for regular rings of mixed characteristic is a primary target.

We next want to record some of the properties of solid closure. We follow the format of Theorem (2.1), including the numbering of the parts.

(6.5) THEOREM (PROPERTIES OF SOLID CLOSURE). *Let R, S be Noetherian rings, $R \to S$ a homomorphism, and let $N \subseteq M \subseteq Q$ be finitely generated R-modules. Solid closures are taken in M unless otherwise specified. u denotes an element of M, and I an ideal of R.*

(a) *$u \in N^\bigstar{}_M$ if and only if $u + N \in 0^\bigstar{}_{M/N}$.*

(b) *$N^\bigstar{}_M \subseteq N^\bigstar{}_Q$; $N^\bigstar{}_Q \subseteq M^\bigstar{}_Q$; $(N^\bigstar)^\bigstar = N$.*

(c) *(Persistence of $\bigstar$.) If $u \in N^\bigstar$ then $1 \otimes u \in \langle N_S\rangle^\bigstar{}_{M_S}$.*

(d) *$u \in N^\bigstar$ iff for every complete local domain B of R, $1 \otimes u \in \langle N_B\rangle^\bigstar{}_{M_B}$.*

(e) *If R is a regular ring of dimension at most two or of characteristic p, $N^\bigstar = N$ for all $N \subseteq M$. (Question: is this true for regular local rings of mixed characteristic?)*

(f) *(Generalized Briançon-Skoda theorem) If R is equicharacteristic and I has at most n generators then for all $k \in \mathbb{N}$, $\overline{I^{n+k}} \subseteq (I^{k+1})^\bigstar$. (Question: is this true in all Noetherian rings?)*

(g) *(Capturing colon ideals) If R is an equicharacteristic and locally excellent ring, $I = (x_1, \dots, x_n)R$ is such that $\operatorname{minheight} I \geq n$, and $I_{n-1} = (x_1, \dots, x_{n-1})R$, then we have $I_{n-1} :_R x_nR \subseteq I^\bigstar_{n-1}$. (Question: is this true for local rings in mixed characteristic and does one get the better result that $I^\bigstar_{n-1} :_R x_nR = I^\bigstar_{n-1}$ in mixed characteristic?)*

(h) *The analogue of (2.1h) is valid for locally excellent rings of equal characteristic with solid closure replacing tight closure.*

(i) *If S is a module-finite extension of R then $u \in N^\bigstar{}_M$ over R if and only if $u_S \in \langle N_S\rangle^\bigstar{}_{M_S}$ over S. Moreover, the contraction of IS to R is contained in $I^\bigstar$.*

PROOF. For (a), (b), (c), (d) we refer the reader to (5.3) and (5.6) of [**Ho8**]. Part (e) is Theorem (7.20) of [**Ho8**], and (f), (g), (h) follow in characteristic p from corresponding facts for tight closure and Theorem (6.4). In equal characteristic 0 they follow from corresponding facts for tight closure and the comparison Theorem (6.6) below. (i) follows from (5.9b) of [**Ho8**]. □

Wherever we have asked a question in the statement of Theorem (6.5) there is work to be done. We now return to the problem of comparing solid closure in equal characteristic zero with tight closure. Our main result along these lines is Theorem (11.4) of [**Ho8**] (part of this was mentioned here earlier, as Theorem (5.6c)), which asserts:

(6.6) THEOREM. *Let R be any Noetherian ring containing $\mathbb{Q}$. Let $N \subseteq M$ be finitely generated R-modules and let $u \in M$.*

(a) *If R is a complete local domain and $u \in N^{*EQ}{}_M$ (hence, if $u \in N^*{}_M$) then there is a big Cohen-Macaulay algebra B over R such that $u_B \in \langle N_B \rangle$.*

(b) *$N^{*EQ}{}_M \subseteq N^{\bigstar}{}_M$. Hence, $N^*{}_M \subseteq N^{\bigstar}{}_M$.*

REMARKS ON THE PROOF. A big Cohen-Macaulay algebra over a complete local domain is automatically solid. Thus, (a) yields (b) at once when R is a complete local domain, and the general case of (b) reduces to the complete local domain case. The proof of (a) is lengthy. □

As mentioned earlier, the result of [**Ro6**] shows that, in general, the inclusion $N^{*EQ}{}_M \subseteq N^{\bigstar}{}_M$ is strict, since in equal characteristic zero every ideal of a regular ring is tightly closed in the sense of ^{*EQ}, but not necessarily solidly closed.

(6.7) BACK TO THE PROLOGUE: A FORMAL POWER SERIES CRITERION. Let $I = (a_1, \ldots, a_n)$ be an ideal of a complete local domain R, and let a be any element of R. It is not hard to see that if there is any solid R-algebra S such that $a \in IS$, then it must be $S = R[X_1, \ldots, X_n]/(F)$ where $F = a - a_1X_1 - \cdots - a_nX_n$, because S maps to any other algebra T such that $a \in IT$ and if S maps to T and T is solid then S is solid. We refer to S as a *generic forcing algebra*. We can study the R-homomorphisms $h : S \to R$ by considering instead the formal power series in auxiliary variables y_i obtained from h by letting

$$g_h = \sum_{\nu \in \mathbb{N}^n} h(X^\nu) y^\nu \in R[[y_1, \ldots, y_n]],$$

where $h(X^\nu)$ represents the value of h on the image of $X_1^{\nu_1} \cdots X_n^{\nu_n}$ in S. Let f be the power series

$$a y_1 \cdots y_n - \sum_{i=1}^{n} a_i y_1 \cdots y_{i-1} y_{i+1} \cdots y_n$$

(labeled (#) in the Prologue). It turns out not to be difficult to see (cf. Theorem (9.3) of [**Ho8**]) that g_h corresponds to a homomorphism that kills the ideal (F) if and only if $g_h f$ is *special* in the sense that it does not contain a term in which all of the y_i occur with positive exponent. It then follows that $a \in I^\bigstar$ if and only if f has a nonzero multiple that is special. This gives yet another characterization of tight closure for complete local domains of characteristic p. The examples considered in the Prologue really ask whether $x_1^2x_2^2x_3^2$ is in the solid closure of (x_1^3, x_2^3, x_3^3) in various regular local rings in which x_1, x_2, x_3 is a regular system of parameters. The answer is not known for the first example ($R = V[[x_2, x_3]]$) mentioned in the Prologue, and is "no" (respectively, "yes") for $K[[x_1, x_2, x_3]]$ when the field K has positive characteristic (respectively, characteristic zero).

Although we have discussed generic forcing algebras and the power series criterion only for ideals, these do extend to pairs of modules. We refer the reader to §9 of [**Ho8**].

(6.8) EXAMPLES IN REGULAR RINGS, AND AN APPLICATION OF THE BRIANÇON-SKODA THEOREM. On the other hand, it is easy to see that $x_1x_2x_3$ is not in the solid closure of (x_1^2, x_2^2, x_3^2) in the ring $K[[x_1, x_2, x_3]]$ (where K has characteristic 0) using the persistence of solid closure. Simply map $K[[[x_1, x_2, x_3]] \to K[[x, y]]$ so as to send x_1, x_2, x_3 to $x+y$, $x-y$, and xy, respectively. If $x_1x_2x_3 \in (x_1^2, x_2^2, x_3^2)^\bigstar$ we find that $(x^2+y^2)(x^2-y^2)xy \in (x^4+y^4, x^2y^2)^\bigstar$ in $K[[x, y]]$, and since we know that every ideal is solidly closed in a regular ring of dimension two, it follows that if $x_1x_2x_3 \in (x_1^2, x_2^2, x_3^2)^\bigstar$ then $(x^2+y^2)(x^2-y^2)xy \in (x^4+y^4, x^2y^2)$ in $K[[x, y]]$, which is false. What is being used here is that $f = x^2 + y^2$, $g = x^2 - y^2$, $h = xy$ gives an example where $fgh \notin (f^2, g^2, h^2)$ in $K[[x, y]]$ when K has characteristic zero. This answers the second part of the "Problem for thought" given at the end of the Prologue.

It is tempting to try to use the same method to show that

$$x_1^2x_2^2x_3^2 \notin (x_1^3, x_2^3, x_3^3)^\bigstar$$

Thus, one winds up seeking f, g, h in a regular ring of dimension two such that $f^2g^2h^2 \notin (f^3, g^3, h^3)$. (This was the first part of the "Problem for thought" following the Prologue.) But this is impossible! I don't know an elementary proof, but one can use the Briançon-Skoda theorem to show that in any regular ring of dimension two, $f^2g^2h^2 \in (f^3, g^3, h^3)$. The issue is local. The ideal (f^3, g^3, h^3) will have a minimal reduction (extend the residue field if necessary: this will not affect any relevant issue) generated by at most two elements, say, (u, v). Now, $(fgh)^3 = f^3g^3h^3$ shows that fgh is integral over (f^3, g^3, h^3) and, hence, over (u, v), so that the Briançon-Skoda theorem implies that $(fgh)^2$ (which is in $\overline{(u, v)^2}$) is in $(u, v) \subseteq (f^3, g^3, h^3)$. (We are using the version of [LS], which is valid for all regular rings.) Similarly, in any regular ring of dimension at most d, the product of the dth powers of any $d+1$ elements is in the ideal generated $(d+1)^{th}$ powers of those elements.

As mentioned earlier, whether $x_1^2x_2^2x_3^2 \in (x_1^3, x_2^3, x_3^3)^\bigstar$ for a regular system of parameters in a regular local ring of mixed characteristic remains open.

★ ★ ★

Epilogue: Twenty Questions

We give here a list of twenty open questions, taken in part from a preliminary version of [**HH12**], connected with tight closure theory.

Throughout these questions, unless otherwise specified, R is a locally excellent Noetherian ring and $N \subseteq M$ are finitely generated R-modules. K, L always denote fields and $K \subseteq L$. If R is local then it has maximal ideal m and residue field K.

If $M = \operatorname{Coker}(a_{ij})$, $\mathbf{F}^e(M) = \operatorname{Coker}(a_{ij}^q)$ with $q = p^e$.

1. Does tight closure commute with localization in characteristic p? In characteristic zero? (Cf. [**AHH**] for a thorough discussion of the localization problem.)

2. Is it true that weakly F-regular rings are F-regular? This is known in the Gorenstein case (and in dimension at most three, using the results of [**Wil**]), but is open in characteristic p even if R is an affine algebra over an algebraically closed field.

3. If R is complete local, is there a positive constant integer b such that for all $e \in \mathbb{N}$, m^{bp^e} kills $H_m^0(\mathbf{F}^e(M))$? Is this true even when R is a complete local weakly F-regular ring and $M = R/I$, where I is primary to a prime P such that $\dim R/P = 1$? An affirmative answer to the second question would yield that weakly F-regular implies F-regular for locally excellent rings R. (Cf. [**AHH**].)

4. In characteristic p, is every F-finite weakly F-regular ring strongly F-regular? (Then weakly F-regular would imply F-regular even without the hypothesis F-finite.) This is known:
 (1) in the Gorenstein case and
 (2) if $\dim R \leq 3$ (cf. [**Wil**]).

5. Let R be a complete local Cohen-Macaulay domain and let J be an ideal of R that is isomorphic as a module with a canonical module for R. Fix a system of parameters $x_1, \dots, x_n$ for R. For every $q = p^e$ and every positive integer t let

$$\theta_{e,t} : J^{[q]}/(x_1^q, \dots, x_n^q) \to J^{[q]}/(x_1^{qt}, \dots, x_n^{qt})$$

be the map induced by mutliplication by $(x_1 \cdots x_n)^{qt-q}$ on $J^{[q]}$. Can one always choose a system of parameters $x_1, \dots, x_n$ and a positive integer t_0 such that $\operatorname{Ker} \theta_{e,t}$ is the same for all $t \geq t_0$? For fixed e, these kernels increase as t increases. Note that t_0 is to be independent of e. This and related problems are studied in [**Wil**]. An affirmative answer in the special case where R is F-finite and weakly F-regular would suffice to show that weakly F-regular is equivalent to strongly F-regular for F-finite rings, and that weakly F-regular is equivalent to F-regular for locally excellent rings of characteristic p.

6. Is the weakly F-regular locus open? (This is an open question for affine rings over algebraically closed fields both in characteristic p and characteristic

0.) This would follow for algebras essentially of finite type over an excellent local ring of characteristic p if weakly F-regular F-finite rings are strongly F-regular, because the strongly F-regular locus is known to be open.

7. Suppose that R has characteristic p. Let $T = \bigcup_e \mathrm{Ass}\,(\mathbf{F}^e(M)/0^*)$. Is T finite? Does it have only finitely many maximal elements? An affirmative answer would reduce the question of whether tight closure commutes with localization to the case where R is local and one is localizing at a prime P with $\dim R/P = 1$. (Cf. [**AHH**].)
(A recent example of M. Katzman shows that $S = \bigcup_e(\mathrm{Ass}\,\mathbf{F}^e(M))$ need not have only finitely many maximal elements. Katzman has also shown that there is a polynomial ring C in one variable over a field of characteristic p, a finitely generated C-algebra R, and a finitely generated R-module M such that there is no element $c \in C^o$ such that all the modules $F^e_{R_c}(M_c)$ are torsion-free over C_c.)

8. Is characteristic p tight closure for locally excellent domains the same as contraction from R^+ ? This is known for parameter ideals ([**Sm1,2**]). An affirmative answer implies that tight closure commutes with localization. This is an open question even in dimension two.

9. Does weak F-regularity deform, i.e., if R is a local domain, $x \neq 0$ and R/xR is weakly F-regular must R be weakly F-regular? (This is known when R is Gorenstein.)

10. Are affine K-algebras with rational singularities necessarily of F-rational type? (The converse is true: [**Sm1,3**].) (This is known in the graded case in dimension two ([**Fe**]) but open in general even in dimension two.)

11. For affine algebras in characteristic 0, does weakly F-regular imply weakly F-regular type? Does F-rational imply F-rational type? Are affine algebras with rational singularities of F-rational type (the converse is known: [**Sm1,3**])?

12. Are direct summands (and pure subrings) of F-rational rings F-rational?

13. Let K have characteristic 0, let A be a finitely generated $\mathbb{Z}$-subalgebra of K, and let (A, R_A, M_A, N_A) be descent data for (K, R, M, N). Suppose that N is K-tightly closed in M. Is it true that for almost all closed fibers, N_κ is tightly closed in M_κ?

14. Is tight closure over a complete local domain, in the equicharacteristic case (0 and p), simply the contracted expansion from a higher level balanced big Cohen-Macaulay algebra?

15. If R is an affine L-algebra, does ${N^{*K}}_M = {N^{*L}}_M$? More generally, if R is an arbitrary Noetherian ring containing $\mathbb{Q}$ does ${N^{*eq}}_M = {N^{*EQ}}_M$? (It would suffice to know this when R is a local ring of a finitely generated $\mathbb{Q}$-algebra.)

16. If a flat homomorphism of rings $R \to S$ has an F-regular base R and geometrically F-regular fibers, is S F-regular? This is known in good cases in characteristic p if the fibers are geometrically regular. (Cf. [**HH9**].)

17. If R is excellent, reduced, characteristic p, and of finite Krull dimension, does R have a test element? (Cf. [**Ab2**].)

18. Under mild conditions on a characteristic p ring R (e.g., if R is reduced and finitely generated as an algebra over an excellent local ring) does formation of the ideal of test elements commute with localization? With completion? With geometrically regular base change (where *geometrically regular* means flat with geometrically regular fibers)?

19. Let R be a characteristic p local ring such that one system of parameters generates a tightly closed ideal. Is R F-rational? This is known ([**HH9**] §4) if R is equidimensional.

20. Is every ideal of a regular ring solidly closed in mixed characteristic? (The mixed characteristic case implies the direct summand conjecture.) The question raised in the prologue is really the question of whether (x_1^3, x_2^3, x_3^3) is solidly closed in $V[[x_2, x_3]]$. Every ideal is solidly closed in regular rings of dimension at most two, even in mixed characteristic.

References

[Ab1] Aberbach, I., *Finite phantom projective dimension*, Amer. J. Math., to appear.

[Ab2] ———, *Test elements in excellent rings with an application to the uniform Artin-Rees property*, Proc. Amer. Math. Soc., to appear.

[Ab3] ———, *Tight closure in F-rational rings*, preprint.

[AHH] Aberbach, I., M. Hochster, and C. Huneke, *Localization of tight closure and modules of finite phantom projective dimension*, J. Reine Angew. Math. (Crelle's Journal) **434** (1993), 67-114.

[Ar1] Artin, M., *Algebraic approximation of structures over complete local rings*, Publ. Math. I.H.E.S. (Paris) **36** (1969), 23-56.

[Ar2] ———, *On the joins of Hensel rings*, Advances in Math. **7** (1971), 282-296.

[ArR] Artin, M. and C. Rotthaus, *A structure theorem for power series rings*, in Algebraic Geometry and Commutative Algebra: in honor of Masayoshi Nagata, Vol. I, Kinokuniya, Tokyo, 1988, 35-44.

[Bou] Boutot, J.-F., *Singularités rationelles et quotients par les groupes réductifs*, Invent. Math. **88** (1987), 65-68.

[BrS] Briancon, J. and H. Skoda, *Sur la clôture intégrale d'un idéal de germes de fonctions holomorphes en un point de* C^n, C. R. Acad. Sci. Paris Sér. A **278** (1974), 949-951.

[BuE] Buchsbaum, D. and D. Eisenbud, *What makes a complex exact*, J. of Algebra **25** (1973), 259-268.

[Ch] Chang, S.-t., *The asymptotic behavior of Hilbert-Kunz functions and their generalizations*, Thesis, University of Michigan, 1993..

[Du] Dutta, S. P., *On the canonical element conjecture*, Trans. Amer. Math. Soc. **299** (1987), 803-811.

[EvG] Evans, E.G. and Griffith P., *The syzygy problem*, Ann. of Math. **114** (1981), 323–333.

[Fe] Fedder, R., *A Frobenius characterization of rational singularity in 2-dimensional graded rings*, Trans. Amer. Math. Soc., to appear.

[FeW] Fedder, R. and K. Watanabe, *A characterization of F-regularity in terms of F-purity*, in Commutative Algebra, Math. Sci. Research Inst. Publ. **15**, Springer-Verlag, New York · Berlin · Heidelberg, 1989, pp. 227–245.

[Gla] Glassbrenner, D. J., *Invariant rings of group actions, determinantal rings, and tight closure*, Thesis, University of Michigan, 1992.

[HaMo] Han, C. and P. Monsky, *Some surprising Hilbert-Kunz functions*, preprint.

[Ho1] Hochster, M., *Contracted ideals from integral extensions of regular rings*, Nagoya Math. J. **51** (1973), 25–43.

[Ho2] ———, *Topics in the homological theory of modules over commutative rings*, CBMS Regional Conf. Ser. in Math. No. **24**, A.M.S., Providence, R.I., 1975.

[Ho3] ———, *Big Cohen-Macaulay modules and algebras and embeddability in rings of Witt vectors*, in Proceedings of the Queen's University Commutative Algebra Conference, Queen's Papers in Pure and Applied Math. **42**, 1975, pp. 106–195.

[Ho4] ———, *Cyclic purity versus purity in excellent Noetherian rings*, Trans. Amer. Math. Soc. **231** (1977), 463–488.

[Ho5] ———, *Some applications of the Frobenius in characteristic* 0, Bull. Amer. Math. Soc. **84** (1978), 886–912.

[Ho6] ———, *Cohen-Macaulay rings and modules*, Proc. of the International Congress of Mathematicians, Helsinki, Finland, Vol. I, Academia Scientarium Fennica, 1980, pp. 291–298.

[Ho7] ———, *Canonical elements in local cohomology modules and the direct summand conjecture*, J. of Algebra **84** (1983), 503–553.

[Ho8] ———, *Solid closure*, in Proc. of the Summer Research Conference on Commutative Algebra (held at Mt. Holyoke College, July, 1992), to appear (in these proceedings).

[HoE] Hochster, M., and J.A. Eagon, *Cohen-Macaulay rings, invariant theory, and the generic perfection of determinantal loci*, Am. J. Math. **93** (1971), 1020–1058.

[HH1] Hochster, M. and C. Huneke, *Tightly closed ideals*, Bull. Amer. Math. Soc. **18** (1988), 45–48.

[HH2] ———, *Tight closure*, in Commutative Algebra, Math. Sci. Research Inst. Publ. **15**, Springer-Verlag, New York · Berlin · Heidelberg, 1989, pp. 305–324.

[HH3] ———, *Tight closure and strong F-regularity*, Mémoires de la Société Mathématique de France, numéro **38** (1989), 119–133.

[HH4] ———, *Tight closure, invariant theory, and the Briançon-Skoda theorem*, J. Amer. Math. Soc. **3** (1990), 31–116.

[HH5] ———, *Absolute integral closures are big Cohen-Macaulay algebras in characteristic p*, Bull. Amer. Math. Soc. (New Series) **24** (1991), 137–143.

[HH6] ———, *Tight closure and elements of small order in integral extensions*, J. of Pure and Appl. Algebra **71** (1991), 233–247.

[HH7] ———, *Infinite integral extensions and big Cohen-Macaulay algebras*, Ann. of Math. **135** (1992), 53–89.

[HH8] ———, *Phantom homology*, Memoirs Amer. Math. Soc. Vol. **103**, No. **490** (1993), 1-91.

[HH9] ———, *F-regularity, test elements, and smooth base change*, preprint.

[HH10] ———, *Tight closures of parameter ideals and splitting in module-finite extensions*, preprint.

[HH11] ———, *Applications of the existence of big Cohen-Macaulay algebras*, preprint.

[HH12] ———, *Tight closure in equal characteristic zero*, in preparation.

[HR1] Hochster, M. and J.L. Roberts, *Rings of invariants of reductive groups acting on regular rings are Cohen-Macaulay*, Advances in Math. **13** (1974), 115–175.

[HR2] ———, *The purity of the Frobenius and local cohomology*, Advances in Math. **21** (1976), 117–172.

[Hu1] Huneke, C., *An algebraist commuting in Berkeley*, Mathematical Intelligencer **11** (1989), 40–52.

[Hu2] ———, *Uniform bounds in Noetherian rings*, Invent. Math. **107** (1992), 203–223.

[Hu3] ———, *Absolute integral closures and big Cohen-Macaulay algebras*, in Proc. of the 1990 Intern. Congress of Mathematicians, Kyoto 1990, Vol. I, Math. Soc. of Japan, Springer-Verlag, New York · Berlin · Heidelberg, 1991, pp. 339–349.

[Ku] Kunz, E., *On Noetherian rings of characteristic p*, Am. J. Math. **98** (1976), 999–1013.

[LS] Lipman J. and A. Sathaye, *Jacobian ideals and a theorem of Briançon-Skoda*, Michigan Math. J. **28** (1981), 199–222.

[LT] Lipman, J. and B. Teissier, *Pseudo-rational local rings and a theorem of Briançon-Skoda about integral closures of ideals*, Michigan Math. J. **28** (1981), 97–116.

[Ma] Ma, F., *Splitting in integral extensions, Cohen-Macaulay modules and algebras*, J. of Algebra **116** (1988), 176–195.

[Mat] Matsumura, H., *Commutative Algebra*, Benjamin, 1970.

[Mo] Monsky, P., *The Hilbert-Kunz function*, Math. Ann. **263** (1983), 43-49.

[PS1] Peskine, C. and L. Szpiro, *Dimension projective finie et cohomologie locale*, I.H.E.S. Publ. Math. (Paris) **42** (1973), 323–395.

[PS2] ———, *Syzygies et multiplicités*, C. R. Acad. Sci. Paris Sér. A **278** (1974), 1421–1424.

[Ro1] Roberts, P., *Two applications of dualizing complexes over local rings*, Ann. Sci. Ec. Norm. Sup. **9** (1976), 103–106.

[Ro2] ———, *Cohen-Macaulay complexes and an analytic proof of the new intersection conjecture*, J. of Algebra **66** (1980), 225–230.

[Ro3] ———, *The vanishing of intersection multiplicities of perfect complexes*, Bull. Amer. Math. Soc. **13** (1985), 127–130.

[Ro4] ———, *Le théorème d'intersection*, C. R. Acad. Sc. Paris Sér. I **304** (1987), 177–180.

[Ro5] ———, *Intersection theorems*, in Commutative Algebra, Math. Sci. Research Inst. Publ. **15**, Springer-Verlag, New York · Berlin · Heidelberg, 1989, pp. 417–436.

[Ro6] ———, *A computation of local cohomology*, in Proc. of the Summer Research Conference on Commutative Algebra (held at Mt. Holyoke College, July, 1992), to appear (in these proceedings).

[Rot] Rotthaus, C., *On the approximation property of excellent rings*, Invent. Math. **88** (1987), 39-63.

[Sh] Sharp, R. Y., *Cohen-Macaulay properties for balanced big Cohen-Macaulay modules*, Math. Proc. Cambridge Philos. Soc. **90** (1981), 229-238.

[Sk] Skoda, H., *Applications des techniques L^2 a la théorie des idéaux d'une algèbre de fonctions holomorphes avec poids*, Ann. Scient. Ec. Norm. Sup. 4ème série t. **5** (1972), 545–579.

[Sm1] Smith, K. E., *Tight closure of parameter ideals and F-rationality*, Thesis, University of Michigan, 1993.

[Sm2] ———, *Tight closure of parameter ideals*, Invent. Math., to appear.

[Sm3] ———, *F-rational rings have rational singularities*, preprint.

[Sw1] Swanson, I., *Joint reductions, tight closure, and the Briançon-Skoda theorem*, J. of Algebra **147** (1992), 128–136.

[Sw2] ———, *Tight closure, joint reductions, and mixed multiplicities*, Thesis, Purdue University, 1992.

[Vel] Velez, J., *Openness of the F-rational locus, smooth base change, and Koh's conjecture*, Thesis, University of Michigan, 1993.

[W1] Watanabe, K.-I., *Study of F-purity in dimension two*, in Algebraic Geometry and Commutative Algebra in honor of Masayoshi Nagata, Vol. II, Kinokuniya, Tokyo, 1988, pp. 791–800.

[W2] ———, *F-regular and F-pure normal graded rings*, J. of Pure and Applied Algebra **71** (1991), 341–350.

[Wil] Williams, L., *Uniform stability of kernels of Koszul cohomology indexed by the Frobenius endomorphism*, Thesis, University of Michigan, 1992; and preprint.

Department of Mathematics, University of Michigan, Ann Arbor, MI 48109-1003, USA

Contemporary Mathematics
Volume **159**, 1994

Indecomposable Canonical Modules and Connectedness

MELVIN HOCHSTER AND CRAIG HUNEKE

1. Introduction

Throughout this paper all rings are commutative, with identity, and Noetherian, unless otherwise specified. In particular, "local ring" always means Noetherian local ring, unless otherwise specified. Our objective is to prove a generalization of Faltings' connectedness theorem [Fal1, Fal2], which asserts that in a complete local domain (R, m, K) of dimension n, if $I \subseteq m$ is an ideal generated by at most $n-2$ elements, then the punctured spectrum of R/I is connected. Our result (see Theorems 3.3 and 3.6) draws the same conclusion without the hypothesis that R be a domain: we assume instead that R is complete, equidimensional (i.e., for every minimal prime p of R, $\dim R/p = \dim R$), and that one of the following conditions, which we shall prove are equivalent, holds:

a) $H^n_m(R)$ (local cohomology with support in m) is indecomposable.

b) The canonical module ω of R is indecomposable.

c) The S_2-ification of R is local.

d) For every ideal I of height two or more, Spec $R - V(I)$ is connected.

e) Given any two distinct minimal primes p, q of R, there is a sequence of minimal primes $p = p_0, \dots, p_i, \dots, p_r = q$ such that for $0 \le i < r$, the height $(p_i + p_{i+1}) \le 1$.

§2 details the properties of canonical modules for a not necessarily Cohen-Macaulay ring, as well as the process of S_2-ification. By and large the results of §2 are known, but in some cases we have not found a convenient reference. We note here only that our definition is that a canonical module ω for (R, m, K) is a finitely generated R-module such that $\mathrm{Hom}_R(\omega, E) \cong H^{\dim R}_m(R)$, where E is an injective hull for K over R. The main results are developed in §3.

1991 *Mathematics Subject Classification*. Primary 13E05, 13C99, 13D45, 13H99, 13J10.
Both authors were supported in part by grants from the National Science Foundation.
This paper is in final form and no version of it will be submitted for publication elsewhere.

2. Canonical modules and S_2-ification

It will be convenient to have a notation for the ideal that turns out to be the annihilator of the canonical module.

(2.1) DEFINITION. If R is a local ring we shall denote by $\mathbf{j}(R)$ the largest ideal which is a submodule of R of dimension smaller than $\dim R$. Then $\mathbf{j}(R)$ is nonzero if and only if some prime P of Ass R is such that $\dim R/P < \dim R$, and then $\mathbf{j}(R) \supseteq \mathrm{Ann}_R P$. Thus, $\mathbf{j}(R) = (0)$ iff R is equidimensional and unmixed (where *unmixed* means that (0) has no embedded primes). Moreover, $\mathbf{j}(R)$ consists of all elements $r \in R$ such that $\dim R/\mathrm{Ann}_R r < \dim R$.

Throughout this section $E = E_R(K)$ denotes an injective hull of the residue field of the local ring (R, m, K) and $^\vee$ denotes the exact functor $\mathrm{Hom}_R(\underline{\quad}, E)$ on R-modules. We begin by summarizing many of the known properties of canonical modules, most of which we shall need in this paper.

(2.2) REMARK. Let (R, m, K) be a local ring with $\dim R = d$.

a) If R is complete, then R has a canonical module, and any canonical module is isomorphic with $H^d_m(R)^\vee$.

b) Any two canonical modules for R are (non-canonically) isomorphic.

c) If R is a homomorphic image of a Gorenstein ring, then R has a canonical module. If $R = S/J$, where S is local, then $\mathrm{Ext}^h_S(R, S)$ is a canonical module for R, where $h = \dim S - \dim R$. More generally, if $S \to R$ is local, R is module-finite over the image of S, S is Cohen-Macaulay with canonical module ω_S, and $h = \dim S - \dim R$, then $\mathrm{Ext}^h_S(R, \omega_S)$ is a canonical module for R. In particular, if R is a module-finite extension of a regular (or Gorenstein) local ring A, then $\mathrm{Hom}_A(R, A)$ is a canonical module for R. (The same holds when R is module-finite over the image of A and the two have the same dimension.)

d) A canonical module for R must be killed by $\mathbf{j}(R)$, and is also a canonical module for $R/\mathbf{j}(R)$, while any canonical module for $R/\mathbf{j}(R)$ is a canonical module for R. Thus, R has a canonical module if and only if $R/\mathbf{j}(R)$ has a canonical module.

For parts e)-i) we let (R, m, K) be a local ring with canonical module ω.

e) The kernel of the map $R \to \mathrm{Hom}_R(\omega, \omega)$ is $\mathbf{j}(R)$. Thus, ω is faithful if and only if R is equidimensional and unmixed.

f) The module ω and its completion are both S_2. Moreover, $\mathrm{Hom}_R(\omega, \omega)$ is a commutative semilocal ring module-finite over the image of R and it is S_2 both as an R-module and as a ring in its own right. It may be identified with a subring of the total quotient ring of $R/\mathbf{j}(R)$. Moreover, its m-adic completion is S_2.

g) For every prime P of R such that $\dim R/P = \dim R$, the ring $(R/P)^\wedge \cong \hat{R}/P\hat{R}$ is equidimensional and unmixed. If $\mathbf{j}(R) = (0)$ then $\mathbf{j}(\hat{R}) = (0)$.

h) $R \to \mathrm{Hom}_R(\omega, \omega)$ is an isomorphism if and only if R is S_2 and equidimensional (the latter condition follows from S_2 if R is catenary), and also iff $\hat{R}$

is S_2. Thus, if R has a canonical module and R is equidimensional and S_2, then $\hat{R}$ is S_2.

i) If R is equidimensional, then for every prime P of R, ω_P is a canonical module for R_P.

For parts j)-l) we suppose that (R, m, K) is local with $\mathbf{j}(R) = (0)$ and let ω be a canonical module for R. Let $S = \mathrm{Hom}_R(\omega, \omega)$. Let $m_1, \ldots, m_s$ denote the maximal ideals of S. Note that ω is an S-module, precisely because $S = \mathrm{Hom}_R(\omega, \omega)$. Then:

j) Every maximal ideal of S has height equal to $\dim R$.

k) When R is complete, so that S is product of local rings S_i, one for every maximal ideal m_i of S, and ω is, correspondingly, a product of modules ω_i over the various S_i, then ω_i is the canonical module for S_i for every i.

l) The module ω is a canonical module for S in the sense that ω_Q is a canonical module for S_Q for every prime ideal Q of S.

PROOF. Parts a) - c) are standard and can be found in [HK]. The rest of the results can be found in either [G] or [A1-2]. In particular, see Theorem 3.2 and Corollary 4.3 in [A2].

(2.3) DISCUSSION. Let (R, m, K) be an equidimensional and unmixed local ring, i.e., such that $\mathbf{j}(R) = (0)$. We shall say that a ring S is an S_2-ification of R if it lies between R and its total quotient ring, is module-finite over R, is S_2 as an R-module, and has the property that for every element $s \in S - R$, the ideal $\mathcal{D}(s)$, defined as $\{r \in R : rs \in R\}$, has height at least two. We are interested in this notion because if $\mathbf{j}(R) = (0)$ and R has a canonical module ω then it has an S_2-ification, to wit, $\mathrm{Hom}_R(\omega, \omega)$ identified with a subring of the total quotient ring of R. Moreover, whenever R has an S_2-ification, it is unique. We prove several propositions in this direction.

(2.4) PROPOSITION. *Let (R, m, K) be a local ring with $\mathbf{j}(R) = (0)$ and let T be its total quotient ring. If $f \in T$ let $\mathcal{D}(f) = \{r \in R : rf \in R\}$. Let S be the subring of T consisting of all elements $f \in T$ such that ht $\mathcal{D}(f) \geq 2$. Then R has an S_2-ification if and only if S is module-finite over R, in which case S is the unique S_2-ification of R.*

PROOF. It is easy to verify that S is a subring of T containing R, since $\mathcal{D}(r) = R$ for $r \in R$ (and the height is $+\infty$), $\mathcal{D}(s \pm s') \supseteq \mathcal{D}(s) \cap \mathcal{D}(s')$, and $\mathcal{D}(ss') \supseteq \mathcal{D}(s)\mathcal{D}(s')$. Moreover, it is immediate from the way that we defined an S_2-ification that it must be contained in T. We next observe that if $S_0 \subseteq S \subseteq T$ with S_0 module-finite over R and S_0 is S_2 as an R-module then $S_0 = S$. To see this, suppose that $f \in S - S_0$. Since $\mathcal{D}(f)$ has height at least two (but cannot be R) and S_0 is a faithful R-module, we must have that there is a regular sequence x, y of length two on S_0 in $\mathcal{D}(f)$. Now $xf, yf \in R$, and so we have that $x(yf) - y(xf) = 0$ is a relation on x, y with coefficients in S_0. It follows that

$xf \in xS$, so that $xf = xs$ with $s \in S_0$. But x is a nonzerodivisor in S_0, hence in R, and so also in T, the total quotient ring of R. Thus, $f = s$, and $f \in S_0$.

Now suppose that S is module-finite over R. We must show that S is S_2. The depth of S on a height one ideal of R is at least one, since the ideal must contain a nonzerodivisor of R (we have that $\mathbf{j}(R) = 0$) and this will be a nonzerodivisor in T. Suppose that I has height at least two. Choose elements x, y in I, nonzerodivisors, such that $(x, y)R$ has height two. We claim that x, y form a regular sequence on S (and this will complete the proof). For suppose that we have a relation $xs = ys'$ with elements s and s' of S. In the total quotient ring let $f = s/y = s'/x$. Choose ideals I, I' of R of height at least two such that $Is \subseteq R$ and $I's' \subseteq R$. Then $II'(x, y)$ multiplies f into R (since $I'xf = I's'$ and $Iyf = Is$), and this ideal has height two. Thus, $f \in S$, and so $s' = xf \in xS$. □

(2.5) REMARKS. Note that if an R-algebra S' is R-isomorphic to the S_2-ification S of R, then there is a unique R-isomorphism $S' \cong S$. (For each element f of S' we can choose a nonzerodivisor $r \in R$ such $fr = r' \in R$. Then if $\phi : S' \to S$ is the isomorphism we must have

$$\phi(r') = \phi(fr) = r\phi(f),$$

which determines $\phi(f)$ uniquely.) Thus, we shall talk about S_2-ifications which are not literally subrings of the total quotient ring of R: they are always, however, canonically identifiable with such a subring.

Note also that if S is an S_2-ification of R, then we can choose finitely many generators for S as an R-module, and for each of these generators an ideal of height at least two in R that multiplies S into R. It follows that there is an ideal of height at least two in R that multiplies S into R: intersect the ideals chosen for the individual generators.

(2.6) PROPOSITION. *If (R, m, K) has an S_2-ification S then for every prime ideal P of R, S_P is an S_2-ification of R_P.*

PROOF. S_P is S_2 over R_P, module-finite, and identifiable with a subring of the total quotient ring. Any element has a unit multiple of the form $s/1$, where s is in S. Then $\mathcal{D}(s/1) \subseteq \mathcal{D}(s)_P \subseteq R_P$ has height at least two. □

(2.7) PROPOSITION. *If (R, m, K) is local, $\mathbf{j}(R) = 0$, and ω is a canonical module for R, then $R \to Hom_R(\omega, \omega)$ is an S_2-ification of R.*

PROOF. We know by 2.2 f) that $\mathrm{Hom}_R(\omega, \omega)$ is a module-finite extension of R that may be identified with a subring of the total quotient ring of R. We also know that it is S_2. Therefore, it will suffice to show that if $s \in \mathrm{Hom}_R(\omega, \omega)$, then $\mathcal{D}(s)$ has height at least two. If not, it will be contained in some height one prime P of R, and P will be in the support of

$$R/\mathcal{D}(s) \cong (R(id_\omega) + Rs)/R(id_\omega) \subseteq \mathrm{Hom}_R(\omega, \omega)/R,$$

and $\mathrm{Hom}_R(\omega, \omega)/R$ is not supported at any height one prime by 2.2 h) and i). □

3. Connectedness theorems

We first give the statement of the Faltings' connectedness theorem for complete domains (see [Fal1], [Fal2]) . In our improvement we will follow the lines of an argument given in [BR]. We will use the following result, the local Hartshorne-Lichtenbaum vanishing theorem (see [Ha], [CS], and [BH]). In its basic form, it asserts that for a complete local domain (R, m, K) of dimension n, if $I \subseteq m$ is not primary to m then $H^n_I(R) = 0$. In a more precise form, it asserts that if (R, m, K) is any complete local ring of dimension n and $I \subseteq m$ is an ideal, then $H^n_I(R) \neq 0$ if and only if there exists a minimal prime ideal p of R such that $\dim R/p = \dim R$ and $I + p$ is primary to m.

(3.1) THEOREM (FALTINGS' CONNECTEDNESS THEOREM). *Let (R, m, K) be an analytically irreducible local ring of dimension n, and let $\mathfrak{A}$ be an ideal of R generated by at most $n-2$ elements. Then the punctured spectrum of Spec $R/\mathfrak{A}$ is connected. (In other words, there do not exist ideals I, J of R such that $\mathrm{Rad}\, I \cap J = \mathrm{Rad}\,\mathfrak{A}$ and $\mathrm{Rad}(I + J) = m$ unless one of the ideals is primary to m and the other has the same radical as $\mathfrak{A}$.)*

(3.2) REMARKS. We do not even need the condition that $\mathfrak{A}$ be generated by $n-2$ or fewer elements: all that is needed is that $H^{n-1}_{\mathfrak{A}}(R) = H^n_{\mathfrak{A}}(R) = 0$, and the second condition is automatic if R is a complete local domain and $\mathfrak{A}$ is not m-primary.

Faltings' original proof was for the equicharacteristic case. In [BR] a much simpler proof was given, whose outline we shall follow here. In [HH2], §6 (see also [HH1]) it is shown that the integral closure R^+ of a complete (or excellent) local domain of positive characteristic p in an algebraic closure of its fraction field is "Cohen-Macaulay " in the sense that every system of parameters in R is a regular sequence in R^+. The characteristic p case of the Faltings' connectedness theorem can be deduced from the Cohen-Macaulay property for R^+, and then the equal characteristic 0 case also follows by the technique of reduction to characteristic p. It was this point of view which led us to suspect that the condition that R is a domain could be weakened.

Our next main objective here is to generalize so that R need not be a domain. We shall state all of our results in the complete case. In each instance, one can achieve the illusion of greater generality by starting with an arbitrary local ring and requiring that its completion satisfy the hypotheses we want.

(3.3) THEOREM. *Let (R, m, K) be a complete equidimensional local ring such that, equivalently, $H^n_m(R)$ is an indecomposable R-module or such that ω_R, its canonical module, is indecomposable. (This is automatic if R is a domain, since ω_R is then an ideal of R.) Let $\mathfrak{A}$ be a proper ideal of R generated by $n-2$ or fewer elements. Then the punctured spectrum of $R/\mathfrak{A}$ is connected.*

PROOF. If not let I, J be ideals which give a disconnection, so that $I \cap J$ has the same radical as $\mathfrak{A}$, $I + J$ is primary to m, but neither I nor J is primary to

m. The Mayer-Vietoris sequence for local cohomology then yields:

$$\cdots \to H^{n-1}_{I\cap J}(R) \to H^n_{I+J}(R) \to H^n_I(R) \oplus H^n_J(R) \to H^n_{I\cap J}(R) \to \cdots$$

and the first and last terms displayed are zero, since $I \cap J$ has the same radical as an ideal with at most $n-2$ generators. Since $I+J$ is primary to m, this yields an isomorphism:

$$H^n_m(R) \cong H^n_I(R) \oplus H^n_J(R).$$

The fact that $H^n_m(R)$ is indecomposable implies that one of the summands, say $H^n_J(R)$, is zero. But then the local Hartshorne-Lichtenbaum vanishing theorem implies that for every prime $P \in \mathcal{P}$, the set of minimal primes of R, $J+P$ is not primary to m. Let P be one of these minimal primes. The intersection of $I+P$ and $J+P$ is still, up to radicals, $\mathfrak{A}+P$, while the sum is still primary to m. Thus, applying the local Hartshorne-Lichtenbaum vanishing theorem to the domain R/P, we see that $I+P$ must be primary to m for every minimal prime P. But this implies that $\bigcap_{P\in\mathcal{P}}(I+P)$ is primary to m, and up to radicals this is the same as $I+\bigcap_{P\in\mathcal{P}} P$. Since $\bigcap_{P\in\mathcal{P}} P$ is the ideal of nilpotents, we find that I itself is primary to m, a contradiction. □

This result motivates a study of when the canonical module of a complete local equidimensional ring is indecomposable. We begin by associating a graph with such a ring.

(3.4) DEFINITION. Let R be an equidimensional local ring. We denote by Γ_R the (undirected) graph whose vertices are the minimal primes of R, and whose edges are determined by the following rule: if P, Q are distinct minimal primes of R, then $\{P,Q\}$ is an edge of Γ_R if and only if $P+Q$ has height one.

We next observe:

(3.5) PROPOSITION. *Let (R,m,K) be a local ring with canonical module ω. Suppose that $\mathfrak{j}(R)=(0)$, and let $S = Hom_R(\omega,\omega)$, the S_2-ification of R. Then:*

a) For every prime ideal P of R and Q of S, if Q lies over P then $ht\ Q = ht\ P$.

b) For every ideal I of R, height IS = height I.

c) Contraction gives a bijection of the minimal primes of S with the minimal primes of R and a bijection of the height one primes of S with the height one primes of R.

PROOF. a) To study the primes of S lying over P, we first replace R, ω, and S, by R_P, ω_P, and S_P. Thus, there is no loss of generality in supposing that P is the maximal ideal of R. But then the result follows from 2.2 j).

b) If P is a prime ideal containing I whose height is the same as that of I, then there is a prime ideal Q of S lying over P. Then Q contains IS, and so $ht\ IS \le ht\ Q = ht\ P = ht\ I$. On the other hand, if Q is a prime ideal containing

IS whose height is the same as that of IS, then Q contracts to a prime ideal P containing I, and so $ht\ IS = ht\ Q = ht\ P \geq ht\ I$.

c) Any height k prime P of R has at least one prime of S lying over it, and all such primes have height k by part a). Moreover, all height k primes of S lie over height k primes of R. Thus, it suffices to show that when $k = 0, 1$, there is at most one prime of S lying over P. But the primes of S lying over P correspond to the primes of S_P lying over PS_P, and S_P is the S_2-ification of R_P. When $\dim R_P \leq 1$, R_P is its own S_2-ification, and the result follows. □

We are now ready to prove a central result:

(3.6) THEOREM. *Let (R, m, K) be a complete local equidimensional ring with* $\dim\ R = n$. *The the following conditions are equivalent:*

a) $H^n_m(R)$ *is indecomposable.*

b) The canonical module $\omega = \omega_R$ of R is indecomposable.

c) The S_2-ification S of $R/\mathbf{j}(R)$ is local.

d) For every ideal J of height at least two, Spec $R - V(J)$ is connected.

e) Γ_R *is connected.*

PROOF. We shall prove that *a)* $\Leftrightarrow$ *b)* $\Leftrightarrow$ *c)* $\Rightarrow$ *d)* $\Leftrightarrow$ *e)* $\Rightarrow$ *c)*. The equivalence of a) and b) is clear. Now assume b). The module ω is also a canonical module for $R/\mathbf{j}(R)$ and, consequently, for the S_2-ification $\mathrm{Hom}_R(\omega, \omega) = \mathrm{Hom}_{R/\mathbf{j}(R)}(\omega, \omega)$ of $R/\mathbf{j}(R)$ as well. If the S is not local, then ω is a product of nonzero factors corresponding to the various factors rings of S, and this will yield a non-trivial direct sum decomposition of ω over R. On the other hand, if $S \cong \mathrm{Hom}_R(\omega, \omega)$ is local, it contains no idempotents other than 0, 1, and this implies that ω is indecomposable. Thus, *c)* $\Rightarrow$ *b)* as well, and we have established the equivalence of the first three conditions.

Now assume that S is local, and that I, $I' \subseteq m$ are such that $I \cap I'$ is nilpotent but $I + I'$ has height at least two. We can replace I, I' by powers and assume that $II' = 0$ but $I + I'$ has height at least two. This situation is preserved when we pass to $R/\mathbf{j}(R)$, for when R is equidimensional $\mathbf{j}(R)$ consists of nilpotents. Thus, we might as well assume that $\mathbf{j}(R) = (0)$. Next, note that $IS + I'S$ has height at least two (since its height is the same as that of $I + I'$, by the preceding proposition). Moreover, if neither I nor I' is primary to m (i.e., if neither has height equal to $\dim R$) then neither IS nor $I'S$ is primary to the maximal ideal of S.

Thus, there is no loss of generality in assuming that R is S_2. The ideal $I + I'$ will contain a regular sequence $u + u'$, $v + v'$ of length two, where u, $v \in I$ and u', $v' \in I'$. The relation $v(u + u') - u(v + v') = 0$ then shows that $u \in (u + u')R$, while $u' \in (u + u')R$ similarly. This yields $u = (u + u')a$, $u' = (u + u')b$, and so $u + u' = (u + u')(a + b)$. Since $u + u'$ is not a zerodivisor, $1 = a + b$, and it follows that at least one of a, b is a unit. Suppose that a is a unit: the other case is similar. Then $u = (u + u')a$ implies that u is a nonzerodivisor, while $uI' \subseteq II' = (0)$ then implies that $I' = 0$, and so I is primary to m. This completes the proof that *c)* $\Rightarrow$ *d)*.

We next want to see that $d) \Leftrightarrow e)$. Suppose that one has ideals I, I' such that $I \cap I'$ is nilpotent. Then we can replace I, I' by their radicals while only increasing $I + I'$. Then each of I, I' is a finite intersection of primes. For each minimal prime p of R, $p \supseteq I \cap I'$, and so p must contain either a minimal prime of I or a minimal prime of I'. Thus, p must be either a minimal prime of I or a minimal prime of I'. If we omit all non-minimal primes from the primary decomposition of I (respectively, I') and intersect the others, we get two larger ideals whose intersection is still $\mathrm{Rad}\,(0)$. Thus, it is possible to give I, I' such that $\mathrm{Rad}\,(I \cap I') = \mathrm{Rad}\,(0)$ and $I + I'$ has height two if and only if one can do this with ideals I, I' coming from a partition of the minimal primes of R into two nonempty sets, with I the intersection of the minimal primes in one set and I' the intesection of the minimal primes in the other set. If one set consists of $\{p_1, \dots, p_h\}$ and the other of $\{q_1, \dots, q_k\}$ we shall have $I = \cap_i p_i$, $I' = \cap_j q_j$, and $I + I'$ will then have the same radical as $\cap_{i,j}(p_i + q_j)$, and will have height at least two if and only if every $p_i + q_j$ has height at least two. Thus, d) fails if and only if the minimal primes can be partitioned into two nonempty sets such that no edge of Γ joins a vertex in one set to a vertex in the other, which is precisely the condition for Γ to be disconnected. Thus, $d \Leftrightarrow e)$.

Finally, we show that $e) \Rightarrow c)$. Suppose that Γ is connected. We want to prove that the S_2-ification of $R/\mathbf{j}(R)$ is local. The graph associated with $R/\mathbf{j}(R)$ is the same as that associated with R, so that we may assume that $\mathbf{j}(R) = (0)$. If the S_2-ification S of R has two or more maximal ideals, say $\mathcal{M}_1, \dots, \mathcal{M}_r$, where $r \geq 2$, for each $\mathcal{M}_j$ let $\mathcal{P}_j$ denote the set of minimal primes of S contained in $\mathcal{M}_j$. Then $\mathcal{P}_j$ is evidently non-empty. There is a bijection between the minimal primes of S and those of R, so that for each $\mathcal{P}_j$ there is a corresponding set of minimal primes $\mathcal{Q}_j$ of R. To complete the argument, it will suffice to show that if i, j are different then it is impossible to have an edge joining a vertex in $\mathcal{Q}_i$ to a vertex in $\mathcal{Q}_j$. If there were such an edge, there would be a height one prime P of R containing both a minimal prime in $\mathcal{Q}_i$ and a minimal prime in $\mathcal{Q}_j$. Then $R_P \cong S_P$, and it follows that the unique prime of S lying over P contains both a prime of $\mathcal{P}_i$ and a prime of $\mathcal{P}_j$. Let $\mathcal{M}$ be a maximal ideal of R containing P. Then $\mathcal{M}$ contains both a prime of $\mathcal{P}_i$ and a prime of $\mathcal{P}_j$, which is impossible: S is a finite product of local rings, and each prime ideal of S is therefore contained in a unique maximal ideal of S, forcing $\mathcal{M}_i = \mathcal{M} = \mathcal{M}_j$. □

When R is not complete, it is necessary to study the graph associated with the minimal primes in the completion: the domain property is frequently lost when one completes. However, the characterization in (3.6c) behaves better, as we see in (3.7) below. First note that in the sequel, if (R, m, K) is equidimensional (but possibly has $\mathbf{j}(R) \neq (0)$), by an S_2-ification of R we mean an S_2-ification for $R/\mathbf{j}(R)$.

(3.7) COROLLARY. *If (R, m, K) is an equidimensional local ring with canonical module ω and $n = \dim R$, then $H^n_m(R)$ is indecomposable if and only if the S_2-ification of R is local. In particular, if R is S_2, then $H^n_m(R)$ is indecompos-*

able.

PROOF. Killing $\mathbf{j}(R)$, if it is not zero, does not affect either issue, and so we mayassume that $\mathbf{j}(R) = (0)$. The S_2-ification of $\hat{R}$ is

$$\mathrm{Hom}_{\hat{R}}(\hat{\omega}, \hat{\omega}) \cong \hat{R} \otimes_R \mathrm{Hom}_R(\omega, \omega).$$

Equidimensionality is preserved by completion here, and the issue of whether a semilocal ring is local is not affected by completing with respect to its Jacobson radical. □

The next Proposition is well-known (e.g. see [B,5.2] which gives a much more general result), but we include a proof as it is fairly short.

(3.8) PROPOSITION. *If (R, m, K) is an excellent local, equidimensional ring then R has an S_2-ification S, and $\hat{S}$ is the S_2-ification of $\hat{R}$.*

PROOF. We may first kill $\mathbf{j}(R)$, and we henceforth suppose that it is 0. Because R is excellent, this is preserved by completion. Let S be the set of elements of the total quotient ring $\mathcal{T}(R)$ of R that are multiplied into R by an ideal of height two or more. It will suffice to show that S is module-finite over R. Note that $\mathcal{T}(R) \subseteq \mathcal{T}(\hat{R})$, since $\hat{R}$ is flat over R. If S is not module-finite over R we can choose a sequence of elements $\{s_i\}$ in S such that the sequence of R-submodules $\sum_{i=1}^{j} Rs_i \subseteq S$ is strictly increasing with j. Clearly, each $s_i \in \mathcal{T}(\hat{R})$. Since $\hat{R}$ has an S_2-ification, we can choose j so large that $\sum_{i=1}^{j} \hat{R}s_i = \sum_{i=1}^{j+1} \hat{R}s_i$. We can choose a nonzerodivisor $a \in R$ such that for $0 \le i \le j+1$, $as_i \in R$. Then $\sum_{i=1}^{j} \hat{R}as_i = \sum_{i=1}^{j+1} \hat{R}as_i$. Since $\hat{R}$ is faithfully flat over R, we find that $\sum_{i=1}^{j} Ras_i = \sum_{i=1}^{j+1} Ras_i$, and so $as_{j+1} \in \sum_{i=1}^{j} Ras_i$. Since a is a nonzerodivisor in R (and hence in S), it follows that $s_{j+1} \in \sum_{i=1}^{j} Rs_i$, a contradiction.

Thus, R has an S_2-ification, S. When we complete, since the fibers of $R \to \hat{R}$ are Cohen-Macaulay , we see that $\hat{S}$ is S_2 as an $\hat{R}$-module. It is clear that it is contained in the total quotient ring of $\hat{R}$: S/R is killed by a nonzerodivisor a in R, and so $\hat{S}/\hat{R}$ is also killed by a. Moreover, if I is an ideal of R of height at least two killing S/R, then $I\hat{R}$ kills $\hat{S}/\hat{R}$. It follows that $\hat{S}$ is the S_2-ification of $\hat{R}$. □

(3.9) PROPOSITION. *Let (R, m, K) be an excellent equidimensional local ring.*

a) The S_2-ification of R is local if and only if the S_2-ification of $\hat{R}$ is local.

b) The S_2-ification of R is local if and only if the S_2-ification of R_{red} is local.

c) If R is S_2 and $x_1, \ldots, x_k$ is a part of a system of parameters, then the S_2-ification of $R/(x_1, \ldots, x_k)R$ is local.

PROOF. a) This is immediate from (3.8).

b) Since R is excellent, $(\hat{R})_{red} \cong (R_{red})^\wedge$. Thus, we may assume that R is complete. We may also assume that $\mathbf{j}(R) = 0$. The result then follows from the fact that the graphs associated with R and R_{red} are the same.

c) The issues are unaffected by completing R and killing the nilpotents. It is easy to see that $B = R/(x_1, \ldots, x_k)R$ is again equidimensional. If the S_2-ification is not local then there is a localization of B at a prime of height at least two such that the punctured spectrum is disconnected, and this ring may be viewed as a quotient of a localization R_Q of R. But R_Q is S_2 and has a canonical module (it is a localization of a complete ring, and so is a homomorphic image of a localization of a complete regular local ring), so that Corollary 3.7 applies. The result now follows from Theorem 3.3. □

(3.10) REMARK. Even for complete domains, the fact that the S_2-ification of the local ring R is local does not imply, in general, that the S_2-ification of every local ring of R is local. Some primes of R may have more than one prime of the S_2-ification lying over them.

For example, consider

$$R = K[x, y, yz, z(z-x), z^2(z-x)] \subseteq S = K[x, y, z].$$

This extension is integral, since z satisfies $Z^2 - xZ - z(z-x) = 0$. The element z is multiplied into R by the height two ideal $(y, z(z-x))$. Now

$$P = (y, yz, z(z-x), z^2(z-x))R \subseteq R$$

is a height two prime, but two prime ideals of S lie over it: $(y, z)S$ and $(y, z-x)S$.

If we complete R, S at their homogeneous maximal ideals both rings remain domains. $\hat{S}$ is the S_2-ification of $\hat{R}$, and is local. However, there are two primes of $\hat{S}$ lying over $\hat{P} = P\hat{R}$, and so the S_2-ification of $\hat{R}_{\hat{P}}$ is not local.

(3.11) REMARK. Suppose that one is trying to give an elementary proof of Faltings' connectedness theorem in the generality we have obtained here, perhaps without using local cohomology. It would suffice to prove that if R is a complete reduced S_2 local ring and x is a single parameter, then the S_2-ification of R/xR is local. The connectedness theorem can be reduced to the case of parameters, and if one knows the single fact stated above, one can carry through an induction on the number of parameters. However, it is quite possible that the case of a single parameter is no easier than the general case. We next note:

(3.12) PROPOSITION. *If (R, m, K), (S, n, K) are two complete equidimensional local rings with algebraically closed coefficient field K and the S_2-ifications of R, S are local then so is the S_2-ification of $T = R\hat{\otimes}_K S$.*

PROOF. We may assume that R, S are reduced, and so is T. We also note that when R, S are domains, then T is a domain. It follows that every minimal prime of T has the form $p\hat{\otimes}_K S + R\hat{\otimes}_K q$ where p is a minimal prime of R and q is a minimal prime of S. (Any prime of T will contract to some prime P of R, and also to some prime Q of S. Hence, it must contain $P\hat{\otimes}_K S + R\hat{\otimes}_K Q$, which is the kernel of the map $R\hat{\otimes}_K S \twoheadrightarrow R/P\hat{\otimes}_K S/Q$. This immediately shows that the minimal primes are a subset of the ideals $p\hat{\otimes}_K S + R\hat{\otimes}_K q$ for p minimal in R and q minimal in S. Since it is easy to see that these ideals are mutually incomparable, they are all minimal primes.) Let Γ, Γ' be the graphs associated with R, S, respectively. The vertices of the graph associated with $R\hat{\otimes}_K S$ are in bijective correspondence with the set $\Gamma \times \Gamma'$. There is an edge from (p, q) to (p', q') if and only if ht $\left((p+p')\hat{\otimes}_K S + R\hat{\otimes}_K (q+q')S\right) = 1$, which happens iff either $\mathrm{ht}(p+p') = 1$ and $q = q'$ or $p = p'$ and $\mathrm{ht}(q+q') = 1$. But then, in the graph associated with $R\hat{\otimes}_K S$, with its vertices identified with $\Gamma \times \Gamma'$, we have that each subgraph $\Gamma \times \{q\}$ is connected for every $q \in \Gamma'$, and each subgraph $\{p\} \times \Gamma'$ is connected for every $p \in \Gamma$. It follows that $\Gamma \times \Gamma'$ is connected. □

(3.13) REMARKS ON THE GRADED CASE. Now suppose that R is a finitely generated $\mathbb{N}$-graded K-algebra with $R_0 = K$. Let m be the homogeneous maximal ideal of R. If R is a domain, so is its completion (with respect to the homogeneous maximal ideal), since its completion has a filtration with respect to which the associated graded ring is R, which is a domain. This implies that the homogeneous primes of R remain prime when we complete. The minimal primes (in fact, all associated primes) of R are homogeneous, and so correspond to the minimal primes of the completion. If R is equidimensional one can check whether $H^n_m(R)$ is indecomposable by checking whether the graph associated with the minimal primes of R is connected: It is not necessary to complete, since the completion will have the same graph. One then gets an immediate family of corollaries of the connectedness theorems given here for intersections of projective varieties. One can also apply the technique of reduction to the diagonal to prove theorems: it may be desirable in that case to assume that the field is algebraically closed, so that products of irreducible components remain irreducible.

It is also worth noting that one can give a graded resolution of R over a polynomial ring. Using Ext to compute the canonical module then produces a graded canonical module. The automorphisms of it yield a "global" S_2-ification S of R. However, while the ring S is a graded module over R, it need not have the property that S_0 is K: when S decomposes, one has non-trivial idempotents in S_0.

REFERENCES

[A1] Y. Aoyama, *On the depth and the projective dimension of the canonical module*, Japan. J. Math. **6** (1980), 61-66.

[A2] ———, *Some basic results on canonical modules*, J. Math. Kyoto Univ. **23** (1983), 85-94.

[B] M. Brodmann, *Finiteness of ideal transforms*, J. Algebra **63** (1980), 162-185.

[BH] M. Brodmann and C. Huneke, *A quick proof of the Hartshorne-Lichtenbaum vanishing theorem*, preprint.

[BR] M. Brodmann and J. Rung, *Local cohomology and the connectedness dimension in algebraic varieties*, Comm. Math. Helv. **61** (1986), 481–490.

[CS] F. W. Call and R. Y. Sharp, *A short proof of the local Lichtenbaum-Hartshorne theorem on the vanishing of local cohomology*, Bull. London Math. Soc. **18** (1986), 261–264.

[EGA] A. Grothendieck (with the collaboration of J. Dieudonné), *Éléments de géometrie algébrique*, Chapitre IV, vol. 24, I.H.E.S. Publ. Math. Paris, 1965, pp. 1–231.

[Fal1] G. Faltings, *A contribution to the theory of formal meromorphic functions*, Nagoya Math. J. **77** (1980), 99–106.

[Fal2] ———, *Some theorems about formal functions*, Publ. of R.I.M.S. Kyoto **16** (1980), 721–737.

[G] A. Grothendieck (notes by R. Hartshorne), *Local Cohomology*, Lect. Notes Math. vol. 41, Springer-Verlag, Berlin, 1967.

[Ha] R. Hartshorne, *Cohomological dimension of algebraic varieties*, Annals of Math **88** (1968), 403–450.

[HK] J. Herzog, E Kunz, et al., *Der kanonische Modul eines Cohen-Macaulay Rings*, Lect. Notes in Math. vol. 238, Springer-Verlag, Berlin, 1971.

[HH1] M. Hochster and C. Huneke, *Absolute integral extensions are big Cohen-Macaulay algebras in characteristic p*, Bull. A.M.S. **24** (1991), 137–143.

[HH2] ———, *Infinite integral extensions and big Cohen-Macaulay algebras*, Annals of Math **135** (1992), 53–89.

[Mat] H. Matsumura, *Commutative Algebra*, W. A. Benjamin, Inc., New York, 1970.

Department of Mathematics, University of Michigan, Ann Arbor, MI 48109-1003, USA

Department of Mathematics, Purdue University, West Lafayette, IN 47907 USA

Contemporary Mathematics
Volume **159**, 1994

Multiplicities in graded rings
I : The general theory.

D. KIRBY AND D. REES

ABSTRACT. This paper is an extension of the paper Kirby[2]. In that paper a multiplicity function, denoted here by $e(G/G'|E)$, was associated with a pair G, G' of finitely generated graded extensions of a local ring Q, both rings being generated by elements of degree $1, G'$ being a sub-ring of G and the Q-module G_1/G'_1 being of finite length. E ranges over finitely generated G-modules and, as a function of E, $e(G/G'|E)$ is an additive function whose values are non-negative integers. In the present paper this is extended by first allowing G, G', E to be graded by sets R of q non-negative integers. Then $e(G/G'|E)$ is replaced by a form $e(G/G'|E|W)$ in a set $W = (w_1, \dots, w_q)$ of q indeterminates, this form being of degree D equal to the dimension of G. $e(G/G'|E|W)$ is as a function of E an additive function and its coefficients are non-negative integers. A degree function $d(G/G'|E|x|W)$ is also introduced, this being a function of E, x, where x ranges over the homogeneous elements x of G which satisfy $\dim(G/xG) = D - 1$. This function is expressible as a linear combination of valuations. These results are used to develop a multiplicity theory of graded ideals of a simply graded extension A of a local ring Q.

Introduction

We consider first a local ring (Q, m, k, d), that is, Q has maximal ideal m, residue field k, and is of Krull dimension d. Suppose that I is an m-primary ideal of Q generated by elements $(a_1, \dots, a_n)$. Finally let E be a finitely generated Q-module. Now let t be an indeterminate over Q and let $R(I)$ denote the graded sub-ring $Q[ta_1, \dots, ta_n]$ of the ring $Q[t]$ graded in the obvious way. Let $E[t]$ denote the graded $Q[t]$-module consisting of all polynomials $e(t) = \sum e_r t^r$ with coefficients in E and $E(I)$ denote the $R(I)$-module contained in $E(I)$ consisting

1991 *Mathematics Subject Classification.* Primary 13A02, 13D25, 13D40.
Key words and phrases. multiplicity, degree function.
This paper is in final form and no version of it will be submitted for publication elsewhere.

of those polynomials $e(t)$ whose coefficients e_r satisfy the stronger condition that $e_r \in I^r E$. Then there exists a polynomial $f(I|r)$ in r of degree at most d, such that, for r sufficiently large, $f(I|r)$ is equal to the length of the Q-module $(E[t]/E(I))_r$, this length being finite. If we now write the coefficient of r^d in $f(I|r)$ as $e(Q|I|E)/d!$, then $e(Q|I|E)$ is the multiplicity of I with respect to E, and, for fixed I, is an additive function of E, when E takes values in the category of finitely generated Q-modules.

In [2], theorem 6, one of us generalised the above approach to multiplicities, by replacing $Q[t], R(I)$ by a pair of finitely generated graded Q-algebras $G, G' \subseteq G$, both these algebras being generated by elements of degree 1, and the rings satisfying the condition that the Q-module G_1/G'_1 has finite length. The dimension $\dim G$ of G is defined to be the maximum of the heights of the *relevant* prime ideals of G, and the spread of G is defined to be $1 + \dim(G/mG)$. They will be denoted by $D, a(G)$ respectively. We impose on G, G' the condition that $a(G') \leq D$. We now consider two graded modules E, E', where E is a finitely generated G-module, and E' is a finitely generated G'-module contained in E. We subject E, E' to the condition that the length $\lambda(E_r/E'_r)$ of the Q-module $\lambda(E_r/E'_r)$ is finite for r large. Then, again, we can find a polynomial $f(E/E'|r)$ in r, of degree at most D, such that $\lambda(E_r/E'_r) = f(E/E'|r)$ if r is large, and we will write the coefficient of r^D as $e(G/G'|E)/D!$. As the notation suggests, this coefficient is independent of the choice of E' and is, further, an additive function of E, when E takes values in the category of finitely generated graded G-modules.

We now come to the main purpose of this paper. This is to extend the results of [2] to the case where the rings G, G' are graded not by a degree taking integer values, but by a degree whose values are sequences of q non-negative integers $R = (r(1), \ldots, r(q))$, i.e., the rings are multigraded. The resulting theory bears a similar relation to the theory of the multiplicities $e(G/G'|E)$ for simply graded rings G, G' as the theory of mixed multiplicities of m-primary ideals initiated by B.Teissier in [11], and developed further in [7], bears to the theory of multiplicities $e(Q|I|E)$. We assume that, if we define the total degree $|R|$ to be $\sum r(i)$, both G and G' are generated by elements of total degree 1, and, further, we impose on G, G' the condition that the Q-modules G_R/G'_R have finite length for all R. We then say that the pair (G, G') is admissible. We can define the dimension $\dim G = D$ and spread $a(G)$ of G in a similar way to the simply graded case. Again we impose on G, G' the additional condition that $D \geq a(G')$, and term the pair acceptable if this condition is satisfied. In fact, we will make more use of a stronger condition, that $\dim G \geq a(G)$, when we say that G is acceptable.

Next, with the assumption that (G, G') is acceptable, we introduce a pair of multigraded modules E and $E' \subseteq E$, E being a finitely generated G-module and E' being a finitely generated G'-module. These two modules are restricted to satisfy the condition that the Q-modules E_R/E'_R have finite length if R is large ($R >> 0$), where this means that $r(1), \ldots, r(q)$ are all sufficiently large. If this condition is satisfied, then we say that the pair of modules (E, E') is (G, G')-admissible.

The main content of the theorem referred to above as extended to the multigraded case is that, if the pair of rings (G, G') is acceptable, and the pair of modules (E, E') is (G, G')-admissible, then there exists a polynomial $f(E/E'|R)$ in R, whose degree is at most equal to D, such that the length $\lambda(E_R/E'_R)$ of E_R/E'_R is equal to $f(E/E'|R)$ if R is sufficiently large. In the present paper, the polynomial $f(E/E'|R)$ is often replaced by a polynomial $H(E/E'|W)$ which is defined as follows. We consider the power series

$$\Lambda(E/E'|Z) = \sum \lambda(E_R/E'_R) Z^R,$$

where $Z^R = z_1^{r(1)} \ldots z_q^{r(q)}$ and the sum is over those R for which $\lambda(E_R/E'_R)$ is finite. Then there is a unique polynomial $H(E/E'|W)$ in the set $W = (w_1, \ldots, w_q)$, where $w_i = (1 - z_i)^{-1}$, such that, when $w_1 \ldots w_q H(E/E'|W)$ is expanded as a power series in Z, the coefficient of Z^R for $R >> 0$ is equal to the coefficient $\lambda(E_R/E'_R)$ of Z^R in $\lambda(E/E'|Z)$. $H(E/E'|W)$ is termed the Hilbert polynomial of the pair E/E'. The degree of this polynomial is at most D, and our main concern is with the sum of the terms of degree $D = \dim G$ in $H(E/E'|W)$. This sum is written as $e(G/G'|E|W)$ and termed the multiplicity form of the G-module E with respect to the pair of rings G, G'. Again, it is independent of the choice of E'. Then $e(G/G'|E|W)$ is, as a function of E, an additive function on the category of finitely generated multigraded G-modules and has other properties. Most of these are contained in the following result, which is part v) of theorem 4.5 below.

Let P range over the set of graded prime ideals of G such that $\dim(G/P) = \dim G$. *Then, for each P, the homogeneous ring of fractions G_P is an artin ring, and the homogeneous module of fractions E_P is a finitely generated G_P-module and so has finite length $\lambda_P(E)$. Then $\lambda_P(E)$ is an additive function on the category of finitely generated multigraded G-modules taking non-negative integer values, and we can write*

$$e(G/G'|E|W) = \sum \lambda_P(E) e(G/G'|G/P|W).$$

The sections earlier than section 4, in which the above result appears, are devoted to the development of the necessary background material. Section 5 is devoted to the related theory of degree functions, which we describe in more detail below, and which is new in the simply graded case of [2]. We conclude, in sections 6 to 8, with the development of a multiplicity theory for suitable ideals in a graded Q-algebra. We now turn to a description of the contents of the individual sections of the paper.

The first section is devoted to the development of the theory of joint reductions of finitely generated multigraded Q-algebras. For more detail the reader is referred to the body of the paper.

The second section develops the power series approach to the theory of Hilbert functions of multi-graded modules which is used in this paper.

After these two preliminary sections, the theory of the multiplicity form referred to above is developed in section 4 using Koszul complexes, for which the necessary results have been described in section 3.

As stated earlier, the fifth section has as its object the development of a theory of degree functions in this situation analogous to the theory of degree functions of ideals in local rings. In general, the degree function $d(G/G'|E|x|W)$ is defined for all x in G satisfying $\dim G/xG = D-1$ by the equation

$$d(G/G'|E|x|W) = e(G_x/G'_x|E/xE|W) - e(G_x/G'_x|0 :_E x|W),$$

where $G_x = G/xG$, $G'_x = G'/G' \cap xG$, and E/xE, $0 :_E x$ are considered as G_x-modules. To develop it properly, we have to assume that $a(G) < D$. This function satisfies an equation

$$d(G/G'|E|x|W) = \sum \lambda_P(E) d(G/G'|G/P|x|W).$$

In addition it satisfies the equation

$$d(G/G'|E|xy|W) = d(G/G'|E|x|W) + d(G/G'|E|y|W).$$

We now, for simplicity of exposition, restrict the further discussion of degree functions to the case where G is a simply graded domain and $E = G$. The restriction to the simply graded case implies that both the multiplicity and the degree function are forms in one variable and hence can be replaced by integers, which we will denote by $e(G/G')$ and $d(G/G'|x)$.

There is an equation relating the degree function to certain valuations on the common homogeneous field of fractions of G and G'. These valuations are integer valued, take non-negative values on Q, and positive values on the maximal ideal m of Q and will be referred to as m-valuations. If v is such a valuation, it determines a canonical graded valuation on G', whose minimum value on G'_r is zero for all r. This has a unique graded extension to G which will also be denoted by v. The equation referred to is then

$$d(G/G'|x) = \delta(x)e(G/G') + \sum d(G/G'|v)v(x),$$

where $\delta(x)$ is the degree of x, and $d(G/G'|v)$ is referred to as a degree coefficient.

As already indicated, the remaining three sections apply the results of the earlier sections to develop a theory of multiplicities of suitable ideals in a finitely generated graded extension A of Q. In sections 6 and 7, the suitable ideals are those termed admissible ideals. An admissible ideal J is one satisfying the two conditions that, a), it is generated by a set of homogeneous elements all of the same degree $\sigma = \sigma(J)$ and b), that A_σ/J_σ has finite length. If J satisfies a) it is said to be level of degree σ. If $D = \dim A$, the theory defines a multiplicity function $e(A|J_1,\ldots,J_D)$ of D admissible ideals, which is symmetric, takes non-negative integer values, and satisfies a logarithmetic equation

$$e(A|J_1,\ldots,J_DJ'_D) = e(A|J_1,\ldots,J_D) + e(A|J_1,\ldots,J'_D).$$

In section 7, the results of section 5 are similarly used to construct a degree function $d(A|J_1,\ldots,J_{D-1}|x)$ of $D-1$ admissible ideals, which satisfies an equation

$$d(A|J_1,\ldots,J_{D-1}|x) = \delta(x)e(A|J_1,\ldots,J_{D-1},\Xi) + \sum d(A|J_1,\ldots,J_{D-1}|v)v(x),$$

where Ξ is the ideal of A generated by the elements of degree 1,v ranges over the set of m-valuations and $d(A|J_1,\ldots,J_{D-1}|v)$ is a symmetric function of the $D-1$ variables $J_1,\ldots,J_{D-1}$, takes non-negative values, and satisfies a logarithmetic equation similar to the one above. A special case is the following. Suppose that we put l of the ideals $J_1,\ldots,J_{D-1}$ equal to Ξ and the rest equal to the same admissible ideal J. Then, if we denote the resulting values of $d(A|J_1,\ldots,J_{D-1}|x)$ and $d(A|J_1,\ldots,J_{D-1}|v)$ by $d_l(A|J|x)$ and $d_l(A|J|v)$, and define $e_l(A|J)$ to be $e(A|J_1,\ldots,J_D)$, if l of the variables equal Ξ and the rest equal J, we obtain

$$d_l(A|J|x) = \delta(x)e_{l+1}(A|J) + \sum_v d_l(A|J|v)v(x).$$

Finally in section 8 we consider as an example of the theory of sections 6 the case where Q is a regular local ring of dimension 1, A is the ring $S = Q[X_1,\ldots,X_m]$ and I is an admissible ideal of degree 1. The basic result relates the sequence $e_l(S|I)(l = 0,1,\ldots)$ to the invariants of the module S_1/I_1, and shows that the only limitations on the sequence $e_l(S|I)(l = 0,1,\ldots)$ in this case is that its members, its forward differences, and its second order forward differences are all non-negative and are ultimately zero.

1.Complete and Joint reductions.

The purpose of this section is to present the theory of complete and joint reductions in a form convenient for applications. We recall that Ooshi, in [5], defined reductions of graded rings, and we follow his example and consider complete and joint reductions of multigraded rings. This approach yields considerable simplifications of the original treatment in Rees[7].

We commence by considering some matters of notation. We will be concerned below with the set $\mathbf{N}^q$ of sequences $R = (r(1),\ldots,r(q))$ of q non-negative integers. We will often be concerned with statements $P(R)$ which hold if "R is sufficiently large", or "for almost all R". The first of these statements will mean that there exists a sequence of integers $r_0(1),\ldots,r_0(q)$ such that $P(R)$ holds if $r(i) \geq r_0(i)$, $(i = 1,\ldots,q)$. We will write $R >> 0$ in place of "R is sufficiently large ". "$P(R)$ holds for almost all R" will mean that that integers $r_0(1),\ldots,r_0(q)$ exist such that $P(R)$ holds when at least one of the conditions $r(i) > r_0(i)$ holds. In this section we will only be concerned with the first of these conditions.

Next, we will write $|R|$ for $\sum r(i)$ and, finally, $\delta(i)$ for the sequence of q integers $(\delta(i,1),...,\delta(i,q))$ defined by $\delta(i,j)=1$ if $j=i$ and 0 if $j\neq i$.

We now describe the multi-graded rings G we will consider. These will be referred to as admissible rings. These rings will be graded by the elements of $\mathbf{N}^q$, that is G is the direct sum of additive groups G_R where R ranges over $\mathbf{N}^q$, and, for each R, R', $G_{R+R'} \supseteq G_R.G_{R'}$. The ring G_0 will be a local ring (Q,m,k,d), where this notation indicates that Q has maximal ideal m, residue field k, and is of Krull dimension $\dim Q = d$. We assume that $G_{\delta(i)}$ is a finitely generated Q-module for each i, and, finally, for each R,

$$G_R = {G_{\delta(1)}}^{r(1)} \cdots {G_{\delta(q)}}^{r(q)}.$$

This implies that, if $\mathbf{y}_i = (y(i,j)|j=1,...,m(i))$ is a set of generators of $G_{\delta(i)}$ as a Q-module, and $\mathbf{y}$ is the union of the sets $\mathbf{y}_i$, then $G = Q[\mathbf{y}]$.

Alternatively, if $\mathbf{Y} = (Y(i,j)|i=1,...,q; j=1,...,m(i))$ is a set of indeterminates over Q, G is a homomorphic image of the multi-graded ring $Q[\mathbf{Y}]$.

We now fix an admissible multigraded ring G. For each i we denote by $\Xi_i(G)$, or simply by Ξ_i, the ideal $\mathbf{y}_i G$ of G. We will term a graded ideal of G irrelevant if its radical contains the product $\Xi_1...\Xi_q$. Alternatively, J is irrelevant if and only if J contains G_R for $R >> 0$.

An ideal which is not irrelevant will be termed a relevant ideal.

We will also consider multigraded G-modules, which we will usually assume to be finitely generated. If E is such a module, then E will be the direct sum of finitely generated G_0-modules E_R satisfying $G_{R'}.E_R \subseteq E_{R+R'}$ for all R, R' while the fact that E is finitely generated can be expressed in the form that for each i, $1 \leq i \leq q$, there exists an integer $l(i)$ such that, if $r(i) \geq l(i)$,

$$E_{R+\delta(i)} = G_{\delta(i)}.E_R,$$

but we will often only need the weaker condition that this holds if $R >> 0$.

It is convenient to allow the degrees $r(i)$ of elements of E to be negative. However, the fact that E is finitely generated will imply that the degrees of non-zero homogeneous elements will be bounded below.

Unless otherwise stated, ideals of G and G-modules considered below will be assumed to be multi-graded. We will also use the term homogeneous instead of multi-graded.

We define the annihilator $\mathrm{Ann}_G E$ as usual, i.e., the set of elements a of G such that $aE = 0$. However, we will occasionally use the ultimate annihilator of E, whose definition now follows.

DEFINITION. The ultimate annihilator $u(E)$ of E is the set of elements a of G such that $aE_R = 0$ if $R >> 0$. Alternatively,

$$u(E) = \mathrm{Ann}_G E : (\Xi_1 ... \Xi_q)^n$$

if n is sufficiently large.

We note that the prime ideals associated with $u(E)$ are precisely the relevant prime ideals associated with $\mathrm{Ann}_G E$.

DEFINITIONS. An ideal J of G is termed irrelevant with respect to E if $J + u(E)$ is an irrelevant ideal of G.

A module E whose annihilator is irrelevant will be termed a trivial module. In particular, if the zero ideal of G is irrelevant, G is said to be trivial.

A straightforward argument shows that J is irrelevant with respect to E if and only if $(JE)_R = E_R$ if $R >> 0$.

LEMMA 1.1. *J is irrelevant with respect to E if and only if $J + P$ is irrelevant for each prime ideal P minimal over $u(E)$.*

PROOF. Let $J' = \cap(J + P)$. Then J' and $J + u(E)$ have the same radical . Hence $J + u(E)$ is irrelevant if and only if J' is irrelevant. But as the intersection of finitely many irrelevant ideals is irrelevant, J' is irrelevant if and only if each of the ideals $J + P$ is irrelevant.

We next turn to modules of fractions E_M, where M is a multiplicatively closed set of homogeneous elements of G. Unless otherwise stated, this will mean the set of fractions x/m ($m \in M$), where x is a homogeneous element of E of the same degree as m. This is not a G-module but a finitely generated G_M-module where G_M is defined in the same way as E_M.

An exception to the above is that, if P is a relevant prime ideal of G, G_P will denote the ring G_M, where M is the set of homogeneous elements in G but not in P. This ring has Krull dimension equal to the height of P and its residue field is $(G/P)_M$ and will be referred to as the homogeneous field of fractions of G/P. It is a finitely generated extension of the field of fractions of $G_0/G_0 \cap P$. Its transcendence degree over this field will be denoted by $t(P)$. Similarly we define E_P to be the module E_M where M is again the set of homogeneous elements of G not in P.

We now record two elementary properties of a prime ideal P of G which is maximal in the set of relevant prime ideals of G. These are that $P \cap G_0 = m$ and that $t(P) = 0$.

Next we consider the projective scheme $\mathrm{Proj}(G)$ determined by G and the sheaf of modules determined by a finitely generated multigraded G-module E. The points of $\mathrm{Proj}(G)$ will be taken to be in 1-1 correspondence with the relevant prime ideals of G, and the stalk of the structure sheaf $O(G)$ of $\mathrm{Proj}(G)$ over the point corresponding to a relevant prime ideal P of G is the local ring G_P. The stalk over P of the sheaf corresponding to the module E is E_P. Note that, if $i(E)$ denotes the sub-module of E consisting of elements whose annihilator is irrelevant, then E and $E/i(E)$ will determine the same sheaf of modules. While we will not make much use of these ideas in the paper, the geometric ideas involved underlie some of the definitions and methods used below, in particular the definition of dimension which follows.

DEFINITIONS.

a) If G is not trivial, the dimension $\dim G$ of G is defined to be $\mathrm{Max}(\mathrm{ht}P)$, where P ranges over the relevant prime ideals of G. It is clear that we can replace relevant by maximal relevant in this definition. If G is trivial, $\dim G = -1$. We define $\dim E$ to be $\dim(G/\mathrm{Ann}_G E)$.
b) The spread $a(E)$ of E is defined to be $\dim(E/mE) + 1$.
c) We define $D(E)$ to be

$$\mathrm{Max}(t(P) + \dim(Q/P \cap Q))$$

where P ranges over all relevant prime ideals minimal over $\mathrm{Ann}_G E$.

We now observe that the definition of $\dim G$ is essentially the definition of the geometric dimension of $\mathrm{Proj}(G)$. It therefore depends on the grading of G, and, in particular, differs from the Krull dimension of the ring derived from G by ignoring the grading. For example, if G is the ring $k[X_1, \dots, X_q]$, where k is a field, and we grade G by sets of q non-negative integers $R = (r(1), \dots, r(q))$ by taking G_R to be the vector space of dimension 1 with basis $X_1^{r(1)} \cdots X_q^{r(q)}$ for all R, then the only relevant graded ideal of G is the zero ideal. Hence $\mathrm{proj}(G)$ will consist of a single point and G has dimension 0.

We now consider some elementary properties of these definitions.

First it is clear that

$$\dim E = \mathrm{Max}(\dim(G/P)),$$

where P ranges over the prime ideals minimal over $\mathrm{Ann}_G E$.

Next, since $\mathrm{Ann}_G(E/mE)$ and $\mathrm{Ann}_G E + mG$ have the same radical, it follows that

$$a(E) = a(G/\mathrm{Ann}_G E).$$

This implies that

$$a(E) = \mathrm{Max}(a(G/P)),$$

where P ranges over the prime ideals P of G minimal over $\mathrm{Ann}_G E$.

A simple consequence of these two statements is that, if

$$0 \to E' \longrightarrow E \longrightarrow E'' \to 0$$

is an exact sequence of G-modules, then

$$\dim E = \mathrm{Max}(\dim E', \dim E'')$$

and

$$a(E) = \mathrm{Max}(a(E'), a(E'')).$$

For

$$\mathrm{Ann}_G E' \cap \mathrm{Ann}_G E'' \supseteq \mathrm{Ann}_G E \supseteq \mathrm{Ann}_G E'.\mathrm{Ann}_G E''.$$

Hence a prime ideal P of G is minimal over $\mathrm{Ann}_G E$ if and only if it is minimal in the set of prime ideals minimal over either $\mathrm{Ann}_G E'$ or $\mathrm{Ann}_G E''$. The two statements are now immediate.

Also, if $p_1, \ldots, p_s$ are the prime ideals minimal over $\mathrm{Ann}_G(E/mE)$,

$$a(E) = \mathrm{Max}(a(G/p_i)).$$

Finally we link the notion of the spread of a graded ring to that of the analytic spread of an ideal of a local ring.

We take $G = G(I)$ to be the Rees ring $Q[ta_1, \ldots, ta_n]$ of the ideal I of Q generated by $(a_1, \ldots, a_n)$, where t is an indeterminate over Q. Then the analytic spread of I is essentially defined in Northcott and Rees, [4], to be the height of the unique irrelevant prime ideal of G/mG and so is equal to $\dim(G/mG) + 1$ according to the definition of dimension in the present paper.

LEMMA 1.2.

a) $\dim E \geq a(E) - 1$, *with strict inequality if no minimal relevant prime ideal of* $u(E)$ *meets* Q *in* m.
b) $\dim E \leq D(E)$.
c) $a(E) = \mathrm{Max}(t(p_i)+1)$, *the maximum being taken over the relevant prime ideals* $p_1, \ldots, p_s$ *of* G *minimal over* $u(E/mE)$.

PROOF.

a) Let $p_1, \ldots, p_s$ be the prime ideals of G minimal over $\mathrm{Ann}_G(E/mE)$. Then, for some i, $a(E)-1 = \dim(G/p_i)$. But, $\mathrm{ht}p_i + \dim(G/p_i) \leq \dim G$. Hence $a(E) - 1 \leq \dim G$ and the equality is strict save in the case where $\dim(G/p_i) = \dim E$, which implies that p_i is a minimal prime ideal over $\mathrm{Ann}_G E$.
b) It will be enough to consider the case where $E = G/P$ where P is a prime ideal of G and to prove in this case that $\dim E \leq t(P) + \dim(Q/P \cap Q)$. Hence we can assume that G, Q are domains and that $E = G$. We denote by K the field of fractions of Q and by F the homogeneous field of fractions of G. Let t be the transcendence degree of F over K. Then we have to prove that

$$\dim G \leq d + t.$$

Now there is a maximal relevant prime ideal p of G whose height is equal to $\dim G$. The local ring G_p is then a localisation of a finitely generated extension of Q whose field of fractions is F and whose maximal ideal meets Q in m. Further its residue field is an algebraic extension of k. The dimension inequality then gives

$$\dim G = \dim G_p \leq d + t,$$

which is the required result.
c) This follows in a similar manner to b) if we replace E by the G/mG-module E/mE and note that, as Q is replaced by k, the dimension inequality is replaced by the equality

$$\dim(E/mE) = \mathrm{Max}(t(p_i)).$$

We now come to the definitions of the types of reduction with which we will be concerned.

DEFINITION.

i) Let $\mathbf{a} = (a(1), \ldots, a(q))$ be a sequence of non-negative integers. Then a joint reduction of type $\mathbf{a}$ of G with respect to E is a set of elements

$$\mathbf{z} = (z(i,j) \in G_{\delta(i)} | i = 1, \ldots, q; j = 1, \ldots, a(i))$$

generating an ideal of G irrelevant with respect to E.

ii) A set of elements $\mathbf{w} = (w(i,j) \in G_{\delta(i)} | i = 1, ..., q; j = 1, ..., a)$ is termed a complete reduction of G (of length a) with respect to E if the elements $w(j) = w(1,j)w(2,j)...w(q,j)$ $(j = 1, \ldots, a)$ generate an ideal of G irrelevant with respect to E.

The main objective of this section is to prove the existence of joint and complete reductions of G with respect to E if $|\mathbf{a}|, a$ are $\geq a(E)$ and k is infinite. We will limit the proof to complete reductions since, if there exists a complete reduction $\mathbf{w}$ of length a and the sequence $\mathbf{a}$ satisfies $|\mathbf{a}| = a$, then we can construct joint reductions $\mathbf{z}$ of type $\mathbf{a}$ as follows.

Express the set of integers $1, ..., a$ as the union of disjoint sets $\sigma_i (i = 1, ..., q)$, where σ_i contains the integers $s(i,1), ..., s(i, a(i))$. Define $z(i,j) = w(i, s(i,j))$ and $\mathbf{z}$ to be the set of elements $z(i,j)$. Then $\mathbf{z}$ generates an ideal of G containing the ideal generated by the elements $w(j)(j = 1, ..., a)$ and so generates an ideal of G irrelevant with respect to E. Hence z is a joint reduction of G with respect to E of type $\mathbf{a}$.

Our first step is to reduce the problem to the case where Q is a field. Now suppose that $\mathbf{w} = (w(i,j) \in G_{\delta(i)} | i = 1, ..., q; j = 1, ..., a)$ is a set of elements of G and $\mathbf{w}' = (w'(i,j) | i = 1, ..., q; j = 1, ..., a)$, where $w'(i,j)$ is the image of $w(i,j)$ in $G' = G/mG$. Then we claim that $\mathbf{w}$ is a complete reduction of G with respect to E if and only if $\mathbf{w}'$ is a complete reduction of G' with respect to $E' = E/mE$. For if $\mathbf{w}'$ is a complete reduction of G' with respect to E' then $((w(1), ..., w(q))E + mE)_R = E_R$ if $R >> 0$, and hence, by Nakayama's Lemma, $((w(1), ..., w(q))E)_R = E_R$ if $R >> 0$, i.e., $\mathbf{w}$ is a complete reduction of G with respect to E. The converse is immediate.

Hence we now replace G by G', E by E' and $w(i,j)$ by $w'(i,j)$.

We now recall the definition of the general extension Q_g of Q. We suppose that $\mathbf{X}$ is a countable set of indeterminates over Q. Then Q_g is the localisation of $Q[\mathbf{X}]$ at the prime ideal $m[\mathbf{X}]$. We also define $G_g = Q_g \otimes_Q G$ and $E_g = G_g \otimes_G E$. Note that, in the case $Q = k$, the associated minimal prime ideals of E_g are then the prime ideals $p_{ig} = p_i G_g$, the field $F(p_{ig}) = k_g \otimes_k F(p_i)$, and $a(E_g) = a(E)$.

We now suppose that a is an integer $\geq a(E)$ and construct a set of elements

$$\mathbf{x} = (x(i,j) | i = 1, ..., q; j = 1, ..., a)$$

of G_g as follows. We choose distinct indeterminates $X(i,j,l)$ from the set $\mathbf{X}$, where i runs from 1 to q, j from 1 to a and for given i, l runs from 1 to $m(i)$. We now define a complete set of general elements of G w.r.t. E to be the set of elements

$$x(i,j) = \sum_{l=1}^{m(i)} X(i,j,l)y(i,l), (i,j = 1, ..., a).$$

Our immediate objective is to prove that $\mathbf{x}$ is a complete reduction of G w.r.t. E. We do this first for the case when $Q = k$ in two lemmas, the first of which is extracted from the proof of theorem 1.4 of Rees and Sally[10].

LEMMA 1.3. *Let E'' be the module $E/\cup_N (0 :_E (\Xi_1 \cdots \Xi_q)^N)$. Then :*

i) *a set of elements*

$$\mathbf{w} = (w(i,j) \in G_{\delta(i)} | i = 1, ..., q; j = 1, ..., a)$$

is a complete reduction of G w.r.t. E if and only if it is a complete reduction of G with respect to E'';

ii) *a set of elements*

$$\mathbf{z} = (z(i,j) \in G_{\delta(i)} | i = 1, ..., q; j = 1, ..., a(i))$$

is a joint reduction of G w.r.t. E if and only if it is a joint reduction of G w.r.t. E''.

PROOF. i) Suppose that $\mathbf{w}$ is a complete reduction of G w.r.t. E''. We can choose an integer n such that

$$\cup_N (0 :_E (\Xi_1 \cdots \Xi_q)^N) = 0 :_E (\Xi_1 \cdots \Xi_q)^n.$$

Now let $[r]$ denote the set of q integers all equal to r . Then by the definition of complete reduction, applied to E'',

$$E_{[r+1]} = w(1,1)...w(q,1)E_{[r]} + ... + w(1,a) \cdots w(q,a)E_{[r]} + (0 :_E (\Xi_1 \cdots \Xi_q)^n)_{[r+1]}$$

if r is large enough. At the cost of possibly increasing r, multiplication of both sides by $(\Xi_1 \cdots \Xi_q)^n$ yields

$$E_{[r+n+1]} = w(1,1)...w(q,1)E_{[r+n]} + \cdots + w(1,a) \cdots w(q,a)E_{[r+n]}$$

and hence $\mathbf{w}$ is a complete reduction of G w.r.t. E. The converse is an immediate consequence of the simple fact that if $\mathbf{w}$ is a complete reduction of G w.r.t.E, it is a complete reduction of G w.r.t. any homomorphic image of E.

The proof of ii) is similar to that of i).

We assume that $Q = k$ in the following lemma also.

LEMMA 1.4. *If* $\mathbf{x}$ *is defined as above, then* $\mathbf{x}$ *is a complete reduction of* G_g *w.r.t.* E_g.

PROOF. It will clearly be sufficient to prove this result in the case $a = a(E)$ and we make this assumption. We proceed by induction on a . We can always impose the restriction on E that $0 :_E (\Xi_1 \cdots \Xi_q) = (0)$, using lemma 1.3 to replace E by E''. This implies that the elements $x(i,j)$ are non-zero-divisors on E_g . First suppose that $a = 1$. This implies that $\dim(G/\mathrm{Ann}_G E) = 0$ and hence that the only relevant prime ideals of G with respect to E are minimal over $\mathrm{Ann}_G E$. Hence, as the elements $x(1,j)$ are non-zero-divisors on E_g, it follows that $\mathrm{Ann}_{G_g} E_g + x(i,1) \cdots x(i,q) G_g$ is irrelevant, and hence that the set $x(i,1), \ldots, x(1,q)$ is a complete reduction of G_g with respect to E_g. .

Now suppose that $a > 1$ and that the result has been proved for smaller values of a without the restriction $0 :_E (\Xi_1 \cdots \Xi_q) = (0)$. We nevertheless impose the restriction in the case under consideration. We now consider the elements $x(1,a), \ldots, x(q,a)$. Their expression involves only a finite number of indeterminates drawn from $\mathbf{X}$. Call this set of indeterminates $\mathbf{X}^*$, and let Q^* be the localisation of $Q[\mathbf{X}^*]$ at $m[\mathbf{X}^*]$. Then we can identify Q_g with $(Q^*)_g$. We also introduce the ring G^* and the module E^* defined in the obvious way. We now consider the module $E\dagger = E^*/x(1,a) \cdots x(q,a)E^*$. Since we are imposing the restriction $0 :_E (\Xi_1 \cdots \Xi_q) = (0)$ on E, the elements $x(1,a), \ldots, x(q,a)$ are not zero divisors on E^* and hence $a(E\dagger) = a(E) - 1$. Hence the result has been proved if E is replaced by $E\dagger$ and so, if r is large enough,

$$((E\dagger)_g)_{[r+1]} = x(1,1) \cdots x(q,1)((E\dagger)_g)_{[r]} + \cdots + x(1,a-1) \cdots x(q,a-1)((E\dagger)_g)_{[r]}$$

which can be written

$$(E_g)_{[r+1]} = x(1,1) \cdots x(q,1)(E_g)_{[r]} + \cdots + x(1,a) \cdots x(q,a)(E_g)_{[r]}$$

which completes the proof in the case $a = a(E)$ and hence for $a \geq a(E)$.

We can now state our basic theorem.

THEOREM 1.5. *Let* (Q, m, k, d) *be a local ring,* $G = Q[\mathbf{y}]$ *be an admissible ring and* E *be a multigraded finitely generated* G*-module. Suppose that* $a \geq a(E)$ *and that*

$$\mathbf{x} = (x(i,j) | i = 1, \ldots, q; j = 1, \ldots, a)$$

is a complete set of general elements of G *w.r.t.* E. *Then the following two statements are true,*

i) $\mathbf{x}$ *is a complete reduction of* G_g *w.r.t.* E_g.
ii) *Let the set of integers* $1, \ldots, a$ *be expressed as the union of disjoint sets* $\sigma_i (i = 1, \ldots, q)$, *where* σ_i *contains the integers* $s(i,1), \ldots, s(i,a(i))$. *Define* $z(i,j) = x(i, s(i,j))$ *and* z *to be the set of elements* $z(i,j)$. *Then* z *is a joint reduction of* G_g *with respect to* E_g *of type* $\mathbf{a} = (a(1), \ldots, a(q))$.

PROOF. The first statement follows from the last lemma using the reduction to the case where Q is a field obtained by replacing Q by $k = Q/m$, G by G/mG and E by E/mE. As indicated earlier, the second statement is an immediate consequence of the first, since the ideal of G_g generated by the set of elements $\mathbf{z}$ contains the ideal generated by the set of elements $\mathbf{x}$.

The set of elements $\mathbf{x}$ is a set of elements of the ring G_g and not of G. However, the above theorem implies

THEOREM 1.6. *If (Q, m, k, d) is a local ring with k infinite and $G, E, a \geq a(E)$ are as in the last theorem, then the following two statements are true.*

i) *There exists a set of elements*

$$\mathbf{w} = (w(i,j)|i = 1, ..., q; j = 1, ..., a)$$

such that $\mathbf{w}$ is a complete reduction of G w.r.t . E.

ii) *If $\mathbf{a} = (a(1), ..., a(q))$ with $\sum a(i) = a$, then there exists a set of elements $\mathbf{z} = (z(i,j)|i = 1, ..., q; j = 1, ..., a(i))$ which is a joint reduction of G w.r.t. E of type $\mathbf{a}$.*

PROOF. i) By the last theorem, there exists an integer r such that

$$(E_g)_{[r+1]} = x(1,1) \cdots x(q,1)(E_g)_{[r]} + \cdots + x(1,a) \cdots x(q,a)(E_g)_{[r]}.$$

Choose a basis $u_1, ..., u_N$ of the Q-module $E_{[r+1]}$. Then,

$$u_s = x(1,1) \cdots x(q,1)v(s,1) + \cdots + x(1,a) \cdots x(q,a)v(s,a), (s = 1, \ldots, N)$$

where $v(s,t)$ belongs to $(E_g)_{[r]}$. Now let $\mathbf{X}'$ be the finite sub-set of the set of indeterminates $\mathbf{X}$ which occur in the set of equations above. Then we can find a polynomial $f(\mathbf{X}')$ in this set of indeterminates, with coefficients in Q and at least one coefficient a unit of Q, such that each of the elements $v(s,t)$ can be written as a finite sum

$$v(s,t) = \sum a(s,t,\mu)\mu/f(\mathbf{X}')$$

with the coefficients $a(s,t,\mu)$ in $E_{[r]}$, and μ denoting a monomial in $\mathbf{X}'$. Multiplying the equations through by $f(\mathbf{X}')$, we obtain, for $s = 1, ..., N$,

$$f(\mathbf{X}')u_s = \sum x(t,1) \cdots x(t,a)a(s,t,\mu)\mu,$$

the sums being over $t = 1$ to q and a finite set of monomials μ. Now, if we substitute for the set of indeterminates $\mathbf{X}'$ a set ξ of elements of Q such that $f(\xi)$ is a unit of Q (this being possible if k is infinite), then the set $\mathbf{w}$ consisting of the elements

$$w(i,j) = \sum_{j=1}^{a} \sum_{i=1}^{q} \xi(i,j,l)y(i,l)$$

where $\xi(i,j,l)$ is the element of Q substituted for $X(i,j,l)$, satisfies the conditions

$$u_s \in w(1,1) \ldots w(q,1)E_{[r]} + \cdots + w(1,a) \cdots w(q,a)E_{[r]}$$

where $s = 1, ..., N$, i.e.,

$$E_{[r+1]} = w(1,1)...w(q,1)E_{[r]} + \cdots + w(1,a)\cdots w(q,a)E_{[r]}$$

for $s = 1, ..., N$ and so is a complete reduction of G w.r.t. E. ii) As in theorem 1.5, this is an immediate consequence of i).

We conclude this section with a definition and a lemma which will be used to reduce the proofs of certain results to the case where G is simply graded (that is, the case where $q = 1$). We refer the reader to the the discussion preceding Lemma 2.1 below.

DEFINITION. Let $C = (c(1), ..., c(q))$ be a sequence of positive integers, and let $G = Q[\mathbf{y}]$ be an admissible ring. Then we will denote by $\Delta_C G$ the subring of G whose homogeneous elements of degree r are the elements of G of degree rC for all non-negative integers r. We consider $\Delta_C G$ as a simply graded domain by defining $(\Delta_C G)_r = G_{rC}$.

Let E be a finitely generated G-module. Then we denote by $\Delta_C E$ the simply graded $\Delta_C G$-module defined by $(\Delta_C E)_r = E_{rC}$, and refer to $\Delta_C E$ as a diagonal of E.

We commence by noting that $\Delta_C G$ is finitely generated over Q, being generated by a basis of G_C. Further $\Delta_C E$ is a finitely generated $\Delta_C G$-module.

LEMMA 1.7. $\dim \Delta_C E = \dim E$; $a(\Delta_C E) = a(E)$.

PROOF. Since $\Delta_C(E/mE) = \Delta_C E/m\Delta_C E$, the second statement will follow from the first applied to E/mE.

We now turn to the first statement and commence with the observation that if P, P' are two relevant prime ideals of G such that P properly contains P', then P, P' have different intersections with G_R if R is sufficiently large, and hence different intersections with $\Delta_C G$.

Let p be a relevant prime ideal of $\Delta_C G$, and consider the set of homogeneous elements x of G with the property that $xG \cap \Delta_C G \subseteq p$. The set of finite sums of such elements form a graded relevant ideal P of G which is prime, since if x, y do not belong to P, there exists z in $xG \cap \Delta_C G$ and w in $yG \cap \Delta_C G$ with neither in p. Then zw is not in p, and so xy is not in P. It is clear that P contains all ideals I of G such that $I \cap \Delta_C G \subseteq p$, in particular, pG, and is therefore the unique prime ideal of G meeting $\Delta_C G$ in p. Hence the correspondence $P \to P \cap \Delta_C G$ is a $1-1$ correspondence between the set of relevant prime ideals of G and the set of prime ideals of $\Delta_C G$ which preserves inclusion. It follows that corresponding prime ideals have the same height and therefore that $\dim \Delta_C G = \dim G$. This completes the proof.

2.The formalism of Hilbert Functions.

In this section we consider Q-modules E which are graded in the sense that E is the direct sum of Q-modules E_R defined for all sequences $R = (r(1), \ldots, r(q))$

of q non-negative integers. We further suppose that E is subject to either the condition that E_R has finite length $\lambda(E_R)$ if $R >> 0$, or to the stronger condition that E_R has finite length for almost all R. Our concern is with the behaviour of the function $\lambda(E_R)$ for large values of R for various choices of E.

We will commence by considering a more general set-up, viz., we will consider integer-valued functions $f(R)$ defined for all $R >> 0$ or for almost all R. In each case we will consider in parallel with such a function the formal power series $F(Z) = \sum f(R)Z^R$, where Z is the set of indeterminates $z_1, \ldots, z_q$, Z^R denotes the monomial ${z_1}^{r(1)} \cdots {z_q}^{r(q)}$, and the sum is over all R for which $f(R)$ is defined. The function $f(R)$ and the power series $F(Z)$ convey the same information and will be freely interchanged below. Any definition applying to either can be translated into an equivalent definition for the other.

If $f(R) = \lambda(E_R)$, then the corresponding power series will be denoted by $\Lambda_E(Z)$.

DEFINITION. Two functions $f(R), g(R)$ defined for $R >> 0$ (resp., for almost all R) will be termed equivalent (resp., strongly equivalent) if $f(R) = g(R)$ for $R >> 0$ (resp., for almost all R), and we will write $f \sim g$ (resp., $f \approx g$).

As indicated above, we will also refer to the power series $F(Z), G(Z)$ which correspond to $f(R), g(R)$ as being equivalent (resp., strongly equivalent) and write $F(Z) \sim G(Z)$ (resp., $F(Z) \approx G(Z)$).

We say that $f(R)$(resp., $F(Z)$) is of polynomial type if there is a polynomial $p(R)$ in R with rational coefficients and taking integer values, such that $f \sim p$ (resp., $F(Z) \sim \sum p(R)Z^R$).

The graded Q-module E is said to be of Hilbert type if $\lambda(E_R)$ is of polynomial type.

We can now state the purpose of this section in a more explicit manner. It is to study the elementary properties of modules of Hilbert type, and of strong Hilbert type, this being defined later. We will first consider functions $f(R)$ of polynomial type, and observe that the polynomial $p(R)$ and the power series $P(Z) = \sum p(R)Z^R$ of the definition are uniquely determined by either $f(R)$ or $F(Z)$ and further each is a complete invariant of f or F to within equivalence. If it is necessary to emphasize the relationship between $f(R)$ and $p(R)$, we will write the latter as $p(f|R)$. Similarly, we will write $P(F|Z)$ instead of $P(Z)$ to emphasize the relationship of the latter to $F(Z)$.

We first characterise the power series of the type $P(Z)$. If the degree of $p(R)$ in $r(i)$ is $m(i) - 1$ for $i = 1, \ldots, q$, then it is clear that the power series

$$(1 - z_1)^{m(1)} \cdots (1 - z_q)^{m(q)} P(Z)$$

is a polynomial in $z_1, \ldots, z_q$ of degree at most $m(i) - 1$ in z_i. By the usual convention the zero polynomial has degree -1. Now we write w_i for the formal power series $\sum z_i^r = (1 - z_i)^{-1}$, and denote the set of variables $w_1, \ldots, w_q$ by

W. W^R is defined in a similar fashion to Z^R, while W^{-1} will denote the set $(w_1^{-1}, \ldots, w_q^{-1})$. Then, since $z_i = 1 - w_i^{-1}$, it follows that $P(Z)$ can be written in the form $w_1 \cdots w_q H(p|W)$, where $H(p|W)$ is a polynomial in W with integer coefficients. Note that $H(0|W)$ is zero. The polynomials $H(p|W)$ and $p(R)$ are related as follows. If $H(p|W) = \sum e(p|K)W^K$, then

$$p(R) = \sum e(p|K)v(R, K)$$

where

$$v(R, K) = \binom{r(1) + k(1)}{k(1)} \cdots \binom{r(q) + k(q)}{k(q)}.$$

Note that the polynomials $p(R)$ and $H(p|W)$ have the same degree and each determines the other.

We now introduce a considerable amount of notation. Let $f(R)$ be of polynomial type, $F(Z) = \sum f(R)Z^R$ and $p(R) = p(f|R)$. Then first we will write either $h(f|R)$ or $h(F|R)$ for $h(p|R)$ and $H(f|W)$ or $H(F|W)$ for $H(p|W)$. $e(f|K)$ or $e(F|K)$ will denote $e(p|K)$. The degree of $p(f|R)$, or, what is the same thing, the degree of $H(f|W) = H(F|W)$ will be denoted either by hdeg$f$ or by hdeg$F$. The sum of the terms of $H(f|W) = H(F|W)$ of degree $h =$ hdeg$f$, will be denoted by $e(f|W)$ or $e(F|W)$, or, if we wish to emphasize the value of h, by $e_h(f|W)$ or $e_h(F|W)$. Finally, we will write the sum of the terms of degree h of $p(f|R)$ in the form $M(f|R)/h!$, or the alternative form $M_h(f|R)/h!$. Note that

$$e_h(f|W) = \sum_{|K|=h} e(f|K)W^K,$$

$$M_h(f|R) = \sum_{|K|=h} \mu(h, K)e(f|K)R^K,$$

where $R = (r(1), \ldots, r(q))$ and $\mu(h, K) = h!/k(1)! \cdots k(q)!$.

DEFINITIONS. When f and F are of polynomial type we will term $h(f|R)$ and $H(F|W)$ the Hilbert polynomials of f and F respectively.

hdeg$f$ = hdeg$F$ will be termed the Hilbert degree of f or F.

The term multiplicity form of f(or F) will be applied to either $e(F|W)$ or $M(f|R)$. Now let E be a module of Hilbert type and consider the case where $f(R)$ above is equal to $\lambda(E_R)$. Then, using the notation and definitions above, we will replace f or F by E. Thus, for example, $e(f|W)$ becomes $e(E|W)$ and $M(f|R)$ becomes $M(E|R)$. These will be referred to as the multiplicity forms of E.

We will, however, usually modify the definitions of the multiplicity forms in the sequel. We consider a class Σ of Q-modules E, graded by sets R of q non-negative integers and which satisfy the following conditions.

i) If $E \in \Sigma$, E is of Hilbert type.

ii) hdeg$E$ is bounded as E varies over Σ.

Let D be the upper bound of hdeg$E$. Then we define the multiplicity forms $e(\Sigma|E|W)$ and $M(\Sigma|E|R)$ of E in Σ to be $e_D(E|W)$ and $M_D(E|R)$ above if hdeg$E = D$, and to be zero if hdeg$E < D$. This definition of multiplicity form depends on the choice of the class of modules Σ. Thus, for all E in Σ, the multiplicity forms of E either have degree D or are zero.

LEMMA 2.1. *Let*

$$0 \longrightarrow E' \longrightarrow E \longrightarrow E'' \longrightarrow 0$$

be an exact sequence of Q-modules preserving degrees. Then

i) *if two of E, E', E'' are of Hilbert type, so is the third;*

ii) *if this is the case,*

$$\text{hdeg}E = \text{Max}(\text{hdeg}E', \text{hdeg}E'');$$

iii) *if E, E', E'' all belong to Σ,*

$$e(\Sigma|E|W) = e(\Sigma|E'|W) + e(\Sigma|E''|W),$$

$$M(\Sigma|E|R) = M(\Sigma|E'|R) + M(\Sigma|E''|R).$$

PROOF. We have $\lambda(E_R) = \lambda(E'_R) + \lambda(E''_R)$. This implies i) and also that, if the conditions of i) hold, then $h(E|R) = h(E'|R) + h(E''|R)$, which implies ii) and iii).

The case $q = 1$ requires special mention since in that case there is only one coefficient $e(E|K)$ to consider, viz., $k(1) = D$. Hence we define the multiplicity $e(E)$ to be $e(E|D)$ and define the multiplicity of E in Σ in a similar manner. The same notational rule will apply to the various multiplicity forms introduced in the later sections.

We have already characterised power series $F(Z)$ of polynomial type as power series equivalent to power series expressible in the form $w_1 \cdots w_q H(W)$, where $H(W)$ is a polynomial in W with integer coefficients. We now give a second characterisation which will be our main tool in proving that modules are of Hilbert type.

LEMMA 2.2. *$F(Z)$ is of polynomial type if and only if there exist integers $m(i) > 0, (i = 1, \ldots, q)$, such that $(1 - z_i)^{m(i)} F(Z) \sim 0$ for $i = 1, \ldots, q$.*

PROOF. Suppose that this condition is satisfied for $F(Z)$. Then we can find a sequence $N = (n(1), \ldots, n(q))$ of non-negative integers and a power series $G(Z)$ such that $F(Z) \sim Z^N G(Z)$ and the coefficient of Z^R in $(1 - z_i)^{m(i)} G(Z)$ is zero if $r(i) \geq m(i)$. It follows that $(1 - z_1)^{m(1)} \cdots (1 - z_q)^{m(q)} G(Z)$ is a polynomial in Z of degree $< m(i)$ in z_i and, as above, $G(Z)$ can be written in the form $w_1 \cdots w_q H(W)$, where $H(W)$ is a polynomial in W. Hence there is a polynomial $p(R)$ taking integer values such that, if $R \geq N$, the coefficient of Z^R in $F(Z)$ is equal to $p(R)$ and $F(Z)$ is of polynomial type.

Conversely, suppose that $f(Z) \sim \sum p(R)Z^R$, where $p(R)$ is a polynomial whose degree in $r(i)$ is $m(i)-1$ for $i=1,\ldots,q$. Then

$$(1-z_i)^{m(i)}F(Z) \sim (1-z_i)^{m(i)}\sum p(R)Z^R \sim 0.$$

At this point we can introduce strong Hilbert type.

DEFINITION. A formal power series $F(Z)$ is said to be of strong polynomial type if there exist pairs of integers $(m(i),n(i))$, $(i=1,\ldots,q)$ such that, for $i=1,\ldots,q$, the coefficient of Z^R in $(1-z_1)^{m(i)}F(Z)$ is zero if $r(i)\geq n(i)$.

A Q-module E is said to be of strong Hilbert type if $\Lambda(E|Z)$ is of strong polynomial type.

It is clear that if $F(Z)$ is of strong polynomial type,

$$(1-z_1)^{m(1)}\cdots(1-z_q)^{m(q)}F(Z)=\phi(Z)$$

where $\phi(Z)$ is a polynomial in Z, and, by substituting $(1-{w_i}^{-1})$ for $z_i(i=1,\ldots,q)$ we can express $F(Z)$ as a polynomial in W,W^{-1}. However it will be more convenient to write $F(Z)$ as a sum $\sum c_K(Z)W^K$, where $c_K(Z)$ is a polynomial involving only those z_i for which $k(i)=0$, and the sum is over those K satisfying $0\leq k(i)\leq m(i)$ for $i=1,\ldots,q$. If $F(Z)=\Lambda_E(Z)$, we will write this polynomial as $L(E|Z,W)$. Note that the sum of the terms of $L(E|Z,W)$ involving all w_i to positive exponent is $w_1\cdots w_qH(E|W)$.

Finally, if $F(Z)$ can be written in the form $\sum c_K(Z)W^K$ as described in the last paragraph then, with $m(i)$ as in the last paragraph, the coefficient of Z_R in $(1-z_i)^{m(i)}F(Z)$ is zero if $r(i)>>0$ for each i, and hence $F(Z)$ is of strong polynomial type. We note that results analogous to lemma 2.1 hold for strong Hilbert type in that, if

$$0\longrightarrow E'\longrightarrow E\longrightarrow E''\longrightarrow 0$$

is an exact sequence of Q-modules preserving degrees, and if two of E,E',E'' are of strong Hilbert type, so is the third, and in this case

$$L(E|Z,W)=L(E'|Z,W)+L(E''|Z,W).$$

We now suppose that E is of Hilbert type and consider two types of operation that can be carried out on E and their effect on $h(E|R)$ and the other functions derived from it.

First we consider diagonalisation, already referred to in the last section.

Suppose that $C=(c(1),\ldots,c(q))$ is a sequence of positive integers and take $R=rC$, where $r>>0$. Then $\lambda(E|rC)$ is a polynomial in r of degree $\leq h$, the coefficient of r^h being $M(E|C)/h!$. Since $\lambda(E_R)\geq 0$ for all R, it follows that $M(E|C)\geq 0$ for all C and is >0 for suitable choice of C.

LEMMA 2.3. *Let $\Delta_C E$ be as defined in the last section. Then, if E is of Hilbert type, so is $\Delta_C E$, and* $\mathrm{hdeg}(\Delta_C E) \leq h = \mathrm{hdeg}E$, *with equality for suitable choice of C.*

PROOF. This is immediate since $\lambda((\Delta_C E)_r)$ is a polynomial of degree $\leq$ hdeg$E$ in r for $r >> 0$, and the coefficient of r^h is $M(E|C)/h!$.

Next we consider the shift operators. Let $K = (k(1), \ldots, k(q))$ be a sequence of integers, not restricted to be non-negative. Then the module $E(K)$ is defined by $E(K)_R = E_{R+K}$ if $r(i) \geq -k(i)$ for $i = 1, \ldots, q$, and $E(K)_R = (0)$ if $r(i) < -k(i)$ for some $i (1 \leq i \leq q)$.

We term $E(K)$ a shift of E.

LEMMA 2.4.

a) If E is of Hilbert type, so is $E(K)$, and

$$i)\mathrm{hdeg}(E(K)) = \mathrm{hdeg}(E),$$

$$ii)M(E(K)|R) = M(E|R),$$

$$iii)e(E(K)|W) = e(E|W).$$

b) E is of Hilbert type if and only if there exists $K \geq 0$ such that $E(K)$ is of strong Hilbert type.

PROOF. a) To prove i), ii), iii), it is sufficient to prove them in the cases where $K = \pm\delta(i)$, since the general shift operation is obtained by iteration of these two cases. As E is of Hilbert type, $\lambda(E_R)$ is a polynomial of degree hdeg$E$ in R for $R >> 0$. Hence, as $\lambda(E(\pm\delta(i))|R) = \lambda(E|R \pm \delta(i))$, if $R >> 0$, i), ii), are immediate, while iii) is an immediate consequence of ii). b) Straightforward.

The following special result will be used in the next section.

THEOREM 2.5. *Let E be a Q-module graded by sets of q non-negative integers R and let*

$$E = E_0 \supseteq E_1 \supseteq \cdots \supseteq E_s \supseteq \ldots$$

be a descending sequence of graded sub-modules of E with the property that the module $(E_s/E_{s+1})_R$ is of finite length for all R, s and the function $f(R,s) = \lambda((E_s/E_{s+1})_R)$ is of strong polynomial type. Let $g(R,s)$ and $g_s(R)$ both denote $\lambda(E/E_{s+1})_R)$, the former as a function of R, s and the latter as a function of R only, s being considered as fixed. Then,

i) *$g(R,s)$ and $g_s(R)$ are of strong polynomial type, the latter for all values of s,*
ii) $\mathrm{hdeg}(g_s)$ *takes a constant value for large values of s, which we will denote by* $\mathrm{limdeg}(g_s)$ *and term the limit degree of g_s,*
iii) $\mathrm{hdeg}(g) = \mathrm{Max}(\mathrm{hdeg}(f) + 1, \mathrm{limdeg}(g_s))$.

PROOF. The condition that f is of strong polynomial type implies that the power series $F(Z,z) = \sum f(R,s)Z^R z^s$ associated with f has a standard representation as a finite sum

$$L(f|Z,z,W,w) = \sum_{K,l} c_{K,l}(Z,z)W^K w^l,$$

where $c_{K,l}(Z,z)$ only involves z_i if $k(i) = 0$ and z if $l = 0$. We write the sum of the terms of $L(f|Z,z,W,w)$ involving $w, w_1, \ldots, w_q$ all to positive exponent as $ww_1 \cdots w_q H(f|W,w)$, so that hdeg(f) is the degree of the polynomial $H(f|W,w)$. We will also write the sum of the terms involving each of $w_1, \ldots, w_q$ to positive exponent but w to exponent zero as the finite sum $w_1 \cdots w_q(\sum_k z^k h_k(f|W))$.

We can write the power series $G(Z,z) = \sum g(R,s)Z^R z^s$ as $(\sum_{i=0}^{\infty} z^i)F(Z,z)$ and hence its standard form $L(g|Z,z,W,w)$ is derived from $L(f|Z,z,W,w)$ by multiplying by w, and replacing any product wz^r which occurs in the product by $w - \sum_{i=0}^{r-1} z^i$. It follows that

$$H(g|W,w) = wH(f|W,w) + \sum_k h_k(f|W).$$

This implies that

$$\text{hdeg}(g) = \text{Max}(\text{hdeg}(f) + 1, \text{hdeg}(\sum_k h_k(f|W))).$$

Next we turn to the functions $g_s(R)$ and the standard form $L(g_s|Z,W)$ of the associated power series $G_s(Z)$. $L(g_s|Z,W)$ is derived from $L(g|Z,z,W,w)$ by replacing w by $\sum_{i=0}^{\infty} z^i$, and picking out the coefficient of z^s in the resulting power series. If the sum of the terms of $L(g_s|Z,W)$ which involve $w_1 \cdots w_q$ to positive exponent is $w_1 \cdots w_q H(g_s|W)$, $H(g_s|W)$ can be expressed as the sum of two polynomials. The first is equal to the coefficient of z^s in $(\sum_{i=0}^{\infty} z^i)(\sum_k z^k h_k(f|W))$, and if we choose s so that $h_k(f|W) = 0$ if $k > s$, is equal to $\sum_k h_k(f|W)$. The second is derived from $w^2 w_1 \cdots w_q H(f|W,w)$ by replacing, in each term, w^n by the coefficient of z^s in $(1-z)^{-n}$, where n is the degree of the term in w, and hence is a polynomial in $w_1, \ldots, w_q$ of degree at most hdeg(f), the coefficients being polynomials in s. Hence for all large s, $H(g_s|W)$ has constant degree, which we denote by limdeg(g_s) and limdeg(g_s) is therefore either $\leq$ hdeg(f) or is equal to the degree of $\sum_k h_k(f|W)$. Hence we have

$$\text{hdeg}(g) = \text{Max}(\text{hdeg}(f) + 1, \text{limdeg}(g_s)).$$

3. Hilbert Functions and Koszul Complexes.

In this section $G = Q[\mathbf{y}]$ will be as in section 1 and E will be a G-module. E, unless otherwise stated, will be assumed to be finitely generated.

The main objective is to prove that various modules are of Hilbert type.

Our basic tool in this section will be the Koszul complex $K(E|\xi)$ of a, not necessarily finitely generated, G-module E with respect to a set of homogeneous elements $\xi = (\xi_1, \ldots, \xi_N)$ of G, ξ_i having degree $\sigma(i)$, where $\sigma(i)(i = 1, \ldots, N)$ is a sequence of q non-negative integers. We recall the definition and some basic properties of $K(E|\xi)$. It is constructed as follows. We first consider the case where $E = G$. We form the exterior algebra $A = \oplus A_j$ of G with respect to a set of indeterminates $u_1, \ldots, u_N$. Then $K(G|\xi)$ is a complex

$$\cdots \xrightarrow{d_{j+1}} K_j(G|\xi) \xrightarrow{d_j} K_{j-1}(G|\xi) \xrightarrow{d_{j-1}} \cdots,$$

the module $K_j(G|\xi) = A_j$ being a free G-module generated by the $\binom{N}{j}$ products

$$u_{i(1)}u_{i(2)}...u_{i(j)}(1 \leq i(1) < i(2) < ... < i(j) \leq N).$$

We grade $K_j(G|\xi)$ by giving $u(i)$ the degree $\sigma(i)$. The map d_j is defined by

$$d_j(u_{i(1)} \cdots u_{i(j)}) = \sum (-1)^k \xi_{i(k)} u_{i(1)} \cdots \hat{u}_{i(k)} \cdots u_{i(j)},$$

where the circumflex denotes a term to be omitted. It is clear that this grading makes the map d_j of degree 0. For general E we take $K(E|\xi) = E \otimes_G K(G|\xi)$. The homology modules of this complex will be denoted by $H_i(E|\xi)$, $(i = 0, \ldots, N)$.

Since we are dealing with graded modules, we note that $K(E|\xi)$ splits up as a direct sum of its graded components $K(E|\xi|R)$ which we write as

$$\cdots \xrightarrow{d_{j+1}(R)} K_j(E|\xi|R) \xrightarrow{d_j(R)} K_{j-1}(E|\xi|R) \xrightarrow{d_{j-1}(R)} \cdots,$$

the jth homology module $H_j(E|\xi|R)$ of $K(E|\xi|R)$ being the component of degree R of $Hj(E|\xi)$. We will require below that the homology modules $H_i(E|\xi)$ of $K(E|\xi)$ satisfy the condition that $\lambda(H_i(E|\xi|R))$ is finite for almost all R. In particular, by taking $i = 0$, we have $\lambda((E/\xi E)_R)$ finite for almost all R.

Given that this condition is satisfied, we can define to within strong equivalence the functions $\lambda((E/\xi E)_R)$ and $\chi(E|\xi|R) = \sum(-1)^i \lambda(H_i(E|\xi E|R))$, and the associated power series

$$\Lambda(E/\xi E|Z) = \sum \lambda((E/\xi E)_R) Z^R \quad \text{and} \quad \chi(E|\xi E|Z) = \sum \chi(E|\xi E|R) Z^R.$$

We note the following result without proof.

If

$$0 \to E' \longrightarrow E \longrightarrow E'' \to 0$$

is an exact sequence of G-modules, and any two of E, E', E'' satisfy the condition above, so does the third, and we have

$$\chi(E|\xi|R) \approx \chi(E'|\xi|R) + \chi(E''|\xi|R) \quad ; \quad \chi(E|\xi|Z) \approx \chi(E'|\xi|Z) + \chi(E''|\xi|Z).$$

In practice, we will be concerned with the following two cases when the above condition is satisfied.

i) *E satisfies the condition that $\lambda(E_R)$ is finite for almost all R.*
ii) *E is finitely generated and satisfies the condition that $\lambda((E/\xi E)_R)$ is finite for almost all R.*

It is the second case that will concern us most, and we will say in that case that E is ξ-admissible.

We now consider one result which concerns case i) above.

LEMMA 3.1. *Let E be as in case i) above. Then, whatever ξ,*

$$\chi(E|\xi|Z) \approx (1 - Z^{\sigma(1)})...(1 - Z^{\sigma(N)})\Lambda(E|Z).$$

PROOF. For almost all R, $K_j(E|\xi|R)$ has finite length for $j = 0, \ldots, N$. Hence, for almost all R,

$$\chi(E|\xi|R) = \sum(-1)^j\lambda(K_j(E|\xi|R)),$$

which yields immediately

$$\chi(E|\xi|Z) \approx \sum(-1)^j\lambda(K_j(E|\xi|Z)).$$

But the right-hand-side is equal to

$$(1 - Z^{\sigma(1)})...(1 - Z^{\sigma(N)})\Lambda(E|Z)$$

by straightforward calculation.

We now assume that E is ξ-admissible, i.e., that ii) holds.

THEOREM 3.2. *Let E be a finitely generated G-module and let E' be a sub-module of E such that $(E/E')_R$ has finite length as a Q-module for almost all R (resp., for all $R >> 0$). Then E/E' is of strong Hilbert type (resp., Hilbert type) and, in either case,* $\mathrm{hdeg}(E/E') \leq a(E) - 1$, *with equality if $E' \subseteq mE$.*

PROOF. If $E' = 0$, so that $E/E' = E$, then the last statement becomes the equality $\mathrm{hdeg}E = a(E) - 1$ under the condition that E_R has finite length for all $R >> 0$. The general case can now be reduced to to the case $E' = 0$, since $a(E/E') \leq a(E) - 1$ with equality holding if $E' \subseteq mE$.

We first assume that $\lambda(E_R)$ is finite for almost all R. Consider the Koszul complex $K(E|\mathbf{y})$. Its homology groups are finitely generated G-modules annihilated by $\mathbf{y}$ and hence for $j = 0, \ldots, m$, where $m = \sum m(i), H_j(E|\mathbf{y})_R = (0)$ for almost all R. Then, by lemma 3.1, because, for $i = 1, \ldots, q$, there are $m(i)$ of the elements of $\mathbf{y}$ of degree $\delta(i)$,

$$(1 - z_1)^{m(1)}...(1 - z_q)^{m(q)}\Lambda(E|Z) \approx \chi(E|\mathbf{y}|Z) \approx 0$$

and this implies that E is of strong Hilbert type.

In view of lemmas 1.7 and 2.3, we can reduce the proof of the last statement to the case $q = 1$ by replacing E by a suitable diagonal. Next, we can find an m-primary ideal J such that J annihilates E_R for almost all R. Hence, for almost all R,

$$\lambda(Q/J)\lambda(E_R/mE_R) \geq \lambda(E_R) \geq \lambda(E_R/mE_R).$$

Hence we can replace G by G/mG, E by E/mE, and thereby reduce the proof to the case where $Q = k$ in addition to the restriction to $q = 1$. We now have $a(E) - 1 = \dim E$ and the result $\mathrm{hdeg} E = \dim E$ is classical in this case.

If we only assume that $\lambda(E_R)$ is finite for $R >> 0$, then we can find $K > 0$, such that $\lambda(E_{R+K})$ is finite for all $R \geq 0$. Hence applying the above to $E(K)$, we see that $E(K)$ is of strong polynomial type and $\mathrm{hdeg} E(K) = a(E(K)) - 1$. The alternative statements now follow by lemma 2.4.

The next lemma requires some preliminary discussion. First suppose that E is a Q-module of strong Hilbert type, and, with the notation of section 2, let

$$L(E|Z, W) = \sum f_\mu(E|Z)\mu,$$

where μ denotes a monomial in W, and $f_\mu(E|Z)$ does not involve z_i if w_i occurs to a positive exponent in μ.

DEFINITION. If E is of strong Hilbert type, a monomial μ in W is said to be critical for E if $f_\mu(E|Z) \neq 0$ and $f_{\mu'}(E|Z) = 0$ for all other monomials μ' divisible by μ.

THEOREM 3.3. *Let E be a finitely generated G-module such that $\lambda(E_R)$ is finite for almost all R. Then if μ is critical for E, $f_\mu(E|Z)$ has non-negative coefficients. In particular, the coefficients of the multiplicity forms $M_D(E|X)$ and $e_D(E|W)$ are non-negative.*

PROOF. We proceed by induction on $m = \sum m(i)$. The result is trivial in the case $m = 0$. Suppose $m > 0$. We suppose $m(i) > 0$, write y for $y(i, m(i))$ and G' for the ring generated over Q by the elements of $\mathbf{y}$ with y deleted. Let n be the least integer such that $(0 : y^n)_E = (0 : y^{n+1})_E$; let $E' = (0 : y^n)_E$ and $E'' = E/E'$. Then the exact sequence

$$0 \to E' \longrightarrow E \longrightarrow E'' \to 0$$

implies that $f_\mu(E|Z) = f_\mu(E'|Z) + f_\mu(E''|Z)$. Now E' has a filtration by the modules $E'_j = (0 : y^j)_E$ and E'_{j+1}/E'_j is a G'- module. Hence, by our inductive hypothesis the result is true for each of the modules E'_{j+1}/E'_j and hence for E'. Further, y is not a zero divisor on E''. Hence we have an exact sequence

$$0 \to E''(\delta(i)) \longrightarrow E'' \longrightarrow E''/yE'' \to 0$$

which implies that $L(E''|Z, W) = w_i L(E''/yE''|Z, W)$. Since E''/yE'' is a G'-module, the result is true for E''/yE'' and hence for E''. Hence, finally, it is true for E. The last remark is immediate.

COROLLARY. *If E is of Hilbert type,the coefficients of the multiplicity form of E are non-negative.*

PROOF. We can replace E by $E(K)$ for a suitable K and assume that E is of strong Hilbert type. Since $M_D(E|R) = M_D(E(K)|R)$, the result follows.

We continue this section by considering the case where Q is a field k and G is a domain $k[\mathbf{y}_1, \ldots, \mathbf{y}_q]$ of dimension D and homogeneous field of fractions F. In what follows, we consider Σ to be the category of finitely generated G-modules, and the multiplicity form $M(E|R)$ below will be the form $M(\Sigma|E|R)$ of section 2 and so has degree D or is zero. The function $\lambda(E_R) = \dim_k(E_R)$ is equal to a polynomial of degree $\leq D$ in R if $R >> 0$, and $M(E|R)/D!$ is the sum of the terms of degree D in this polynomial.

We recall for reference the following elementary results concerning $M(E|R)$.

a) If

$$0 \to E' \longrightarrow E \longrightarrow E'' \to 0$$

is an exact sequence of f.g. graded G-modules,

$$M(E|R) = M(E'|R) + M(E''|R).$$

b) If $E(C)$ is a shift of E,

$$M(E(C)|R) = M(E|R).$$

c) If E is annihilated by a non-zero divisor of G,

$$M(E|R) = 0.$$

Next we introduce some more notation. We fix an arbitrary sub-set σ of $(1, \ldots, q)$ and denote its complement by σ'. However, we can always renumber $(1, \ldots, q)$ so that $\sigma = (1, \ldots, p)$, where $p = |\sigma|$, and we will restrict attention to this case, and take p as fixed. Further, if $A = (a(1), \ldots, a(q))$ is any sequence indexed by $(1, ..., q)$ we will write A' for $(a(1), \ldots, a(p))$ and A'' for $(a(p+1), \ldots, a(q))$.

Further fix a sequence of q non-negative integers $K = (k(1), \ldots, k(q))$ such that $|K| = D$. Let G' denote the sub-domain $k[\mathbf{y}_1, \ldots, \mathbf{y}_p]$ of G, $D' = \dim G'$ and F' be the homogeneous field of fractions of G'. The ring G' is graded by sequences of p non-negative integers R' and the function $\lambda(R') = \dim_k G'_{R'}$ is of polynomial type of degree D'. We denote its multiplicity form by $M(G'|R')$, and write the coefficient of $R^{K'}$ in $M(G'|R')$ as $\binom{D'}{K'}e(G'|K')$. Let G'' denote the ring $F'[\mathbf{y}_{p+1}, \ldots, \mathbf{y}_q]$ which is graded by sequences R'' of $q-p$ non-negative integers and the function $\lambda(R'') = \dim_{F'} G''_{R''}$ is of polynomial type of degree $D'' = D - D'$. We denote its multiplicity form by $M(G''|R'')$, and write the coefficient of $(R'')^{K''}$ in $M(G''|R'')$ as $\binom{D''}{K''}e(G''|K)$.

THEOREM 3.4. *Let G be as above. Then, if $|K| = D$, the coefficient of R^K in $M(G|R)$ is zero if $|K'| > D'$ while if $|K'| = D'$,*

$$e(G|K) = e(G'|K')e(G''|K'').$$

PROOF. Fix $R'' = (r''(p+1), \ldots, r''(q))$ and let $G(R'') = \oplus G_R$, where R is restricted to satisfy $r(p+j) = r''(p+j)$ for $j = 1, \ldots, q-p$. Then $G(R'')$ is a finitely generated torsion-free G'-module graded by the sequences of p integers $R' = (r(1), \ldots, r(p))$. Now $G(R'')$, being torsion-free, contains a sub-module L

which is the direct sum of shifts of G', the number of summands being $\lambda(R'') = \dim_{F'}(G''_{R''})$, and, also, $G(R'')/L$ is annihilated by a non-zero-divisor of G'. By a)-c) above, it follows that

$$M(G(R'')|R') = \lambda(R'')M(G'|R')$$

and hence, for large R', $\lambda(G_{R',R''})$ is equal to a polynomial of degree at most D' in R' in which the sum of terms of degree D' is $\lambda(R'')M(G'|R')/D'!$. If we now let R'' vary, $\lambda(R'')$ is, for $R'' >> 0$, equal to a polynomial in R'' in which the sum of the terms of degree D'' is $M(G''|R'')/D''!$. It follows, first, that any term in $M(G|R)$ of degree $> D'$ in $r(1), \dots, r(p)$ has zero coefficient. Furthermore, the sum of the terms of degree D' in $r(1), \dots, r(p)$ and degree D'' in $r(p+1), \dots, r(q)$ in $M(G|R)$ is $\binom{D}{D'}M(G''|R'')M(G'|R')$. This completes the proof.

COROLLARY. *If $D' = D$ and $K'' = 0$, then $e(G|K) = [F : F']e(G'|K')$.*

PROOF. This is a special case of the above, since

$$e(G''|0) = [F : F'].$$

For the final result of this section, we first introduce some notation. We consider a G-module E and an m-primary ideal I of Q which is generated by the set of elements $(a_1, \dots, a_n)$. We will write $G(I)$ for the Rees ring $G[ta_1, \dots, ta_n]$ of the ideal IG of G, t being an indeterminate. $G(I)$ is graded by sets R, s of $q + 1$ non-negative integers. $G(I)_{R,s}$ consists of the elements of the form xt^s, where x belongs to I^sG_R. We will write $E(I)$ for the graded $G(I)$-module defined by $E(I)_{R,s} = I^sE_R$. If J is any ideal of G, we will write $J*$ for the ideal of $G(I)$ consisting of all finite sums $\sum c_st^s$ with c_s in $J \cap I^s$. Note that, if $Q' = Q/J \cap Q$, I' is the ideal $I + J \cap Q$ of Q', and $G' = G/J$, $J*$ is the kernel of the homomorphism of $G(I)$ onto $G'(I')$. We also note the following elementary results.

(1) $\mathrm{Ann}_{G(I)}E(I) = (\mathrm{Ann}_G E)*$,
(2) the relevant prime ideals of $G(I)$ minimal over $\mathrm{Ann}_{G(I)}E(I)$, are the prime ideals $P*$, where P ranges over the relevant prime ideals of G minimal over $(\mathrm{Ann}_G E)$ which do not contain mG. (Note that, if P contains mG, $P*$ is an irrelevant prime ideal of $G(I)$).

We note that 2) implies that no relevant minimal prime ideal of $G(I)$ over $\mathrm{Ann}_{G(I)}E(I)$ contains $mG(I)$, since if $P*$ contains $mG(I), P$ contains mG.

LEMMA 3.5. $\dim(E(I)) \leq \dim E$.

PROOF. We can write

$$\dim(E(I)) = \mathrm{Max}(\dim(G(I)/P*)),$$

the maximum being taken over all relevant minimal prime ideals P of G over $\mathrm{Ann}_G E$ not containing mG. We also have

$$\dim(E) = \mathrm{Max}(\dim(G/P)),$$

the maximum being taken over all P minimal over $\mathrm{Ann}_G E$ without the restriction to those not containing mG. Hence it will be sufficient to prove that $\dim(G(I)) \leq \dim G$ in the case where G is a domain.

Now $\dim(G(I))$ is the maximum of the Krull dimensions of the rings $G(I)_P$, where P ranges over the relevant prime ideals of $G(I)$. If $p = P \cap G$, then the local rings $G(I)_P$ and G_p have the same field of fractions, (the homogeneous field of fractions of G), and $G(I)_P$ dominates G_p. It follows that the Krull dimension of $G(I)_P$ is not greater than the Krull dimension of G_p, which in turn is $\leq \dim G$. Hence the lemma follows.

We now come to the last theorem of this section. $D(E)$ is as defined in section 1.

THEOREM 3.6.

i) *If* $g(I|E|R,s) = \lambda(E_R/I^s E_R)$, $g(I|E|R,s)$ *is of strong polynomial type,*
ii) $\mathrm{hdeg}(g) = \dim E$, *where* $\mathrm{hdeg}(g)$ *is the Hilbert degree of* $g(I|E|R,s)$ *as a function of R,s,*
iii) $\dim E = D(E)$.

PROOF. I) Consider the $G(I)$-module $E(I)/IE(I)$. By theorem 3.2, this is of strong Hilbert type. Now write $f(I|E|R,s)$ for $\lambda((E(I)/IE(I))_{R,s}) = \lambda(I^s E_R/I^{s+1}E_R)$. The function $f(I|E|R,s)$ is then of strong polynomial type and, by theorem 2.5, g is also of strong polynomial type.

The statements of the remaining two parts will follow if we prove *a*) $\mathrm{hdeg}(g) \leq \dim E$, and *b*) $\mathrm{hdeg}(g) \geq D(E)$, since, by lemma 1.2, $D(E) \geq \dim E$.

Proof of *a*). We introduce the functions $g_s(I|E|R) = \lambda(E_R/I^{s+1}E_R)$ for $s = 0, 1, \ldots$. By theorem 2.5, each of these functions is of strong polynomial type. In this case, in view of the inequalities

$$\lambda(E_R/mE_R) \leq \lambda(E_R/I^{s+1}E_R) \leq \lambda(Q/I^s)\lambda(E_R/mE_R),$$

they all have Hilbert degree $a(E) - 1$, and hence $\mathrm{limdeg}(g_s) = a(E) - 1$. Again invoking Theorem 2.5, and writing $\mathrm{hdeg}(f)$ for the degree of $f(I|E|R,s)$ as a function of R, s, we therefore have

$$\mathrm{hdeg}(g) = \mathrm{Max}(\mathrm{hdeg}(f) + 1, a(E) - 1).$$

But, as $f(I|E|R,s) = \lambda((E(I)/IE(I))_{R,s})$, a similar argument to the above shows that $\mathrm{hdeg}(f) = a(E(I)) - 1$ and hence that

$$\mathrm{hdeg}(g) = \mathrm{Max}(a(E(I)), a(E) - 1).$$

Since no relevant prime ideal of $G(I)$ minimal over $\mathrm{Ann}_{G(I)}E(I)$ contains $mG(I)$, it follows that $a(E(I)) \leq \dim E(I) \leq \dim E$ while $a(E) - 1 \leq \dim E$. Hence the proof of a) is complete.

We now turn to the proof of b). This will follow if we prove that, for each relevant prime ideal P minimal over $\mathrm{Ann}_G E$,

$$\mathrm{hdeg}(g) \geq \dim(Q/Q \cap P) + t(P).$$

Now there is a non-zero quotient module E_1 of E which is annihilated by P, and which, as a G/P-module, is torsion-free. Since $\lambda((E_1)_R/(I^s E_1)_R) \leq \lambda(E_R/(I^s E_R))$ for all R, s, it follows that we can replace E by E_1, G by G/P, Q by $Q/Q \cap P$, and reduce the proof to the case where G is a domain and E is a torsion-free G-module. We then have to prove that

$$\mathrm{hdeg}(g) \geq t + d,$$

where t is the transcendence degree of the homogeneous field of fractions F of G over the field of fractions K of Q.

We now fix R and consider $(E_R/I^s E_R)$ as a function of s only. It is a classical result that, for large s this is equal to a polynomial in s of degree d, the coefficient of s^d being $v(R)e(I)/d!$, where $e(I)$ is the multiplicity of I, and $v(R)$ is the dimension of the vector space $K \otimes_Q E_R$ over K. But for R large, $v(R)$ is equal to a polynomial in R of degree t. Hence the polynomial in R, s equal to $g(I|E|R, s)$ for large R, s contains non-zero terms of degree t in R and d in s and so has degree at least $d + t$. Hence we have proved b) and thus completed the proof of parts ii) and iii) of theorem 3.6.

4. Admissible Pairs of Rings and Modules.

As in section 1, G will denote the multi-graded ring $Q[\mathbf{y}]$, where

$$\mathbf{y} = (y(i,j) \in G_{\delta(i)} | i = 1, \ldots, q; j = 1, \ldots, m(i)).$$

We further remind the reader that a power series $\sum c_R Z^R$ is of polynomial type (resp., strong polynomial type) if c_R is a polynomial in R for R large (resp., for almost all R), and a module E graded by the sets R is of Hilbert type (resp., strong Hilbert type) if $\lambda(E_R)$ is finite if R is large (for almost all R) and the power series $\sum \lambda(E_R) Z^R$ is of polynomial type (resp., strong polynomial type).

LEMMA 4.1. *Let*

$$\mathbf{v} = (v(i,j) \in G_{\delta(i)} | i = 1, \ldots, q; j = 1, \ldots, n(i)).$$

satisfy the condition that $\lambda(G_R/(\mathbf{v}G)_R) < \infty$ *for almost all* R *(for* $R >> 0$*). Then, if* E *is any finitely generated* G*-module, the homology groups* $H_i(E|\mathbf{v})$ *of the Koszul complex* $K(E|\mathbf{v})$ *of* E *with respect to* $\mathbf{v}$ *satisfy* $\lambda((H_i(E|\mathbf{v})_R) < \infty$ *for almost all* R *(resp., for* $R >> 0$*), and the functions* $\Lambda(H_i(E|\mathbf{v})|Z)$ *and* $\chi(E|\mathbf{v}|Z) = \sum(-1)^i \Lambda(H_i(E|\mathbf{v})|Z)$ *are of strong polynomial type (resp., polynomial type).*

PROOF. The first statement follows since the modules $H_i(E|\mathbf{v})$ are finitely generated and annihilated by the elements of $\mathbf{v}$. It now follows from theorem 3.2 that $\Lambda(H_i(E|\mathbf{v})|Z)$ is of strong polynomial type (polynomial type) for each i and hence $\chi(E|\mathbf{v}|Z)$ is of strong polynomial type (polynomial type).

DEFINITIONS. Let $G' = Q[\mathbf{u}]$, where

$$\mathbf{u} = (u(i,j) \in G_{\delta(i)} | i = 1, \dots, q; j = 1, \dots, n(i)),$$

be a graded sub-ring of G. We will say that the pair (G, G') is admissible if G' is subject to the condition that $\lambda(G_{\delta(i)}/G'_{\delta(i)})$ is finite for each i, and hence $\lambda(G_R/G'_R)$ is finite for all R.

Let E be a finitely generated G-module and E' be a finitely generated G'-module contained in E. We will say that the pair (E, E') is (G, G')-admissible (or simply admissible) if $\lambda(E_R/E'_R)$ is finite when $R >> 0$.

THEOREM 4.2. *Let (G, G') be admissible and (E, E') be (G, G')-admissible. Let $b \geq a(E')$. Then, if*

$$\mathbf{v} = (v(i,j) | i = 1, \dots, q; j = 1, \dots, b(i))$$

is a joint reduction of G' of type $B = (b(1), \dots, b(q))$, where $|B| = b$,

i) *$\lambda(H_i(E|\mathbf{v})_R)$ is finite if $R >> 0$ for $i = 0, \dots, b$;*
ii) *$\chi(E|\mathbf{v}|Z) = \sum(-1)^i \lambda(H_i(E|\mathbf{v})|Z)$ is of polynomial type;*
iii)
$$(1 - Z)^B \Lambda(E/E'|Z) \sim \chi(E|\mathbf{v}|Z),$$
where
$$(1 - Z)^B = (1 - z_1)^{b(1)} \cdots (1 - z_q)^{b(q)};$$
iv) *E/E' is of Hilbert type. Further, if $H(E/E'|W)$ is the Hilbert polynomial of E/E', and we write $H_b(G/G'|E|W)$ for the sum of the terms of $H(E/E'|W)$ of degree greater than or equal to b, then $H_b(G/G'|E|W)$ is independent of the choice of E', i.e., depends only on G, G', E and b.*

PROOF. First we note that $(\mathbf{v}E')_R = E'_R$ if $R >> 0$ for any finitely generated G'-module E'. Hence $(\mathbf{v}E)_R \supseteq E'_R$ if $R >> 0$, implying that $(E/\mathbf{v}E)_R$ has finite length if $R >> 0$. Parts i), ii) now follow by lemma 4.1. We now consider the Koszul complexes of the modules $E, E', E'' = E/E'$ with respect to the set of elements $\mathbf{v}$, E' and E'' being considered as G'-modules, while E is considered as either a G'-module or as a G-module. First consider $K(E'|\mathbf{v})$. As E' is finitely generated as a G'-module the homology modules of $K(E'|\mathbf{v})$ are annihilated by $\mathbf{v}G'$ and hence $\lambda(H_i(E'|\mathbf{v})_R) = 0$ if $R >> 0$. It follows that $\chi(E'|\mathbf{v}|R) = 0$ if $R >> 0$, and $\chi(E'|\mathbf{v}|Z) \sim 0$.

Next, we have an exact sequence

$$0 \to K(E'|\mathbf{v}) \to K(E|\mathbf{v}) \to K(E''|\mathbf{v}) \to 0,$$

and this implies that

$$\chi(E|\mathbf{v}|Z) \sim \chi(E'|\mathbf{v}|Z) + \chi(E''|\mathbf{v}|Z) \sim \chi(E''|\mathbf{v}|Z).$$

Now $\chi(E''|\mathbf{v}|Z) \approx (1 - Z)^B \Lambda(E''|Z)$ by lemma 3.1, and hence we have proved iii). It now follows that E/E' is of Hilbert type, since $\chi(E|\mathbf{v}|Z)$ is of polynomial type. Further, if K is such that $|K| \geq b$, then we can choose B such that

$|B| = b$ and $b(i) \leq k(i)$ for each i. If we now choose $\mathbf{v}$ as in the statement of the theorem, then the coefficient of W^K in $H(E/E'|W)$ is the same as the coefficient of W^{K-B} in the Hilbert polynomial of $\chi(E|\mathbf{v}|Z)$ and so depends only on E and $\mathbf{v}$. But the coefficient of W^K in $H(E/E'|W)$ does not depend on the particular choice of B and hence not on the choice of the joint reduction $\mathbf{v}$ of G'. Hence $H_b(G/G'|E|W)$ depends only on G, G', E and b.

In our next theorem, we wish to relate $\operatorname{hdeg}(E/E')$ to $\dim E$, and to this end we make use of a suitable diagonal of G to reduce the proof to the case $q = 1$. To be precise, suppose that $K = (k(1), \ldots, k(q))$ is a set of positive integers. Then, if we consider the rings $\Delta_K G, \Delta_K G'$ and the modules $\Delta_K E, \Delta_K E'$ in place of G, G' and E, E', we do not alter $\dim E$, $\dim E'$, while we can choose K so that $\operatorname{hdeg}(\Delta_K E/\Delta_K E') = \operatorname{hdeg}(E/E')$. We may further assume that $(\Delta_K E)_{r+1} = (\Delta_K G)_r.(\Delta_K E)_1$ and $(\Delta_K E')_{r+1} = (\Delta_K G')_r.(\Delta_K E')_1$ for all $r \geq 0$, and, finally, that $\lambda((\Delta_K E)_1/(\Delta_K E')_1)$ is finite. The notation of the theorem following is as in theorem 4.2.

THEOREM 4.3.

$$\operatorname{hdeg}(E/E') \leq \dim E.$$

PROOF. By the remarks above we can assume that $q = 1$, and the following conditions hold

$$a) E = E_0 + GE_1, \quad b) E' = E'_0 + G'E'_1, \quad c) \lambda((E/E')_r) < \infty \quad \text{if} \quad r > 0.$$

We now introduce the G'-module E'' defined as $E'_0 + G'E_1$, so that $E \supseteq E'' \supseteq E'$. By c), we can find an m-primary ideal I annihilating both E_1/E'_1 and E_2/E'_2. Hence

$$i) IE_1 \subseteq E'_1 \subseteq E''_1, \quad ii) IG_1E_1 = IE_2 \subseteq E'_2 \subseteq E''_2 = G'_1E_1.$$

But then, if $r \geq 0$,

$$I^r E_{r+1} = I^r (G_1)^r E_1 \subseteq (G'_1)^r E_1 = E''_{r+1} \quad \text{and} \quad IE''_{r+1} = (G'_1)^r.IE_1 \subseteq E'_{r+1}.$$

Hence $I^r E_r \subseteq E'_r$ for all $r > 0$, implying that $\lambda(E_r/E'_r) \leq \lambda(E_r/I^r E_r)$ for all $r > 0$. By theorem 3.5, the function $\lambda(E_r/I^s E_r)$ is, as a function of r, s, of strong polynomial type, its degree being $\dim E$. Hence, the function $\lambda(E_r/I^r E_r)$ is, as a function of r for r large, equal to a polynomial in r of degree at most $\dim E$, and the same will be true of the function $\lambda(E_r/E'_r)$.

We now come to the main object of this section. This is to associate with each pair of admissible graded Q-algebras (G, G') an additive function $e(G/G'|E|W)$ on the category of finitely generated graded G-modules E whose values are forms of degree $D = \dim G$ in $W = (w_1, \ldots, w_q)$ with integer coefficients. To define $e(G/G'|E|W)$ we choose, as is always possible, a G'-module $E' \subseteq E$ such that (E, E') is (G, G')-admissible. For example, if Σ is a set of homogeneous generators of E as a G-module, we could choose E' to be the G'-module with the same set of generators. It follows from theorem 4.2 that the power series

$\Lambda(E/E'|Z) = \sum \lambda(E_R/E'_R)Z^R$ is of polynomial type and from theorem 4.3 that its Hilbert polynomial $H(E/E'|W)$ has degree at most D.

DEFINITION. $e(G/G'|E|W)$ is the sum of the terms of $H(E/E'|W)$ of degree D.

If K is a set of non-negative integers such that $|K| = D$, $e(G/G'|E|K)$ is defined to be the coefficient of W^K in $e(G/G'|E|W)$. Finally we will write $e(G/G'|W)$ for $e(G/G'|G|W)$.

If we are to make full use of theorem 4.2, it is necessary that we should be able to take the integer b of that theorem equal to $D = \dim G$, which would require that $D \geq a(E')$. If, now, we replace B in that theorem by a set K of non-negative integers such that $|K| = D$, and choose a joint reduction $\mathbf{v}$ of G' of type K, we would have

$$\chi(E|\mathbf{v}|Z) \sim e(G/G'|E|K)w_1 \cdots w_q.$$

The condition $D \geq a(E')$ will hold for all choices of the G'-module E' if it holds when $E' = G'$. We therefore introduce the definitions.

DEFINITIONS. The admissible pair of graded rings (G, G') will be termed acceptable if they satisfy the condition that $\dim G \geq a(G')$. The ring G is termed acceptable if $\dim G \geq a(G)$.

There are two conditions, $a), b)$ below, which will individually imply that the admissible pair (G, G') is acceptable.

a) *The ring G is acceptable.*

b) $G'_{\delta(i)} \subseteq mG_{\delta(i)}$ *for at least one i.*

We will prove these implications in theorem 4.4 below.

Once $e(G/G'|E|W)$ and its coefficients $e(G/G'|E|K)$ are defined, we can immediately define the form $M(G/G'|E|R)$. In the present section, we will find it more convenient to use $e(G/G'|E|W)$ in the proofs.

THEOREM 4.4. *Let (G, G') be an admissible pair of graded rings and let (E, E') be a (G, G')-admissible pair of modules. Then the following statements hold.*

i) $\dim G' \leq \dim G$.

ii) *If either of the conditions a) or b) are satisfied by G, G', then (G, G') is acceptable.*

iii) *If (G, G') is acceptable, $|K| = D$ and $\mathbf{v}_K$ is a joint reduction of G' of type K, the coefficient $\chi(E|\mathbf{v}_K|R)$ of Z^R in $\chi(E|\mathbf{v}_K|Z)$ for $R >> 0$ is $e(G/G'|E|K)$ and does not depend on the choice of $\mathbf{v}_K$.*

iv) *The terms of degree D in $H(E/E'|W)$ do not depend on the choice of E'.*

PROOF. i) We will use the alternative dimension function $D(G)$ which is equal to $\dim G$ by theorem 3.5, and for the ring G of the theorem, we will write D for $D(G) = \dim G$.

The radical of G' is the intersection of the radical of G with G'. Hence, if p is any relevant minimal prime ideal of G', there is a relevant minimal prime ideal P of G such that $p = P \cap G'$. If we choose p such that $D(G') = D(G'/p)$, then it will be sufficient to prove that $D(G/P) \geq D(G'/p)$, since $D(G/P) \leq D(G)$. Hence we may assume that G is a domain and prove that $D(G) \geq D(G')$. Let F, F' be the homogeneous fields of fractions of G, G' and K be the field of fractions of Q. Then $F \supseteq F'$ and

$$D(G) = \text{trans.deg}_K F + d \geq \text{trans.deg}_K F' + d = D(G')$$

and the result is proved.

ii) Assume first that a) holds and that $a(G') > D = \dim G$. Then, by i), $a(G') > \dim G'$, and hence, by lemma 1.2, there is a prime ideal p of G' meeting Q in m such that $\dim(G'/p) = D$ and, therefore, $a(G'/p) = D + 1$. If P is a minimal prime ideal of G meeting G' in p, then $\dim G/P$ must also equal D and, as $P \cap Q = m$, $a(G/P) = D + 1$ which contradicts a).

Now assume that b) holds. If possible, let p be a minimal relevant prime ideal of G' meeting Q in m, and let P be a minimal prime ideal of G meeting G' in p. By the remarks in i) above, we know such a P exists. But then, as P contains m, p contains $G'_{\delta(i)}$ for some i and so is irrelevant. Hence no minimal relevant prime ideal of G' meets Q in m and $a(G') \leq \dim G' \leq D$.

We now impose on the pair (G, G') the restriction that it be acceptable.

iii) We note that the degree of $H(E/E'|W)$ is $\leq \dim E$, and hence if $\dim E < D$ then $\chi(E|\mathbf{v}_K|Z) \sim 0$ while $e(G/G'|E|W) = 0$ whatever E'. We therefore assume that $\dim E = D$. Since $D \geq a(E')$, the result is an immediate consequence of 4.2 iv).

iv) follows from iii).

In our next theorem we list some elementary properties of $e(G/G'|E|W)$.

THEOREM 4.5. *Let the admissible pair of rings (G, G') be acceptable and let $D = \dim G$. Then the following statements hold.*

i) *If*

$$0 \to E_1 \to E \to E_2 \to 0$$

is an exact sequence of G-modules,

$$e(G/G'|E|W) = e(G/G'|E_1|W) + e(G/G'|E_2|W).$$

ii)

$$e(G/G'|E(K)|W) = e(G/G'|E|W),$$

where $E(K)$ is a shift of E.

iii) *Let $G \supseteq G' \supseteq G''$, and $\lambda(G_R/G''_R)$ be finite for $R >> 0$. Assume that the pairs (G, G') and (G', G'') are acceptable and, in addition, that $\dim G' = \dim G = D$. Then*

$$e(G/G''|W) = e(G/G'|W) + e(G'/G''|W).$$

iv) *If G is a finite G'-module, then*

$$e(G/G'|E|W) = 0$$

for all E.

v)

$$e(G/G'|E|W) = \sum \lambda(E_P)e(G/G'|G/P|W)$$

the sum being over the finite set of prime ideals P of G such that $\dim G/P = D$.

PROOF. i) With the notation of theorem 4.4,

$$\chi(E|\mathbf{v}_K|Z) = \chi(E_1|\mathbf{v}_K|Z) + \chi(E_2|\mathbf{v}_K|Z)$$

by the additive property of the Euler-Poincaré characteristic and the result follows from 4.4 applied to all K such that $|K| = D$.

ii) This follows immediately from lemma 2.4.

iii) We first observe that the conditions imply that (G, G'') is also admissible and, further, acceptable. Since both $e(G/G'|W)$ and and $e(G'/G''|W)$ are sums of terms of degree D, the result now follows from the equation

$$\lambda(G_R/G''_R) = \lambda(G_R/G'_R) + \lambda(G'_R/G''_R).$$

iv) For E is then a finitely generated G'-module and can be taken as E'.

v) This is a standard consequence of i), ii) and the fact that $e(G/G'|E|W) = 0$ if $\dim E < D$. We recall the argument. The module E has a filtration

$$E = E_0 \supseteq E_1 \supseteq \ldots \supseteq E_N = (0),$$

where $E_i/E_{i+1} \cong (G/P_i)(K_i)$ for some prime ideal P_i of G and some shift K_i. Application of i), ii) now implies that $e(G/G'|E|W) = \sum e(G/G'|G/P_i|W)$. We next observe that if $\dim(G/P_i) < D$, then $e(G/G'|G/P_i|W) = 0$. Hence we can restrict attention to those i such that $\dim(G/P_i) = D$. In this case P_i is a minimal prime ideal. Let P be a minimal prime ideal of G. Then the homogeneous localisation G_P of G at P is an artin ring and E_P is a finite G_P-module. E_P will have a filtration

$$E_P = (E_0)_P \supseteq (E_1)_P \supseteq \ldots \supseteq (E_N)_P = (0),$$

and, if $E_i/E_{i+1} \cong (G/P_i)(K_i)$ with $P_i \neq P$, then $(E_i)_P/(E_{i+1})_P = (0)$, while, if $P_i = P$, $(E_i)_P/(E_{i+1})_P$ is of length 1 as a G_P-module. Hence $\lambda(E_P)$ is equal to the number of values of i for which $P_i = P$, and

$$e(G/G'|E|W) = \sum \lambda(E_P)e(G/G'|G/P|W)$$

the sum being over the finite set of prime ideals P of G with $\dim G/P = D$.

The result of v) above indicates that the case where G is a domain is of particular importance. We now consider this case in more detail. Condition a) is then satisfied save in the case $\dim Q = 0$, or, in other words, when Q is a field k. The condition that (G, G') is admissible is then satisfied if G' is any graded sub-domain of G generated over k by elements of total degree 1, and if E is a finitely generated G-module, (E, E') is admissible for any finitely generated G'-module E'. We note further that E and E' are themselves of Hilbert type, and the definition of $e(G/G'|E|W)$ can be written as the sum of the terms of degree D in $H(E|W) - H(E'|W)$. Further, since $a(G') = \dim G' + 1$, it follows that (G, G') is acceptable except in the case when $\dim G = \dim G'$. If $\dim G = \dim G'$, it is not true that $e(G/G'|E|W)$ as defined, is independent of the choice of E'. For example, if $E = G$, both $E' = G'$ and $E' = 0$ are valid choices of E', but give different values to $e(G/G'|E|W)$. Hence the theory developed above is not valid in this case.

We now consider an alternative approach to the form $e(G/G'|E|W)$ which will give us extra information.

DEFINITION. We will term G/G' simple if $G = G'[y]$, where the element y is of degree $\delta(i)$ for some i and satisfies the condition that $ym \subseteq G'_{\delta(i)}$.

LEMMA 4.6. *If G, G' is an admissible pair of rings, we can find a sequence of rings $G = G(0) \supseteq G(1) \supseteq \cdots \supseteq G(N) = G'$ such that $G(i)/G(i+1)$ is simple for $i = 0, \ldots, N-1$.*

PROOF. We proceed by induction on $l(G/G') = \sum \lambda(G_{\delta(i)}/G'_{\delta(i)})$, the sum being from $i = 1, \ldots, q$. If $l(G/G') = 1$, G/G' is simple. If $l(G/G') > 1$, we can choose j such that $\lambda(G_{\delta(j)}/G'_{\delta(j)}) > 0$. Let r be the least integer such that $m^r G_{\delta(j)} \subseteq G'_{\delta(j)}$ and choose y in $m^{r-1}G_{\delta(j)}$ and not in $G'_{\delta(j)}$. Let G'' be $G'[y]$. Then $G \supseteq G'' \supseteq G'$, G''/G' is simple, and $l(G/G'') = l(G/G') - 1$.

THEOREM 4.7. *Let G, G' be acceptable and let $E, E' \subseteq E$ be, respectively, a finitely generated G-module and a finitely generated G'-module with $E = GE'$. Then:*

i) *the power series $\Lambda(E/E'|Z) = \sum \lambda(E_R/E'_R)Z^R$ is of strong polynomial type;*
ii) *the coefficients of $e(G/G'|E|W)$ are non-negative.*

PROOF. Suppose we choose a sequence of rings $G_{(i)}$ as in lemma 4.6. Then if we write $E_{(i)} = G_{(i)}E'$, we have

$$\Lambda(E/E'|Z) = \sum \Lambda(E_{(i)}/E_{(i+1)}|Z).$$

It follows that to prove both i) and ii) it is sufficient to consider the case where G/G' is simple. Suppose therefore that $G = G'[y]$ with y of degree $\delta(i)$ and that $ym \subseteq G'_{\delta(i)}$. We can express E as the union of the ascending sequence of finitely generated G'-modules $E' + yE' + ... + y^r E'$. Write M_r for the quotient module

$$(E' + yE' + ... + y^r E')/(E' + yE' + ... + y^{r-1}E').$$

Each of the modules M_r is annihilated by m and so $\lambda((M_r)_R)$ is finite for all r, R. Hence, by theorem 3.2, M_r is of strong Hilbert type. Multiplication by y

induces a homomorphism of degree $\delta(i)$ of M_r onto M_{r+1} for all r. It follows that we can choose an integer r' such that, if we write M' for $M_{r'}$, $M_{r'+s}$ is isomorphic to $M'(s\delta(i))$ for all $s \geq 0$. Hence, if $\Lambda(M_r|Z) = \sum \lambda((M_r)_R)Z^R$, and $\Lambda(M'|Z) = \sum \lambda(M'_R)Z^R$,

$$\Lambda(E/E'|Z) = \sum_r \Lambda(M_r|Z) + \sum_s z_i^s \Lambda(M'|Z) \quad (*),$$

the first sum being from $r = 1$ to $r' - 1$ and the second from $s = 0$ to infinity, the infinite sum having a meaning since only a finite number of terms in it can make a contribution to the coefficient of Z^R for any given R. i) is now immediate. Hence we can write $\Lambda(E/E'|Z), \Lambda(M_r|Z), \Lambda(M'|Z)$ as polynomials $L(E/E'|Z,W), L(M_r|Z,W), L(M'|Z,W)$ in Z, W, and we can rewrite $(*)$ as

$$L(E/E'|Z,W) = \sum_r L(M_r|Z,W) + w_i L(M'|Z,W) \quad (**).$$

ii) If $D = \dim G$, the sum of the terms of $L(E/E'|Z,W)$ of degree $D + q$ in $w_1, \ldots, w_q$ which involve each of $w_1, \ldots, w_q$ to a positive exponent is equal to $w_1 ... w_q e(G/G'|E|W)$. Since hdeg$M_r \leq a(E') - 1 < D$, and hdeg$M' \leq a(E') - 1 < D$, it follows that the terms of this sum can only occur in the last term $w_i L(M'|Z,W)$ on the right-hand side of $(**)$. Such a term can arise in two ways. First it can arise from a term $c_\mu w_1 \cdots w_q \mu$ of $L(M'|Z,W)$, where μ is a monomial of degree $D - 1$ in W, so that the coefficient of $w_i \mu$ in $e(G/G'|E|W)$ is c_μ. Now the monomial $w_1 \cdots w_q \mu$ is clearly critical for M', implying that $c_\mu \geq 0$. Next it can arise from a term $f(z_i) w_1 \cdots \hat{w}_i \cdots w_q \mu'$ of $L(M'|Z,W)$, where μ' is a monomial in $w_1, \ldots, w_{i-1}, w_{i+1}, \ldots, w_q$ of degree D. Again the monomial $w_1 ... \hat{w}_i ... w_q \mu'$ is critical for M', and hence the coefficients of $f(z_i)$ are non-negative. Using the identity $z_i w_i = w_i - 1$, we see that the coefficient of μ' in $e(G/G'|E|W)$ is $f(1) \geq 0$.

We now return to theorem 3.6 and the function $g(I|E|R,s) = \lambda(E_R/I^s E_R)$, where E is any finitely generated graded G-module and I is an m-primary ideal of Q. This function was proved in theorem 3.6 to be of strong Hilbert type and to have degree equal to $\dim E \leq \dim G$. Let $p(R,s)$ be the polynomial equal to $g(I|E|R,s)$ for almost all R, s and let $H(p|W,w)$ be the associated form in W, w as in section 2.

DEFINITION. $e(I|E|W,w)$ is the sum of the terms of $H(p|W,w)$ of degree $D = \dim G$. This, if not zero, will be a form of degree D in W, w.

THEOREM 4.8.

i) *$e(I|E|W,w)$ can be expressed in the form*

$$e(I|E|W,w) = \sum \lambda(E_P) e(I|G/P|W,w)$$

where P ranges over all prime ideals of G such that $\dim(G/P) = D$.

ii) *The coefficients of $e(I|E|W,w)$ are non-negative.*

PROOF. As in theorem 3.6, we consider the Rees ring $G(I) = G[ta_1, \dots, ta_n]$, where t is an indeterminate over G, and, for any finitely generated G-module E, the module $E(I)$ similarly defined to be the set of finite sums $\sum c_r t^r$ with $c_r \in I^r E$. Both $G(I)$ and $E(I)$ are graded by sequences of $q+1$ integers $(r(1), \dots, r(q), s)$ We abbreviate this sequence as R, s. We now introduce the pair of rings $G[t]$ and $G(I)$. We grade $G[t]$ by taking $G_R t^s$ as the module of elements of $G[t]$ of degree (R, s). Then the grading of $G(I)$ in which the module of elements of degree (R, s) is $I^s G_R t^s$ is the same as that induced by considering $G(I)$ as a graded Q-sub-module of $G[t]$. We define $E[t]$ to be the $G[t]$-module consisting of all sums $\sum c_s t^s$ with c_s in E, and graded in a similar manner to that adopted for $G[t]$. $E(I)$ is then a graded Q-sub-module of $E[t]$. We note that $(G[t], G(I))$ is an admissible pair of graded rings, and that $(E[t], E(I))$ is a $(G[t], G(I))$-admissible pair of modules. We also note that, for all $s > 0$, $G(I)_{R,s} \subseteq mG[t]$, so that the pair $G[t], G(I)$ satisfies condition b) and hence is acceptable. Finally $E[t] = G[t]E(I)$. Now $E[t]_{R,s}/E(I)_{R,s}$ is isomorphic to $E_R/I^s E_R$. Hence the series $\sum \lambda((E[t]/E(I))_{R,s}) Z^R z^s$ and $\sum g(I|E|R, s) Z^R z^s$ are the same. The two parts of the theorem will now follow if we apply theorem 4.5 v) and theorem 4.7 ii) with the rôle of G being played by $G[t]$, that of G' by $G(I)$, that of E by $E[t]$ and that of E' by $E(I)$.

COROLLARY. *If $G'_{\delta(i)} \subseteq mG_{\delta(i)}$ for $i = 1, \dots, q$, then $e(G/G'|G|W) \neq 0$.*

PROOF. The hypotheses imply that $G'_R \subseteq m^{|R|} G_R$, which in turn implies that $\lambda(G_R/G'_R) \geq \lambda(G_R/m^{|R|}G_R)$. But the above shows that $\lambda(G_R/m^{|R|}G_R)$ is equal to a polynomial in R of degree exactly $\dim G$ if $R \geq 0$ and hence the same is true of $\lambda(G_R/G'_R)$. It follows that $H(G/G'|G|W)$ is of degree exactly $\dim G$, and hence $e(G/G'|G|W)$ is not zero.

THEOREM 4.9. *Write $e(I|E|W, w)$ in the form*

$$e(I|E|W, w) = \sum e_l(W) w^l.$$

Then $e_l(W) = 0$ if either $l < D - a(E) + 1$ or $l > d = \dim Q$.

PROOF. We replace $e(I|E|W, w)$ by $M(R, s)$ The latter only differs from $e(I|E|W, w)$ by replacing W, w by R, s and multiplying the coefficients by certain non-zero multinomial coefficients. Now we can find integers $r_0(i) (i = 1, \dots, q)$, s_0 such that, if $r(i) \geq r_0(i) (i = 1, \dots, q), s \geq s_0$, then $\lambda(E_R/I^s E_R)$ is equal to the coefficient of $Z^R z^s$ in the expansion of the corresponding Hilbert polynomial $H(W, w)$ as a power series in Z, z. This coefficient is a polynomial in R, s of degree D, the sum of the terms of degree D being $M(R, s)/D!$. Fix $s \geq s_0$. Then by theorem 3.2, the resulting polynomial in R has degree at most $a(E) - 1$. Since this is true for all $s \geq s_0$, $e_l(W) = 0$ if $l < D - a(E) + 1$.

Next, fix R subject to the conditions $r(i) \geq r_0(i)$. Then the resulting polynomial is $\lambda(E_R/I^s E_R)$ which is a polynomial in s of degree at most d. Hence $e_l(W) = 0$ if $l > d$.

To simplify the statement of the theorem which follows we commence by introducing some notation. Let σ be a sub-set of $(1, \dots, q)$ containing t elements and σ' be its complement. If X denotes a sequence $(x(1), \dots, x(q))$, we define $X(\sigma) = (x(j)|j \in \sigma)$ and $X(\sigma') = X - X(\sigma)$.

In order to further simplify the notation, we will find it convenient in what follows to renumber $(1, \dots, q)$ so that σ is $(1, \dots, t)$.

Let $G = Q[\mathbf{y}_1, \dots, \mathbf{y}_q]$ be a graded domain such that $\dim G = D$. Let $G(\sigma)$ denote the sub-ring $Q[\mathbf{y}_1, \dots, \mathbf{y}_t]$, $D(\sigma)$ its dimension and $F(\sigma)$ its homogeneous field of fractions. We now consider the ring $A(\sigma) = F(\sigma)[\mathbf{y}(\sigma')]$ as graded by sets $R(\sigma')$. If we denote the dimension of the vector space of elements of $A(\sigma)$ of degree R' by $\lambda(R')$, the series $\sum \lambda(R(\sigma'))Z(\sigma')^{R(\sigma')}$ is of polynomial type and we can define its multiplicity form which we will write as $e_\sigma(G|W(\sigma'))$. $e_\sigma(G|W(\sigma'))$ has degree $D - D(\sigma)$.

THEOREM 4.10.

i) *Let G, G' be an acceptable pair of rings. Then, if σ is any sub-set of $(1, \dots, q)$, K is a set of q non-negative integers such that $|K| = D$ and $|K(\sigma)| > D(\sigma)$, the coefficient $e(G/G'|E|K)$ of W^K in $e(G/G'|E|W)$ is zero for all finitely generated G-modules E.*

ii) *The sum of the terms of $e(G/G'|G|W)$ of degree $D(\sigma)$ in $W(\sigma)$ and $D - D(\sigma)$ in $W(\sigma')$ is $e(G(\sigma)/G'(\sigma)|G(\sigma)|W(\sigma))e_\sigma(G|W(\sigma'))$.*

PROOF. Using theorem 4.5(v), we can reduce the proof of i) to the case where G is a domain and $E = G$. We will therefore consider i) and ii) together, and assume that G is a domain and $E = G$. We will again find it convenient to renumber $(1, \dots, q)$ so that σ is $(1, \dots, t)$. We now fix $L = (l(1), \dots, l(t))$ such that $|L| = D(\sigma)$ and choose a joint reduction $\mathbf{v}$ of $G'(\sigma)$ of type L. Note that $\mathbf{v}$ will then also be a joint reduction of G'.

We now consider the Koszul complex $K(G|\mathbf{v})$ and, more particularly, its component $K(G|\mathbf{v}|R)$ of degree R for a value of $R >> 0$. Its homology modules are of finite length. Hence we can define $\chi(G|\mathbf{v}|R)$ and, therefore, to within equivalence, the power series $\chi(G|\mathbf{v}|Z) = \sum \chi(G|\mathbf{v}|R)Z^R$ which is of Hilbert type, and is equivalent to the series $(1 - Z)^L \Lambda(G/G'|Z)$. It follows that its multiplicity form is $e_L(G/G'|G|W)$, defined as the quotient when $e(G/G'|G|W)$ is divided by W^L. We can also calculate $\chi(G|\mathbf{v}|Z)$ by considering G as the direct sum of finitely generated $G(\sigma)$-modules $G_{R'}$, where R' ranges over all sequences of integers $(r(t+1), \dots, r(q))$ and $G_{R'}$ is the direct sum of the components G_R of G for which $R(\sigma') = R'$. $G_{R'}$ is a torsion free $G(\sigma)$-module of rank $\lambda(R')$. It follows that the multiplicity form of $\chi(G_{R'}|\mathbf{v}|Z(\sigma))$ is $\lambda(R')e(G(\sigma)/G'(\sigma)|G(\sigma)|L)$; hence

$$e_L(G/G'|G|W) = e(G(\sigma)/G'(\sigma)|G(\sigma)|L)e_\sigma(G|W(\sigma')),$$

and so does not involve $W(\sigma)$. This implies i), and by multiplying by W^L and summing over L, we obtain ii).

Our final result is limited to the graded case i.e., the case $q = 1$. In this case we note that $e(G/G'|W)$ takes the form $e(G/G')w^D$, and we will consider the non-negative integer $e(G/G')$ in place of $e(G/G'|W)$.

THEOREM 4.11. *Let (Q, m, k, d) be a quasi-unmixed local ring. Further, let $G = Q[y_1, \dots, y_m]$ be a finitely generated graded Q-algebra, where $y_1, \dots, y_m$ have degree 1 and which satisfies the condition that for every minimal prime ideal P of G, $\dim(G/P) = \dim G$. Let $G' = Q[u_1, \dots, u_n]$ be a graded sub-algebra of G, where $u_1, \dots, u_n$ have degree 1 and $\lambda(G_1/G'_1)$ is finite. Then, if $e(G/G') = 0$, G is a finite G'-module.*

PROOF. We commence by reducing the proof to the case where G is a domain and G/G' is simple.

If we denote by J the ideal $mG + (u_1, \dots, u_n)$ of G, the condition that G is a finite G'-module is equivalent to the condition that J is irrelevant.

Now suppose that $P_1, \dots, P_q$ are the minimal prime ideals of G. Then J will be an irrelevant ideal of G if and only if $J + P_i/P_i$ is an irrelevant ideal of G/P_i for each i. Hence we may assume that G is a domain and we will denote $\dim G$ by D.

The reduction to the case where G/G' is simple follows immediately from lemma 4.6.

Since G/G' is simple, $G_r = J_r + Qy^r$, and $my^r \subseteq J_r$ for all $r > 0$. We will suppose that J is relevant and prove that this implies that $e(G/G') > 0$, which will prove the theorem. Since, for all r, $J_r \subseteq G_r$, $G/J \cong k[X]$, where X is an indeterminate and hence J is a prime ideal. We will therefore now write it as p. We now recall that if the set of elements $\mathbf{v} = (v_1, \dots, v_D)$ is a reduction of G', or what is the same thing, the ideal of G generated by $\mathbf{v}$ is a reduction of p, then by theorem 4.2 iii), $\chi(G|\mathbf{v}|Z) \approx e(G/G')w$. Since $\mathbf{v}$ is a reduction of p, p is the only relevant prime ideal of G containing $\mathbf{v}G$.

Now suppose we form the homogeneous ring of fractions L of G with respect to the set S of homogeneous elements of G not in p. The maximal ideal n of the local ring L is generated by the elements $\xi_i = u_i/y$, $(i = 1, \dots, n)$, since $ym \subseteq G'_1$, and meets Q in m. Hence its residue field is isomorphic to k. Since Q is quasi-unmixed, the dimension inequality becomes an equality and

$$\dim L = t(G) + d = D.$$

Now we consider the complete ring of fractions of G with respect to S. This will be isomorphic to $L[Y, Y^{-1}]$, where Y is an indeterminate of degree 1 and is the image of y.

Next suppose that E is any finitely generated G-module, such that the only relevant prime ideal of G containing $\mathrm{Ann}E$ is p. Then $(G/\mathrm{Ann}E)_r$ will have constant length if $r >> 0$, and hence E_r will have constant length if $r >> 0$. Denote this constant length by $\Lambda(E)$. If we now form the complete module of fractions of E with respect to S, the result will be of the form $E''[Y, Y^{-1}]$, where

E'' is the homogeneous module of fractions of E with respect to S and is an L-module of length $\Lambda(E)$.

Now suppose we apply the same procedure of forming modules of fractions to the Koszul complex $K(G|\mathbf{v})$. The result will be the Koszul complex of $L[Y, Y^{-1}]$ with respect to the elements $\mathbf{v}$ considered as elements of degree 1 of $L[Y, Y^{-1}]$. The graded components of this complex are all isomorphic to the complex $K(L|\eta)$, where η denotes the sequence of elements $\eta_i = v_i/y$, $i = 1, \ldots, n$. Note that this set of elements is a reduction of the maximal ideal n of L. Since the modules $H_i(G|\mathbf{v})$ are annihilated by $\mathbf{v}$, it follows that $H_i(L|\eta)$ has length $\Lambda(H_i(G|\mathbf{v}))$. But $e(G/G') = \sum(-1)^i\Lambda(H_i(G|\mathbf{v}))$ and hence $e(G/G') = \chi(L|\eta)$. But, since $\chi(L|\eta)$ is equal to the multiplicity of the ideal $(\eta_1, \ldots, \eta_D)$ of L and so is positive, this implies that $e(G/G')$ is positive as required.

5.Degree Functions.

In this section we consider a companion to the multiplicity form $e(G/G'|E|W)$ associated with the admissible pair of graded rings G, G' of the last section. This is the degree function $d(G/G'|E|x|W)$, which is a form of degree $D - 1$, where $D = \dim G$. We will need to impose a stronger restriction on the pair (G, G') than the restriction to acceptable pairs imposed in the last section.

Before defining degree functions, and the conditions under which they are defined, we must introduce some notation. In what follows S will be the set of homogeneous elements of G which are not in any prime ideal P of G satisfying the condition that $\dim(G/P) = D$. The set of such prime ideals is finite. If $x \in S$, we will denote by G_x the ring G/xG, but G'_x will denote the image of G' under the natural homomorphism of G onto G_x and not G'/xG'. It is clear that the pair of rings (G_x, G'_x) is admissible, and that $\dim(G/xG) = D - 1$. Hence there are prime ideals P containing xG such that $\dim(G/P) = D - 1$. In order to define degree functions, we have to consider multiplicity forms $e(G_x/G'_x|E_x|W)$, where E_x denotes a finitely generated graded G_x-module and x is an arbitrary element of S. In order to use the properties of such multiplicity forms proved in section 4, we would require that, for all $x \in S$, the pairs of rings (G_x, G'_x) should be acceptable in the sense of section 4. This is most conveniently ensured if we impose on G the condition that it be fully acceptable in the sense of the following definition.

DEFINITION. We say that the graded ring G is fully acceptable if $a(G) < \dim G$.

It is clear that, as G_x is a homomorphic image of G, this will imply that $a(G_x) \leq D - 1$. This, in turn, implies that G_x itself is acceptable and hence that the pair (G_x, G'_x) is acceptable by theorem 4.4.

We now come to the definition of degree functions.

DEFINITION. Let G, G' be an admissible pair of rings, with G fully acceptable. Let $x \in S$, and E be a finitely generated graded G-module. Let

$$d(G/G'|E|x|W) = e(G_x/G'_x|E/xE|W) - e(G_x/G'_x|0 :_E x|W)$$

where E/xE, $0 :_E x$ are both considered as G_x-modules. Then $d(G/G'|E|x|W)$, considered as a function of x, is termed the degree function of the module E with respect to G, G'. We will refer to it as the degree function of E, if the values of G, G' are clear from the context,

It is clear that the form $d(G/G'|E|x|W)$ has degree $D-1$. If $|K| = D-1$, the coefficient of W^K in $d(G/G'|E|x|W)$ will be denoted by $d(G/G'|E|x|K)$.

In our next theorem, we show that the degree function has many properties analogous to those of the function $e(G/G'|E|W)$ listed in the theorems of the last section.

THEOREM 5.1. *Let G, G' be an admissible pair of graded rings, G being fully acceptable and satisfying* $\dim G = D$. *Let x be a homogeneous element of G not in any graded prime ideal P of G satisfying* $\dim(G/P) = D$. *Then the following statements hold.*

i) *Let $|K| = D-1$, and $\mathbf{v} = (v_1, \ldots, v_D)$, where $v_1 = x$ and $v_2, \ldots, v_D$ are elements of G' whose images in G'_x form a joint reduction of G'_x of type K. Then the coefficient $\chi(E|\mathbf{v}|R)$ of Z^R in the Euler-Poincaré characteristic $\chi(E|\mathbf{v}|Z)$ of $K(E|\mathbf{v})$ is equal to $d(G/G'|E|x|K)$ if $R >> 0$.*

ii) *If*

$$0 \to E' \to E \to E'' \to 0$$

is an exact sequence of G-modules,

$$d(G/G'|E|x|W) = d(G/G'|E'|x|W) + d(G/G'|E''|x|W).$$

iii)

$$d(G/G'|E(K)|x|W) = d(G/G'|E|x|W),$$

where $E(K)$ is a shift of E.

iv) *If G is a finite G'- module, then, for all E,*

$$d(G/G'|E|x|W) = 0.$$

v)

$$d(G/G'|E|x|W) = \sum \lambda(E_P) d(G/G'|G/P|x|W)$$

the sum being over the set of prime ideals P of G such that $\dim G/P = D$, *this set being finite.*

vi) *The coefficients of $d(G/G'|E|x|W)$ are non-negative.*

vii) *If $x = x'x''$, then*

$$d(G/G'|E|x|W) = d(G/G'|E|x'|W) + d(G/G'|E|x''|W).$$

PROOF.

i) Let $\mathbf{v}' = (v_2', \dots, v_D')$, where v_i' is the image of v_i in G_x'. Then, by definition and theorem 4.4,

$$d(G/G'|E|x|K) = \chi(E/xE|\mathbf{v}'|R) - \chi(0 :_E x|\mathbf{v}'|R)$$

if $R >> 0$. Since the terms on the right-hand side are the same for all $R >> 0$, it follows from the multigraded version of proposition 3 ii) of Kirby [2] that the right-hand side is equal to $\chi(E|\mathbf{v}|R)$ for $R >> 0$.

ii) Since $\chi(E|\mathbf{v}|Z) = \chi(E'|\mathbf{v}|Z) + \chi(E''|\mathbf{v}|Z)$,

$$d(G/G'|E|x|K) = d(G/G'|E'|x|K) + d(G/G'|E''|x|K)$$

for all K such that $|K| = D - 1$, and the result is immediate.

iii) Since $0 :_{E(K)} x = (0 :_E x)(K)$ and $E(K)/xE(K) = (E/xE)(K)$, this is a consequence of 4.5 (ii).

iv) As G_x is a finite G_x'-module, this follows from 4.5 iv).

v) We first prove that $d(G/G'|G/P|x|W) = 0$ if $\dim(G/P) < D$. First suppose that x is not in P. Then $0 :_{G/P} x = (0)$ and the G_x-module E/xE has a filtration whose quotients are shifts of modules G/p, where $p \supseteq xG + P$ and so $\dim G/p < D - 1$. Hence the result follows in this case. If $x \in P$, and $E = G/P$, then $E/xE = E = 0 :_E x$ and the result is immediate. We can now complete the proof in the same way as in the proof of theorem 4.5(v).

vi) In view of v), it is enough to prove this in the case where E is of the form G/P. If x is not in P then $0 :_E x = (0)$, and $e(G_x/G_x'|E/xE|W) \geq 0$, while if $x \in P$ we have $d(G/G'|E|x|W) = 0$.

vii) Again, it will be enough to consider the case where $E = G$ and G is a domain. This implies that

$$d(G/G'|G|x|W) = e(G_x/G_x'|G/xG|W)$$

and this equation remains valid if we replace x throughout by x' and, also, if we replace x by x'' throughout. We also have the following exact sequence of G_x-modules.

$$0 \to (G/x''G)(\Delta') \to G/xG \to G/x'G \to 0,$$

where Δ' is the degree of x'.

This yields the equation

$$e(G_x/G_x'|G/xG|W) = e(G_x/G_x'|G/x'G|W) + e(G_x/G_x'|G/x''G|W).$$

The right-hand side will be unaltered if we replace G_x/G_x' by $G_{x'}/G_{x'}'$ in the first term and by $G_{x''}/G_{x''}'$ in the second term. The required equation now follows immediately.

For the remainder of this section, the assumption that G is fully acceptable will be in force.

Using vii), we can define $d(G/G'|E|x|W)$ over a larger set of elements x. This can be described as follows. Let H denote the ring of all fractions x/s, where $x \in G$ and $s \in S$. The units of this ring are then the elements of the form ss'^{-1} where $s, s' \in S$ and we can define

$$d(G/G'|E|ss'^{-1}|W) = d(G/G'|E|s|W) - d(G/G'|E|s'|W).$$

Hence we can define $d(G/G'|E|x|W)$ when x is a unit of H.

Our next objective is to express the degree function $d(G/G'|E|x|W)$ in terms of valuations. This will generalize the main result of Rees [6]. For this purpose, it is convenient to restrict attention initially to the case where G is a domain and $E = G$. In this case we will drop reference to E and write $d(G/G'|x|W)$ in place of $d(G/G'|E|x|W)$. We also recall that the homogeneous fields of fractions of G and G' coincide, and we denote this field by F. Note that the ring H above can then be described as $F[T, T^{-1}]$, where T is a set $(t_1, \ldots, t_q)$ of q indeterminates.

DEFINITION. If B denotes either G or G', a good m-prime of B is a prime ideal p of B containing mB which satisfies $\dim(B/p) = D - 1$. It is clear that these prime ideals are necessarily of height 1 and hence minimal over mB. Hence they are finite in number.

Next we consider the ring B/p, where p is a good m-prime of B. This is a graded extension of k, and we will write $\Lambda(p|Z)$ for $\sum \lambda((B/p)_R)Z^R$. This is of polynomial type of degree $D - 1$ and we will write its multiplicity form as $e(p|W)$.

If p is a good m-prime of $B = G$ or G', we denote by $L(p)$ the homogeneous localisation of B at p. This will be a 1-dimensional local domain and its residue field will be denoted by $K(p)$.

For future reference we note that, if M is a finitely generated B-module such that $\lambda(M_R)$ is finite when $R >> 0$, then M_p is an $L(p)$-module of finite length for each p. Further the power series $\sum \lambda(M_R)Z^R$ is of polynomial type of degree $D - 1$ and the sum of the terms of degree $D - 1$ in its Hilbert polynomial is $\sum_p \lambda(M_p)e(p|W)$, the sum being over the good m-primes p of B. Next we turn to valuations. We describe the valuations we shall consider in the following definition.

DEFINITION. A valuation v on the common homogeneous field of fractions F of G, G' will be termed an m-valuation of G or G' if it satisfies conditions i) and ii) below and a good m-valuation if it also satisfies iii).

i) It takes integer values together with infinity.
ii) It takes non-negative values on Q and positive values on m.
iii) Its residue field K_v is a finitely generated field extension of k of transcendence degree $D - 1$.

Let B denote either of the domains G, G'. In addition to the homogeneous field of fractions F of B, we will consider the graded ring of all fractions x/y, where $x, y \neq 0$ are homogeneous elements of B, not necessarily of the same degree. This ring can be written as $F[\xi, \xi^{-1}]$, where $\xi = (\xi_1, \ldots, \xi_q)$, with $\xi_i \in B_{\delta(i)}$ for $i = 1, \ldots, q$, and $\xi^{-1} = (\xi_1^{-1}, \ldots, \xi_q^{-1})$. If v is any integer-valued valuation on F, and $a_1, \ldots, a_q$ are arbitrary integers, there is a unique graded extension w of v to $F[\xi, \xi^{-1}]$ in which $w(\xi_i) = a_i (i = 1, \ldots, q)$. We will use v itself to denote the extension corresponding to $a_i = -\text{Min}(v(x/\xi_i))$ for $i = 1, \ldots, q$. This extension is characterised by the two properties that it takes non-negative values on B and its centre on B is a relevant prime ideal.

It will often be more convenient below to consider this graded valuation on B rather than the valuation on F from which it arises

If p is any good m-prime of B, we define $V(p)$ to be the finite set of valuations v on F whose centre on B is p. These valuations are necessarily good m-valuations of B. The integral closure $L^*(p)$ of $L(p)$ is then $\cap O_v$, v ranging over $V(p)$, and O_v is the valuation ring of v.

We now require some results concerning 1-dimensional local domains. We first recall some results which are essentially contained in Northcott[3].

If L is a local domain of dimension 1, and $\hat{L}$ is its completion, then there is a 1-1 correspondence between the minimal prime ideals $\hat{P}$ of $\hat{L}$ and the valuations v on the field of fractions F of L which take non-negative values on L. This correspondence arises as follows. The integral closure of $\hat{L}/\hat{P}$ in its field of fractions $\hat{F}$ is a discrete valuation ring defining a valuation $\hat{v}$ on $\hat{F}$, and v is the restriction of $\hat{v}$ to F. This implies that the set of valuations v non-negative on L is finite and, if v is such a valuation and K, K_v denote the residue fields of L and the valuation ring of v, then $[K_v : K]$ is finite.

DEFINITIONS. The prime ideal $\hat{P}$ of $\hat{L}$ corresponding to v will be denoted by $\hat{P}_v$ and the length of the $\hat{P}_v$-primary component of zero in $\hat{L}$ will be denoted by $\delta(v, L)$.

We now prove a lemma, essentially theorem 6 of Northcott[3].

LEMMA 5.2. *Let L be a 1-dimensional local domain with field of fractions F, and residue field K. Let v range over the finite set of valuations on F non-negative on L and let K_v be the residue field of v. Then, if $x \neq 0$ is an element of L,*

$$\lambda(L/xL) = \sum_v \delta(v, L)[K_v : K]v(x).$$

PROOF. Since L is a 1-dimensional local domain, $\lambda(L/xL)$ is equal to the multiplicity of the ideal xL of L and hence to the multiplicity of the ideal $x\hat{L}$ of $\hat{L}$. Hence we have

$$\lambda(L/xL) = \sum_v \delta(v, L)e(x\hat{L}|\hat{L}/\hat{P}_v)$$

where v ranges over the valuations ≥ 0 on L. Now consider the integral closure L^* of $\hat{L}/\hat{P}_v$ in its field of fractions. This is the valuation ring of $\hat{v}$, and is a finite $(\hat{L}/\hat{P})$-module of rank 1. Hence $e(x\hat{L}|L^*) = e(x\hat{L}|\hat{L}/\hat{P}_v)$, while, as

$$\lambda(L^*/x^n L^*) = n[K_v : K]v(x),$$

$e(x\hat{L}|L^*) = [K_v : K']v(x)$, and the result follows.

DEFINITIONS. Let v be an m-valuation of G'. If v is good and $p(v) = p$, we define the degree coefficient form $d(G'|v|W)$ by the equation

$$d(G'|v|W) = \delta(v, L(p))[K_v : K(p)]e(p|W).$$

If v is not good we define $d(G'|v|W)$ to be zero. If $|K| = D - 1$, the coefficient of W^K in $d(G'|v|W)$ will be denoted by $d(G'|v|K)$.

If $1 \leq i \leq q$ we define $e_{(i)}(G/G'|E|W)$ to be the quotient when $e(G/G'|E|W)$ is divided by w_i, with the consequent definition of $e_{(i)}(G/G'|E|K)$ if $|K| = D-1$.

THEOREM 5.3. *Let x be a non-zero homogeneous element of G of degree $\Delta(x) = (\delta_1(x), \ldots, \delta_q(x))$. Then,*

$$d(G/G'|x|W) = \sum_i \delta_i(x)e_{(i)}(G/G'|W) + \sum_v d(G'|v|W)v(x),$$

where the first sum is summed over $i = 1, \ldots, q$ and the second sum is over the set of m-valuations of G', each v being extended to G as indicated above.

PROOF. It will be sufficient to prove this result under the additional hypothesis that x belongs to G'. For we can write $x = y/z$, with y, z in G', and if the above equation holds when x is replaced by y or z, it holds for x in view of 5.1(vii). $d(G/G'|x|W)$ is the multiplicity form of the power series $\Lambda(Z) = \sum \lambda(G_R/xG_{R-\Delta(x)} + G'_R)Z^R$. Now

$$\begin{aligned}\lambda(G_R/xG_{R-\Delta(x)} + G'_R) &= \lambda(G_R/G'_R) - \lambda(xG_{R-\Delta(x)} + G'_R/G'_R)\\ &= \lambda(G_R/G'_R) - \lambda(xG_{R-\Delta(x)}/G'_R \cap xG_{R-\Delta(x)})\\ &= \lambda(G_R/G'_R) - \lambda(xG_{R-\Delta(x)}/xG'_{R-\Delta(x)})\\ &\quad + \lambda(G'_R \cap xG_{R-\Delta(x)}/xG'_{R-\Delta(x)}).\end{aligned}$$

Now the power series $\sum(\lambda(G_R/G'_R) - \lambda(xG_{R-\Delta(x)}/xG'_{R-\Delta(x)}))Z^R$ can be written to within equivalence as $(1-Z^{\Delta(x)})\sum \lambda(G_R/G'_R)Z^R$, and a straight-forward calculation shows that this is of polynomial type of degree $D-1$ with multiplicity form

$$\sum_i \delta_i(x)e_{(i)}(G/G'|W).$$

To complete the proof, we consider the power series

$$\Lambda''(Z) = \sum \lambda(G'_R \cap xG_{R-\Delta(x)}/xG'_{R-\Delta(x)})Z^R.$$

Let I be the ideal $xG \cap G'$ of G' and $M = I/xG'$. Then $\lambda(M_R)$ is finite when $R >> 0$, and the series above can be written $\sum \lambda(M_R)Z^R$. This series has

Hilbert degree at most $D-1$ and the sum of the terms of degree $D-1$ in its Hilbert polynomial is $\sum_p \lambda(M_p)e(p|W)$, where p ranges over the prime ideals of G' minimal over Ann$M$ which satisfy $\dim(G'/p) = D-1$. Since $\lambda(M_p) \leq \lambda(L(p)/xL(p))$ for each p, we have the inequality

$$d(G/G'|x|W) \leq \sum_i \delta(x)e_{(i)}(G/G'|W) + \sum_v d(G'|v|W)v(x)$$

by the definition of $d(G'|v|W)$ and lemma 5.2.

Equality will hold if I is not contained in any good prime ideal p of G' minimal over over mG'. Now suppose p' is a prime ideal of G' minimal over I. Then, as the radical of I is the intersection of the radical of xG with G', it follows that $p' = P \cap G'$, where P is a prime ideal of G minimal over xG. If, in addition, p' contains mG', P must contain mG. Since G is fully acceptable, $a(G) < D$, which implies that $\dim(G/P) < D-1$, and, in turn, that $\dim(G'/p') < D-1$. It follows that p' cannot be contained in any good prime ideal p minimal over mG' and hence that the same is true of I. Hence equality holds.

We have remarked earlier that the function $d(G/G'|x|W)$ of x can be defined for all $x \neq 0$ in F. In this case, by the theorem above, we can write

$$(*) \qquad d(G/G'|x|W) = \sum_v d(G'|v|W)v(x).$$

The form $d(G'|v|W)$,defined for each good m-valuations v, is zero for all save a finite set of v. Hence, by the theorem of independence of valuations, it is uniquely determined by the fact that (*) holds for all $x \neq 0$ in F.

We now conclude this section, by considering the removal of the restriction to the case where G is a domain and $E = G$. Let $P_1, \ldots, P_h$ be the minimal prime ideals of G satisfying the condition that $\dim(G/P_j) = D$.

Let $G_j = G/P_j, G'_j = G'/P_j \cap G'$ and T, T' be the sets of homogeneous elements of G, (resp. G'), not in any of the prime ideals P_j. The homogeneous localisations of G by the set T and G' by the set T' then coincide, their common value being a semi-local artin ring which we will denote by Φ. Its maximal ideals will be denoted by Π_j where j takes the values $1, \ldots, h$. Π_j consists of the fractions y/z with y in P_j and $z \in T$. We write F_j for Φ/Π_j, so that F_j is the homogeneous field of fractions of either G_j or G'_j. We now define m-valuations of G' as follows. Let v' be an m-valuation of G'_j, and let θ_j be the homomorphism of Φ onto F_j. Then the function $v(x) = v'(\theta_j(x))$ is termed an m-valuation of G' and is said to be good if v' is good. It takes the value infinity on Π_j and it can be extended to G' and hence to G in an analogous manner to that described in the domain case; this extension is infinite on P_j. We will also write $\Pi(v), P(v)$ for Π_j, P_j where j is determined by the condition that $v(x)$ is infinite on Π_j. Finally if $P(v) = P_j$, we define

$$d(G'|v|W) = d(G'_j|v|W).$$

We can now state the following general form of theorem 5.3.

THEOREM 5.4.

$$d(G/G'|E|x|W) = \sum_i \delta_i(x)e_{(i)}(G/G'|E|W) + \sum_v \lambda(E_{P(v)})d(G'|v|W)v(x),$$

the first sum being over $i = 1, \ldots, q$ and the second over all m-valuations of G'.

PROOF. This follows from 5.3 and 5.1(v).

We conclude this section by considering the case $q = 1$, when the result above takes a simpler form. In this case $d(G/G'|E|x|W)$ involves only one term which we will write $d(G/G'|E|x)w_1^{D-1}$, $e_{(1)}(G/G'|E|W)$ becomes $e(G/G'|E)w_1^{D-1}$, and finally we will write $d(G'|v|W)$ as $d(G'|v)w_1^{D-1}$. We will only consider the coefficents $d(G/G'|E|x)$, $e(G/G'|E)$ and $d(G'|v)$. The coefficients $d(G'|v)$ take a simple form if G' is assumed to be basic, i.e., generated by D elements. In this case G'/mG' is isomorphic to $k[X_1, \ldots, X_D]$, implying that mG' is a prime p and also that $d(p|W) = w_1^{D-1}$. Since only one prime ideal p is involved, we will write L for $L(p)$ and K for $K(p)$, noting that K is isomorphic to $k(X_1, \ldots, X_{D-1})$. Finally, we may write $d(G'|v) = \delta(v, L)[K_v : K]$, where $\delta(v, L)$ and K_v are as defined earlier.

We collect these statements together as a theorem.

THEOREM 5.5. *Let $G = Q[y_1, \ldots, y_m]$, $G' = Q[u_1, \ldots, u_n]$, where the sets of generators $y_1, \ldots, y_m$, $u_1, \ldots, u_n$ are all homogeneous elements of degree 1 of the graded ring G, and the pair of rings G, G' is admissible. Then, if E is any finitely generated graded G-module, and x is a homogeneous element of G of degree δ ,*

$$d(G/G'|E|x) = \delta e(G/G'|E) + \sum_v \lambda(E_{P(v)})d(G'|v)v(x),$$

the sum being over all m-valuations of G'. If, further, G' is basic, then

$$d(G'|v) = \delta(v, L)[K_v : K].$$

6. Multiplicities of ideals in graded rings.

In this section we consider a situation which can be described as follows. We suppose that $(Q, m, k, d), d \geq 0$, is a local ring and that $A = Q[y_1, \ldots, y_m]$ is an extension of Q graded by a single integer s, the elements y_i being of degree 1. We will impose on A the restriction that it is acceptable, that is, that it satisfies the condition $a(A) \leq \dim A$. We will write D for $\dim A$. Finally we denote by Ξ the ideal $(y_1, \ldots, y_m)$ of A.

We now introduce some definitions.

DEFINITION. An ideal J of A is said to be level if it possesses a set of generators all of the same degree.If this degree is equal to σ, we will say that J is of degree σ. A level ideal J of degree σ is termed admissible if $\lambda(A_\sigma/J_\sigma)$ is finite (so that $\lambda(A_r/J_r)$ is finite for all $r \geq \sigma$). It is weakly admissible if the weaker condition $\lambda(A_r/J_r)$ is finite for $r >> 0$ is satisfied.

It is clear that the product of two admissible ideals is admissible. In the theorems which follow, various ideals which occur will be assumed to be admissible. Nevertheless, this condition can often be replaced by weak admissibility in view of the following lemma.

LEMMA 6.1. *Let J be a level ideal of A of degree σ and let $J\dagger$ denote the ideal generated by the elements of A_σ integrally dependent on J. Assume that no minimal prime ideal of A is irrelevant. Then, if J is weakly admissible, there is an admissible ideal J' of degree σ such that $J\dagger \supseteq J' \supseteq J$ and $J'_r = J_r$ if $r >> 0$.*

PROOF. We take J' to be the ideal generated by the elements x of A_σ which satisfy $xA_s \subseteq J_{s+\sigma}$ for $s >> 0$. It then follows that, if $s >> 0$, $J'_\sigma A_s \subseteq J_{s+\sigma} = J_\sigma A_s$. But the restriction that no minimal prime ideal of A is irrelevant implies that, for all s, the annihilator of A_s is nilpotent, and hence the condition $J'_\sigma A_s \subseteq J_\sigma A_s$ implies that $J' \subseteq J\dagger$. It is also clear that $J'_{s+\sigma} = J_{s+\sigma}$ if $s >> 0$. Now there exists an m-primary ideal I such that $IA_{s+\sigma} \subseteq J_{s+\sigma}$ for all $s >> 0$. Then $IA_\sigma.A_s \subseteq J_{s+\sigma} = J_\sigma.A_s$ and hence $IA_\sigma \subseteq J'_\sigma$, proving that A_σ/J'_σ is finite.

Let $\sigma = (\sigma(1), \ldots, \sigma(q))$ be a sequence of non-negative integers and let $\mathbf{J}$ denote a sequence $(J_1, \ldots, J_q)$ of admissible ideals of A. Let J_i be of degree $\sigma(i)$, so that it is generated by elements $u_{ij}(j = 1, \ldots, n(i))$ of degree $\sigma(i)$. We now associate with the sequence of ideals $\mathbf{J}$ an admissible pair of graded rings $G(\sigma, \Xi)$ and $G(\mathbf{J})$ graded by sequences of $q+1$ non-negative integers $(R, s) = (r(1), \ldots, r(q), s)$.

For each i, we choose a set of generators a_{ij} of $A_{\sigma(i)}$, so that $\Xi^{\sigma(i)}$ is generated by the elements a_{ij}. Let $\mathbf{t} = (t_1, \ldots, t_q)$ be a set of indeterminates over A. We then write $a(i,j)$ for $t_i a_{ij}$, $u(i,j)$ for $t_i u_{ij}$ and write $\mathbf{a}, \mathbf{u}$ for the sets of elements $a(i,j), u(i,j)$. Then $G(\sigma, \Xi) = A[\mathbf{a}]$ and $G[\mathbf{J}] = A[\mathbf{u}]$. It follows that both these rings are graded by the sets (R, s), and, if $\sigma(R) = \sum \sigma(i) r(i)$, $G(\sigma, \Xi)_{R,s} = \mathbf{t}^R A_{\sigma(R)+s}$, while $G(\mathbf{J})_{R,s} = \mathbf{t}^R(\mathbf{J}^R)_{\sigma(R)+s}$. If $\mathbf{J}$ consists of a single level ideal J of degree $\sigma(1)$, we will write $G(\sigma(1), \Xi)$ for $G(\sigma, \Xi)$ and $G(J)$ for $G(\mathbf{J})$. Note that, if $C = (c(1), \ldots, c(q))$ is a sequence of positive integers, $\Delta_C(G(\mathbf{J}))$ is isomorphic to $G(\mathbf{J}^C)$

Next let E be any finitely generated A-module. Then we write $E(\sigma, \Xi)$ for the $G(\sigma, \Xi)$-module defined by $E(\sigma, \Xi)_{R,s} = \mathbf{t}^R E_{\sigma(R)+s}$ and $E(\mathbf{J})$ for the $G(\mathbf{J})$-module defined by $E(\mathbf{J})_{R,s} = \mathbf{t}^R(\mathbf{J}^R E)_{\sigma(R)+s}$. It follows that

$$(E(\sigma, \Xi)/E(\mathbf{J}))_{R,s} \cong (E/\mathbf{J}^R E)_{\sigma(R)+s}.$$

Let I be a graded ideal of A. Then we define the graded ideal $I\ddagger$ of $G(\sigma, \Xi)$ to be the ideal of $G(\sigma, \Xi)$ whose homogeneous elements of degree (R, s) are the elements $a\mathbf{t}^R$ where $a \in I \cap A_{\sigma(R)+s}$. In particular, if p is a graded prime ideal of A, $p\ddagger$ is prime and the homogeneous rings of fractions A_p and $G(\sigma, \Xi)_{p\ddagger}$ coincide. Hence if p is minimal over I, $p\ddagger$ is minimal over $IG(\sigma, \Xi) = IA[\mathbf{a}]$. Conversely, if x belongs to the intersection of the prime ideals p minimal over

I, and $x\mathbf{t}^R \in G(\sigma, \Xi)$, $x\mathbf{t}^R$ belongs to the radical of $I\ddagger$. Hence a minimal prime ideal of $G(\sigma, \Xi)$ is of the form $p\ddagger$, where p is a minimal prime ideal of A, and therefore $\dim(G(\sigma, \Xi)) = \dim A = D$. Similarly, if p ranges over the prime ideals of A minimal over mA, $p\ddagger$ ranges over the prime ideals of $G(\sigma, \Xi)$ minimal over $mG(\sigma, \Xi)$, implying that $a(G(\sigma, \Xi)) = a(A)$. It follows that the ring $G(\sigma, \Xi)$ is acceptable.

There is a 1-1 correspondence $P \longleftrightarrow p$ between the prime ideals P of $G(\sigma, \Xi)$ which do not contain m and the prime ideals p of $G(\mathbf{J})$ which do not contain m. This is defined by $p = P \cap G(\mathbf{J})$ and $P = G(\mathbf{J})_p[\mathbf{t}] \cap G(\sigma, \Xi)$, where $G(\mathbf{J})_p$ is the homogeneous localisation of $G(\mathbf{J})$ at p. Since $G(\mathbf{J})_p = G(\sigma, \Xi)_P$, it follows that $G(\sigma, \Xi)/P$ and $G(\mathbf{J})/p$ have the same dimension. It also follows that the minimal prime ideals of $G(\mathbf{J})$ are the prime ideals $p\ddagger \cap G(\mathbf{J})$, where p ranges over the minimal prime ideals of A. Next $\dim(G(\mathbf{J})) = D$. Finally we note that if $C = (c(1), ..., c(q))$ is a set of positive integers, $a(G(\mathbf{J})) = a(G(\mathbf{J}^C)$, by lemma 1.7.

We now take $G = G(\sigma, \Xi)$ and $G' = G(\mathbf{J})$ and apply the results of section 4 to the pair of admissible rings G, G' and the (G, G')-admissible modules $(E(\sigma, \Xi), E(\mathbf{J}))$.

In what follows, we will consider the function

$$\lambda(R, s) = \lambda((E(\sigma, \Xi)/E(\mathbf{J}))_{R,s}) = \lambda((E/\mathbf{J}^R E)_{\sigma(R)+s})$$

and the power series $\sum \lambda(R, s) Z^R z^s$ which is of polynomial type by theorem 4.2, and let $e(\mathbf{J}|E|W, w) = e(G(\sigma, \Xi)/G(\mathbf{J})|E(\sigma, \Xi)|W, w)$ and $M(\mathbf{J}|E|R, s)$ be its multiplicity forms. Then, if $(K, l) = (k(1), \ldots, k(q), l)$ is a sequence of $q+1$ non-negative integers satisfying $|K| + l = D$, we will now write $e(\mathbf{J}|E|K, l)$ for the coefficient of $W^K w^l$ in $e(\mathbf{J}|E|W, w)$.

LEMMA 6.2. *Let ξ be a joint reduction of the set of q+1 ideals $\mathbf{J}, \Xi$ of A of type (K, l), where $|K| + l = D$. Then $e(\mathbf{J}|E|K, l)$ is equal to the Euler-Poincaré characteristic $\chi(E|\xi|r)$ of the component of degree r of the Koszul complex $K(E|\xi)$ for $r >> 0$.*

PROOF. Let ξ consist of elements $\xi_{ij} \in (J_i)_{\sigma(i)} (i = 1, \ldots, q; j = 1, \ldots, k(i))$ together with elements $\xi_1, \ldots, \xi_l \in A_1$. Let $\mathbf{X}$ denote the set consisting of the elements $t_i \xi_{ij} (i = 1, \ldots, q; j = 1, \ldots, k(i))$ together with the elements $\xi_1, \ldots, \xi_l$. The condition that $\mathbf{X}$ is a joint reduction of type (K, l) of $G(\mathbf{J})$ is equivalent to the condition that ξ is a joint reduction of the set of ideals $(\mathbf{J}, \Xi)$ of A of type K, l. Hence it follows from theorem 4.4 that $e(\mathbf{J}|E|K, l)$ is equal to the Euler-Poincaré characteristic $\chi(E(\Xi)|\mathbf{X}|R, s)$ of the component of degree (R, s) of $K(E(\sigma, \Xi)|\mathbf{X})$ for $(R, s) >> 0$. But we can identify this component with the component of degree $\sigma(R) + s$ of $K(E|\xi)$ and hence the result follows.

The above lemma gives an expression for $e(\mathbf{J}|E|K, l)$ in which the rings $G(\sigma, \Xi)$ and $G(\mathbf{J})$ do not enter. We carry this further in the sets of definitions which follow.

DEFINITIONS. Let $\mathbf{I} = (I_1, \dots, I_D)$ be a sequence of D admissible ideals of A which need not be distinct or different from Ξ. Then $e(A|\mathbf{I}|E)$ is defined to be $\chi(E|\mathbf{v}|R)$ for $R >> 0$, where $\mathbf{v} = (v_1, \dots, v_D)$ is a joint reduction of $\mathbf{I}$ of type $(1, \dots, 1)$. We term $e(A|\mathbf{I}|E)$ the mixed multiplicity of the set of ideals $\mathbf{I} = (I_1, \dots, I_D)$ with respect to E. If $I_1 = I_2 = \cdots = I_D = I$, we will write $e(A|I|E)$ for $e(A|\mathbf{I}|E)$. If $I_j = \Xi$ for $j > D - l$, and $\mathbf{I}' = (I_1, \dots, I_{D-l})$, we will also write $e_l(A|\mathbf{I}'|E)$ for $e(A|\mathbf{I}|E)$.

Lemma 6.2 implies that $e(\mathbf{J}|E|K, l)$ is equal to $e(A|\mathbf{I}|E)$ if we take $\mathbf{I}$ to consist of $k(1)$ copies of J_1, ... , $k(q)$ copies of J_q and l copies of Ξ . In particular if we take $\mathbf{J} = \mathbf{I}$, and $k(i) = 1 (i = 1, \dots, D)$, $e(A|\mathbf{I}|E)$ is equal to $e(\mathbf{J}|E|K, 0)$. This implies that $e(A|\mathbf{I}|E)$ is independent of the choice of $\mathbf{v}$.

We now consider the elementary properties of the function $e(A|\mathbf{I}|E)$, the results being for the most part translations of parts of theorems 4.4, 4.5 or of Proposition 3 of Kirby[2].

THEOREM 6.3. *Let* $\mathbf{I} = (I_1, \dots, I_D)$ *be a sequence of admissible ideals of* A, I_j *being level of degree* $\sigma(j)$. *Then the following statements hold.*

i) $e(A|\mathbf{I}|E)$ *is a symmetric function of* $I_1, \dots, I_D$ *and, as a function of* E, *is an additive function on the category of finitely generated graded* A*-modules, the values being non-negative integers.*

ii) *If* $E(c)$ *is a shift of* E,

$$e(A|\mathbf{I}|E(c)) = e(A|\mathbf{I}|E).$$

iii) *If* P *ranges over the set of prime ideals* P *of* A *such that* $\dim(A/P) = D$, *and, for each* P, $\mathbf{I}(P)$ *denotes the set of ideals* $I_j + P/P$, $(j = 1, \dots, D)$ *of* A/P, *then*

$$e(A|\mathbf{I}|E)) = \sum_P \lambda(E_P) e(A/P|\mathbf{I}(P)|A/P).$$

iv) *Let* $\mathbf{x} = (x_1, \dots, x_D)$ *be a joint reduction of* $\mathbf{I}$ *of type* $(1, \dots, 1)$, $A' = A/x_DA$, $\mathbf{I}'$ *be the set of ideals* $I_j + x_DA/x_DA$ $(j = 1, \dots, D-1)$ *of* A' *and* E_1, E_2 *be the* A'*-modules* E/x_DE, $0 :_E x_D$. *Then*

$$e(A|\mathbf{I}|E) = e(A'|\mathbf{I}'|E_1) - e(A'|\mathbf{I}'|E_2).$$

v) *If* $I_D = I'I''$, *then*

$$e(A|\mathbf{I}|E) = e(A|I_1, \dots, I_{D-1}, I'|E) + e(A|I_1, \dots, I_{D-1}, I''|E).$$

vi) *If* $\mathbf{J} = (J_1, \dots, J_q)$ *is a sequence of admissible ideals of* A, *then,*

$$e(A|J^R\Xi^s|E) = M(\mathbf{J}|E|R, s)/d!.$$

vii) *If* $a = a(\Xi)$, $e_l(A|\mathbf{I}'|E) = 0$ *if* $l \geq a$ *for all choices of* $\mathbf{I}' = (I_1, \dots, I_{D-l})$ *and* E.

PROOF. i) Since $\chi(E|x_1,\ldots,x_D|r)$ is a symmetric function of $x_1,\ldots,x_D$, $e(A|\mathbf{I}|E)$ is a symmetric function of $I_1,\ldots,I_D$. the additivity of $e(A|\mathbf{I}|E)$ follows from theorem 4.5 i) and the fact that it is non-negative follows from 4.6 ii).

Parts ii) and iii) follow from 4.5 ii) and v).

Parts iv) and v) follow from Kirby[2] proposition 3. First we note that $e(A|\mathbf{I}|E)$ is the value of the symbol $e_t(x_1,\ldots,x_D|E)$ of Kirby[2] for $t >> 0$, this being constant in the present context. Hence iv) is a consequence of part ii) and v) is a consequence of part iii) of proposition 3. Part vi) follows by iteration of v).

If $I_j = \Xi$ for $j > D-a$ we can choose $\mathbf{x}$ so that $x_{D-a+1},\ldots,x_D$ is a reduction of Ξ. Then the homology modules $H_i(E|\mathbf{x})$ will be annihilated by a power of Ξ for all E and $i = 0,\ldots,D$. Hence, if $l \geq a, e_l(A|\mathbf{I}'|E) = 0$, for all choices of I' and E which proves vii).

THEOREM 6.4. *Let $\mathbf{I}, \mathbf{I}'$ be two sequences of D admissible ideals such that I_j, I'_j have the same degree for each j and $I_j \subseteq I'_j$ for $j = 1,\ldots,D$. Then, for all E*

$$e(A|\mathbf{I}|E) \geq e(A|\mathbf{I}'|E),$$

with equality if $I'_j \subseteq I_j\dagger$ where $I_j\dagger$ is as defined in lemma 6.1.

PROOF. We can reduce the proof to the case where $E = A$ by using 6.3 iii). We now use the alternative definition of $e(A|\mathbf{I}|E)$ and note that if $G' = G(\mathbf{I})$ and $G = G(\mathbf{I}')$ then $e(A|\mathbf{I}|A) - e(A|\mathbf{I}'|A)$ is the coefficient of $w_1\cdots w_D$ in $e(G/G'|G|W,w)$ and so is non-negative. The second part follows from 4.5 iv), since its hypotheses imply that $G(\mathbf{I}')$ is a finite $G(\mathbf{I})$-module.

This theorem enables us to define $e(A|\mathbf{I}|E)$ not merely for $\mathbf{I}$ a sequence of admissible ideals but for $\mathbf{I}$ a sequence of weakly admissible ideals by choosing I'_j as in lemma 6.1 and then defining $e(A|\mathbf{I}|E)$ to be $e(A|\mathbf{I}'|E)$. This is independent of the choice of the ideals I'_j in view of the above theorem.

We now deduce a reduction criterion from theorem 4.10. Note that I is a reduction of I' is equivalent to $I \subseteq I' \subseteq I\dagger$, where $I\dagger$ is as in lemma 6.1.

THEOREM 6.5. *Let Q be quasi-unmixed, and let A satisfy the condition that* $\dim(A/P) = D$ *for all minimal prime ideals of A. Then, if $I, I' \subseteq I$ are admissible ideals of A of the same degree which satisfy $e(A|I|A) = e(A|I'|A)$, I' is a reduction of I.*

PROOF. Let I be generated by $(u_1,\ldots,u_k)$ and I' by $(u'_1,\ldots,u'_l)$. Let G, G' be the sub-rings $Q[u_1,\ldots,u_k], Q[u'_1,\cdots,u'_l]$ of A. It then follows that $e(G/G'|G)$ is equal to $e(A|I'|A) - e(A|I|A)$ and so is zero. Therefore we can apply theorem 4.11, and deduce that G is a finite G'-module, and hence that I' is a reduction of I.

We now come to the last theorem of this section. To simplify the statement, we will impose the restriction that Q is a domain. We will also introduce some notation. We will denote the field of fractions of Q by F, write A' for the graded domain A_S, where S is the set of non-zero elements of Q, and, more generally, if E is any finitely generated graded A-module, we denote by E' the graded A'-module E_S. Then A' is a finitely generated graded extension of F, and E' is a finitely generated A'-module. Hence there is a polynomial $P(E'|n)$ such that $\dim_F E'_n = P(E'|n)$ for n large. The degree of $P(E'|n)$ is at most $D-d$ and we will write the coefficient of n^{D-d} in $P(E'|n)$ as $b(E)/(D-d)!$.

THEOREM 6.6. *Assume that A is a domain, and let $\mathbf{I} = (I_1, \ldots, I_q)$ be a set of admissible ideals of A, where I_j is level of degree $\sigma(j)$, not all $\sigma(j)$ being zero. Let J be an $\mathfrak{m}$-primary ideal of Q, so that JA is an admissible ideal of A of degree 0. If $R = (r(1), \ldots, r(q))$, let*

$$M(\mathbf{I}, JA|E|R, s) = \sum_{|K|+l=D} \binom{D}{K, l} e(\mathbf{I}, JA|E|K, l) R^K s^l$$

be the multiplicity form of the set of admissible ideals $\mathbf{I}, JA$ of A with respect to the module E. Then $e(\mathbf{I}, JA|E|K, l) = 0$ if $l > d$, while the sum of the terms in which s is raised to the power d is

$$\binom{D}{d} b(E) e(Q|J) \sigma(R)^{D-d} s^d,$$

where $\sigma(R) = \sum \sigma(j) r(j)$ and $e(Q|J)$ is the multiplicity of the ideal J of Q.

PROOF. If $r(1), \ldots, r(q), s$ are all sufficiently large, $\lambda(E_{\sigma(R)}/(\mathbf{I}^R J^s E)_{\sigma(R)})$ is equal to a polynomial in R, s of degree at most D. The sum of the terms of degree D in this polynomial is $M(\mathbf{I}, JA|E|R, s)/D!$. Fix large values of $r(1), \ldots, r(q)$ and allow s to vary. Now,

$$\lambda(E_{\sigma(R)}/(\mathbf{I}^R J^s E)_{\sigma(R)}) = \lambda(E_{\sigma(R)}/(\mathbf{I}^R E)_{\sigma(R)}) + \lambda((\mathbf{I}^R E)_{\sigma(R)}/(\mathbf{I}^R J^s E)_{\sigma(R)})$$

and the first term is independent of s, while, for s large, the second is a polynomial in s of degree at most d, the coefficient of s^d being

$$e(Q|J) \dim_F(F \otimes_Q (\mathbf{I}^R E)_{\sigma(R)})/d!.$$

Since the ideals of $\mathbf{I}$ are admissible, we can write the second factor as $\dim_F E'_{\sigma(R)}$ and it is therefore equal to $P(E'|\sigma(R))$ if $r(1), \ldots, r(q)$ are large. Hence the sum of the terms of degree d in s and $D-d$ in $r(1), \ldots, r(q)$ in $M(\mathbf{I}, JA|E|R, s)/D!$ and $e(Q|J)P(E'|\sigma(R))$ are equal. But the latter is equal to

$$b(E) e(Q|J) \sigma(R)^{D-d} s^d/(D-d)!d!$$

and the result is proved.

7. Degree Functions of ideals in graded rings.

We now consider degree functions, associated with ideals rather than pairs of rings. We commence by considering matters of notation.

Suppose that A is as in the last section and that x is an element of A such that $\dim(A/xA) = D - 1$. We define A_x to be the ring A/xA, and if J is any ideal of A, we write J_x for the ideal $(J + xA)/xA$ of A_x.

We now suppose that $\mathbf{J} = (J_1, \ldots, J_q)$ is a sequence of admissible ideals of A, the sequence of degrees of $\mathbf{J}$ being $\sigma = (\sigma(1), \ldots, \sigma(q))$. We will denote by $\mathbf{J}_x$ the sequence of ideals $(J_i)_x, (i = 1, \ldots, q)$. Finally, we will use the notation of the last section without further comment. In particular, we will now write G for the ring $G(\sigma, \Xi)$ and G' for the ring $G(\mathbf{J})$. If E is any A-module, GE will denote $G \otimes_A E = E(\sigma, \Xi)$.

We now write $d(\mathbf{J}|E|x|W, w)$ for $d(G/G'|GE|x|W, w)$, and, if v is any good m-valuation of A, it has a natural graded extension to G which we will also denote by v. We will now write $d(\mathbf{J}|v|W, w)$ for $d(G/G'|v|W, w)$, while if $|K| + l = D - 1, d(\mathbf{J}|E|x|K, l)$ and $d(\mathbf{J}|v|K, l)$ will denote the coefficients of $W^K w^l$ in $d(\mathbf{J}|E|x|W, w)$ and $d(\mathbf{J}|v|W, w)$ respectively. Note that we could also define

$$d(\mathbf{J}|E|x|W, w) = e(Ax|\mathbf{J}_x|E/xE) - e(A_x|\mathbf{J}_x|0 :_E x).$$

We now introduce a degree function analogous to $e(A|\mathbf{I}|E)$. This will require that A satisfies the condition that $a(A) < D$, that is, A is fully acceptable.

DEFINITIONS. Let $\mathbf{I} = (I_1, \ldots, I_{D-1})$ be a sequence of $D - 1$ admissible ideals of A. Then if x is any element of A such that $\dim(A/xA) = D - 1$, E is any finitely generated graded A-module, and $E/xE, 0 :_E x$ are considered as A_x-modules, we define

$$d(A|\mathbf{I}|E|x) = e(A_x|\mathbf{I}_x|E/xE) - e(A_x|\mathbf{I}_x|0 :_E x).$$

If $I_1 = I_2 = \cdots = I_{D-1} = I$, we will write $d(A|I|E|x)$ for $d(A|\mathbf{I}|E|x)$. If $I_j = \Xi$ for $j > D - l - 1$, and $I' = (I_1, \ldots, I_{D-l-1})$, we will also write $d_l(A|\mathbf{I}'|E|x)$ for $d(A|\mathbf{I}|E|x)$.

We can obtain an expression for $d(A|\mathbf{I}|E|x)$ as an Euler-Poincaré characteristic as follows. Let $\eta = (\eta_1, \ldots, \eta_{D-1})$ be a sequence of elements of A whose image η_x in A_x is a joint reduction of the sequence of ideals $\mathbf{I}_x$ of type $(1, \ldots, 1)$. Then

$$d(A|\mathbf{I}|E|x) = \chi(E/xE|\eta_x|r) - \chi(0 :_E x|\eta_x|r)$$

for $r >> 0$ and so is equal to $\chi(E|x, \eta|r)$ for $r >> 0$.

We now translate the results of section 5 into the language of this section. First we consider theorem 5.1.

THEOREM 7.1. *Let* $\mathbf{I} = (I_1, \ldots, I_{D-1})$ *be a set of admissible ideals of* A, I_j *having degree* $\sigma(j)$ *and let* x *be a homogeneous element of* A *satisfying the condition* $\dim(A/xA) = D - 1$. *Then the following statements hold.*

i) *If*

$$0 \longrightarrow E' \longrightarrow E \longrightarrow E'' \longrightarrow 0$$

is an exact sequence of A*-modules,*

$$d(A|\mathbf{I}|E|x) = d(A|\mathbf{I}|E'|x) + d(A|\mathbf{I}|E''|x).$$

ii) *If* $E(c)$ *is a shift of* E, $d(A|\mathbf{I}|E(c)|x) = d(A|\mathbf{I}|E|x)$.

iii)

$$d(A|\mathbf{I}|E|x) = \sum \lambda(E_P) d(A|\mathbf{I}|A/P|x)),$$

the sum being over the set of prime ideals P *of* G *such that* $\dim G/P = D$.

iv) $d(A|\mathbf{I}|E|x)$ *takes non-negative values.*

v) *If* $x = x'x''$, *then*

$$d(A|\mathbf{I}|E|x) = d(A|\mathbf{I}|E|x') + d(A|\mathbf{I}|E|x'').$$

vi) *If* $I_{D-1} = I'I'', \mathbf{I}' = (I_1, \ldots, I_{d-2}, I'), \mathbf{I}'' = (I_1, \ldots, I_{d-2}, I'')$

$$d(A|\mathbf{I}|E|x) = d(A|\mathbf{I}'|E|x) + d(A|\mathbf{I}''|E|x).$$

PROOF. We consider the rings $G = G(\sigma, \Xi)$ and $G' = G(\mathbf{I})$ and commence with the observation that, if

$$0 \longrightarrow E' \longrightarrow E \longrightarrow E'' \longrightarrow 0$$

is an exact sequence of A-modules, then the sequence

$$0 \longrightarrow GE' \longrightarrow GE \longrightarrow GE'' \longrightarrow 0$$

is an exact sequence of G-modules. i) now follows from 5.1(ii).

ii) follows from 5.1(iii), applied to GE, if we note that a shift of E induces a shift of GE.

iii) follows from 5.1 v) if we note first that as P ranges over the prime ideals of A such that $\dim(A/P) = D$, then, if $P\ddagger$ is defined as in the last section, $P\ddagger$ ranges over the prime ideals of G of dimension D and that, further, $\lambda(E_P) = \lambda((GE)_{P\ddagger})$.

iv) and v) are translations of 5.1 vi) and vii).

To prove vi), we choose $\eta = (\eta_1, \ldots, \eta_{D-1})$ to satisfy, for $r >> 0$, $d(A|\mathbf{I}|E|x) = \chi(E|x, \eta|r)$, and the two similar equations obtained by replacing x by x' and by x''. The result then follows from the equation

$$\chi(E|x, \eta|r) = \chi(E|x', \eta|r) + \chi(E|x'', \eta|r),$$

which holds for $r >> 0$.

Next we consider the analogue of theorem 6.4 for degree functions.

THEOREM 7.2. *Let* $\mathbf{I}, \mathbf{I}'$ *be two sequences of* $D-1$ *admissible ideals such that* I_j, I'_j *have the same degree for each* j *and* $I_j \subseteq I'_j$ *for* $j = 1, \ldots, D-1$. *Then, for all* E *and all* x *such that* $\dim A/xA = D-1$,

$$d(A|\mathbf{I}|E|x) \geq d(A|\mathbf{I}'|E|x)$$

with equality if $I'_j \subseteq I_j\dagger$ *for each* j *where* $I_j\dagger$ *is defined as in 6.1.*

PROOF. As in the proof of 6.4 we can assume that $E = A$ and further we may assume that A is a domain. Hence we only have to prove that

$$e(A_x|\mathbf{I}_x|A_x) \geq e(A_x|\mathbf{I}'_x|A_x)$$

and that equality holds in this inequality if $I_j\dagger \supseteq I'_j \supseteq I_j$ for each j. Both these statements follow if we apply 6.4 to the ring A_x and the sequences of ideals $\mathbf{I}_x$ and $\mathbf{I}'_x$.

We now consider the translation of theorem 5.3. We therefore restrict attention to the case where A is a domain, and $E = A$, and assume that $\mathbf{I} = (I_1, \ldots, I_{D-1})$ is a set of $D-1$ admissible ideals of A. We will write $d(A|\mathbf{I}|x)$ in place of $d(A|\mathbf{I}|E|x)$. As observed earlier, if $G = G(\sigma, \Xi), a(G) = a(A)$ and so is $< D$, by our restriction on A.

The term m-valuation of A used below requires elaboration. We use the definition given in section 5, with A replacing G or G'. Then, initially, it is defined to mean an integer-valued valuation v on the homogeneous field of fractions of A which takes non-negative values on Q and positive values on m, and we term it a good m-valuation if its residue field K_v has transcendence degree $D-1$ over k. As in section 5, there is a unique extension of v to the complete field of fractions of A which has the properties that it is graded and takes non-negative integer values on A, and its centre is a relevant prime ideal of A. This extension will also be denoted by v.

THEOREM 7.3. *Let* $\mathbf{I} = (I_1, \ldots, I_{D-1})$ *be a set of* $D-1$ *admissible ideals of the domain* A. *Then we can define for each* m*-valuation* v *of* A *a unique non-negative integer* $d(A|\mathbf{I}|v)$ *satisfying the conditions i), ii) below, and which also satisfies iii) to v).*

i) $d(A|\mathbf{I}|v) = 0$ *for all save a finite set of good* m*-valuations* v.

ii) *If* x *is a homogeneous element of* A *of degree* $\delta(x)$ *satisfying the condition that* $\dim(A/xA) = D-1$, *then,*

$$d(A|\mathbf{I}|x) = \delta(x).e(A|\mathbf{I}, \Xi) + \sum_v d(A|\mathbf{I}|v)v(x).$$

iii) $d(A|\mathbf{I}|v)$ *is symmetric in* $I_1, \ldots, I_{D-1}$.

iv) *If* $I_{D-1} = I'I'', \mathbf{I}' = (I_1, ..., I_{D-2}, I'), \mathbf{I}'' = (I_1, ..., I_{D-2}, I'')$,

$$d(A|\mathbf{I}|v) = d(A|\mathbf{I}'|v) + d(A|\mathbf{I}''|v).$$

v) *Let $I' \supseteq I_{D-1}$ and be level of the same degree as I_{D-1}. Further suppose that I' is contained in $I_{D-1}\dagger$. Then*

$$d(A|I_1, \ldots, I_{D-2}, I'|v) = d(A|\mathbf{I}|v)$$

for all v.

PROOF. We suppose that $\mathbf{J} = (J_1, \ldots, J_q)$ is a sequence of admissible ideals of A and $K = (k(1), \ldots, k(q)), l$ are such that $\mathbf{I}$ consists of $k(1)$ copies of $J_1, \ldots, k(q)$ copies of J_q and l copies of Ξ, so that $|K| + l = D - 1$.

Now take $G = G(\sigma, \Xi)$ and $G' = G(\mathbf{J})$, these rings being graded by sets of $q+1$ integers $R = (r(1), \ldots, r(q), s)$, the last integer s corresponding to the degree in A itself. Further, $d(A|\mathbf{I}|x) = d(G/G'|x|K, l)$ and $e(A|\mathbf{I}, \Xi) = e(G/G'|K, l+1)$. If we consider x as an element of G', it has its first q degrees equal to zero and its final degree equal to $\delta(x)$. We can apply theorem 5.3, which takes the form

$$d(G/G'|x|W, w) = \delta(x) e_{(0)}(G/G'|W, w) + \sum_v d(G'|v|W) v(x),$$

where $e_{(0)}(G/G'|W, w)$ denotes the quotient when $e(G/G'|W, w)$ is divided by w. By equating the coefficients of $W^K w^l$ on the two sides of the equation we obtain

$$d(A|\mathbf{I}|x) = d(G/G'|x|K, l) = \delta(x) e(G/G'|K, l+1) + \sum_v d(G'|v|K, l) v(x),$$

which holds for all homogeneous x in A. Hence we can define $d(A|\mathbf{I}|v) = d(G'|v|K, l)$ and these values will satisfy i) and ii) as required.

The function $d(A|\mathbf{I}|x)$ defined on A can be defined for all x of the form y/z where y, z are non-zero homogeneous elements of A, by the equation

$$d(A|\mathbf{I}|x) = d(A|\mathbf{I}|y) - d(A|\mathbf{I}|z).$$

If x belongs to the homogeneous field of fractions F of A , then we have

$$d(A|\mathbf{I}|x) = \sum_v d(A|\mathbf{I}|v) v(x).$$

The theorem of independence of valuations then implies that, if $d(A|\mathbf{I}|v) = 0$ for all save a finite set of values of v, then the truth of this equation for all non-zero x determines $d(A|\mathbf{I}|v)$ uniquely for all v. This justifies the adjective "unique" as applied to $d(A|\mathbf{I}|v)$, and shows that it is only necessary to consider ii) as holding for $E = A/P$ where P ranges over those prime ideals P of A such that $\dim(A/P) = D$.

To prove iii), suppose that $\mathbf{I}'$ is a permutation of $\mathbf{I}$. Then, since by its definition, $d(A|\mathbf{I}|x)$ is symmetric in $\mathbf{I}$, it follows that

$$\sum_v d(A|\mathbf{I}|v) v(x) = \sum_v d(A|\mathbf{I}'|v) v(x)$$

for all x in F, and this implies that $d(A|\mathbf{I}|v) = d(A|\mathbf{I}'|v)$ for all v by the theorem of independence of valuations.

By 7.1, vi), we have the equation,

$$d(A|\mathbf{I}|x) - d(A|\mathbf{I}'|x) - d(A|\mathbf{I}''|x) = 0,$$

and hence that

$$\sum_v (d(A|\mathbf{I}|v) - d(A|\mathbf{I}'|v) - d(A|\mathbf{I}''|v))v(x) = 0,$$

for all non-zero x in F. iv) then follows by the theorem of independence of valuations.

To prove v) we note that, for r large enough, $(I')^{r+1} = (I')^r.I_{D-1}$, since I_{D-1} is a reduction of I'. Hence v) follows from iv).

THEOREM 7.4. *Let* $A, \mathbf{I} = (I_1, \ldots, I_{D-1})$ *be as in the last theorem, and further suppose that* I_j *is of degree* $\sigma(j)(j = 1, \ldots, D-1)$. *Let* J *be a another admissible ideal of* A *of degree* σ. *Then*

$$e(A|\mathbf{I}, J) = \sigma e(A|\mathbf{I}, \Xi) + \sum d(A|\mathbf{I}|v)v(J).$$

PROOF. By the usual device of adjoining an indeterminate to the residue field k of Q, we may assume that k is infinite. Under this assumption, we can now choose elements $x_j \in I_j, (j = 1, \ldots, D-1)$ and $x \in J$ such that

a) $(x_1, \ldots, x_{D-1}, x)$ is a joint reduction of $\mathbf{I}$,J of type (1, ... , 1),

b) $v(x) = v(J)$ for the good m-valuations associated with $\mathbf{I}$.

It follows from 6.3(iv) that

$$e(A|\mathbf{I}, J) = d(A|\mathbf{I}|x).$$

But the right-hand side is equal to $\sigma e(A|\mathbf{I}, \Xi) + \sum_v d(A|\mathbf{I}|v)v(x)$ by 7.3(ii) and hence is equal to

$$\sigma e(A|\mathbf{I}, \Xi) + \sum_v d(A|\mathbf{I}|v)v(J).$$

To end this section we consider a single admissible ideal J of A of degree σ, generated by elements $u_1, \ldots, u_n$. We will write $G(J)$ for the ring $A[tu_1, \ldots, tu_n]$ and B for its sub-ring $Q[tu_1, \ldots, tu_n]$ where t is an indeterminate over A. We remind the reader that the condition that $a(A) \leq D-1$ is in force. We now define $e_l(A|J)$ to be $e(A|\mathbf{I})$ where $\mathbf{I}$ consists of l copies of Ξ and $D-l$ copies of J, and the functions $d_l(A|J|x)$ and $d_l(A|J|v)$ to be $d(A|\mathbf{I}|x)$ and $d(A|\mathbf{I}|v)$ when $\mathbf{I}$ consists of l copies of Ξ and $D-1-l$ copies of J. The valuations v such that $d_l(A|J|v) \neq 0$ for some l will be referred to as the good m-valuations associated with J.

DEFINITION. Let v be a good m-valuation associated with J, let $P(v)$ be the centre of the canonical extension of v to $G(J)$, and write $p(v)$ for $P(v) \cap A$ and $p'(v)$ for $P(v) \cap B$. Then we define $c(J|v)$ to be $\dim(A/p(v))$ and $c'(J|v)$ to be $\dim(B/p'(v))$.

THEOREM 7.5.
a) If $p \supseteq J$ is a homogeneous prime ideal of A meeting Q in m, then there exists a good m-valuation v associated with J such that $p(v) = p$ if and only if

$$i) a(J_p) = \text{ht}p, \quad ii) \text{ht}p + \text{tr.deg}_k k(p) = D.$$

b) If v is an m-valuation associated with J, then if $d_l(A|J|v) \neq 0$

$$c(J|v) \geq l \geq D - 1 - c'(J|v).$$

PROOF. a) If $p = p(v)$, then $p = P(v) \cap A$. The prime ideals $P(v)$ are characterised by the properties that they contain $mG(J)$ and satisfy $\text{tr.deg}_k k(P) = D - 1$.

Next we refer back to the discussion of analytic spread of ideals preceding lemma 1.2, to obtain the expression $\dim(G(J_p)/pG(J_p)) + 1$ for the analytic spread $a(J_p)$ of the ideal J_p of A_p, where $G(J_p)$ is the Rees ring of J_p. If we now apply part iii) of theorem 3.6 to the ring $G(J_p)/pG(J_p)$, we can write this as $\text{Max}(1 + \text{tr.deg}_{k(p)} k(P))$, taken over all prime ideals P of $G(J)$ meeting A in p.

We now apply the dimension inequality to $P = P(v)$ and $p = p(v)$ to obtain

$$\text{ht}P + \text{tr.deg}_k k(P) = \text{ht}P + \text{tr.deg}_{k(p)} k(P) + \text{tr.deg}_k k(p) \leq \text{ht}p + \text{tr.deg}_k k(p) \leq D.$$

As $\text{tr.deg}_k k(P) = D - 1$, then we must have $\text{ht}P = 1$ and both inequalities must be equalities. Hence, if $p = p(v)$, i) and ii) both hold.

If i) holds, we can choose P such that $1 + \text{tr.deg}_{k(p)} k(P) = \text{ht}p$, implying that $\text{ht}P = 1$ and the first inequality is an equality.

If ii) holds the second inequality is an equality.

Hence our chosen P satisfies the equation $\text{tr.deg}_k k(P) = D - 1$, and there exists a good m-valuation v associated with J, such that $P = P(v)$ and hence $p = p(v)$.

b) Consider the bigraded domain $G(J)/P(v)$ over k, whose dimension is $D-1$. It is graded by pairs of integers (r, s), the subring consisting of the elements whose first degree is zero being $A/p(v)$ and the subring consisting of the elements whose second degree is zero being $B/p'(v)$. But, apart from a non-zero multiple, $d_l(A|J|v)$ is the coefficient of $r^{D-l-1}s^l$ in $M(G(J)/P(v)|r, s)$. Since $A/p(v)$ has dimension $c(J|v)$, it follows from theorem 3.4 that $d_l(A|J|v) = 0$ if $l > c(J|v)$, while since $B/p'(v)$ has dimension $c'(J|v)$, it follows that $d_l(A|J|v) = 0$ if $D - l - 1 > c'(J|v)$.

THEOREM 7.6. *Let J be an admissible ideal of A of degree σ. Then:*
a) if x is a homogeneous element of A of degree $\delta(x)$,

$$d_l(A|J|x) = \delta(x).e_{l+1}(A|J) + \sum d_l(A|J|v)v(x);$$

b)

$$e_l(A|J) = \sigma.e_{l+1}(A|J) + \sum d_l(A|J|v)v(J);$$

the last sum in each case being over the m-valuations v associated with J satisfying $c(J|v) \geq l \geq D-1-c'(J|v)$.

PROOF. The two parts follow from 7.3 ii), theorem 7.4 and theorem 7.5, if we take the set I of the first two theorems to consist of $D-l-1$ copies of J and l copies of Ξ.

THEOREM 7.7. *Let $I, J \subseteq I$ be admissible ideals of A of the same degree σ, and suppose that I/J is generated by a single element. Then,*

$$e_l(A|I) = \sigma e_{l+1}(A|J) + \sum_v d_l(A|J|v)v(I),$$

the sum being over all good valuations associated with J. As a consequence,

$$e_l(A|I) \geq \sigma e_{l+1}(A|J) \quad (l = 0, 1, \dots),$$

with equality if $v(I) = 0$ for all good m-valuations v associated with J which satisfy $c(J|v) \geq l \geq D-1-c'(J|v)$.

PROOF. We can reduce the proof to the case when k is infinite by the device of adjoining an indeterminate to k. If k is infinite, we can find an element a of degree σ in I such that $I = J + aA$, and $v(a) = v(I)$ for all good valuations v associated with J or I. We next note that, as $I + aA = J + aA$, $d_l(A|I|a) = d_l(A|J|a)$ for all l. Applying theorem 7.6, we have

$$\begin{aligned} e_l(A|I) &= \sigma e_{l+1}(A|I) + \sum_v d_l(A|I|v)v(I) \\ &= \sigma e_{l+1}(A|I) + \sum_v d_l(A|I|v)v(a) \\ &= d_l(A|I|a) = d_l(A|J|a) \\ &= \sigma e_{l+1}(A|J) + \sum_v d_l(A|J|v)v(a) \\ &= \sigma e_{l+1}(A|J) + \sum_v d_l(A|J|v)v(I) \end{aligned}$$

for all l. This proves the first statement in the theorem, and the second is an immediate consequence of it. If I satisfies $v(I) = 0$ for all valuations v associated with J for which $d_l(A|J|v) \neq 0$, then it is clear that $e_l(A|I) = \sigma e_{l+1}(A|J)$. But such valuations satisfy $c(J|v) \geq l \geq D-1-c'(J|v)$ by 7.5 b). (If k is infinite we can chose I to satisfy this condition.)

8. An Example.

We finish this paper with a simple example of the theory developed in sections 6 and 7.

We will suppose that Q is a regular local ring of dimension 1 whose maximal ideal is generated by the element x. We take A to be the ring $S = Q[X_1, \dots, X_m]$, where $X_1, \dots, X_m$ are indeterminates over Q and we will write S instead of

A in what follows. Note that this implies that S has dimension m. I will be an admissible ideal of S of degree 1. Then, by applying a suitable linear transformation to the set of generators $X_1, \ldots, X_m$ of S, we can take I to be generated by elements $x^{a(i)}X_i (i = 1, \ldots, m)$ where $a(1) \geq a(2) \geq \ldots \geq a(m) \geq 0$. We will denote the sequence $a(1), \ldots, a(m)$ by $\mathbf{a}$. We now have the following theorem.

THEOREM 8.1.

$$e_l(S|I) = a(l+1) + \ldots + a(m),$$

where $l = 0, \ldots, m-1$ *and is zero if* $l \geq m$.

PROOF. We write $\Lambda(\mathbf{a}, r, s)$ for the length $\lambda(S_{r+s}/(J^r)_{r+s})$ which is equal, for $r, s >> 0$, to a polynomial in r, s of degree m, in which the terms of degree m can be written as

$$(e_0(S|I)r^m + \ldots + \binom{m}{l} e_l(S|I)r^{m-l}s^l + \ldots + me_{m-1}(S|I)rs^{m-1} + e_m(S|I)s^m)/m!.$$

We will now prove that it is a polynomial in r, s of degree m in which the terms of degree m are

$$(s(0)r^m + \ldots + \binom{m}{l} s(l)r^{m-l}s^l + \ldots + ma(m)rs^{m-1})/m!,$$

where $s(l) = a(l+1) + \ldots + a(m)$, which will prove the theorem.

We proceed by induction on m, the case $m = 0$ being trivial. We will also write $\mu(m, t)$ for the number of monomials in m variables of degree t, i.e. $(t+1)\ldots(t+m-1)/(m-1)!$, which is a polynomial in t of degree $m-1$. Now we can write

$$\Lambda(\mathbf{a}, r, s) = \lambda(S_{r+s}/x^{ra(m)}S_{r+s}) + \lambda(x^{ra(m)}S_{r+s}/(I^r)_{r+s}).$$

The first term is equal to $a(m)r\mu(m, r+s)$ and so is a polynomial in r, s of degree m, and the terms of degree m can be written

$$a(m)(mr^m + \cdots + \binom{m}{l}(m-l)r^{m-l}s^l + \cdots + mrs^{m-1}))/m!.$$

The second term can be written as $\sum \Lambda(\mathbf{a}', t, s)$, where $\mathbf{a}'$ is the set of $m-1$ integers $a'(l) = a(l) - a(m), (l = 1, \ldots, m-1)$ and the sum is from $t = 0$ to $t = r$. By our inductive hypothesis,

$$\Lambda(\mathbf{a}', t, s) = ((s(0) - ma(m))t^{m-1} + \ldots$$
$$+ \binom{m-1}{l}(s(l) - (m-l)a(m))t^{m-l-1}s^l + \ldots)/(m-1)!,$$

and since $\sum t^{m-l-1}$ (summed from $t = 0$ to r) is a polynomial in r of degree $m-l$, the term of degree $m-l$ being $r^{m-l}/(m-l)$, $\Lambda(\mathbf{a}, r, s)$ is a polynomial in r, s of degree m and the terms of degree m can be written as

$$(s(0)r^m + ... + \binom{m}{l} s(l)r^{m-l}s^l + ... + ms(m-1)rs^{m-1})/m!$$

which completes the proof.

If we consider the ring S for all values of m, and the ideal I to be an arbitrary ideal of S generated by elements of degree 1, the above shows that the sequence of numbers $e_l(S|I)(i = 0, 1, \dots)$ is only subject to the conditions that $e_l(S|I)$, the forward differences $e_l(S|I) - e_{l+1}(S|I)$ and second differences $e_l(S|I) - 2e_{l+1}(S|I) + e_{l+2}(S|I)$ are non-negative for all l and zero for large l. The statements concerning $e_l(S|I)$ and the forward differences $e_l(S|I) - e_{l+1}(S|I)$ hold in general in view of theorem 7.6 b), but whether the result concerning the second difference holds in general is still open.

REFERENCES

1. J-P. Henry and M. Merle, *Conormal Space and Jacobian Module:Integral Closure of a module and related questions.*, Preprint (1991).
2. D. Kirby, *Graded Multiplicity Theory and Hilbert Functions*, J.London.Math.Soc.(2) **36** (1987), 16-22.
3. D.G. Northcott, *A general theory of one-dimensional local rings.*, Proc.Glasgow Math.Ass. **II** (1956), 159-169.
4. D.G. Northcott and D. Rees, *Reductions of Ideals in local rings.*, Proc.Camb.Phil.Soc. **50** (1954), 145-158..
5. A. Ooshi, *Reductions of graded rings and psuedo-flat graded modules.*, Hiroshima Math.J. **18** (1988), 463-477.
6. D. Rees, *Degree Functions.*, Proc.Camb.Phil.Soc. **57** (1961), 1-7.
7. ———, *Generalisations of reductions and mixed multiplicities.*, J.London.Math.Soc,(2) **29** (1984), 397-414.
8. ———, *Amao's Theorem and Reduction Criteria*, J.London.Math.Soc.(2) **32** (1985), 404-410.
9. ———, *Reductions of Modules.*, Math.Proc.Camb.Phil.Soc. **101** (1987), 431-449.
10. D. Rees and J. Sally, *General elements and joint reductions.*, Michigan J.Math. **35** (1988), 241-254..
11. B.Teissier, *Cycles evanscents, sections planes, et conditions de Whitney.*, Asterique **7/8** (1973), 285-362.

FACULTY OF MATHEMATICAL STUDIES, THE UNIVERSITY, SOUTHAMPTON, SO9 5NH, U.K.

6 HILLCREST PARK, EXETER EX4 4SH, DEVON, U.K.

Contemporary Mathematics
Volume **159**, 1994

Pfaffian Identities, with Applications to Free Resolutions, DG–Algebras, and Algebras with Straightening Law

ANDREW R. KUSTIN

ABSTRACT. A pfaffian identity plays a central role in the proof that the minimal resolution of a Huneke-Ulrich deviation two Gorenstein ring is a DGΓ–algebra. A second pfaffian identity is the patch which holds together two strands of the minimal resolution of a residual intersection of a grade three Gorenstein ideal. In this paper, we establish a family of pfaffian identities from which the two previously mentioned identities can be deduced as special cases. Our proof is by induction: the base case is an identity of binomial coefficients and the inductive step is a calculation from multilinear algebra.

Two unrelated problems in commutative algebra have recently been solved by finding the appropriate pfaffian identity. Fix a commutative noetherian ring R. For the first problem, let $X_{2n\times 2n}$ and $Y_{1\times 2n}$ be matrices with entries from R. Assume that X is an alternating matrix. Huneke and Ulrich [**10**] showed that if the ideal $I = I_1(YX) + \operatorname{Pf}(X)$ has the maximum possible grade (namely, $2n-1$), then I is a perfect Gorenstein ideal. The minimal R–resolution $\mathbb{M}$ of $A = R/I$ was found in [**11**]. Srinivasan [**15**] proved $\mathbb{M}$ is a DGΓ–algebra. The pfaffian identity [**15, 4.3**] is the key to showing that the proposed multiplication satisfies the differential property. The algebra structure on $\mathbb{M}$ was used in [**12**] to show that if $(R, \mathfrak{m}, k)$ is a regular local ring of equicharacteristic zero, then the Poincaré series

$$P_A^M(z) = \sum_{i=0}^{\infty} \dim_k \operatorname{Tor}_i^A(M,k) z^i$$

1991 *Mathematics Subject Classification.* 15A69, 13C40, 13D25, 13F50.

Key words and phrases. algebras with straightening law, deviation two Gorenstein ideal, DG-algebras, pfaffian, residual intersection.

The author was supported in part by the National Science Foundation DMS–9106329.

This paper is in final form and no version of it will be submitted for publication elsewhere.

is a rational function for all finitely generated A–modules M.

For the second problem, let J be a residual intersection of a grade three Gorenstein ideal I. If the data is sufficiently generic, then the generators of J, the minimal resolution of R/J, the minimal resolution of "half" of the divisor class group of R/J, and the minimal resolution of each power I^k have all been found in [**14**]. Each resolution in [**14**] is obtained by patching together two complexes all of whose maps are linear and well understood. The patch involves maps of many different degrees and is much more difficult to understand. Pfaffian identity [**14, 3.2**] shows that patched object is a complex. (Some of the results from [**14**] may be obtained using other techniques: the generators of J are also calculated in [**13**] and the ideals I^k are also resolved in [**2**].)

The above mentioned pfaffian identities (see Propositions 5.1 and 4.1 in the present paper) appear to be somewhat similar, but the exact relationship between them is not immediately obvious. The existing proofs of these identities are ad hoc and unpleasant. In the present paper, we derive the two identities from a common result: identity [**15, 4.3**] is the left side of Theorem 3.1 when the data is arranged so that the right side consists of only one term; and [**14, 3.2**] is the right side of Theorem 3.1 when the left side has become trivial. The proof of Theorem 3.1 is an induction which depends on an identity of binomial coefficients (Lemma 1.3) and a fact from multilinear algebra (Lemma 2.4).

In section 6 we apply our techniques to derive a straightening formula for pfaffians. We conclude with a brief section which shows that many familiar pfaffian identities are special cases of Theorem 3.1.

1. Binomial coefficients

We often use binomial coefficients with negative parameters; consequently, we now recall the standard definition and properties of these objects.

DEFINITION 1.1. For integers i and m, the binomial coefficient $\binom{m}{i}$ is defined to be

$$\binom{m}{i} = \begin{cases} \dfrac{m(m-1)\cdots(m-i+1)}{i!} & \text{if } 0 < i, \\ 1 & \text{if } 0 = i, \text{ and} \\ 0 & \text{if } i < 0. \end{cases}$$

OBSERVATION 1.2. (*a*) *If* $0 \le m < i$, *then* $\binom{m}{i} = 0$.

(b) *For all integers* i *and* m,

$$\binom{m}{i-1} + \binom{m}{i} = \binom{m+1}{i}.$$

(c) *If* i *and* m *are integers with* $0 \le m$, *then* $\binom{m}{i} = \binom{m}{m-i}$.

(d) *If* i *is a nonnegative integer, then* $\binom{-1}{i} = (-1)^i$.

(e) *For all integers a and b,*

$$\binom{a}{b} = (-1)^b \binom{b-a-1}{b}.$$

The identities in Lemma 1.3 form the base step in the proof of our main result. Corollary 1.4 is well known. It may be proved directly using an argument similar to the proof of Lemma 1.3 or it may be deduced from Lemma 1.3 after two applications of Observation 1.2 (e).

LEMMA 1.3. *Let A, B, and C be integers. If $0 \le A$, then*

(a)
$$\sum_{k\in\mathbb{Z}} (-1)^k \binom{B+k}{C+k}\binom{A}{k} = (-1)^A \binom{B}{A+C}, \quad \textit{and}$$

(b)
$$\sum_{k\in\mathbb{Z}} (-1)^k \binom{B-k}{C-k}\binom{A}{k} = \binom{B-A}{C}.$$

PROOF. The proof of (a) proceeds by induction on A. If $A = 0$, then the only nonzero term on the left side occurs when $k = 0$, and this term is $\binom{B}{C}$, which is equal to the right side. We now suppose that the result holds for a fixed A for all values of B and C. Let

$$X = \sum_{k\in\mathbb{Z}} (-1)^k \binom{B+k}{C+k}\binom{A+1}{k}.$$

Observation 1.2 (b) gives $X = X_1 + X_2$, where

$$X_1 = \sum_{k\in\mathbb{Z}} (-1)^k \binom{B+k}{C+k}\binom{A}{k} \quad \text{and} \quad X_2 = \sum_{k\in\mathbb{Z}} (-1)^k \binom{B+k}{C+k}\binom{A}{k-1}.$$

The induction hypothesis gives

$$X_1 = (-1)^A \binom{B}{A+C} \quad \text{and} \quad X_2 = (-1)^{A+1} \binom{B+1}{A+C+1}.$$

We conclude that

$$X = X_1 + X_2 = (-1)^{A+1}\left[\binom{B+1}{A+1+C} - \binom{B}{A+C}\right] = (-1)^{A+1}\binom{B}{A+1+C},$$

as desired. Replace k with $A-k$ in order to deduce (b) from (a). □

COROLLARY 1.4. *Let A, B, and C be integers. If $0 \le A$, then*

$$\sum_{k\in\mathbb{Z}} \binom{B}{C-k}\binom{A}{k} = \binom{A+B}{C}. \quad \square$$

The following generalization of Observation 1.2 (c) is used in the proof of Corollary 3.3.

LEMMA 1.5. *If A and B are integers, then*

$$\binom{A}{A-B} = \binom{A}{B} + (-1)^{A+B}\binom{-B-1}{-A-1}.$$

PROOF. Let

$$T_1 = \binom{A}{A-B}, \quad T_2 = \binom{A}{B} \quad \text{and} \quad T_3 = (-1)^{A+B}\binom{-B-1}{-A-1}.$$

If $0 \le A$, then $T_1 = T_2$ and $T_3 = 0$. Henceforth, we assume that $A \le -1$. If $0 \le B$, then $A - B \le A \le -1$ and $T_1 = 0$. Furthermore, in this case, $0 \le B - A - 1$; consequently,

$$T_2 = \binom{A}{B} = (-1)^B\binom{B-A-1}{B} = (-1)^B\binom{B-A-1}{-A-1} = (-1)^{A+B+1}\binom{-B-1}{-A-1} = -T_3.$$

Finally, we assume that $A \le -1$ and $B \le -1$. In this case, $T_2 = 0$ and

$$T_1 = \binom{A}{A-B} = (-1)^{A+B}\binom{-B-1}{A-B} = (-1)^{A+B}\binom{-B-1}{-A-1} = T_3. \quad \square$$

2. Multilinear algebra

DATA 2.1. *Let R be a commutative noetherian ring, F be a free R-module of finite rank, and φ be an element of $\bigwedge^2 F$.*

We make much use of the $\bigwedge^\bullet F^*$-module structure on $\bigwedge^\bullet F$, as well as the $\bigwedge^\bullet F$-module structure on $\bigwedge^\bullet F^*$. In particular, if $a_i \in \bigwedge^i F$ and $b_j \in \bigwedge^j F^*$, then

$$a_i(b_j) \in \bigwedge^{j-i} F^* \quad \text{and} \quad b_j(a_i) \in \bigwedge^{i-j} F.$$

The following formulas are well known and not difficult to establish; see, for example, [**4, Proposition A.3**].

PROPOSITION 2.2. *Adopt Data 2.1. Let $a, b \in \bigwedge^\bullet F$ and $c \in \bigwedge^\bullet F^*$ be homogeneous elements.*

(a) *If $a \in \bigwedge^1 F$, then*

$$(a(c))(b) = a \wedge (c(b)) + (-1)^{1+\deg c} c(a \wedge b).$$

(b) *If $c \in \bigwedge^{\operatorname{rank} F} F^*$, then*

$$(a(c))(b) = (-1)^\nu (b(c))(a),$$

where $\nu = (\operatorname{rank} F - \deg a)(\operatorname{rank} F - \deg b)$. $\square$

NOTE. The value for ν which is given above is correct and is different than the value given in [**4**].

We also make heavy use of the divided power structure on $\bigwedge^\bullet F$. If a is a homogeneous element of $\bigwedge^\bullet F$ of even degree, then, for each integer k, $a^{(k)}$ is an element of $\bigwedge^\ell F$, where $\ell = k \cdot \deg a$. (Of course, if $\ell < 0$ or $\operatorname{rank} F < \ell$, then $\bigwedge^\ell F = 0$.) Some of the properties of divided powers are collected below; more information may be found in [**4**].

PROPOSITION 2.3. *Adopt Data 2.1.*

(a) *If k and ℓ are integers, then*

$$\varphi^{(k)} \wedge \varphi^{(\ell)} = \binom{k+\ell}{k} \varphi^{(k+\ell)}.$$

(b) *If $b \in F^*$ and k is an integer, then $b\left(\varphi^{(k)}\right) = b(\varphi) \wedge \varphi^{(k-1)}$.*

The following multilinear algebra calculation is the key to the induction step in the proof of our main result.

LEMMA 2.4. *Adopt Data 2.1. Let k and ℓ be integers. If b_1 and b are homogeneous elements of $\bigwedge^\bullet F^*$ with $\deg b_1 = 1$, then*

$$b_1\left[\left(\varphi^{(k)}(b)\right)(\varphi^{(\ell)})\right] + b_1\left[\left(\varphi^{(k-1)}(b)\right)(\varphi^{(\ell+1)})\right] - b_1(\varphi^{(1)}) \wedge \left[\left(\varphi^{(k-1)}(b)\right)(\varphi^{(\ell)})\right]$$
$$= \left[\varphi^{(k)}(b_1 \wedge b)\right](\varphi^{(\ell)}).$$

PROOF. We expand $S_1 = \left[(b_1(\varphi^{(1)}))\left(\varphi^{(k-1)}(b)\right)\right](\varphi^{(\ell)})$ two different ways. On the one hand, Proposition 2.2 gives $S_1 = S_2 + S_3$, where

$$S_2 = b_1(\varphi^{(1)}) \wedge \left(\varphi^{(k-1)}(b)\right)(\varphi^{(\ell)}) \text{ and } S_3 = (-1)^{1+\deg b}\left(\varphi^{(k-1)}(b)\right)\left(b_1(\varphi^{(1)}) \wedge \varphi^{(\ell)}\right).$$

Proposition 2.3, together with the module action of $\bigwedge^\bullet F^*$ on $\bigwedge^\bullet F$, yields

$$S_3 = (-1)^{1+\deg b}\left(\varphi^{(k-1)}(b)\right)\left(b_1(\varphi^{(\ell+1)})\right) = (-1)^{1+\deg b}\left(\varphi^{(k-1)}(b) \wedge b_1\right)(\varphi^{(\ell+1)})$$
$$= -\left[b_1 \wedge \varphi^{(k-1)}(b)\right](\varphi^{(\ell+1)}) = -b_1\left[\left(\varphi^{(k-1)}(b)\right)(\varphi^{(\ell+1)})\right].$$

On the other hand, Proposition 2.2 gives

$$\left(b_1(\varphi^{(k)})\right)(b) = b_1 \wedge \varphi^{(k)}(b) - \varphi^{(k)}(b_1 \wedge b);$$

and therefore,

$$S_1 = \left(\left[b_1(\varphi^{(1)}) \wedge \varphi^{(k-1)}\right](b)\right)(\varphi^{(\ell)}) = \left[\left(b_1(\varphi^{(k)})\right)(b)\right](\varphi^{(\ell)})$$
$$= b_1\left[\left(\varphi^{(k)}(b)\right)(\varphi^{(\ell)})\right] - \left(\varphi^{(k)}(b_1 \wedge b)\right)(\varphi^{(\ell)}).$$

The proof is completed by equating the two values for S_1. □

The identities in section 3 are stated and proved in a coordinate free manner; however, some of the applications in sections 4 through 7 are derived from these identites by using bases. The hypothesis in these applications is that X is an alternating matrix with entries from the commutative noetherian ring R. We recover Data 2.1 as follows. Suppose that the matrix $X = (x_{ij})$ has N rows and columns. Let $e_1, \ldots, e_N$ be a basis for a free R–module F; let $\varepsilon_1, \ldots, \varepsilon_N$ be the corresponding dual basis for F^*; and let φ be the element

$$\varphi = \sum_{1 \le i < j \le N} x_{ij}\, e_i \wedge e_j$$

of $\bigwedge^2 F$. If $I = (n_1, \dots, n_i)$ is an i-tuple of integers, then define $e_I \in \bigwedge^i F$ and $\varepsilon_I \in \bigwedge^i F^*$ by

$$e_I = e_{n_1} \wedge e_{n_2} \wedge \dots \wedge e_{n_i} \qquad \text{and} \qquad \varepsilon_I = \varepsilon_{n_1} \wedge \varepsilon_{n_2} \wedge \dots \wedge \varepsilon_{n_i}.$$

CALCULATION 2.5. *If I is an i-tuple of integers, then*

$$e_I(\varepsilon_I) = \varepsilon_I(e_I) = (-1)^{\frac{i(i-1)}{2}}.$$

PROOF. Let $f_i = (e_1 \wedge e_2 \wedge \dots \wedge e_i)(\varepsilon_1 \wedge \varepsilon_2 \wedge \dots \wedge \varepsilon_i)$. It is clear that $f_1 = 1$ and $f_{i+1} = (-1)^i f_i$. It is also clear that $f_i = (-1)^{\frac{i(i-1)}{2}}$ solves this recurrence relation. □

Let $I = (n_1, \dots, n_i)$ be a fixed i-tuple of integers. If the integers $n_1, \dots, n_i$ are distinct, then $\sigma(I)$ is the sign of the permutation which rearranges $n_1, \dots, n_i$ into ascending order. If there is a repeat among the n's, then $\sigma(I) = 0$. We say that I is *an index set of size i* if $n_1 < n_2 < \dots < n_i$. Sometimes, we write $|I|$ for the size of the index set I. Let S be a fixed finite set of distinct integers. We write

$$\sum_{\substack{J_1 \cup J_2 = S \\ |J_1| = j_1}}$$

to mean that the sum is taken over all j_1-element subsets J_1 of S, and the complement of J_1 in S is denoted J_2. In the above sum, J_1 and J_2 are both taken to be index sets. If $b \in \bigwedge^r F^*$, then $b(e_I)$ is the element

$$b(e_I) = \sum_{\substack{I_1 \cup I_2 = I \\ |I_1| = r}} \sigma(I_1 I_2) b(e_{I_1}) e_{I_2} \tag{2.6}$$

of $\bigwedge^{i-r} F$. If $Y = (y_{ij})$ is a matrix and I_1 and J are j-tuples of integers, then we take $Y(I_1; J)$ to be the determinant of the submatrix of Y which consists of rows I_1 and columns J. In particular, if $f\colon F^* \to F$ is a map with $f(\varepsilon_j) = \sum_i y_{ij} e_i$ and J is a j-tuple of integers, then

$$(\textstyle\bigwedge^j f)(\varepsilon_J) = \displaystyle\sum_{\substack{I_1 \cup I_2 = \{1, \dots, N\} \\ |I_1| = j}} Y(I_1; J) e_{I_1}.$$

Thus, Calculation 2.5 shows that if $|I| = |J| = j$, then

$$\varepsilon_I\left((\textstyle\bigwedge^j f)(\varepsilon_J)\right) = \varepsilon_I(e_I) Y(I, J) = (-1)^{\frac{j(j-1)}{2}} Y(I; J). \tag{2.7}$$

We use the notation "X_I" to denote pfaffians; in particular,

$$X_I = \begin{cases} \text{the pfaffian of the principal submatrix of } X \text{ consisting} \\ \text{of rows and columns } n_1, \dots, n_i \text{ (in the given order)} & \text{if } 0 < i, \\ 1 & \text{if } i = 0, \text{ and} \\ 0 & \text{if } i < 0. \end{cases}$$

It follows that if i is an even integer, then

$$\varphi^{(i/2)} = \sum_{\substack{J_1 \cup J_2 = \{1,\dots,N\} \\ |J_1| = i}} X_{J_1} e_{J_1};$$

and therefore, Calculation 2.5 shows that

$$\varepsilon_I\left(\varphi^{(i/2)}\right) = \varepsilon_I(e_I) X_I = (-1)^{i/2} X_I. \tag{2.8}$$

3. The main results

The first main result in this paper is Theorem 3.1. Everything in sections 4 through 7, with one exception, follows from Theorem 3.1 or one of its five Corollaries. The one exception is Proposition 5.4, and this result is a consequence of Theorem 3.7, which is the other main result in the paper.

THEOREM 3.1. *Adopt Data 2.1. Let A, B, C, and d be integers. If $b \in \bigwedge^d F^*$, then*

(a) $$\sum_{k\in\mathbb{Z}} (-1)^k \binom{B+k}{C+k} b(\varphi^{(k)}) \wedge \varphi^{(A-k)} = \sum_{k\in\mathbb{Z}} (-1)^{A+k} \binom{B+d-k}{A+C-k} \left(\varphi^{(k)}(b)\right)(\varphi^{(A-k)}),$$

and

(b) $$\sum_{k\in\mathbb{Z}} (-1)^k \binom{B-k}{C-k} b(\varphi^{(k)}) \wedge \varphi^{(A-k)} = \sum_{k\in\mathbb{Z}} (-1)^{d+k} \binom{B-A+k}{C-d+k} \left(\varphi^{(k)}(b)\right)(\varphi^{(A-k)}).$$

NOTE. If $k < 0$ or $A < k$, then the corresponding terms in each of the above summations is zero.

PROOF. The proof of (a) proceeds by induction on d. We first suppose that $d = 0$. In this case, the left hand side is $Xb(\varphi^{(A)})$, where

$$X = \sum_{k\in\mathbb{Z}} (-1)^k \binom{B+k}{C+k}\binom{A}{k}.$$

The only non-zero term on the right side occurs when $k = 0$, and this term is $Yb(\varphi^{(A)})$, where

$$Y = (-1)^A \binom{B}{A+C}.$$

If $A \le -1$, then $\varphi^{(A)} = 0$. If $0 \le A$, then Lemma 1.3 shows that $X = Y$.

Henceforth, we assume that $1 \le d$. We take $b = b_1 \wedge b'$, with $\deg b_1 = 1$ and $\deg b' = d - 1$. Let

$$X = \sum_{k\in\mathbb{Z}} (-1)^k \binom{B+k}{C+k} b(\varphi^{(k)}) \wedge \varphi^{(A-k)}.$$

The element b_1 of F^* acts like a graded derivation on $\bigwedge^\bullet F$; therefore,

$$X = b_1(S_1) - b_1(\varphi^{(1)}) \wedge S_2,$$

where

$$S_1 = \sum_{k\in\mathbb{Z}} (-1)^k \binom{B+k}{C+k} b'(\varphi^{(k)}) \wedge \varphi^{(A-k)} \quad \text{and}$$

$$S_2 = \sum_{k\in\mathbb{Z}} (-1)^k \binom{B+k}{C+k} b'(\varphi^{(k)}) \wedge \varphi^{(A-1-k)}.$$

The induction hypothesis gives

$$S_1 = (-1)^A \sum_{k\in\mathbb{Z}} (-1)^k \binom{B+d-1-k}{A+C-k} \left(\varphi^{(k)}(b')\right)(\varphi^{(A-k)}) \quad \text{and}$$

$$S_2 = (-1)^A \sum_{k\in\mathbb{Z}} (-1)^{k+1} \binom{B+d-1-k}{A-1+C-k} \left(\varphi^{(k)}(b')\right)(\varphi^{(A-1-k)}).$$

Use Observation 1.2 (b) in order to write $S_1 = S_1' + S_1''$, where

$$S_1' = (-1)^A \sum_{k\in\mathbb{Z}} (-1)^k \binom{B+d-k}{A+C-k} \left(\varphi^{(k)}(b')\right)(\varphi^{(A-k)}) \quad \text{and}$$

$$S_1'' = (-1)^A \sum_{k\in\mathbb{Z}} (-1)^{k+1} \binom{B+d-1-k}{A+C-k-1} \left(\varphi^{(k)}(b')\right)(\varphi^{(A-k)}).$$

An index shift yields

$$S_1'' = (-1)^A \sum_{k\in\mathbb{Z}} (-1)^k \binom{B+d-k}{A+C-k} \left(\varphi^{(k-1)}(b')\right)(\varphi^{(A-k+1)}) \quad \text{and}$$

$$S_2 = (-1)^A \sum_{k\in\mathbb{Z}} (-1)^k \binom{B+d-k}{A+C-k} \left(\varphi^{(k-1)}(b')\right)(\varphi^{(A-k)}).$$

We now have

$$X = b_1(S_1' + S_1'') - b_1(\varphi^{(1)}) \wedge S_2$$

$$= (-1)^A \sum_{k\in\mathbb{Z}} (-1)^k \binom{B+d-k}{A+C-k} \left[\begin{array}{c} b_1\left[(\varphi^{(k)}(b'))(\varphi^{(A-k)})\right] \\ +b_1\left[(\varphi^{(k-1)}(b'))(\varphi^{(A-k+1)})\right] \\ -\left(b_1(\varphi^{(1)})\right) \wedge \left[(\varphi^{(k-1)}(b'))(\varphi^{(A-k)})\right] \end{array} \right].$$

Apply Lemma 2.4 to complete the proof of (a). The proof of (b) is completely analogous to the proof of (a); we omit the details. □

In our first variation of Theorem 3.1, we arrange the data so that only one term on the right side of the identity survives.

COROLLARY 3.2. *Adopt Data 2.1. Let p, q, and d be integers. If $b \in \bigwedge^d F^*$, then*

(a) $$\sum_{k\in\mathbb{Z}} (-1)^{k+p} \binom{q-d-1+k}{k-p} b(\varphi^{(k)}) \wedge \varphi^{(p+q-k)} = \left(\varphi^{(q)}(b)\right)(\varphi^{(p)}), \quad \textit{and}$$

(b) $$\sum_{k\in\mathbb{Z}} (-1)^{k+d+q} \binom{p-1-k}{d-q-k} b(\varphi^{(k)}) \wedge \varphi^{(p+q-k)} = \left(\varphi^{(q)}(b)\right)(\varphi^{(p)}).$$

PROOF. Apply Theorem 3.1 (a) with A replaced by $p+q$, B replaced by $q-d-1$, and C replaced by $-p$, in order to see that

$$\sum_{k\in\mathbb{Z}}(-1)^k\binom{q-d-1+k}{k-p}b(\varphi^{(k)})\wedge\varphi^{(p+q-k)}=\sum_{k\in\mathbb{Z}}(-1)^{q+p+k}\binom{q-1-k}{q-k}\Big(\varphi^{(k)}(b)\Big)(\varphi^{(p+q-k)}).$$

The proof of (a) is complete because right side of the above sum is zero unless $q=k$. Assertion (b) follows from Theorem 3.1 (b) with A replaced by $p+q$, B replaced by $p-1$ and C replaced with $d-q$. An alternate proof of (b) can be obtained from (a) by way of Lemma 1.5 and Corollary 3.4. $\square$

In our second variation of Theorem 3.1, we remove the "k" from the bottom part of the binomial coefficients.

COROLLARY 3.3. *Adopt Data 2.1. Let A, B, L, and d be integers. If b is an element of $\bigwedge^d F^*$, then*

$$\text{(a)}\quad \sum_{k\in\mathbb{Z}}(-1)^k\binom{B-k}{L}b(\varphi^{(k)})\wedge\varphi^{(A-k)}=\sum_{k\in\mathbb{Z}}(-1)^{d+k}\binom{B-A+k}{L+d-A}\Big(\varphi^{(k)}(b)\Big)(\varphi^{(A-k)}),$$

and

$$\text{(b)}\quad \sum_{k\in\mathbb{Z}}(-1)^k\binom{B+k}{L}b(\varphi^{(k)})\wedge\varphi^{(A-k)}=\sum_{k\in\mathbb{Z}}(-1)^{A+k}\binom{d+B-k}{L+d-A}\Big(\varphi^{(k)}(b)\Big)(\varphi^{(A-k)}).$$

PROOF. We first prove (a). Apply (b) of Theorem 3.1, with C replaced by $B-L$, in order to see that $L_1=R_1$, where

$$L_1=\sum_{k\in\mathbb{Z}}(-1)^k\binom{B-k}{B-L-k}b(\varphi^{(k)})\wedge\varphi^{(A-k)}\quad\text{and}$$

$$R_1=(-1)^d\sum_{k\in\mathbb{Z}}(-1)^k\binom{B-A+k}{B-L-d+k}\Big(\varphi^{(k)}(b)\Big)(\varphi^{(A-k)}).$$

Apply (a) of Theorem 3.1, with B replaced by $L-B-1$ and C replaced by $-B-1$, in order to see that $L_2=R_2$, where

$$L_2=\sum_{k\in\mathbb{Z}}(-1)^k\binom{L-B-1+k}{-B-1+k}b(\varphi^{(k)})\wedge\varphi^{(A-k)}\quad\text{and}$$

$$R_2=\sum_{k\in\mathbb{Z}}(-1)^{A+k}\binom{L-B-1+d-k}{-B-1+A-k}\Big(\varphi^{(k)}(b)\Big)(\varphi^{(A-k)}).$$

Apply Lemma 1.5 (with A replaced by $B-k$ and B replaced by $B-L-k$) in order to see that

$$L_1+(-1)^L L_2=\sum_{k\in\mathbb{Z}}(-1)^k\binom{B-k}{L}b(\varphi^{(k)})\wedge\varphi^{(A-k)}.$$

Lemma 1.5 (with A replaced by $B-A+k$ and B replaced by $B-L-d+k$) shows that

$$R_1+(-1)^L R_2=\sum_{k\in\mathbb{Z}}(-1)^{k+d}\binom{B-A+k}{d+L-A}\Big(\varphi^{(k)}(b)\Big)(\varphi^{(A-k)}).$$

The proof of (a) is complete. Identity (b) follows from (a), by way of Observation 1.2 (e); replace B with $L-B-1$. (An alternate proof of Corollary 3.3 can be obtained by mimicking the proof of Theorem 3.1.) □

In our third variation of Theorem 3.1, we impose hypotheses which allow us to set the identities in Corollary 3.3 to zero, one side at a time.

COROLLARY 3.4. *Adopt Data 2.1. Let A, B, d, and L be integers, and let b be an element of $\bigwedge^d F^*$.*

(i) *If $A+L+1 \leq d$, then*

$$\sum_{k\in Z}(-1)^k\binom{B+k}{L}(\varphi^{(k)}(b))(\varphi^{(A-k)})=0, \quad \textit{and}$$

$$\sum_{k\in Z}(-1)^k\binom{B-k}{L}(\varphi^{(k)}(b))(\varphi^{(A-k)})=0.$$

(ii) *If $d \leq A-L-1$, then*

$$\sum_{k\in Z}(-1)^k\binom{B-k}{L}b(\varphi^{(k)})\wedge\varphi^{(A-k)}=0, \quad \textit{and}$$

$$\sum_{k\in Z}(-1)^k\binom{B+k}{L}b(\varphi^{(k)})\wedge\varphi^{(A-k)}=0.$$

PROOF. Identity (a) from Corollary 3.3 gives

$$\sum_{k\in\mathbb{Z}}(-1)^{d+k}\binom{A+B-k}{L-d+A}b(\varphi^{(k)})\wedge\varphi^{(A-k)}=\sum_{k\in Z}(-1)^k\binom{B+k}{L}(\varphi^{(k)}(b))(\varphi^{(A-k)});$$

and the hypothesis ensures that $\binom{A+B-k}{L-d+A}=0$. The other three assertions are established in the same way. □

In a similar manner, one can find hypotheses which set each side of the identities in Theorem 3.1 to zero. Corollary 3.5, which plays a central role in section 6, is an example of this technique.

COROLLARY 3.5. *Adopt Data 2.1. Let A, B, C, and d be integers. Assume that $b\in\bigwedge^d F^*$. If $\frac{-1-d}{2}\leq B\leq A+C-d-1$, then*

$$\sum_{k\in\mathbb{Z}}(-1)^k\binom{B+k}{C+k}b(\varphi^{(k)})\wedge\varphi^{(A-k)}=0.$$

PROOF. If S is the left side of the identity, then Theorem 3.1 shows that

$$S=\sum_{k\in\mathbb{Z}}(-1)^{A+k}\binom{B+d-k}{A+C-k}\left(\varphi^{(k)}(b)\right)(\varphi^{(A-k)}).$$

If $0\leq B+d-k$, then the hypothesis guarantees that $0\leq B+d-k\leq A+C-k-1$, and the binomial coefficient is zero. If $B+d-k\leq -1$, then the hypothesis also guarantees that $\deg\varphi^{(k)}(b)\leq -1$; thus, $\varphi^{(k)}(b)=0$. It follows that $S=0$. □

Our fifth variation of Theorem 3.1 involves two elements of $\bigwedge^\bullet F^*$.

COROLLARY 3.6. *Adopt Data 2.1. Let A, B, d, and L be integers, and let b and b' be homogeneous elements of $\bigwedge^\bullet F^*$ with $b \in \bigwedge^d F^*$. If $A+L+1 \le d$, then*

$$\sum_{k\in\mathbb{Z}}(-1)^k\binom{B+k}{L}\left(\left(b'(\varphi^{(k)})\right)(b)\right)(\varphi^{(A-k)}) = 0.$$

PROOF. The proof is by induction on the degree of b'. If $\deg b' = 0$, then the result is contained in Corollary 3.4. Henceforth, we assume that $b' = b_1 \wedge b''$, where $\deg b_1 = 1$ and $\deg b'' = \deg b' - 1$. Apply Proposition 2.2 in order to see that $\left(b'(\varphi^{(k)})\right)(b)$ is equal to

$$(b_1\left(b''(\varphi^{(k)})\right))(b) = b_1 \wedge \left(\left(b''(\varphi^{(k)})\right)(b)\right) + (-1)^{\deg b''+1}\left(b''(\varphi^{(k)})\right)(b_1\wedge b).$$

When this identity is applied to $\varphi^{(A-k)}$, we see that $((b'(\varphi^{(k)}))(b))(\varphi^{(A-k)})$ is equal to

$$b_1\left(\left(\left(b''(\varphi^{(k)})\right)(b)\right)(\varphi^{(A-k)})\right) + (-1)^{\deg b''+1}\left(\left(b''(\varphi^{(k)})\right)(b_1\wedge b)\right)(\varphi^{(A-k)}).$$

It follows that the left side of the identity is equal to $b_1(S_1) + (-1)^{\deg b''+1}S_2$, where

$$S_1 = \sum_{k\in\mathbb{Z}}(-1)^k\binom{B+k}{L}\left(\left(b''(\varphi^{(k)})\right)(b)\right)(\varphi^{(A-k)}) \quad \text{and}$$

$$S_2 = \sum_{k\in\mathbb{Z}}(-1)^k\binom{B+k}{L}\left(\left(b''(\varphi^{(k)})\right)(b_1\wedge b)\right)(\varphi^{(A-k)}).$$

The induction hypothesis yields $S_1 = S_2 = 0$, because

$$1+A+L \le \deg b < \deg(b_1\wedge b) \qquad \text{and} \qquad \deg b'' < \deg b'. \quad \square$$

The second main result in this paper is concerned with the data of Corollary 3.6 when the hypothesis $A+L+1\le d$ is not satisfied. Our proof of Theorem 3.7 is much like an argument from a class in Differential Equations: we show that two expressions are equal by verifying that they both are solutions of the same "Initial Value Problem".

THEOREM 3.7. *Adopt Data 2.1. Let $\widetilde{\varphi}\colon F^* \to F$ be the homomorphism which is given by $\widetilde{\varphi}(b_1) = b_1(\varphi)$ for all $b_1 \in F^*$. Let d and A be integers and let b and b' be homogeneous elements of $\bigwedge^\bullet F^*$. If $\deg b = d$, then*

$$\sum_{k\in\mathbb{Z}}(-1)^k\left(\left(b'(\varphi^{(k)})\right)(b)\right)(\varphi^{(A-k)}) = \sum_{k\in\mathbb{Z}}(-1)^k\varphi^{(A-d-k)}\wedge\left(\varphi^{(k)}(b')\right)\left((\textstyle\bigwedge^d\widetilde{\varphi})(b)\right).$$

PROOF. For each fixed integer A, we define bilinear maps

$$\Phi_A\colon \textstyle\bigwedge^\bullet F^*\oplus\bigwedge^\bullet F^* \to \bigwedge^\bullet F \qquad \text{and} \qquad \Psi_A\colon \bigwedge^\bullet F^*\oplus\bigwedge^\bullet F^*\to\bigwedge^\bullet F.$$

If b and b' are homogeneous elements of $\bigwedge^\bullet F^*$, then

$$\Phi_A(b',b)=\sum_{k\in\mathbb{Z}}(-1)^k\left(\left(b'(\varphi^{(k)})\right)(b)\right)(\varphi^{(A-k)})\qquad\text{and}$$

$$\Psi_A(b',b)=\sum_{k\in\mathbb{Z}}(-1)^k\varphi^{(A-k-\deg b)}\wedge\left(\varphi^{(k)}(b')\right)\left((\textstyle\bigwedge^{\deg b}\widetilde{\varphi})(b)\right).$$

We prove that $\Phi_A(b',b)=\Psi_A(b',b)$ by induction on $\deg b'$. We must show

$$\Phi_A(1,b)=\varphi^{(A-\deg b)}\wedge\left(\textstyle\bigwedge^{\deg b}\widetilde{\varphi}\right)(b),\tag{3.8}$$

$$\Psi_A(1,b)=\varphi^{(A-\deg b)}\wedge\left(\textstyle\bigwedge^{\deg b}\widetilde{\varphi}\right)(b),\tag{3.9}$$

$$\Phi_A(b_1\wedge b',b)=b_1(\Phi_A(b',b))+(-1)^{\deg b'+1}\Phi_A(b',b_1\wedge b),\qquad\text{and}\tag{3.10}$$

$$\Psi_A(b_1\wedge b',b)=b_1(\Psi_A(b',b))+(-1)^{\deg b'+1}\Psi_A(b',b_1\wedge b)\tag{3.11}$$

for all $b_1\in F^*$.

It is clear that $\Phi_A(1,1)=\varphi^{(A)}$. If $b_1\in F^*$, then Lemma 2.4 shows that

$$\begin{aligned}\Phi_A(1,b_1\wedge b)&=b_1(\Phi_A(1,b))-b_1(\Phi_A(1,b))+b_1(\varphi)\wedge\Phi_{A-1}(1,b)\\&=b_1(\varphi)\wedge\Phi_{A-1}(1,b).\end{aligned}$$

We conclude that (3.8) holds. Assertion (3.9) is obvious. The proof of Corollary 3.6 shows that $\left[((b_1\wedge b')(\varphi^{(k)}))\,(b)\right](\varphi^{(A-k)})$ is equal to

$$b_1\left(\left(\left(b'(\varphi^{(k)})\right)(b)\right)(\varphi^{(A-k)})\right)+(-1)^{\deg b'+1}\left(\left(b'(\varphi^{(k)})\right)(b_1\wedge b)\right)(\varphi^{(A-k)})$$

and (3.10) holds.

Now we prove (3.11). Let $d=\deg b$, and let $X=\left((\bigwedge^d\widetilde{\varphi})(b)\right)$. We expand

$$Y=\left[\left(b_1(\varphi^{(k)})\right)(b')\right](X)$$

two different ways. On the one hand, Proposition 2.2 gives

$$Y=\left[b_1\wedge\varphi^{(k)}(b')\right](X)-\left[\varphi^{(k)}(b_1\wedge b')\right](X)=b_1\left(\left[\varphi^{(k)}(b')\right](X)\right)-\left[\varphi^{(k)}(b_1\wedge b')\right](X).$$

On the other hand,

$$Y=\left[\left(b_1(\varphi)\wedge\varphi^{(k-1)}\right)(b')\right](X)=\left[(b_1(\varphi))\left(\varphi^{(k-1)}(b')\right)\right](X).$$

Apply Proposition 2.2, once again, in order to see that

$$Y=b_1(\varphi)\wedge\left(\varphi^{(k-1)}(b')\right)(X)+(-1)^{\deg b'+1}\left(\varphi^{(k-1)}(b')\right)\left(b_1(\varphi)\wedge X\right).$$

Combine the two expansions of Y in order to obtain that $\left[\varphi^{(k)}(b_1\wedge b')\right](X)$ is equal to

$$b_1\left(\left[\varphi^{(k)}(b')\right](X)\right)-b_1(\varphi)\wedge\left(\varphi^{(k-1)}(b')\right)(X)-(-1)^{\deg b'+1}\left[\varphi^{(k-1)}(b')\right]\left(b_1(\varphi)\wedge X\right).$$

Thus, $S_1 = S_2 + S_3 + (-1)^{\deg b'+1}S_4$, where

$$S_1 = \sum_{k\in\mathbb{Z}}(-1)^k\varphi^{(A-d-k)} \wedge \left[\varphi^{(k)}(b_1 \wedge b')\right](X) = \Psi_A(b_1 \wedge b', b),$$

$$S_2 = \sum_{k\in\mathbb{Z}}(-1)^k\varphi^{(A-d-k)} \wedge b_1\left(\left[\varphi^{(k)}(b')\right](X)\right),$$

$$\begin{aligned} S_3 &= -\sum_{k\in\mathbb{Z}}(-1)^k\varphi^{(A-d-k)} \wedge b_1(\varphi) \wedge \left(\varphi^{(k-1)}(b')\right)(X) \\ &= \sum_{k\in\mathbb{Z}}(-1)^k b_1(\varphi^{(A-d-k)}) \wedge \left(\varphi^{(k)}(b')\right)(X), \quad \text{and} \end{aligned}$$

$$\begin{aligned} S_4 &= -\sum_{k\in\mathbb{Z}}(-1)^k\varphi^{(A-d-k)} \wedge \left(\varphi^{(k-1)}(b')\right)(b_1(\varphi) \wedge X) \\ &= \sum_{k\in\mathbb{Z}}(-1)^k\varphi^{(A-d-1-k)} \wedge \left(\varphi^{(k)}(b')\right)(b_1(\varphi) \wedge X). \end{aligned}$$

Observe that $S_2 + S_3 = b_1\left(\Psi_A(b', b)\right)$. Observe also that

$$b_1(\varphi) \wedge X = b_1(\varphi) \wedge \left((\textstyle\bigwedge^d \widetilde{\varphi})(b)\right) = (\textstyle\bigwedge^{d+1} \widetilde{\varphi})(b_1 \wedge b).$$

It follows that $S_4 = \Psi(b', b_1 \wedge b)$. Assertion (3.11) has been established and the proof is complete. □

4. A new proof of the Kustin-Ulrich pfaffian identity

We give a new proof of [**14, 3.2**]. The notation is explained in section 2.

PROPOSITION 4.1. *Let X be an alternating matrix with entries from the commutative noetherian ring R. If A, B, C, and D are index sets of size a, b, c, and d, respectively, with $a + b + d$ even and $a + b \le d - 2$, then*

$$\sum_{\substack{t\in\mathbb{Z} \\ a+c+t \text{ even}}} \sum_{\substack{E\cup F=D \\ |E|=t}} (-1)^{(a+c+t)/2}\sigma(EF)X_{ACE}X_{BCF} = 0.$$

PROOF. If M is an integer and β, β', and β'' are homogeneous elements of $\bigwedge^\bullet F^*$, then Corollary 3.6 shows that

$$\sum_{k\in\mathbb{Z}}(-1)^k \left[\left(\left(\beta'(\varphi^{(k)})\right)(\beta)\right)(\varphi^{(M-k)})\right](\beta'')$$

is the zero element of the ring R, provided

$$\deg\beta'' = 2M - \deg\beta - \deg\beta' \quad \text{and} \tag{4.2}$$

$$1 + M \le \deg\beta. \tag{4.3}$$

We may apply Corollary 3.6 with $M = c + \frac{a+b+d}{2}$, $\beta' = \varepsilon_A$, $\beta = \varepsilon_C \wedge \varepsilon_D$, and $\beta'' = \varepsilon_B \wedge \varepsilon_C$. Indeed, (4.2) is satisfied because $b + c = 2M - (a) - (c+d)$; and (4.3) is equivalent to the hypothesis $a + b + 2 \leq d$. It follows that

$$\sum_{k \in \mathbb{Z}} (-1)^k \left[\left(\left(\varepsilon_A(\varphi^{(k)}) \right) (\varepsilon_C \wedge \varepsilon_D) \right) (\varphi^{(M-k)}) \right] (\varepsilon_B \wedge \varepsilon_C) = 0. \tag{4.4}$$

We complete the proof by translating (4.4) back into the language of the original statement. For the time being, consider k to be fixed. Let $r = 2k - a$ and let Y represent the element $\varepsilon_A(\varphi^{(k)})$ of $\bigwedge^r F$. We evaluate

$$\left[(Y(\varepsilon_C \wedge \varepsilon_D))(\varphi^{(M-k)}) \right] (\varepsilon_B \wedge \varepsilon_C). \tag{4.5}$$

Recall, from (2.6), that

$$Y(\varepsilon_C \wedge \varepsilon_D) = \sum_{j=0}^{r} \sum_{\substack{C_1 \cup C_2 = C \\ |C_1| = j}} \sum_{\substack{E \cup F = D \\ |E| = r-j}} (-1)^{|C_2| \cdot |E|} \sigma(C_1 C_2) \sigma(EF) Y(\varepsilon_{C_1} \wedge \varepsilon_E) \wedge \varepsilon_{C_2} \wedge \varepsilon_F. \tag{4.6}$$

Furthermore, in (4.6), $2(M-k) - |C_2| - |F| = |B| + |C|$; consequently, when (4.6) is inserted into (4.4), we have

$$\left[(\varepsilon_{C_2} \wedge \varepsilon_F)(\varphi^{(M-k)}) \right] (\varepsilon_B \wedge \varepsilon_C) = (\varepsilon_B \wedge \varepsilon_C \wedge \varepsilon_{C_2} \wedge \varepsilon_F)(\varphi^{(M-k)}).$$

If C_2 is nonempty, then $\varepsilon_C \wedge \varepsilon_{C_2} = 0$. It follows that (4.5) is equal to

$$\sum_{\substack{E \cup F = D \\ |E| = r-c}} \sigma(EF) Y(\varepsilon_C \wedge \varepsilon_E) \cdot (\varepsilon_B \wedge \varepsilon_C \wedge \varepsilon_F)(\varphi^{(M-k)}). \tag{4.7}$$

Furthermore, in (4.7), we have $\deg Y = \deg(\varepsilon_C \wedge \varepsilon_E)$; therefore,

$$Y(\varepsilon_C \wedge \varepsilon_E) = (\varepsilon_C \wedge \varepsilon_E)Y = (\varepsilon_C \wedge \varepsilon_E)\left(\varepsilon_A(\varphi^{(k)}) \right) = (-1)^a (\varepsilon_A \wedge \varepsilon_C \wedge \varepsilon_E)(\varphi^{(k)}).$$

We now see that (4.4) is

$$\sum_{k \in \mathbb{Z}} (-1)^k \sum_{\substack{E \cup F = D \\ |E| = 2k-a-c}} \sigma(EF)(\varepsilon_A \wedge \varepsilon_C \wedge \varepsilon_E)(\varphi^{(k)}) \cdot (\varepsilon_B \wedge \varepsilon_C \wedge \varepsilon_F)(\varphi^{(M-k)}) = 0.$$

Apply (2.8) to see that

$$\sum_{k \in \mathbb{Z}} (-1)^k \sum_{\substack{E \cup F = D \\ |E| = 2k-a-c}} \sigma(EF) X_{ACE} X_{BCF} = 0.$$

Replace k by $\frac{t+a+c}{2}$ in order to complete the proof. □

5. Srinivasan's identities

In this section we show that the identities of **[15]** and **[16]** are special cases of our results. We begin by deducing **[15, 4.3]** from Theorem 3.1. Proposition 4.4 in **[15]** is an immediate consequence of Corollary 3.3; results 4.5, 4.6, 4.7, and 4.8 in **[15]** are immediate consequences of Lemma 1.3. Proposition 5.3 provides a new proof of **[15, 4.2]**. In Proposition 5.4, we give a new proof of Srinivasan's decomposition **[15, page 447]** of the pfaffian X_{IJ} into products of the form $X_{IJ_1} \cdot X_{J_2}$ and $X_{I_1} \cdot X(I_2; J)$. Proposition 5.5 is a reformulation of identity (b) from **[15, page 447]**.

PROPOSITION 5.1. *Adopt Data 2.1. Let F have rank $2n$, and let i, j, and t be integers with $t + j$ even. If ξ is an element of $\bigwedge^{2n} F^*$ and a is an element of $\bigwedge^t F$, then*

$$\left(\left(\varphi^{(n-\frac{t+j}{2})} \wedge a\right)(\xi)\right)(\varphi^{([[\frac{i+j}{2}]])})$$

is equal to

$$\sum_{\substack{w=0 \\ j-w \text{ even}}}^{j} \binom{n - [[\frac{i+t}{2}]]}{\frac{j-w}{2}} \left[\left(\varphi^{(n-\frac{t-w}{2})}(\xi)\right)(a)\right] \wedge \varphi^{([[\frac{i-w}{2}]])}.$$

PROOF. Write $i = i' + \varepsilon$, where $i' + t$ is even and ε is either 0 or 1. Notice that

$$\left[\left[\frac{i+j}{2}\right]\right] = \frac{i'+j}{2}, \quad \left[\left[\frac{i+t}{2}\right]\right] = \frac{i'+t}{2}, \quad \text{and} \quad \left[\left[\frac{i-w}{2}\right]\right] = \frac{i'-w}{2}. \tag{5.2}$$

In identity (b) from Corollary 3.2, replace p with $\frac{i'+j}{2}$, q with $n - \frac{t+j}{2}$, d with $2n - t$, and k with $n + \frac{w-t}{2}$, in order to see that

$$\left(\varphi^{(n-\frac{t+j}{2})}(b)\right)(\varphi^{(\frac{i'+j}{2})})$$

is equal to

$$\sum_{\substack{w \in \mathbf{Z} \\ j-w \text{ even}}} (-1)^{\frac{w-j}{2}} \binom{\frac{j-w}{2} - n + \frac{i'+t}{2} - 1}{\frac{j-w}{2}} b(\varphi^{(n+\frac{w-t}{2})}) \wedge \varphi^{(\frac{i'-w}{2})}.$$

Observation 1.2 (e) yields

$$\binom{\frac{j-w}{2} - n + \frac{i'+t}{2} - 1}{\frac{j-w}{2}} = (-1)^{\frac{j-w}{2}} \binom{n - \frac{i'+t}{2}}{\frac{j-w}{2}};$$

and therefore,

$$\left(\varphi^{(n-\frac{t+j}{2})}(b)\right)(\varphi^{(\frac{i'+j}{2})}) = \sum_{\substack{w \in \mathbf{Z} \\ j-w \text{ even}}} \binom{n - \frac{i'+t}{2}}{\frac{j-w}{2}} b(\varphi^{(n+\frac{w-t}{2})}) \wedge \varphi^{(\frac{i'-w}{2})}.$$

If $j < w$, then the binomial coefficient

$$\binom{n - \frac{i'+t}{2}}{\frac{j-w}{2}}$$

is equal to zero. If $w < 0$, then

$$b(\varphi^{(n+\frac{w-t}{2})}) \in \bigwedge^w F = 0.$$

Thus,

$$\left(\varphi^{(n-\frac{t+j}{2})}(b)\right)(\varphi^{(\frac{i'+j}{2})}) = \sum_{\substack{0\le w\le j\\ j-w \text{ even}}} \binom{n - \frac{i'+t}{2}}{\frac{j-w}{2}} b(\varphi^{(n+\frac{w-t}{2})}) \wedge \varphi^{(\frac{i'-w}{2})}.$$

In light of (5.2), we see that

$$\left(\varphi^{(n-\frac{t+j}{2})}(b)\right)(\varphi^{([[\frac{i+j}{2}]])}) = \sum_{\substack{0\le w\le j\\ j-w \text{ even}}} \binom{n - [[\frac{i+t}{2}]]}{\frac{j-w}{2}} b(\varphi^{(n-\frac{t-w}{2})}) \wedge \varphi^{([[\frac{i-w}{2}]])}.$$

To complete the proof, replace b with the element $a(\xi)$ of $\bigwedge^{2n-t} F^*$. The module action of $\bigwedge^\bullet F$ on $\bigwedge^\bullet F^*$ gives

$$\varphi^{(r)}(b) = \varphi^{(r)}(a(\xi)) = \left(\varphi^{(r)} \wedge a\right)(\xi)$$

and Proposition 2.2 (b) gives

$$b(\varphi^{(r)}) = (a(\xi))\left(\varphi^{(r)}\right) = \left(\varphi^{(r)}(\xi)\right)(a). \quad \square$$

PROPOSITION 5.3. *Adopt Data 2.1. Let $\widetilde{\varphi}\colon F^* \to F$ be the homomorphism which is given by $\widetilde{\varphi}(b_1) = b_1(\varphi)$ for all $b_1 \in F^*$, and let i and d be integers. If $b \in \bigwedge^d F^*$, then*

$$b(\varphi^{(i)}) = \sum_{k=1}^{[[i-\frac{d}{2}]]} (-1)^{k+1} b(\varphi^{(i-k)}) \wedge \varphi^{(k)} + \left(\bigwedge^{2i-d} \widetilde{\varphi}\right)\left(\varphi^{(d-i)}(b)\right).$$

PROOF. In the notation of (3.9), observe that

$$\left(\bigwedge^{2i-d} \widetilde{\varphi}\right)\left(\varphi^{(d-i)}(b)\right) = \Psi_{2i-d}(1, \varphi^{(d-i)}(b));$$

therefore, Theorem 3.7 and Proposition 2.3 yield

$$\begin{aligned}\left(\bigwedge^{2i-d} \widetilde{\varphi}\right)\left(\varphi^{(d-i)}(b)\right) &= \sum_{k\in\mathbb{Z}} (-1)^k \left(\varphi^{(k)}(\varphi^{(d-i)}(b))\right)(\varphi^{(2i-d-k)})\\ &= \sum_{k\in\mathbb{Z}} (-1)^k \binom{k+d-i}{d-i} \left(\varphi^{(k+d-i)}(b)\right)(\varphi^{(2i-d-k)})\\ &= (-1)^i \sum_{\ell\in\mathbb{Z}} (-1)^{d+\ell} \binom{\ell}{d-i} \left(\varphi^{(\ell)}(b)\right)(\varphi^{(i-\ell)}).\end{aligned}$$

Apply Corollary 3.3, with $L=0$ and $B=A=i$, in order to see that

$$\begin{aligned}\left(\bigwedge^{2i-d}\widetilde{\varphi}\right)\left(\varphi^{(d-i)}(b)\right) &= (-1)^i\sum_{\ell\in\mathbb{Z}}(-1)^\ell b(\varphi^{(\ell)})\wedge\varphi^{(i-\ell)}\\ &= \sum_{k\in\mathbb{Z}}(-1)^k b(\varphi^{(i-k)})\wedge\varphi^{(k)}\\ &= \sum_{k=0}^{[[i-\frac{d}{2}]]}(-1)^k b(\varphi^{(i-k)})\wedge\varphi^{(k)}.\end{aligned}$$

The last equality holds because if $k<0$, then $\varphi^{(k)}=0$; and if $\left[\left[i-\frac{d}{2}\right]\right]<k$, then $b(\varphi^{(i-k)})=0$. □

Recall, from section 2, that $X(I_2;J)$ is a minor of X.

PROPOSITION 5.4. *Let X be an alternating matrix with entries from the commutative noetherian ring R. Let I and J be index sets of size i and j, respectively. If $i+j$ is even, then*

$$\sum_{\substack{t\in\mathbb{Z}\\ t+i \text{ even}}}(-1)^{\frac{t+j}{2}}\sum_{\substack{J_1\cup J_2=J\\ |J_1|=t}}\sigma(J_1J_2)X_{IJ_1}\cdot X_{J_2}=(-1)^{\frac{j(j+1)}{2}}\sum_{\substack{I_1\cup I_2=I\\ |I_1|=i-j}}\sigma(I_1I_2)X_{I_1}\cdot X(I_2;J).$$

PROOF. Evaluate Theorem 3.7 at $b'=\varepsilon_I$, $b=\varepsilon_J$, and $A=\frac{i+j}{2}$, in order to see that $L=R$, where

$$L=\sum_{k\in\mathbb{Z}}(-1)^k\left(\left(\varepsilon_I(\varphi^{(k)})\right)(\varepsilon_J)\right)\left(\varphi^{(\frac{i+j}{2}-k)}\right)\quad\text{and}$$

$$R=\sum_{k\in\mathbb{Z}}(-1)^k\varphi^{(\frac{i-j}{2}-k)}\wedge\left(\varphi^{(k)}(\varepsilon_I)\right)\left((\textstyle\bigwedge^j\widetilde{\varphi})(\varepsilon_J)\right).$$

Apply (2.6) to see that

$$L=\sum_{k\in\mathbb{Z}}(-1)^k\sum_{\substack{J_1\cup J_2=J\\ |J_1|=2k-i}}\sigma(J_1J_2)\left(\varepsilon_I(\varphi^{(k)})\right)(\varepsilon_{J_1})\cdot\varepsilon_{J_2}\left(\varphi^{(\frac{i+j}{2}-k)}\right).$$

We know, from (2.8), that

$$\left(\varepsilon_I(\varphi^{(k)})\right)(\varepsilon_{J_1})=(-1)^kX_{J_1I}=(-1)^{k+i}X_{IJ_1}.$$

It follows that

$$L=\sum_{k\in\mathbb{Z}}(-1)^{k+i+\frac{i+j}{2}}\sum_{\substack{J_1\cup J_2=J\\ |J_1|=2k-i}}\sigma(J_1J_2)X_{IJ_1}\cdot X_{J_2}.$$

Let $t=2k-i$ in order to see that L is equal to the left side of the announced identity.

Notice that

$$\frac{i-j}{2}<k\implies\varphi^{(\frac{i-j}{2}-k)}=0\text{ and }k<\frac{i-j}{2}\implies\left(\varphi^{(k)}(\varepsilon_I)\right)\left((\textstyle\bigwedge^j\widetilde{\varphi})(\varepsilon_J)\right)=0.$$

Thus, there is only one non-zero term in R. If we write k for $\frac{i-j}{2}$, then R is equal to

$$\begin{aligned}&(-1)^k\left(\varphi^{(k)}(\varepsilon_I)\right)\left((\textstyle\bigwedge^j\widetilde{\varphi})(\varepsilon_J)\right)\\ &=(-1)^k\sum_{\substack{I_1\cup I_2=I\\|I_1|=2k}}\sigma(I_1I_2)\left(\varphi^{(k)}(\varepsilon_{I_1})\right)\wedge\varepsilon_{I_2}\left((\textstyle\bigwedge^j\widetilde{\varphi})(\varepsilon_J)\right).\end{aligned}$$

Apply (2.8) to see that the

$$R=\sum_{\substack{I_1\cup I_2=I\\|I_1|=2k}}\sigma(I_1I_2)X_{I_1}\cdot\varepsilon_{I_2}\left((\textstyle\bigwedge^j\widetilde{\varphi})(\varepsilon_J)\right).$$

We know that $\widetilde{\varphi}\colon F^*\to F$ is the map

$$\widetilde{\varphi}(\varepsilon_j)=\varepsilon_j(\varphi)=\varepsilon_j\left(\sum_{p<q}x_{pq}e_p\wedge e_q\right)=\sum_i -x_{ij}e_i.$$

It follows, from (2.7), that, if A and B are index sets with $|A|=|B|=r$, then

$$\varepsilon_A\left((\textstyle\bigwedge^r\widetilde{\varphi})(\varepsilon_B)\right)=(-1)^{\frac{r(r+1)}{2}}X(A;B).$$

Thus,

$$R=(-1)^{\frac{j(j+1)}{2}}\sum_{\substack{I_1\cup I_2=I\\|I_1|=i-j}}\sigma(I_1I_2)X_{I_1}\cdot X(I_2;J).\quad\square$$

If X has size $N\times N$ and $\{1,\dots,N\}$ is the disjoint union of A, B, C, and D, then one obtains Srinivasan's version [**15, page 447**] of the following identity by letting $J=B\cup C$ and $I=C\cup D$.

PROPOSITION 5.5. *Let X be an alternating matrix with entries from the commutative noetherian ring R, t be an integer, and A, B, and C be index sets of size a, b, and c, respectively. If $t+b$ is even and $a+c$ is even, then*

$$\sum_{\substack{A_1\cup A_2=A\\|A_1|=t}}\sigma(A_1A_2)X_{A_1B}\cdot X_{A_2BC}=\sum_{\substack{s\in\mathbb{Z}\\ s+a+b\text{ even}}}\binom{\frac{a-c}{2}}{\frac{s+t-c}{2}}\sum_{\substack{C_1\cup C_2=C\\|C_1|=s}}\sigma(C_1C_2)X_{ABC_1}\cdot X_{BC_2}.$$

PROOF. Apply Corollary 3.2 (a) to the element $\varepsilon_B\wedge\varepsilon_C$. Replace the element b of $\bigwedge^d F^*$ with $\varepsilon_A\wedge\varepsilon_B$. If p and q are integers which satisfy $2q+2p=a+2b+c$, then

$$\left(\left(\varphi^{(q)}(\varepsilon_A\wedge\varepsilon_B)\right)(\varphi^{(p)})\right)(\varepsilon_B\wedge\varepsilon_C)\quad\text{and}$$

$$\sum_{k\in\mathbb{Z}}(-1)^{k+p}\binom{q-a-b-1+k}{k-p}\left((\varepsilon_A\wedge\varepsilon_B)(\varphi^{(k)})\wedge\varphi^{(q+p-k)}\right)(\varepsilon_B\wedge\varepsilon_C)$$

are equal elements of R. The techniques which are employed in the proof of Proposition 4.1 yield that

$$\sum_{\substack{A_1\cup A_2=A\\|A_1|+b=2q}}\sigma(A_1A_2)X_{A_1B}\cdot X_{A_2BC}$$

is equal to

$$\sum_{k\in\mathbb{Z}}(-1)^{k+p}\binom{q-a-b-1+k}{k-p}\sum_{\substack{C_1\cup C_2=C\\|C_1|+a+b=2k}}\sigma(C_1C_2)X_{ABC_1}\cdot X_{BC_2}.$$

Replace k with $\frac{s+a+b}{2}$, q with $\frac{t+b}{2}$, and p with $\frac{a+b+c-t}{2}$. Use Observation 1.2 (e) to complete the proof. □

6. An application to Algebras with Straightening Law

Let X be an $N\times N$ alternating matrix with indeterminate entries. Consider the poset $(P,\le)$ of pfaffians of X. The order on P is given as follows. If r and s are even integers, $a_1<\cdots<a_r$, and $b_1<\cdots<b_s$, then

$$X_{a_1\ldots a_r}\le X_{b_1\ldots b_s} \text{ provided } s\le r \text{ and } a_i\le b_i \text{ for } 1\le i\le s.$$

It is well known (see, for example, [**7**]) that the polynomial ring $\mathbb{Z}[X]$ is an algebra with straightening law on P over $\mathbb{Z}$. (We use the language of [**3**].) In particular, for each pair of index sets A and B, there exist integers r_{CD} such that

$$X_AX_B=\sum_{C,D}r_{CD}X_CX_D, \tag{6.1}$$

where the sum is taken over all pairs of index sets C and D with

$$X_C\le X_D \quad\text{and}\quad X_C\le X_A. \tag{6.2}$$

Most, but not all, of the above "straightening formula" is proved in [**8, Lemma 6.2**]. Indeed, [**8, Lemma 6.2**] establishes (6.1), where C and D vary over all pairs of index sets with $X_C\le X_D$. One must modify the argument of [**8**] in order to obtain (6.1), where C and D fulfill both of the requirements of (6.2). One version of this modification occurs in [**6**]. Suppose that $|A|$ and $|B|$ are even integers with $|B|\le|A|$. Let $i(A,B)$ be the largest index i for which $a_k\le b_k$ for all k with $1\le k\le i$. If $|A|=N$ or $i(A,B)=|B|$, then $X_A\le X_B$ and $X_AX_B=X_AX_B$ satisfies both (6.1) and (6.2). Day's argument proceeds by induction on $|A|$ and $i(A,B)$. Let $i=i(A,B)$. It is convenient to write $A=A'\cup C_1'$ and $B=C_2'\cup B'$, where A' is $a_1<\cdots<a_i$, and C_1' is $c_1<\cdots<c_\lambda$, with $a_i<c_1$; and C_2' is $c_{\lambda+1}<\cdots<c_{\lambda+i+1}$, and B' is $b_1<\cdots<b_t$, with $c_{\lambda+i+1}<b_1$. The information about the indices is summarized as

$$\begin{array}{rccccccccccc}
B: & c_{\lambda+1} & < & \ldots & < & c_{\lambda+i} & < & c_{\lambda+i+1} & < & b_1 & < \ \ldots\ < & b_t\\
 & \veebar & & & & \veebar & & \wedge & & & & \\
A: & a_1 & < & \ldots & < & a_i & < & c_1 & < & \ldots & \ldots\ < & c_\lambda.
\end{array}$$

Let C represent the index set $c_{\lambda+1}<\cdots<c_{\lambda+i+1}<c_1\cdots<c_\lambda$. Day proves that there exist integers r_{C_1} and $r_{B_1C_1}$ such that

$$X_{A'C_1'}X_{C_2'B'}=\sum_{\substack{C_1\cup C_2=C\\|C_1|=\lambda\\C_1\ne C_1'}}r_{C_1}X_{A'C_1}X_{C_2B'}+\sum_{\substack{B_1\cup B_2=B'\\0\le|B_1|\le t}}\ \sum_{\substack{C_1\cup C_2=C\\\lambda+1\le|C_1|}}r_{C_1B_1}X_{A'C_1B_1}X_{C_2B_2}. \tag{6.3}$$

It is clear that $X_{A'C_1} \le X_A$ and $X_{A'C_1B_1} \le X_A$. Furthermore,

$$i(A,B) < i(A'C_1, C_2B') \quad \text{and} \quad |A| < |A'C_2B_2|;$$

so, (6.1) with hypotheses (6.2) follows by induction.

In Proposition 6.6 we generalize (6.3) and produce explicit values for the integers r_{C_1} and $r_{B_1C_1}$; indeed, (6.3) is the special case $M = \lambda$, $Q = t-1$. Corollary 6.5, which is the base step in our proof of Proposition 6.6, is a translation of Corollary 3.5 from the language of multilinear algebra to the language of pfaffians. The following conventions are in effect throughout the rest of this section.

DATA 6.4. *Let X be an alternating matrix with entries from the commutative noetherian ring R, and let A, B, and C be index sets of size a, b, and c, respectively. Assume that $a+b+c$ is even. For fixed values of ℓ and r, let*

$$X(\ell,r) = \sum_{\substack{B_1 \cup B_2 = B \\ |B_1| = r}} \sum_{\substack{C_1 \cup C_2 = C \\ |C_1| = \ell}} \sigma(B_1B_2)\sigma(C_1C_2) X_{AB_1C_1} X_{B_2C_2}.$$

COROLLARY 6.5. *Adopt Data 6.4. If M and Q are integers with $Q+M$ even, $M+a$ even, $-1 \le Q$, and $Q+M \le b+c-a-2$, then*

$$\sum_{\substack{r,\ell \in \mathbf{Z} \\ \ell+r+a \text{ even}}} (-1)^{\ell c + \frac{r+a-\ell}{2}} \binom{\frac{\ell+r+Q}{2}}{\frac{\ell+r-M}{2}} X(\ell,r) = 0.$$

PROOF. Apply Corollary 3.5 with A replaced by $\frac{a+b+c}{2}$, B replaced by $\frac{Q-a}{2}$, C replaced by $-\frac{M+a}{2}$, d replaced by a, and b replaced by ε_A. Since the hypothesis of Corollary 3.5 is equivalent to the present hypothesis, we conclude that

$$\sum_{k\in\mathbb{Z}} (-1)^k \binom{\frac{Q-a}{2}+k}{\frac{-M-a}{2}+k} \left[\varepsilon_A(\varphi^{(k)}) \wedge \varphi^{(\frac{c+a+b}{2}-k)}\right] (\varepsilon_B \wedge \varepsilon_C)$$

is the zero element of R. Calculations similar to those in section 4 yield

$$\left[\varepsilon_A(\varphi^{(k)}) \wedge \varphi^{(\frac{c+a+b}{2}-k)}\right] (\varepsilon_B \wedge \varepsilon_C) = \sum_{\substack{r,\,\ell\in\mathbf{Z} \\ r+\ell+a=2k}} (-1)^{\ell c+\ell+a+\frac{c+a+b}{2}} X(\ell,r).$$

Replace k by $\frac{r+\ell+a}{2}$ to see that

$$(-1)^{a+\frac{c+a+b}{2}} \sum_{\substack{r,\,\ell\in\mathbf{Z} \\ r+\ell+a \text{ even}}} (-1)^{\ell c+\frac{r+a-\ell}{2}} \binom{\frac{r+\ell+Q}{2}}{\frac{r+\ell-M}{2}} X(\ell,r) = 0. \quad \square$$

PROPOSITION 6.6. *Adopt Data 6.4. If Q and M are integers with $Q+M$ even, $M+a$ even, $b-1 \le Q$, and $Q+M \le b+c-a-2$, then*

$$\sum_{\substack{r,\ell \in \mathbf{Z} \\ \ell+r+a \text{ even}}} (-1)^{\ell c + \frac{r+a-\ell}{2}} \binom{\frac{\ell-r+Q}{2}}{\frac{\ell-r-M}{2}} X(\ell,r) = 0.$$

REMARK. We note, for the purpose of induction, that Proposition 6.6 says that $X(M,0)$ is equal to

$$\sum_{\substack{M+1\le \ell\\ \ell+a \text{ even}}} (-1)^{1+\frac{M-\ell}{2}} \binom{\frac{\ell+Q}{2}}{\frac{\ell-M}{2}} X(\ell,0) + \sum_{r=1}^{b} \sum_{\substack{\ell\in\mathbf{Z}\\ \ell+r+a \text{ even}}} (-1)^{rc+1+\frac{M+r-\ell}{2}} \binom{\frac{\ell-r+Q}{2}}{\frac{\ell-r-M}{2}} X(\ell,r).$$

PROOF. For each integer r, with $0\le r\le b$, let

$$T_r = \sum_{\substack{\ell\in\mathbf{Z}\\ \ell+r+a \text{ even}}} (-1)^{\ell c+\frac{r+a-\ell}{2}} \binom{\frac{\ell+r+Q}{2}}{\frac{\ell+r-M}{2}} X(\ell,r).$$

Recall, from Corollary 6.5, that $\sum_{r=0}^{b} T_r = 0$. If $b=0$, then the result holds. The proof proceeds by induction on b; henceforth, we assume that $0<b$. Write $T_r = T_r' + T_r''$, with

$$T_r' = (-1)^{Mc+rc+r+\frac{a-M}{2}} X(M-r,r)$$

and

$$T_r'' = \sum_{\substack{M+1-r\le \ell\\ \ell+r+a \text{ even}}} (-1)^{\ell c+\frac{r+a-\ell}{2}} \binom{\frac{\ell+r+Q}{2}}{\frac{\ell+r-M}{2}} X(\ell,r).$$

For each r with $1\le r\le b$, we apply the induction hypothesis with A replaced by AB_1, a replaced by $a+r$, B replaced by B_2, b replaced by $b-r$, M replaced by $M-r$, and Q replaced by $Q-r$. Notice that all of the hypotheses are still satisfied. When the induction hypothesis is applied to

$$T_r' = (-1)^{Mc+rc+r+\frac{a-M}{2}} \sum_{\substack{B_1\cup B_2=B\\ |B_1|=r}} \sigma(B_1B_2) \sum_{\substack{C_1\cup C_2=C\\ |C_1|=M-r}} \sigma(C_1C_2) X_{AB_1C_1} X_{B_2C_2},$$

we see that $T_r' = V_r + V_r'$, where

$$V_r = \sum_{\substack{M-r+1\le \ell\\ \ell+a+r \text{ even}}} (-1)^{\ell c+1+\frac{a+r-\ell}{2}} \binom{\frac{\ell-r+Q}{2}}{\frac{\ell+r-M}{2}} X(\ell,r)$$

and

$$V_r' = \sum_{r'=1}^{b-r} \sum_{\substack{\ell\in\mathbf{Z}\\ \ell+r'+a+r \text{ even}}} (-1)^{\ell c+1+\frac{a+r+r'-\ell}{2}} \binom{\frac{\ell-r'-r+Q}{2}}{\frac{\ell-r'+r-M}{2}} X(\ell;r;r'),$$

for $X(\ell;r;r')$ equal to

$$\sum_{\substack{C_1\cup C_2=C\\ |C_1|=\ell}} \sigma(C_1C_2) \sum_{\substack{B_1\cup B_2=B\\ |B_1|=r}} \sum_{\substack{B_1'\cup B_2'=B_2\\ |B_1'|=r'}} \sigma(B_1'B_2')\sigma(B_1B_2) X_{AB_1B_1'C_1} X_{B_2'C_2}.$$

Co-associativity in the co-algebra $\bigwedge^\bullet F^*$, together with the fact that the composition

$$\textstyle\bigwedge^{r+r'} F^* \xrightarrow{\Delta} \bigwedge^{r} F^* \otimes \bigwedge^{r'} F^* \xrightarrow{\wedge} \bigwedge^{r+r'} F^*$$

is equal to multiplication by $\binom{r+r'}{r'}$, yields that

$$X(\ell;r;r') = \binom{r+r'}{r'} X(\ell, r+r').$$

Notice that

$$V_r' = \sum_{r'=1}^{b-r} \sum_{\substack{M-r-r'+1\le\ell \\ \ell+r'+a+r \text{ even}}} (-1)^{\ell c+1+\frac{a+r+r'-\ell}{2}} \binom{\frac{\ell-r'-r+Q}{2}}{\frac{\ell-r'+r-M}{2}} \binom{r+r'}{r'} X(\ell, r+r');$$

because, if $\ell < M - r' - r + 1$, then $\ell - r' + r - M < 1 - 2r' < 0$, and the first binomial coefficient is zero. It follows that T_r', which is equal to $V_r + V_r'$, is equal to

$$\sum_{r'=0}^{b-r} \sum_{\substack{M-r-r'+1\le\ell \\ \ell+r'+a+r \text{ even}}} (-1)^{\ell c+1+\frac{a+r+r'-\ell}{2}} \binom{\frac{\ell-r'-r+Q}{2}}{\frac{\ell-r'+r-M}{2}} \binom{r+r'}{r'} X(\ell, r+r').$$

Replace r with $r - r'$ to see that

$$\sum_{r=1}^{b} T_r' = \sum_{r=1}^{b} \sum_{\substack{M-r+1\le\ell \\ \ell+r+a \text{ even}}} (-1)^{\ell c+1+\frac{a+r-\ell}{2}} \sum_{r'=0}^{r-1} \binom{\frac{\ell-r+Q}{2}}{\frac{\ell+r-M}{2} - r'} \binom{r}{r'} X(\ell, r).$$

Recall that $T_r = T_r' + T_r''$. We now see that is equal to

$$\sum_{r=1}^{b} T_r = \sum_{r=1}^{b} \sum_{\substack{M-r+1\le\ell \\ \ell+r+a \text{ even}}} (-1)^{\ell c+\frac{a+r-\ell}{2}} \left[\binom{\frac{\ell+r+Q}{2}}{\frac{\ell+r-M}{2}} - \sum_{r'=0}^{r-1} \binom{\frac{\ell-r+Q}{2}}{\frac{\ell+r-M}{2} - r'} \binom{r}{r'} \right] X(\ell, r).$$

Use Corollary 1.4 to see that

$$\sum_{r=1}^{b} T_r = \sum_{r=1}^{b} \sum_{\substack{M-r+1\le\ell \\ \ell+r+a \text{ even}}} (-1)^{\ell c+\frac{a+r-\ell}{2}} \binom{\frac{\ell-r+Q}{2}}{\frac{\ell-r-M}{2}} X(\ell, r).$$

The constraint $M - r + 1 \le \ell$ is irrelevant; thus,

$$\sum_{r=1}^{b} T_r = \sum_{r=1}^{b} \sum_{\substack{\ell\in\mathbf{Z} \\ \ell+r+a \text{ even}}} (-1)^{\ell c+\frac{a+r-\ell}{2}} \binom{\frac{\ell-r+Q}{2}}{\frac{\ell-r-M}{2}} X(\ell, r).$$

The proof is complete because $\sum_{r=0}^{b} T_r = 0$. $\square$

7. Other Applications

In this section we compare our results to a few familiar pfaffian identities. We begin with the "Laplace expansion" of a pfaffian.

PROPOSITION 7.1. *Let X be an alternating matrix with entries from the commutative noetherian ring R. If N is an even integer, then*

$$\sum_{k=2}^{N}(-1)^k X_{1k}X_{2\cdots\widehat{k}\cdots N} = X_{12\cdots N}.$$

PROOF. Apply Proposition 5.5 with $C = \{1\}$, $A = \{2,\ldots,N\}$, and B empty. □

The next two examples are Lemmas 2.3 and 2.4 of [**4**].

PROPOSITION 7.2. *Adopt Data 2.1.*

(a) *Let $\widetilde{\varphi}\colon F^* \to F$ be the homomorphism which is given by $\widetilde{\varphi}(b_1) = b_1(\varphi)$ for all $b_1 \in F^*$. If $b \in \bigwedge^d F^*$, then*

$$(\textstyle\bigwedge^d \widetilde{\varphi})(b) = \displaystyle\sum_{k\in\mathbb{Z}}(-1)^k\left(\varphi^{(k)}(b)\right)(\varphi^{(d-k)}).$$

(b) *If $b \in \bigwedge^{2m+1} F^*$, then $b(\varphi^{(m+1)}) = \left(\varphi^{(m)}(b)\right)(\varphi)$.*

PROOF. Apply Theorem 3.7 with $A = d$ and $b' = 1$ to prove (a). Assertion (b) is Corollary 3.3 (a) with $L = 0$ and $A = B = m+1$. □

The following identities played a crucial role in preliminary versions of [**2**]; see, for example, [**1**]. Other proofs may be found in [**5**] and [**9, Relations 2.23 and 2.24**].

PROPOSITION 7.3. *Let X be an alternating matrix with entries from the commutative noetherian ring R. If $p, q, p_1, \ldots, p_\ell$ are integers, then*

$$X(p,p_1,\ldots,p_\ell;q,p_1,\ldots,p_\ell) = \begin{cases} X_{p_1,\ldots,p_\ell}X_{p,q,p_1,\ldots,p_\ell} & \text{if } \ell \text{ is even, and} \\ X_{p,p_1,\ldots,p_\ell}X_{q,p_1,\ldots,p_\ell} & \text{if } \ell \text{ is odd.} \end{cases}$$

PROOF. Apply Proposition 5.4 with $I = \{p,p_1,\ldots,p_\ell\}$ and $J = \{q,p_1,\ldots,p_\ell\}$. □

REFERENCES

1. G. Boffi and R. Sánchez, *Some classical formulas and a determinantal ideal*, Geometry Seminars, 1988–1991, Univ. Stud. Bologna, Bologna, 1991, pp. 27–35.
2. G. Boffi and R. Sánchez, *On the resolutions of the powers of the pfaffian ideal*, preprint (1990).
3. W. Bruns and U. Vetter, *Determinantal rings*, Lecture Notes in Mathematics **1327**, Springer Verlag, Berlin Heidelberg New York, 1988.
4. D. Buchsbaum and D. Eisenbud, *Algebra structures for finite free resolutions, and some structure theorems for ideals of codimension 3*, Amer. J. Math. **99** (1977), 447–485.
5. A. Cayley, *Sur les déterminants gauche*, J. reine angew. Math. **38** (1849), 93–96.

6. C. Day, *Ring around the poset: an introduction to Algebras with Straightening Law*, M. A. thesis, University of South Carolina, 1989.
7. C. De Concini, D. Eisenbud, and C. Procesi, *Hodge algebras*, Astérisque **91** (1982), 1–87.
8. C. De Concini and C. Procesi, *A characteristic free approach to invariant theory*, Advances Math. **21** (1976), 330–354.
9. P. Heymans, *Pfaffians and skew-symmetric matrices*, Proc. Lond. Math. Soc. **19** (1969), 730–768.
10. C. Huneke and B. Ulrich, *Divisor class groups and deformations*, Amer. J. Math. **107** (1985), 1265–1303.
11. A. Kustin, *The minimal free resolutions of the Huneke-Ulrich deviation two Gorenstein ideals*, J. Alg. **100** (1986), 265–304.
12. A. Kustin, *The deviation two Gorenstein rings of Huneke and Ulrich*, preprint, (University of South Carolina, 1992).
13. A. Kustin and B. Ulrich, *If the socle fits*, J. Alg. **147** (1992), 63–80.
14. A. Kustin and B. Ulrich, *A family of complexes associated to an almost alternating map, with applications to residual intersections*, Mem. Amer. Math. Soc. **95** (1992), 1–94.
15. H. Srinivasan, *Minimal algebra resolutions for cyclic modules defined by Huneke-Ulrich ideals*, J. Alg. **137** (1991), 433–472.
16. H. Srinivasan, *Decomposition formulas for pfaffians*, J. Alg. (to appear).

MATHEMATICS DEPARTMENT, UNIVERSITY OF SOUTH CAROLINA, COLUMBIA, SC 29208

E-mail address: kustin@milo.math.scarolina.edu

Contemporary Mathematics
Volume **159**, 1994

Proximity inequalities for complete ideals in two-dimensional regular local rings

JOSEPH LIPMAN

0. Introduction
1. Preliminaries
2. Proximity inequalities
3. Unique factorization
4. Simple complete ideals
5. Valuations and proximity

Introduction. Among the various algebraic approaches to the classification of singularities of irreducible plane curves (via characteristic pairs, multiplicity sequence, value semigroup, etc.) a particularly attractive one, based on the idea of *proximity,* was developed by Enriques [**4**, book 4]. His analysis can be adapted to arbitrary valuations birationally dominating two-dimensional regular local rings, and also to complete—i.e., integrally closed—ideals in such rings.

While some motivational material on valuations appears in §5, this paper deals mainly with complete ideals. The principal result, Theorem (2.1), provides a necessary and sufficient condition, the *proximity inequalities,* for the existence of a complete ideal having a given "point basis." This is the ideal-theoretic version of an old result on the existence of plane curves with given effective multiplicities at infinitely near points, cf. [**4**, p. 392, p. 427], [**13**, p. 196, Thm. 14], [**8**, p. 49]. Unique factorization for complete ideals and some basic properties of simple complete ideals fall out as corollaries, more or less. Theorem (4.11), on the *predecessor* of a simple complete ideal, is inspired by [**5**, §3], but says more.

Basically then, our purpose is to publicize the efficacy of the notion of proximity, and in particular to redo some of Zariski's theory of complete ideals[1] so as to expose further its roots in the classical treatment of the local behavior of linear systems of curves on smooth surfaces .

1991 *Mathematics Subject Classification.* 13H05, 14H20.
Partially supported by the National Security Agency.

[1]cf. [**15**, Appendix 5], [**7**, chaps. II, V], [**9**], [**6**].

This paper is in final form and no version of it will be submitted for publication elsewhere.

1. Preliminaries. (1.1) Fix a field K, and denote by α, β, γ, ... two-dimensional regular local rings—which we call "points"—having fraction field K.

To connect with classical language, we say that a point β is "infinitely near" to a point α if $\beta \supset \alpha$. Then the maximal ideal $\mathfrak{m}_\beta$ of β intersects α in $\mathfrak{m}_\alpha$. Moreover, a factorization theorem of Zariski and Abhyankar [**1**, p. 343, Thm. 3] gives the existence of a finite sequence—clearly unique—

$$\alpha = \alpha_0 \subset \alpha_1 \subset \cdots \subset \alpha_n = \beta \tag{1.1.1}$$

such that for $0 \le i < n$, α_{i+1} is a *quadratic transform* of α_i, i.e., a localization at a maximal ideal of a ring $\alpha_i[x^{-1}\mathfrak{m}_{\alpha_i}]$ with $x \in \mathfrak{m}_{\alpha_i}$, $x \notin \mathfrak{m}^2_{\alpha_i}$. So the residue field extension $\alpha/\mathfrak{m}_\alpha \subset \beta/\mathfrak{m}_\beta$ is *finite;* we denote its degree by $[\beta : \alpha]$.

(1.2) To each point β associate the unique valuation ord_β of K such that

$$\operatorname{ord}_\beta(x) = \max\{\, n \mid x \in \mathfrak{m}^n_\beta \,\} \qquad (0 \ne x \in \beta).$$

This association is a one-one correspondence between points infinitely near to a given point α and valuations v of K dominating and residually transcendental over α, the point associated to such a v being the largest one containing α and dominated by v [**1**, p. 336, Prop. 3].

(1.3) We say that $\beta \supsetneq \alpha$ is *proximate to* α, and write $\beta \succ \alpha$, if the valuation ring of $\operatorname{ord}_\alpha$ contains β—and hence is of the form $\beta_\mathfrak{p}$ where $\mathfrak{p}$ is a height one prime ideal in β containing $\mathfrak{m}_\alpha$.

An easy induction on the length n of the sequence (1.1.1) shows that if $\beta \supsetneq \alpha$ then $\mathfrak{m}_\alpha\beta = t^a u^b \beta$, where $t\beta = \mathfrak{m}_{\alpha_{n-1}}\beta$, $(t,u)\beta = \mathfrak{m}_\beta$, $a > 0$ and $b \ge 0$. Consequently β is proximate to α_{n-1} and to *at most one other point* in (1.1.1).

Thinking geometrically, consider a map $f\colon X \to Y$ of smooth surfaces, let α be the local ring of a point $y \in Y$, and let β be the local ring of a point $x \in f^{-1}(y)$; then the points to which β is proximate correspond to the components of $f^{-1}(y)$ through x—at most two, since $f^{-1}(y)$ is a normal-crossing divisor.

(1.4) The integral closure $\bar{I}$ of an ideal I in α satisfies

$$\bar{I} = \{\, x \in \alpha \mid \operatorname{ord}_\beta(x) \ge \operatorname{ord}_\beta(I) \text{ for all } \beta \supset \alpha \,\}.$$

It is actually enough here to consider only those β such that $\operatorname{ord}_\beta \in \mathbf{R}(I)$, the set of Rees valuations of I, i.e., those valuations which correspond to the components of the closed fiber on the normalized blowup of I, cf. e.g., [**15**, p. 354, Lemma].

Conforming with Zariski's terminology, we say that I is *complete* if $I = \bar{I}$.

The product of any two complete ideals is complete [**15**, p. 385, Thm. 2′].[2]

(1.5) Let I be an α-ideal of finite colength, i.e., $\lambda_\alpha(\alpha/I) < \infty$ (where λ_α denotes the length of an α-module). The *transform of* I in a point $\beta \supset \alpha$ is the finite-colength β-ideal $I^\beta := I(I\beta)^{-1}$. Note that $I\beta = t^c u^d I^\beta$ where t and u are as in (1.3), $c = \operatorname{ord}_{\gamma_1}(I)$ with γ_1 the predecessor of β in the sequence (1.1.1), and $d = \operatorname{ord}_{\gamma_2}(I)$ with γ_2 the other point in (1.1.1) to which β is proximate (if there is one; otherwise $d = 0$).

[2] For a generalization to rational singularities, cf. [**7**, p. 209, Thm. (7.1)].

Transform preserves products: $(IJ)^\beta = I^\beta J^\beta$.

The transform operation is transitive: if $\alpha \subset \beta \subset \gamma$ then $(I^\beta)^\gamma = I^\gamma$.

(1.6) $\beta \supset \alpha$ is a *base point* of I if $\mathrm{ord}_\beta(I^\beta) \neq 0$, i.e., if $I\beta$ is not a principal ideal. A given α-ideal I has only finitely many base points, since any such β is dominated by a Rees valuation of I (as follows from Zariski's Main Theorem, because β does not contain any local ring on the normalized blowup of I).

The *point basis* of I is the family of nonnegative integers

$$\mathbf{B}(I) := (\mathrm{ord}_\beta(I^\beta))_{\beta \supset \alpha}.$$

For any two finite-colength α-ideals I, J, we have:

(i) $\mathbf{B}(IJ) = \mathbf{B}(I) + \mathbf{B}(J)$ (since $(IJ)^\beta = I^\beta J^\beta$)

(ii) $\bar{I} = \bar{J} \iff \mathbf{B}(I) = \mathbf{B}(J)$ ([**9**, p. 209, (1.10)]).

Thus $\mathbf{B}$ maps the multiplicative monoid $\mathcal{M}_\alpha$ of finite-colength complete α-ideals *isomorphically* onto a submonoid $\mathbf{B}(\mathcal{M}_\alpha)$ of the free commutative monoid generated by all the points infinitely near to α.

The following central result describes $\mathbf{B}(\mathcal{M}_\alpha)$.

2. Proximity inequalities.

THEOREM (2.1). *Let $(r_\beta)_{\beta\supset\alpha}$ be a family of nonnegative integers, with $r_\beta = 0$ for all but finitely many β. Then there exists a finite-colength α-ideal I with $\mathbf{B}(I) = (r_\beta)$ iff the following* proximity inequality *holds for each $\beta \supset \alpha$:*

$$r_\beta \geq \sum_{\gamma \succ \beta} [\gamma : \beta] r_\gamma.$$

And if there is such an I then there is one and only one which is complete.

The second assertion follows from (1.6)(ii); the proof of the first takes up the rest of this section. (See also the remark at the very end of the paper.) We'll say that an α-ideal I is divisible by an α-ideal J if $I = JJ'$ for some α-ideal J' (or, equivalently, if $I = J(I : J)$). A key point is:

LEMMA (2.2). *Let I be a finite-colength complete α-ideal, and let $\nu \geq 0$ be the integer such that I is divisible by $\mathfrak{m}_\alpha^\nu$ but not by $\mathfrak{m}_\alpha^{\nu+1}$. Then*

$$\nu = \mathrm{ord}_\alpha(I) - \sum_{\gamma \succ \alpha} [\gamma : \alpha] \mathrm{ord}_\gamma(I^\gamma).$$

Assuming (2.2), whose proof will be given below, we can prove (2.1) as follows.

Suppose that $(r_\beta) = \mathbf{B}(I)$. To prove the inequality for $r_\beta := \mathrm{ord}_\beta(I^\beta)$, we can simply replace α by β and I by I^β in (2.2) and apply transitivity of transform (1.5).

Suppose conversely that the family (r_β) satisfies the proximity inequalities. Let $\beta_1, \beta_2, \ldots, \beta_n$ $(n \geq 0)$ be all those quadratic transforms of α whose corresponding r doesn't vanish. Inducting on the number of β such that $r_\beta \neq 0$, we may assume that there exists a finite-colength complete β_j-ideal I_j with point basis $(r_\beta)_{\beta\supset\beta_j}$. By [**9**, p. 217, Lemma (2.3)] there is a finite-colength complete α-ideal I', not divisible by $\mathfrak{m}_\alpha$, whose transform in β_j is I_j $(1 \le j \le n)$ and whose transform in every other quadratic transform of α is the unit ideal. By assumption,

$$0 \le r_\alpha - \sum_{\gamma\succ\alpha}[\gamma:\alpha]r_\gamma \overset{(2.2)}{=} r_\alpha - \operatorname{ord}_\alpha(I') =: a,$$

so we may set $I := \mathfrak{m}_\alpha^a I'$; and then $\mathbf{B}(I) = (r_\beta)$. □

Now here is the proof of (2.2). Set $\mathfrak{m} := \mathfrak{m}_\alpha$, $k := \alpha/\mathfrak{m}$. If $I = \mathfrak{m}^e J$ $(e \ge 0)$ then $I^\gamma = J^\gamma$ for all $\gamma \supsetneqq \alpha$, so that the sum $\sum_{\gamma\succ\alpha}[\gamma:\alpha]\operatorname{ord}_\gamma(I^\gamma)$ does not change when I is replaced by J; and it follows that for (2.2) it suffices to treat the case where $\nu = 0$.

Let $r := \operatorname{ord}_\alpha(I)$, so that $I \subset \mathfrak{m}^r$, $I \not\subset \mathfrak{m}^{r+1}$. Let $s(I)$ be the degree of the greatest common divisor $c(I)$ (in the graded UFD $\operatorname{gr}(\alpha) := \oplus_{n\ge0}\mathfrak{m}^n/\mathfrak{m}^{n+1}$) of all the elements in the initial form vector space

$$\operatorname{in}(I) := (I + \mathfrak{m}^{r+1})/\mathfrak{m}^{r+1} \subset \mathfrak{m}^r/\mathfrak{m}^{r+1} \subset \operatorname{gr}(\alpha).$$

A basic result of Zariski [**15**, p. 368, Prop. 3], [**6**, p. 327, Prop. 2.5] is that *if I is not divisible by $\mathfrak{m}$ then $s(I) = \operatorname{ord}_\alpha(I)$*. Thus (2.2) (for $\nu = 0$, and hence for all ν) follows from the next result, which is an ideal-theoretic analog of [**13**, p. 191, Thm. 11]:

PROPOSITION (2.3). *For any finite-colength α-ideal I, we have*

$$s(I) = \sum_{\gamma\succ\alpha}[\gamma:\alpha]\operatorname{ord}_\gamma(I^\gamma).$$

PROOF. The proof uses a family v_p of valuations of K, one for each homogeneous height one prime ideal p in $\operatorname{gr}(\alpha)$. For any nonzero $x \in \alpha$, let

$$\bar{x} := x + \mathfrak{m}^{b+1} \in (\mathfrak{m}^b/\mathfrak{m}^{b+1}) \subset \operatorname{gr}(\alpha) \qquad (b := \operatorname{ord}_\alpha(x))$$

be the initial form of x. Since $\operatorname{gr}(\alpha)$ is isomorphic to a polynomial ring over the residue field k, we can factor the principal ideal $(\bar{x})$ uniquely in the form

$$(\bar{x}) = \prod_p p^{n_p(x)}.$$

It is then easily checked that the mapping $x \mapsto \big(\operatorname{ord}_\alpha(x), n_p(x)\big)$ of α into the lexicographically ordered group $\mathbb{Z} \times \mathbb{Z}$ gives rise to a valuation v_p of K which dominates α and is composite with $\operatorname{ord}_\alpha$ (i.e., its valuation ring is contained in that of $\operatorname{ord}_\alpha$). There is a unique quadratic transform β_p of α which is dominated by v_p; and one verifies that $p \mapsto \beta_p$ is a one-one correspondence between the set of homogeneous height one primes p and the set of quadratic transforms of α (both sets corresponding to the set of closed points in the closed fiber of the blowup of $\mathfrak{m}$.) Note that if a point γ is contained in the valuation ring V_p of v_p, then v_p *dominates* γ: the center of v_p in γ—i.e., the intersection $\mathfrak{q}$ of γ with the maximal ideal of V_p—is $\mathfrak{m}_\gamma$ (because V_p contains the localization $\gamma_\mathfrak{q}$).

The significance of the v_p with respect to proximity is given by:

LEMMA (2.4). (i) *If* $\alpha \subset \gamma \subset V_p$, *then* γ *contains* β_p *and is proximate to* α.

(ii) *Conversely, if* γ *contains* β_p *and is proximate to* α, *then* $\gamma \subset V_p$. *In fact* V_p *is the unique valuation ring dominating* γ *and composite with* $\operatorname{ord}_\alpha$, *i.e., contained in the valuation ring* R *of* $\operatorname{ord}_\alpha$. *Moreover, the inclusion* $\gamma \hookrightarrow V_p$ *induces an isomorphism of residue fields, and* $v_p(\mathfrak{m}_\gamma) = (0,1)$.

Before proving (2.4), let us deduce (2.3). Factoring the $\operatorname{gr}(\alpha)$-ideal $\big(c(I)\big)$ as

$$\big(c(I)\big) = \prod_p p^{n_p(I)}$$

and observing that p is generated by a homogeneous element of degree $[\beta_p : \alpha]$, we reduce to showing that

$$n_p(I) = \sum_{\substack{\gamma \supset \beta_p \\ \gamma \succ \alpha}} [\gamma : \beta_p] \operatorname{ord}_\gamma(I^\gamma).$$

In view of (2.4), which implies in particular that γ and β_p have the same residue field, i.e., $[\gamma : \beta_p] = 1$, the right hand sum becomes $\sum_{V_p \supset \gamma \supsetneqq \alpha} \operatorname{ord}_\gamma(I^\gamma)$. But

$$\big(\operatorname{ord}_\alpha(I), n_p(I)\big) = v_p(I) = \sum_{V_p \supset \gamma \supset \alpha} \operatorname{ord}_\gamma(I^\gamma) v_p(\mathfrak{m}_\gamma).$$

(For the second equality use, repeatedly, the fact that with δ the unique quadratic transform of γ dominated by v_p, we have $I^\gamma\delta = \mathfrak{m}_\gamma^{\operatorname{ord}_\gamma(I^\gamma)} I^\delta$, so that

$$v_p(I^\gamma) = v_p(I^\delta) + \operatorname{ord}_\gamma(I^\gamma) v_p(\mathfrak{m}_\gamma),$$

cf. [**9**, p. 209, Lemma (1.11)].) Since $v_p(\mathfrak{m}_\alpha) = (1,0)$ and since (by (2.4)) $v_p(\mathfrak{m}_\gamma) = (0,1)$ for all $\gamma \supsetneqq \alpha$, the desired conclusion follows.

It remains to prove (2.4). Assertion (i) is obvious, by the definition of β_p and since v_p is composite with $\operatorname{ord}_\alpha$. As for (ii), note first that if $\mathfrak{p}$ (resp. $\mathfrak{q}$) is the center of $\operatorname{ord}_\alpha$ in β_p (resp. γ), then $\mathfrak{p}$ is a height one prime ideal, whence so is $\mathfrak{q}$ (since $\mathfrak{q} \cap \beta_p = \mathfrak{p}$). Let $h\colon R \to \kappa$ be the canonical map of R onto its residue field κ. Then $h(\beta_p) \subset h(\gamma) \subset \kappa$. Since $h(\beta_p) \cong \beta_p/\mathfrak{p}$ is a discrete valuation ring with fraction field κ, and $h(\gamma) \cong \gamma/\mathfrak{q} \neq \kappa$, therefore $h(\gamma) = h(\beta_p)$.

Now the valuation rings V which dominate γ and are contained in R are in one-one correspondence with the valuation rings in κ which dominate $h(\gamma)$, the correspondence being $V \leftrightarrow h(V)$ [**2**, p. 111, §4.1, Prop. 2]; and since $h(\gamma)$ is a discrete valuation ring, there is a unique such V, namely $h^{-1}h(\gamma)$. The corresponding valuation of K dominates β_p too, and so by uniqueness it must be v_p. Thus $\gamma \subset V_p = V$.

Moreover, $V = h^{-1}h(\gamma)$ has the same residue field as γ. Also, since $h(\mathfrak{m}_\gamma V)$ is the maximal ideal of $h(\gamma)$ and since the ideals in V are totally ordered, so that $\mathfrak{m}_\gamma V$ contains the kernel of $V \twoheadrightarrow \gamma$, therefore $\mathfrak{m}_\gamma V$ is the maximal ideal of V, i.e., $v_p(\mathfrak{m}_\gamma) = (0,1)$. $\square$

3. Unique factorization. Before proceeding, we recall a useful device, due to Du Val [**3**], for representing proximity relations. Fixing a point α, consider the *proximity matrix*

$$\mathbf{p} = \mathbf{p}(\alpha) = (p_{\beta\gamma})_{\beta\supset\alpha,\,\gamma\supset\alpha}$$

given by

$$\begin{aligned} p_{\beta\gamma} &= 1 && \text{if } \beta = \gamma \\ &= -1 && \text{if } \beta \prec \gamma \\ &= 0 && \text{otherwise.} \end{aligned}$$

Consider also the diagonal matrix $\mathbf{d} = \mathbf{d}(\alpha) = (d_{\beta\gamma})_{\beta\supset\alpha,\,\gamma\supset\alpha}$ given by

$$\begin{aligned} d_{\beta\gamma} &= 0 && \text{if } \beta \neq \gamma \\ d_{\beta\beta} &= [\beta : \alpha] \end{aligned}$$

and the *refined proximity matrix*

$$\mathbf{d}^{-1}\mathbf{p}\mathbf{d} =: \mathbf{P} = \mathbf{P}(\alpha) = (P_{\beta\gamma})_{\beta\supset\alpha,\,\gamma\supset\alpha}$$

given by

$$\begin{aligned} P_{\beta\gamma} &= 1 && \text{if } \beta = \gamma \\ &= -[\gamma : \beta] && \text{if } \beta \prec \gamma \\ &= 0 && \text{otherwise.} \end{aligned}$$

One checks via (1.3) that the matrices $\mathbf{p}$ and $\mathbf{P}$ are column-finite and *invertible*, and that the entries of $\mathbf{p}^{-1}$ and $\mathbf{P}^{-1}$ are nonnegative integers (cf. also (3.1) and (4.6)).

REMARK. The proximity inequalities for a "column vector" $\mathbf{B} = (r_\beta)$ as in (2.1) can be expressed as $\mathbf{PB} \geq 0$.

COROLLARY (3.1) (Unique Factorization). *For each $\gamma \supset \alpha$ let $\mathbf{e}_\gamma$ be the "unit column vector" corresponding to γ, i.e., the family $(e_{\gamma\delta})_{\delta\supset\alpha}$ with $e_{\gamma\delta} = 0$ if $\delta \neq \gamma$ and $e_{\gamma\gamma} = 1$. Then:*

(i) *There is a unique $\mathfrak{m}_\alpha$-primary complete ideal $\wp_\gamma$ with point basis $\mathbf{P}^{-1}\mathbf{e}_\gamma$.*

(ii) *Any $\mathfrak{m}_\alpha$-primary complete ideal I can be factored uniquely as*

$$I = \prod_{\gamma\supset\alpha} \wp_\gamma^{a_\gamma} \quad (a_\gamma \geq 0).$$

The factorization vector $\mathbf{F}(I) := (a_\gamma)$ *is related to the point basis $\mathbf{B}(I)$ of I by*

$$\mathbf{F}(I) = \mathbf{PB}(I).$$

PROOF. Since $\mathbf{PP}^{-1}\mathbf{e}_\gamma = \mathbf{e}_\gamma \geq 0$, the existence of $\wp_\gamma$ is given by (2.1) (see preceding remark), and its uniqueness by (1.6)(ii).

As mentioned in (1.4), the ideal $\prod \wp_\gamma^{a_\gamma}$ is complete; and by (1.6)(i), its point basis is $\mathbf{P}^{-1}(a_\gamma)$, which equals $\mathbf{B}(I)$ iff $(a_\gamma) = \mathbf{PB}(I)$, whence the conclusion. □

EXAMPLE (3.2). To each of the ideals $\wp_\delta$ associate an *adjoint* (or *conductor-*) ideal $\mathfrak{C}_\delta$, as follows.

Let $(r_\beta) := \mathbf{P}^{-1}\mathbf{e}_\delta$ be the point basis of $\wp_\delta$, cf. (3.1). The equation $\mathbf{P}(r_\beta) = \mathbf{e}_\delta$ gives, for $\beta \neq \delta$,

(3.2.1)
$$r_\beta = \sum_{\gamma \succ \beta} [\gamma : \beta] r_\gamma .$$

Hence,

(3.2.2)
$$\beta \subset \gamma \implies r_\beta \geq r_\gamma .$$

(By induction on the number of points between β and γ we reduce to where γ is a quadratic transform of β, so that $\gamma \succ \beta$ and we can apply (3.2.1).) Moreover,

(3.2.3)
$$r_\delta = 1, \quad \text{and} \quad r_\gamma = 0 \quad \text{if } \gamma \not\subseteq \delta .$$

This is because $\mathbf{P}$ is "upper triangular" ($\mathbf{P}_{\beta\gamma} = 0$ unless $\beta \subset \gamma$) with 1's on the diagonal, so the same is true of $\mathbf{P}^{-1}$, whose δ-column is $(r_\gamma)_{\gamma \supset \alpha}$.

Now consider the family (r'_β) given by

$$\begin{aligned} r'_\beta &= r_\beta - 1 & &\text{if } r_\beta > 0 \\ &= 0 & &\text{if } r_\beta = 0; \end{aligned}$$

and set

$$c_\beta := r'_\beta - \sum_{\gamma \succ \beta} [\gamma : \beta] r'_\gamma .$$

By (3.2.3) and (3.2.2), $r'_\gamma = 0$ unless $\gamma \subsetneqq \delta$, and $r'_\gamma = r_\gamma - 1$ if $\gamma \subset \delta$. So if $r_\beta > 1$ then $\beta \neq \delta$ and

$$\begin{aligned} c_\beta &= (r_\beta - 1) - \sum_{\delta \supset \gamma \succ \beta} [\gamma : \beta](r_\gamma - 1) \\ &= \Big(r_\beta - \sum_{\gamma \succ \beta} [\gamma : \beta] r_\gamma\Big) - 1 + \sum_{\delta \supset \gamma \succ \beta} [\gamma : \beta] \\ &\overset{(3.2.1)}{=} -1 + \sum_{\delta \supset \gamma \succ \beta} [\gamma : \beta] \geq 0; \end{aligned}$$ [3]

while if $r_\beta \leq 1$ then (3.2.2) implies that $c_\beta = 0$.

By (2.1) then, there exists a unique complete ideal $\mathfrak{C}_\delta$ with point basis (r'_β), and by (3.1) that ideal is

$$\mathfrak{C}_\delta := \prod_{\beta \supset \alpha} \wp_\beta^{c_\beta} .$$

This $\mathfrak{C}_\delta$ has a number of interesting properties, cf. [**10**], of which we mention only one (conductor property): for every integer $n \geq \operatorname{ord}_\delta(\mathfrak{C}_\delta)$, there is a $z \in \alpha$ with $\operatorname{ord}_\delta(z) = n$; and if $[\delta : \alpha] = 1$ (but not otherwise), then there is no $z \in \alpha$ with $\operatorname{ord}_\delta(z) = \operatorname{ord}_\delta(\mathfrak{C}_\delta) - 1$.

---

[3] Incidentally, (2.4)(ii) yields that $[\gamma : \beta]$ has the same value for all γ such that $\delta \supset \gamma \succ \beta$.

4. Simple complete ideals. A simple ideal in α is, by definition, one which cannot be factored non-trivially. A complete α-ideal I is simple iff it is not a product of two other complete ideals (for, if $I = JL$ then $I = \bar{J}\bar{L}$).

COROLLARY (4.1). *The map $\gamma \mapsto \wp_\gamma$ is a one-one correspondence between points infinitely near to α and simple $\mathfrak{m}_\alpha$-primary complete ideals. The inverse map takes such a simple ideal to its unique largest base point.*

PROOF. The first statement follows from (3.1); the second from (3.2.3). □

Now we want to vary α, so we will write $\wp_{\alpha\gamma}$ instead of $\wp_\gamma$. We also set $\wp_{\alpha\delta} = \delta$ whenever $\alpha \not\subset \delta$. Note then that for any three points $\alpha \subset \beta$ and γ,

$$\operatorname{ord}_\gamma(\wp^\beta_{\alpha\gamma}) = \operatorname{ord}_\gamma(\wp_{\beta\gamma}).$$

This is clear, from the last assertion in (4.1), if $\beta \not\subset \gamma$. Otherwise it just says that the $\beta\gamma$-entries in the inverse refined proximity matrices $\mathbf{P}(\alpha)^{-1}$ and $\mathbf{P}(\beta)^{-1}$ are the same, which holds because $\mathbf{P}(\beta)$ is obtained from $\mathbf{P}(\alpha)$ by chopping off all rows and columns indexed by points not containing β, so that a similar relation holds between the inverse matrices. (Note that in the calculation of column γ of $\mathbf{P}^{-1}$, only those $P_{\beta\gamma}$ for which $\beta \subset \gamma$ come into play, so that in essence we are working with finite upper triangular matrices.) By (1.4), then, $\wp_{\beta\gamma}$ is the integral closure of the transform $\wp^\beta_{\alpha\gamma}$. If we use the result that *transforms of complete ideals are complete,* ([**15**, p. 381, Prop. 5], [**7**, p. 209, Prop. (6.5)]), then we can conclude that in fact $\wp_{\beta\gamma} = \wp^\beta_{\alpha\gamma}$. But just for variety, let us take a brief stroll along another logical path ((4.2) and (4.3)).

PROPOSITION (4.2). *For any points $\alpha \subset \beta$ and γ, the β-transform of $\wp_{\alpha\gamma}$ is $\wp_{\beta\gamma}$.*

PROOF. In view of the last assertion in (4.1), we need only consider the case where $\beta \subset \gamma$. Using transitivity of transform (1.5) to induct on the number of points between α and β, we reduce to where β is a quadratic transform of α. Then by [**9**, p. 217, Lemma (2.3)], there exists a simple $\mathfrak{m}_\alpha$-primary complete ideal $\wp$ whose point basis outside of α is the same as that of its β-transform $\wp_{\beta\gamma}$. It follows at once from (4.1) that $\wp = \wp_{\alpha\gamma}$. □

COROLLARY (4.3). *For any two points $\alpha \subset \beta$, the β-transform of any complete finite-colength α-ideal is again complete.*

PROOF. Since transform respects products (1.5), therefore (3.1) reduces us to the case of simple ideals, given by (4.2). □

PROPOSITION (4.4). *A complete $\mathfrak{m}_\alpha$-primary ideal I is divisible by $\wp_\gamma$ iff $\operatorname{ord}_\gamma$ is a Rees valuation of I.*

PROOF. Using the fact that a local ring dominates the blowup of a product of ideals iff it dominates the blowup of each of the factors, we reduce readily to where $I = \wp_\gamma$, in which case the assertion is that (*): ord_γ *is the unique Rees valuation of* $\wp_\gamma$, which is shown in [**7**, p. 245, Prop. (21.3)], or, in a more elementary way, in [**6**, p. 333, Thm. 4.2].)[4] □

We define the *valuation vector* $\mathbf{V}(I)$ of an α-ideal I to be the column vector $(\mathrm{ord}_\beta(I))_{\beta\supset\alpha}$. Though $\mathbf{V}(I)$ has infinitely many nonzero entries, it can still be premultiplied by a row-finite matrix, for example by the transpose $\mathbf{p^t}$ of $\mathbf{p}$.

PROPOSITION (4.5). *For any α-ideal I,*

$$\mathbf{B}(I) = \mathbf{p^t}\mathbf{V}(I).$$

PROOF. The proposition states that for any $\beta \supset \alpha$,

$$\mathrm{ord}_\beta(I^\beta) = \mathrm{ord}_\beta(I) - \sum_{\alpha\subset\gamma\prec\beta} \mathrm{ord}_\gamma(I),$$

which is an immediate consequence of the relation $I\beta = t^c u^d I^\beta$ in (1.5). □

COROLLARY (4.6). *The entries of the matrix* $\mathbf{p}^{-1}$ *are*

$$\begin{aligned}(\mathbf{p}^{-1})_{\beta\gamma} &= \mathrm{ord}_\gamma(\mathfrak{m}_\beta) && \text{if } \beta\subset\gamma\\ &= 0 && \text{otherwise.}\end{aligned}$$

PROOF. If $\beta \subset \gamma$, then after chopping off some rows and columns from $\mathbf{p}$ we may, as in the remarks preceding (4.2), assume that $\beta = \alpha$, and then just take $I = \mathfrak{m}_\alpha$ in (4.5) (or at least in the equivalent relation $\mathbf{V}(I) = (\mathbf{p}^{-1})^{\mathbf{t}}\mathbf{B}(I)$). The second equality results from the corresponding property of $\mathbf{p}$. □

REMARKS. (1) The formulation $\mathbf{V}(I) = (\mathbf{p}^{-1})^{\mathbf{t}}\mathbf{B}(I)$ of (4.5), i.e., by (4.6),

$$\mathrm{ord}_\beta(I) = \sum_{\alpha\subset\gamma\subset\beta} \mathrm{ord}_\beta(\mathfrak{m}_\gamma)\mathrm{ord}_\gamma(I^\gamma),$$

is a special case of [**9**, p. 209, Lemma (1.11)], cf. proof of (2.3) above.

[4] I learned only recently (April, 1993) that (*) is essentially contained in Hironaka's 1960 Harvard thesis (Chapter 1, §4, Thm. 10). When I mentioned (*) to Zariski in the early 1970's, he seemed unfamiliar with it, but quickly came up with the following proof. Proceed by induction on the number of base points of γ, the assertion being obvious when that number is 1 (i.e., $\gamma = \alpha$ and $\wp_\gamma = \mathfrak{m}_\alpha$). So assume $\wp := \wp_\gamma \ne \mathfrak{m}_\alpha$, and let Y be the scheme over $\mathrm{Spec}(\alpha)$ obtained by first blowing up $\mathfrak{m}_\alpha$ and then blowing up the transform $\wp_{\beta\gamma}$ of $\wp_\gamma$ in the unique quadratic transform β of α dominated by γ, cf. (4.2). It results from the inductive hypothesis that the only valuations which dominate α and whose centers on Y are one-dimensional are ord_α and ord_γ. It suffices therefore to show that the center of ord_α on the blowup X of $\wp$ is 0-dimensional. But the equality $s(\wp) = \mathrm{ord}_\alpha(\wp)$ preceding (2.3) implies that $\wp$ has a basis $(x_0, x_1, \dots, x_n)$ with $\mathrm{ord}_\alpha(x_i) > \mathrm{ord}_\alpha(x_0)$ for all $i > 0$, i.e., x_i/x_0 lies in the maximal ideal of the valuation ring of ord_α, so that the local ring on X dominated by ord_α has the same residue field as α. □

(2) The fact that the identity matrix $\mathbf{p}\mathbf{p}^{-1}$ has zero entries off the diagonal translates to the relation

$$\operatorname{ord}_\gamma(\mathfrak{m}_\beta) = \sum_{\beta \prec \delta \subset \gamma} \operatorname{ord}_\gamma(\mathfrak{m}_\delta) \qquad (\beta \subsetneqq \gamma).$$

(3) Replacing $\mathbf{p}$ by $\mathbf{P}$, we find similarly that

$$\operatorname{ord}_\beta(\wp_{\beta\gamma}) = \sum_{\beta \prec \delta \subset \gamma} [\gamma : \beta] \operatorname{ord}_\delta(\wp_{\delta\gamma}) \qquad (\beta \subsetneqq \gamma).$$

(These last two equations transform into each other by reciprocity, cf. (4.8).)

Next we define the *intersection number* $(I \cdot J) = (J \cdot I)$ of two finite-colength α-ideals to be

$$(I \cdot J) := \sum_{\beta \supset \alpha} [\beta : \alpha] \operatorname{ord}_\beta(I^\beta) \operatorname{ord}_\beta(J^\beta) = \mathbf{B}(I)^{\mathbf{t}} \mathbf{d} \mathbf{B}(J).$$

COROLLARY (4.7). *For any α-ideal I and any $\beta \supset \alpha$,*

$$(I \cdot \wp_\beta) = [\beta : \alpha] \operatorname{ord}_\beta(I).$$

PROOF. $\mathbf{B}(\wp_\beta)$ is the β-column of $\mathbf{P}^{-1} = \mathbf{d}^{-1}\mathbf{p}^{-1}\mathbf{d}$, so we have equal row vectors

$$\big((I \cdot \wp_\beta)\big)_{\beta \supset \alpha} = \mathbf{B}(I)^{\mathbf{t}} \mathbf{d} \mathbf{P}^{-1} = \mathbf{B}(I)^{\mathbf{t}} \mathbf{p}^{-1} \mathbf{d} \overset{(4.5)}{=} \mathbf{V}(I)^{\mathbf{t}} \mathbf{p}\mathbf{p}^{-1} \mathbf{d} = \big(\mathbf{d}\mathbf{V}(I)\big)^{\mathbf{t}},$$

whence the conclusion. □

COROLLARY (4.8) (Reciprocity).[5] *For any $\beta \supset \alpha$, $\gamma \supset \alpha$,*

$$[\beta : \alpha] \operatorname{ord}_\beta(\wp_\gamma) = [\gamma : \alpha] \operatorname{ord}_\gamma(\wp_\beta).$$

PROOF. $(\wp_\gamma \cdot \wp_\beta) = (\wp_\beta \cdot \wp_\gamma)$. □

Given two vectors $\mathbf{B} = (r_\beta)_{\beta \supset \alpha}$, $\mathbf{B}' = (r'_\beta)_{\beta \supset \alpha}$, we write $\mathbf{B} \geq \mathbf{B}'$ to signify that $r_\beta \geq r'_\beta$ for all β.

COROLLARY (4.9). *For any two finite-colength complete α-ideals I and J,*

$$\mathbf{B}(I) \geq \mathbf{B}(J) \implies J \supset I.$$

PROOF. According to (4.6), the entries of $(\mathbf{p}^{\mathbf{t}})^{-1}$ are all ≥ 0, so by (4.5), $\mathbf{B}(I) \geq \mathbf{B}(J) \implies \mathbf{V}(I) \geq \mathbf{V}(J)$, whence the conclusion. □

---

[5] [**7**, p. 247, Prop. (21.4)], [**6**, p. 334, Thm. 4.3].

COROLLARY (4.10).[6] *If $\delta \subset \gamma$ then $\wp_\delta \supset \wp_\gamma$.*

PROOF. By an obvious induction, we may assume that γ is a quadratic transform of δ, so that $\delta \prec \gamma$. Then the matrix equation $\mathbf{P}^{-1}\mathbf{P} = \mathbf{1}$ yields, for $\alpha \subset \beta \subsetneq \gamma$,

$$\mathrm{ord}_\beta(\wp_\gamma^\beta) = \sum_{\alpha \subset \delta' \prec \gamma} [\gamma : \delta']\mathrm{ord}_\beta(\wp_{\delta'}^\beta) \geq \mathrm{ord}_\beta(\wp_\delta^\beta),$$

whence by (3.2.3), $\mathbf{B}(\wp_\gamma) \geq \mathbf{B}(\wp_\delta)$, and we can apply (4.9). □

The next result gives a characterization of proximity for points containing α in terms of their corresponding simple complete ideals—a characterization found by D. B. Scott and used by Hoskin in [**5**, §3]—namely $\beta \succ \gamma \Leftrightarrow \wp_\gamma$ divides $\mathfrak{q}_\beta$, the smallest ord_β-ideal strictly containing $\wp_\beta$. Note however that our result says more; and besides being valid in arbitrary two-dimensional regular local rings, it also shows that $\mathfrak{q}_\beta$ is actually the smallest *complete* ideal strictly containing $\wp_\beta$. (This has been proved previously by Noh [**12**, Thm. 3.1], at least when the residue field of α is algebraically closed.)

THEOREM (4.11). *Let $\beta \supset \alpha$. Among complete ideals I in α strictly containing $\wp_\beta$, there is a smallest, viz.*

$$\mathfrak{q}_\beta := \prod_{\alpha \subset \gamma \prec \beta} \wp_\gamma^{[\beta:\gamma]}.$$

For any valuation v dominating β, $\mathfrak{q}_\beta$ is even a v-ideal ($v(x) \geq v(\mathfrak{q}_\beta) \Rightarrow x \in \mathfrak{q}_\beta$).

PROOF. To prove that $\mathfrak{q}_\beta \subset I$, it's enough to show that $w(\mathfrak{q}_\beta) \geq w(I)$ for every Rees valuation $w = \mathrm{ord}_\delta$ of I, cf. (1.4); so it's certainly enough that $w(\mathfrak{q}_\beta) \geq w(\wp_\beta)$. Now every $\wp_\gamma$ dividing I strictly contains $\wp_\beta$, and so by (4.10), $\beta \not\subset \gamma$, and by (4.1), β is not a base point of $\wp_\gamma$, i.e., $(\wp_\gamma)\beta$ is a principal ideal. Thus $I\beta$ is a principal ideal, so β dominates a local ring S on the blowup of I, and consequently $\delta \not\supset \beta$ (otherwise S, being dominated by the valuation ring of ord_δ—which is also a local ring on the blowup—would have to be that valuation ring, and so could not be contained in β). The first assertion follows then from:

LEMMA (4.12). *For any $\delta \supset \alpha$, we have*

$$\begin{aligned} \mathrm{ord}_\delta(\mathfrak{q}_\beta) &= \mathrm{ord}_\delta(\mathfrak{p}_\beta) && \text{if } \delta \not\supset \beta \\ &= \mathrm{ord}_\delta(\mathfrak{p}_\beta) - \mathrm{ord}_\delta(\mathfrak{m}_\beta) && \text{if } \delta \supset \beta. \end{aligned}$$

In particular, $\mathrm{ord}_\delta(\mathfrak{q}_\beta) \leq \mathrm{ord}_\delta(\mathfrak{p}_\beta)$ for all δ, and so $\mathfrak{q}_\beta \supsetneq \wp_\beta$.

PROOF. By definition, $\mathbf{F}(\wp_\beta) - \mathbf{F}(\mathfrak{q}_\beta)$ is column β of the matrix $\mathbf{P}$. Hence by (3.1), $\mathbf{B}(\wp_\beta) - \mathbf{B}(\mathfrak{q}_\beta)$ is column β of $\mathbf{P}^{-1}\mathbf{P} = \mathbf{1}$, and so by (4.5), $\mathbf{V}(\wp_\beta) - \mathbf{V}(\mathfrak{q}_\beta)$ is column β of $(\mathbf{p^t})^{-1}$. The conclusion results then from (4.6). □

[6]Cf. [**15**, p. 392, (F)].

For the second assertion in (4.11), set

$$I := \{\, x \in \alpha \mid v(x) \geq v(\mathfrak{q}_\beta)\,\} \supset \mathfrak{q}_\beta \supsetneqq \wp_\beta,$$

so that as above $I\beta$ is principal, say $I\beta = z\beta$. The kernel J of the homomorphism taking $x \in I$ to $(xz^{-1} + \mathfrak{m}_\beta) \in \beta/\mathfrak{m}_\beta$ consists of all $x \in I$ such that $v(xz^{-1}) > 0$, i.e., $v(x) > v(\mathfrak{q}_\beta)$; so J is a complete ideal not containing $\mathfrak{q}_\beta$. But J does contain $\mathfrak{p}_\beta$, because $z^{-1}\mathfrak{p}_\beta\beta$ is a non-principal β-ideal, so that $z^{-1}\mathfrak{p}_\beta\beta \subset \mathfrak{m}_\beta$; and so by the first part of (4.11), $J = \wp_\beta$. Thus we have an injective $\alpha/\mathfrak{m}_\alpha$-linear map $I/\mathfrak{p}_\beta \hookrightarrow \beta/\mathfrak{m}_\beta$, whence

$$\lambda_\alpha(I/\wp_\beta) \leq [\beta : \alpha].$$

(Recall that λ denotes length.) But the point bases of $\wp_\beta$ and $\mathfrak{q}_\beta$ are identical except for a 1 at β, cf. proof of (4.12). So by the Hoskin-Deligne length formula [**9**, p. 222, Thm. (3.1)],

$$\lambda_\alpha(\mathfrak{q}_\beta/\wp_\beta) = [\beta : \alpha].$$

Since $\mathfrak{q}_\beta \subset I$, we conclude that $\mathfrak{q}_\beta = I$. □

5. Valuations and proximity. To a valuation v dominating α, associate the sequence

$$(5.1)_v \qquad \alpha = \alpha_0 \subset \alpha_1 \subset \cdots \subset \alpha_i \subset \alpha_{i+1} \subset \cdots$$

where for each $i \geq 0$, α_{i+1} is the unique quadratic transform of α_i dominated by v. As in (1.2), the sequence $(5.1)_v$ terminates after finitely many steps iff v is residually transcendental over α. Assume from here on that the sequence is infinite, i.e., v is residually algebraic (or "zero-dimensional") over α. Note, conversely, that *any* infinite sequence $(\beta_i)_{i\geq 0}$ of successive quadratic transforms is associated to a unique zero-dimensional valuation, namely the one having valuation ring $\cup_i \beta_i$. Thus a zero-dimensional valuation dominating α can be identified with a maximal branch running through the tree of points infinitely near to α.

In some sense, v is the *limit* of the valuations ord_{α_i}. And, as will now be briefly discussed, *the proximity relations among the α_i*—as encoded, say, in the matrix $\mathbf{P}_v$ obtained from the refined proximity matrix $\mathbf{P}(\alpha)$ by restricting to those entries $P_{\beta\gamma}$ for which both β and γ are in $(5.1)_v$—*determine many of the basic properties of v.*

For instance, $\mathbf{P}_v$ determines the rank and rational rank of v,[7] and in case of rational rank 1, whether v is discrete or not. In essence, this is shown in [**14**, §9].[8]

[7]except when α is not complete, v is residually finite over α, and there are only finitely many "satellite" points in $(5.1)_v$, i.e., points proximate to two others in $(5.1)_v$: the problem is that a rank-two discrete valuation of the completion of α can restrict to a rank-one discrete valuation of α with the same proximity matrix, cf. [**14**, p. 118, Example 3.5].

[8]Spivakovsky represents proximity relations via dual graphs of closed fibers on smooth birational Spec(α)-schemes. (See also [**3**] for this technique.) An equivalent, but more direct and very effective graphical representation was invented by Enriques: it is the "Enriques diagram," a structured version of $(5.1)_v$ obtained by grouping maximal sets of points proximate

(5.2). So let us indicate how, in analogy with the classical theory of plane curve singularities, we can use $\mathbf{P}_v$ to analyze v more closely.

EXAMPLE. Suppose that α is complete (or at least henselian). Let f_0 be an irreducible element in $\alpha_0 := \alpha$; and having inductively defined $f_i \in \alpha_i$, let α_{i+1} be the unique quadratic transform of α_i in which the proper transform f_{i+1} of f_i is a non-unit. The α_i are "the points infinitely near to α lying on f."

This sequence of points determines a rank-two composed valuation $v = v_1 \circ v_2$, where the valuation ring of v_1 is the integral closure of α/f_0. The *singularity invariants* of f_0 (multiplicity sequence, value semigroup, ...) are all calculable from $\mathbf{P}_v$. Here $\mathbf{P}_v$ carries only a finite amount of information, because for large i, f_i is of order one in α_i, $[\alpha_{i+1} : \alpha_i] = 1$, and there is just one α_j (namely α_{i+1}) proximate to α_i.

Now consider an arbitrary zero-dimensional valuation v, let $(5.1)_v$ be the corresponding quadratic sequence, and set

$$\mathfrak{m}_i := \mathfrak{m}_{\alpha_i}, \qquad e_i := [\alpha_i : \alpha].$$

Consider also the sequence of v-ideals defined inductively by

$$J_0 = \alpha, \quad J_{i+1} = \{\, x \in \alpha \mid v(x) > v(J_i) \,\}.$$

These J_i are all the finite-colength v-ideals. We have then four sequences:

(1) The *multiplicity sequence* $(v(\mathfrak{m}_0)/v(\mathfrak{m}_i), e_i)_{1 \le i < \infty}$.
[The quotients $(v(\mathfrak{m}_0)/v(\mathfrak{m}_i))$ are rational numbers in the interval $[1, \infty)$.]

(2) The *semigroup-length sequence* $(v(J_i)/v(\mathfrak{m}_0), \lambda_\alpha(\alpha/J_i))_{1 \le i < \infty}$.
[The quotients $(v(J_i)/v(\mathfrak{m}_0))$ are rational numbers in the interval $[1, \infty)$.]

(3) The *point basis sequence* $\mathbf{B}(J_i)_{1 \le i < \infty}$.
[The base points of any J_i are among the α_j, so that $\mathbf{B}(J_i)$ may be represented in the form $(b_{ij})_{j \ge 0}$.]

(4) The *factorization sequence* $\mathbf{F}(J_i)_{1 \le i < \infty}$.
[The simple complete ideals dividing any J_i are among the $\wp_j$ corresponding to the points α_j in $(5.1)_v$—and for each $j \ge 0$, $\wp_j$ *is* a v-ideal—so that $\mathbf{F}(J_i)$ may be represented in the form $(a_{ij})_{j \ge 0}$.]

THEOREM. *The proximity matrix $\mathbf{P}_v$ determines each one of the preceding four sequences, and vice-versa.*

The *proof* begins with the observation that v can be replaced by its "approximations" $\operatorname{ord}(\alpha_i)$ $(i \to \infty)$. The analysis for $\operatorname{ord}(\alpha_i)$ is in many respects closely related to that of the singularity at the origin of the "plane curve" $f_i = 0$, where f_i is a sufficiently general element of the corresponding simple complete ideal $\wp_i$. (For some results along these lines, cf. [**11**] and [**14**].)

Details may appear elsewhere.

---

to the same one (when those sets contain more than one member) successively along alternating horizontal and vertical lines, cf. [**4**, Chap. 1, §8, pp. 374–381].

Remark (added in proof). Here is another more geometric way of looking at the proximity inequalities of Theorem (2.1).

Let I be a finite-colength α-ideal, and let $f\colon X \to \operatorname{Spec}(\alpha)$ be a proper birational map such that X is non-singular and $\mathcal{I} := I\mathcal{O}_X$ is invertible. Then each irreducible component E of the closed fibre $f^{-1}(\mathfrak{m}_\alpha)$ is a projective line over the field $\alpha/\mathfrak{m}_\alpha$, and the restriction $\mathcal{I}|_E$ is an invertible $\mathcal{O}_E$-module generated by global sections; so the intersection number $(\mathcal{I}\cdot E)$ (i.e., the degree of $\mathcal{I}|_E$) is ≥ 0. But *this intersection number is precisely* $[\beta : \alpha](r_\beta - \sum_{\gamma \succ \beta}[\gamma : \beta]r_\gamma)$ where $\beta \supset \alpha$ is the unique point such that the valuation ring of ord_β is the local ring of the generic point of E on X. The proof, suitably generalized, leads further to higher-dimensional proximity inequalities, cf. [**16**, p. 988, Cor. 4].

The converse ("if") part of Theorem (2.1) also results from the preceding interpretation, basically because any invertible $\mathcal{O}_X$-module $\mathcal{J}$ with $(\mathcal{J}\cdot E) \geq 0$ for all E must be of the form $J\mathcal{O}_X$ for some α-ideal J [**7**, p. 210, Thm. (12.1)].

References

1. S. S. Abhyankar, *On the valuations centered in a local domain*, Amer. J. Math **78** (1956), 321–348.
2. N. Bourbaki, *Algèbre Commutative, chap. 5-6*, Act. Sci. et Ind., no. 1308, Hermann, Paris, 1964.
3. P. Du Val, *On isolated singularities of surfaces which do not affect the conditions of adjunction*, Proc. Cambridge Phil. Soc. **30** (1934), 453–459.
4. F. Enriques and O. Chisini, *Lezioni sulla teoria geometrica delle equazioni e delle funzioni algebriche, vol. II*, N. Zanichelli, Bologna, 1918.
5. M. A. Hoskin, *Zero-dimensional valuation ideals associated with plane curve branches*, Proc. London Phil. Soc. (3) **6** (1956), 70–99.
6. C. Huneke, *Complete ideals in two-dimensional regular local rings*, Commutative Algebra, Proceedings of a Microprogram held June 15–July 2, 1987, Springer-Verlag, New York, 1989, pp. 325–337.
7. J. Lipman, *Rational singularities, with applications to algebraic surfaces and unique factorization*, Publ. Math. Inst. Hautes Études Sci. **36** (1969), 195–279.
8. ———, *Appendix to Chapter II*, Algebraic Surfaces (2nd edition), by O. Zariski, Springer-Verlag, New York, 1971, pp. 45–50.
9. ———, *On complete ideals in regular local rings*, Algebraic Geometry and Commutative Algebra, vol. I, in honor of Masayoshi Nagata, Kinokuniya, Tokyo, 1988, pp. 203–231.
10. ———, *Adjoints and polars of simple complete ideals in two-dimensional regular local rings*, Algebra and Algebraic Geometry, Tenerife, June 1992, Proceedings (to appear).
11. S. Noh, *Sequence of valuation ideals of prime divisors of the second kind in 2-dimensional regular local rings*, J. Algebra (to appear).
12. S. Noh, *Adjacent integrally closed ideals in dimension two*, J. Pure and Applied Algebra (to appear).
13. D. G. Northcott, *Abstract dilatations and infinitely near points*, Proc. Cambridge Phil. Soc. **52**, Part 2 (1956), 176–197.
14. M. Spivakovsky, *Valuations in function fields of surfaces*, Amer. J. Math **112** (1990), 107–156.
15. O. Zariski and P. Samuel, *Commutative Algebra, vol. 2*, D. van Nostrand, Princeton, 1960.
16. A. Campillo, G. Gonzalez-Sprinberg, and M. Lejeune-Jalbert, *Amas, idéaux à support fini et chaînes toriques*, C. R. Acad. Sci. Paris, Série I, **315** (1992), 987–990.

Department of Mathematics, Purdue University, W. Lafayette, IN 47907, USA

E-mail address: lipman@math.purdue.edu

Contemporary Mathematics
Volume **159**, 1994

Cohomological annihilators and Castelnuovo-Mumford regularity

UWE NAGEL AND PETER SCHENZEL

ABSTRACT. For a finitely generated module M over a local Noetherian ring $(R, \mathfrak{m})$ we relate the cohomological annihilators of M to those of M modulo a bunch of hypersurface sections. These results are used in order to get bounds on the Castelnuovo-Mumford regularity for locally Cohen-Macaulay schemes. This implies contributions to the Eisenbud-Gôto conjecture, in particular for 1-Buchsbaum schemes.

1. Introduction

Castelnuovo-Mumford regularity was first defined by D. Mumford [19], who attributes the idea to G. Castelnuovo. Let $X \subset \mathbb{P}^n_K$ be a projective scheme over an algebraically closed field K with ideal sheaf $\mathcal{I}_X$. Then X is said to be m-regular if $H^i(\mathbb{P}^n_K, \mathcal{I}_X(m-i)) = 0$ for all $i \geq 1$. The Castelnuovo-Mumford regularity $\operatorname{reg}(X)$ of X is the least such integer m. The interest in an upper bound for $\operatorname{reg}(X)$ stems partly from the fact that X is m-regular iff for every $j \geq 0$ the minimal generators of the j-th syzygy module of the saturated defining ideal $\mathcal{I}(X)$ of X have degree $\leq m - j$ (cf. [6]). Moreover, $\operatorname{reg}(X)$ provides an upper bound on the complexity of algorithms computing all syzygies of $\mathcal{I}(X)$ (cf. [2]).

Recently upper bounds for the Castelnuovo-Mumford regularity of locally Cohen-Macaulay schemes have attracted some attention (cf., e. g., [5], [14], [18], [23], [32], [34]). This latter notion has been refined by introducing k-Buchsbaum schemes. A projective scheme $X \subset \mathbb{P}^n_K$ is said to be k-Buchsbaum if $\oplus_{j \in \mathbb{Z}} H^i(\mathbb{P}^n_K, \mathcal{I}_X(j))$ is annihilated by $(x_0, \ldots, x_n)^k$ for all i with $1 \leq i \leq$

1991 *Mathematics Subject Classification.* 14 B 15; Secondary 13 H 10, 14 M 05.
Key words and phrases. local cohomology, cohomological annihilator, Castelnuovo-Mumford regularity, Cohen-Macaulay scheme.
The first author was supported by the DFG during the final preparation of the paper. The second author is grateful to the NSF for a support in order to participate at this meeting.
This paper is in final form and no version of it will be submitted elsewhere.

$\dim X$. For example, if X is arithmetically Buchsbaum then X and all its successive hyperplane sections are 1-Buchsbaum. If we assume only X to be 1-Buchsbaum then a general hyperplane section $X \cap H$ does not need to be 1-Buchsbaum. Note that any locally Cohen-Macaulay scheme is k-Buchsbaum for some k. Moreover, X is arithmetically Cohen-Macaulay if and only if it is a 0-Buchsbaum scheme. A 1-Buchsbaum scheme is often called quasi-Buchsbaum.

Let $X \subset \mathbb{P}^n_K$ be an irreducible, reduced and nondegenerate scheme of dimension d which is k-Buchsbaum for some $k \geq 1$. Then there are bounds (cf., e. g., [31], [13])

$$\operatorname{reg}(X) \leq \left\lceil \frac{\deg(X) - 1}{\operatorname{codim}(X)} \right\rceil + 1 - d + C(k),$$

where $C = C(k)$ is a certain integer depending on k. Here $\lceil a \rceil$ denotes the least integer $\geq a$ for $a \in \mathbb{R}$. Of course the question is how large C has to be. From [31], Theorem 1, it follows that we may take $C = (2^d - 1)k + d$. This was improved by [13], Theorem 3.1, allowing $C = (2^d - 1)k$. A big progress was achieved in [14], Theorem 3.4, implying that the choice of $C = \binom{d+1}{2}k$ is possible. One goal of the present paper is to show a further improvement, namely that the above estimation remains true if we put $C = (2d - 1)k$. This provides for example:

COROLLARY 1.1. *Let X be an irreducible, reduced and nondegenerate subvariety of $\mathbb{P}^n_K$ which is 1-Buchsbaum. Then we have*

$$\operatorname{reg}(X) \leq \left\lceil \frac{\deg(X) - 1}{\operatorname{codim}(X)} \right\rceil + \dim(X).$$

In particular, if $\deg(X) \geq n + \dim(X) - 1$ and $\operatorname{codim} X \geq 2$ we get

$$\operatorname{reg}(X) \leq \left\lceil \frac{\deg(X) - 1}{\operatorname{codim}(X)} \right\rceil + \dim(X) \leq \deg(X) - \operatorname{codim}(X) + 1.$$

Note that a conjecture of D. Eisenbud and S. Gôto (see [6]) states that

$$\operatorname{reg}(X) \leq \deg(X) - \operatorname{codim}(X) + 1$$

for an arbitrary irreducible, reduced and nondegenerate subvariety of $\mathbb{P}^n_K$. Hence our results prove this conjecture for most of the 1-Buchsbaum subvarieties. Moreover, we get often better bounds than expected by the Eisenbud conjecture.

The key for our results is the careful study of the cohomological annihilators of general hypersurface sections $X \cap F_1 \cap \cdots \cap F_r$ where $1 \leq r \leq \dim(X)$. Here it is essential that we consider not only hypersurfaces of degree 1 as usual (cf., e. g., [3], [4], [24]) but also hypersurfaces F_j of degree > 1. After some preliminaries we begin these investigations in Section 3. The local ring at the vertex of the affine cone of a projective variety X covers a lot of properties of X. So we work there in the more general framework of a finitely generated module M over a Noetherian local ring $(R, \mathfrak{m})$. Using spectral sequences we relate the cohomological annihilators $\mathfrak{a}_i(M) = \operatorname{Ann}_R H^i_{\mathfrak{m}}(M)$ of M to those of $M/\underline{x}M$ where $\underline{x} = \{x_1, \ldots, x_r\}$ is an M-filter regular sequence and $1 \leq r \leq \dim M$.

According to our applications we are interested in showing that we may choose $\underline{x}$ in such a way that $\mathfrak{a}_i(M/\underline{x}M)$ becomes large. This will be done in Section 4 where the crucial concept of an (r,i)-standard sequence is introduced. This is a generalization of the well-known standard systems of parameters (cf., e.g., [28], [33]). In Section 5 we restrict ourselves to modules with finite local cohomology. In particular we derive a new characterization of k-Buchsbaum modules with the help of standard systems of parameters. Finally, in Section 6 we apply the previous results in order to obtain the upper bounds for the Castelnuovo-Mumford regularity of projective k-Buchsbaum schemes.

In [13] and [14] also (k,s)-Buchsbaum schemes were considered. This concept requires not only information on the cohomological annihilators of the scheme itself but also of its successive hypersurface sections. Having this in mind we think that it is clear how our results specialize to results on (k,s)-Buchsbaum schemes and modules. Hence we state them only for k-Buchsbaum schemes with one exception in Section 5 in order to allow a comparison with the main result of [14].

2. Notations and Preliminary Results

In the sequel we have to use a certain generalization of the notion of a regular sequence. To this end let us summarize basic facts about filter regular sequences introduced in [29]. Let M denote a finitely generated R-module over $(R,\mathfrak{m})$, a local Noetherian ring.

DEFINITION 2.1. A system of elements $\{x_1,\dots,x_r\} \subseteq \mathfrak{m}$ is called a filter regular sequence of M (or M-filter regular sequence), if

$$x_i \notin \mathfrak{p} \quad \text{for all } \mathfrak{p} \in (\operatorname{Ass} M/(x_1,\dots,x_{i-1})M) \setminus \{\mathfrak{m}\}$$

for all $i = 1,\dots,r$.

This is equivalent to saying that the R-modules

$$(x_1,\dots,x_{i-1})M : x_i/(x_1,\dots,x_{i-1})M, \quad i = 1,\dots,r,$$

are of finite length. In particular, an M-regular sequence is an M-filter regular sequence. For an ideal $\mathfrak{a}$ of R and a submodule N of M set $N :_M \langle \mathfrak{a} \rangle = \{m \in M : \mathfrak{a}^i m \subseteq N \text{ for some } i \geq 1\}$. Then $\{x_1,\dots,x_r\} \subseteq \mathfrak{m}$ is an M-filter regular sequence if and only if $(x_1,\dots,x_{i-1})M : x_i \subseteq (x_1,\dots,x_{i-1})M : \langle \mathfrak{m} \rangle$ for all $i = 1,\dots,r$.

PROPOSITION 2.2. *Let $\underline{x} = \{x_1,\dots,x_r\}$ denote a system of non-units. Then the following conditions are equivalent:*

(i) *$\underline{x}$ is an M-filter regular sequence.*
(ii) *$\{\frac{x_1}{1},\dots,\frac{x_i}{1}\} \in R_\mathfrak{p}$ is an $M_\mathfrak{p}$-regular sequence for all $\mathfrak{p} \in (V(x_1,\dots,x_i) \cap \operatorname{Supp} M) \setminus \{\mathfrak{m}\}$ and $i = 1,\dots,r$.*
(iii) *$\{x_1^{k_1},\dots,x_r^{k_r}\}$ is M-filter regular for some r-tuple $(k_1,\dots,k_r) \in \mathbb{N}_+^r$.*
(iv) *$\{x_1^{k_1},\dots,x_r^{k_r}\}$ is M-filter regular for every r-tuple $(k_1,\dots,k_r) \in \mathbb{N}_+^r$.*

PROOF. The equivalence of the first two conditions is an easy consequence of the fact that (i) is equivalent to the finite length of the finitely generated R-module $(x_1, \dots, x_{i-1})M : x_i/(x_1, \dots, x_{i-1})M$, $i = 1, \dots, r$. The equivalence of the first with the third (resp. fourth) condition is now a consequence of elementary properties of regular sequences, see e.g. [16], Exercise 12, p. 102. □

While regular sequences are stable under permutations this is no longer true for M-filter regular sequences. Put $R = K[x,y,z]_{(x,y,z)}/(x) \cap (x^2, y)$, K a field. Then $\{z, y\}$ is R-filter regular, while this does not hold for $\{y, z\}$. In [29] it is shown that any system of parameters of M is an M-filter regular sequence if and only if depth $M_{\mathfrak{p}} + \dim R/\mathfrak{p} = \dim M$ for all $\mathfrak{p} \in \operatorname{Supp} M \setminus \{\mathfrak{m}\}$.

COROLLARY 2.3. *Let $\{x_1, \dots, x_r\}$, $\{y_1, \dots, y_r\}$ be two systems of non-units of R. Suppose that $\{x_1y_1, \dots, x_ry_r\}$ is an M-filter regular sequence. Then this is true for $\{x_1, \dots, x_r\}$ and $\{y_1, \dots, y_r\}$.*

PROOF. It is enough to show that $\{x_1, \dots, x_r\}$ is M-filter regular. Assume the contrary. Then there exists a non-maximal prime ideal

$$\mathfrak{p} \in \operatorname{Ass}((x_1, \dots, x_{i-1})M : x_i/(x_1, \dots, x_{i-1})M),$$

i.e., $\mathfrak{p} \in \operatorname{Supp} M/(x_1, \dots, x_i)M$. Therefore

$$(x_1, \dots, x_i)M_{\mathfrak{p}} \neq M_{\mathfrak{p}} \text{ and } (x_1y_1, \dots, x_iy_i)M_{\mathfrak{p}} \neq M_{\mathfrak{p}}.$$

By (2.2) it is known that $\{\frac{x_1y_1}{1}, \dots, \frac{x_iy_i}{1}\} \subseteq \mathfrak{p}R_{\mathfrak{p}}$ forms an $M_{\mathfrak{p}}$-regular sequence. Then $\{\frac{x_1}{1}, \dots, \frac{x_i}{1}\}$ forms an $M_{\mathfrak{p}}$-regular sequence, see [16], Exercise 12, p. 102. But this contradicts the choice of $\mathfrak{p}$. □

If $\{x_1, \dots, x_r\}$ is an M-filter regular sequence, then $\{x_i, \dots, x_r\}$ forms a filter regular sequence of $M/(x_1, \dots, x_{i-1})M$ for $i = 1, \dots, r$. For our purposes here we need an observation on the Koszul cohomology. For a system of elements $\{x_1, \dots, x_r\} = \underline{x}$ let $H^i(\underline{x}; M)$, $i \in \mathbb{Z}$, denote the Koszul cohomology of M with respect to $\underline{x}$, see [10] for the definitions.

COROLLARY 2.4. *If the system $\underline{x} = \{x_1, \dots .x_r\} \subseteq \mathfrak{m}$ is an M-filter regular sequence, then $H^i(\underline{x}; M)$ is an R-module of finite length for $i = r - 1$ (resp. all $i < r$).*

PROOF. First note that

$$\operatorname{Supp} H^i(\underline{x}; M) \subseteq V(\underline{x}) \cap \operatorname{Supp} M, \quad i \in \mathbb{Z}.$$

On the other hand $\underline{x}$ is an M-regular sequence if and only if $H^i(\underline{x}; M) = 0$ for $i = r - 1$ (resp. all $i < r$) see [10]. Then the result follows by Proposition 2.2 and the localization of Koszul complexes. □

By the above example it follows that the converse of Corollary 2.4 does not hold. In algebraic geometry the notion of filter regular sequences is closely related to generic chosen hypersurfaces. Let $R = \oplus_{i \geq 0} R_i$ denote a non-negatively graded Noetherian ring with $(R_0, \mathfrak{n})$ a local ring. By [12], (22.4), a sequence of homogeneous elements $\underline{x} = \{x_1, \ldots, x_r\}$ is called M-weakly regular, M a finitely generated graded R-module, if

$$[(x_1, \ldots, x_{i-1})M : x_i]_n = [(x_1, \ldots, x_{i-1})M]_n, \quad \text{for all} \quad n \gg 0$$

and $i = 1, \ldots, r,$. (For a graded R-module N the R_0-module $[N]_n$, $n \in \mathbb{Z}$, denotes the n-th graded component.) This is equivalent to saying that

$$x_i \notin \mathfrak{p} \quad \text{for all } \mathfrak{p} \in (\operatorname{Ass}_R M/(x_1, \ldots, x_{i-1})M) \setminus V(R_+)$$

for $i = 1, \ldots, r$, where $R_+ = \oplus_{i>0} R_i$ denotes the irrelevant ideal of R. In the case of R_0 an Artinian local ring this coincides with our definition of M-filter regular sequences.

Moreover, for a graded R-module M and $k \in \mathbb{Z}$ let $M(k)$ denote the R-module M whose grading is given by $[M(k)]_i = [M]_{k+i}$, $i \in \mathbb{Z}$.

Now let us return to the situation of a local ring $(R, \mathfrak{m})$. For an ideal $\mathfrak{a}$ of R the local cohomology modules $H^i_{\mathfrak{a}}(M)$, $i \in \mathbb{Z}$, are defined as in [10]. There is a functorial isomorphism

$$\varinjlim H^i(\underline{x}^{(k)}; M) \simeq H^i_{\mathfrak{a}}(M), \quad i \in \mathbb{Z},$$

where $\mathfrak{a} = (x_1, \ldots, x_r)R$ and $\underline{x}^{(k)} = \{x_1^k, \ldots, x_r^k\}$. For the definition of the Čech complex $K^{\bullet}_{\underline{x}}$ of $\underline{x} = \{x_1, \ldots, x_r\}$, see [10]. There are functorial isomorphisms

$$H^i(K^{\bullet}_{\underline{x}} \otimes_R M) \simeq H^i_{\mathfrak{a}}(M), \quad i \in \mathbb{Z},$$

see [10] for the details.

PROPOSITION 2.5. *Let $\underline{x} = \{x_1, \ldots, x_r\}$ be an M-filter regular sequence. Then*

$$\operatorname{Supp} H^i_{\mathfrak{a}}(M) \subseteq V(\mathfrak{m}),$$

for all $0 \leq i < r$, where $\mathfrak{a} = (x_1, \ldots, x_r)R$.

PROOF. It is obviously true that $\operatorname{Supp} H^i_{\mathfrak{a}}(M) \subseteq V(\underline{x}) \cap \operatorname{Supp} M$. Let $\mathfrak{p}$ be a non-maximal prime ideal $\mathfrak{p} \in V(\underline{x}) \cap \operatorname{Supp} M$. By a localization argument it follows that

$$H^i_{\mathfrak{a}}(M) \otimes_R R_{\mathfrak{p}} \simeq H^i_{\mathfrak{a}R_{\mathfrak{p}}}(M_{\mathfrak{p}}) = 0 \quad \text{for } i < r,$$

since $\{\frac{x_1}{1}, \ldots, \frac{x_r}{1}\}$ is an $M_{\mathfrak{p}}$-regular sequence. $\square$

For some applications we need the technique of spectral sequences. To this end we refer to the presentation in [17], Theorem 4. Let $K^\bullet$, $L^\bullet$ be two complexes of R-modules. Put $C^\bullet = K^\bullet \otimes L^\bullet$ the tensor product of these complexes. For computing the hypercohomology of $C^\bullet$ there are two spectral sequences

$$\begin{array}{rcll} E_2^{ij} & = & H^i(K^\bullet \otimes H^j(L^\bullet)) & \Rightarrow \quad E^{i+j} = H^{i+j}(C^\bullet) \quad \text{and} \\ 'E_2^{ij} & = & H^j(H^i(K^\bullet) \otimes L^\bullet) & \Rightarrow \quad 'E^{i+j} = H^{i+j}(C^\bullet), \end{array}$$

see [17]. For two ideals $\mathfrak{a}, \mathfrak{b}$ of R there is – as a particular case of the Grothendieck spectral sequence for the composite of two functors – the spectral sequence

$$E_2^{ij} = H^i_{\mathfrak{a}}(H^j_{\mathfrak{b}}(M)) \Rightarrow E^{i+j} = H^{i+j}_{(\mathfrak{a},\mathfrak{b})}(M),$$

functorial on the category of R-modules, see [17].

In the case of R a graded ring and M a graded R-module the local cohomology modules admit a natural grading. For further results in this situation we refer to [8].

3. Cohomological Annihilators

Let M denote a finitely generated R-module over a local ring $(R, \mathfrak{m})$. The vanishing resp. non-vanishing of the local cohomology modules $H^i_{\mathfrak{m}}(M)$ provides useful local information on M, see [10]. For a more subtle consideration the annihilators of $H^i_{\mathfrak{m}}(M)$ were considered by P. Roberts in [25], see also [26].

DEFINITION 3.1. For a finitely generated R-module M let

$$\mathfrak{a}_i(M) := \operatorname{Ann}_R H^i_{\mathfrak{m}}(M), \quad i \in \mathbb{Z},$$

denote the i-th cohomological annihilator of M.

For a summary of basic properties of $\mathfrak{a}_i(M)$, see [26]. It is a well-known fact that $\prod_{i=0}^{d-1} \mathfrak{a}_i(M), d = \dim_R M$, is an $\mathfrak{m}$-primary ideal if and only if $H^i_{\mathfrak{m}}(M), i \neq d$, are finitely generated R-modules. The study of this class of generalized Cohen-Macaulay modules was initiated in [27]. In particular, if M is a Buchsbaum R-module, then $\mathfrak{m}H^i_{\mathfrak{m}}(M) = 0$, for all $i \neq d$, i.e., $H^i_{\mathfrak{m}}(M)$ is a finite dimensional $R/\mathfrak{m}$-vector space. More generally we define:

DEFINITION 3.2. Let M be a finitely generated R-module such that $H^i_{\mathfrak{m}}(M)$ is of finite length for all $i \neq d$. Set

$$\lambda_i := \lambda_i(M) := \min\{\lambda \in \mathbb{N} : \mathfrak{m}^\lambda \subseteq \mathfrak{a}_i(M)\}$$

for $0 \leq i < d$. Moreover, M is said to be a k-Buchsbaum module, if $k \geq \lambda_i(M)$ for all $i \neq d$.

It is known, see [30], that a Buchsbaum module is 1-Buchsbaum, while the converse does not hold in general, see e.g. [7], where the ubiquity of non-Buchsbaum but 1-Buchsbaum rings is established. In [18], Corollary 2.10, some example of curves are given for which the integer k such that the curves are k-Buchsbaum but not $(k-1)$-Buchsbaum is a priori known.

Now we relate the cohomological annihilators of M to those of M modulo a bunch of generic hyperplane sections.

THEOREM 3.3. *Let $\underline{x} = \{x_1, \dots, x_r\}$ denote an M-filter regular sequence. Then*

$$\mathfrak{a}_i(M) \cdot \ldots \cdot \mathfrak{a}_{i+r}(M) \subseteq \mathfrak{a}_i(M/\underline{x}M)$$

for all integers i.

PROOF. Let $K^\bullet := K^\bullet_{\underline{y}}$ denote the Čech complex of R for a system of parameters $\underline{y} = \{y_1, \dots, y_d\}$, $d = \dim_R M$, of M as considered in Section 2. Let $K^\bullet(\underline{x}; R) \otimes_R M$ be the Koszul co-complex of M with respect to $\underline{x}$. Put

$$C^\bullet := (K^\bullet \otimes_R M) \otimes_R K^\bullet(\underline{x}; R) \simeq K^\bullet \otimes_R K^\bullet(\underline{x}; M).$$

As mentioned in Section 2 there are two spectral sequences for computing the cohomology of $C^\bullet$. First consider

$$E_2^{ij} = H^i(K^\bullet \otimes_R H^j(\underline{x}; M)) \Rightarrow E^{i+j} = H^{i+j}(C^\bullet).$$

Note that $H^i(K^\bullet \otimes_R N) \simeq H^i_{\mathfrak{m}}(N)$, $i \in \mathbb{Z}$, for a finitely generated R-module N. Therefore

$$E_2^{ij} \simeq H^i_{\mathfrak{m}}(H^j(\underline{x}; M)) \quad \text{for all } i, j \in \mathbb{N}.$$

By Corollary 2.3 the R-modules $H^i(\underline{x}; M)$, $i < r$, are of finite length. So there are the following isomorphisms

$$E_2^{ij} \simeq \begin{cases} 0 & \text{for } i \neq 0 \quad \text{and } j \neq r, \\ H^j(\underline{x}; M) & \text{for } i = 0 \quad \text{and } j \neq r, \\ H^i_{\mathfrak{m}}(M/\underline{x}M) & \text{for } j = r. \end{cases}$$

To this end recall that $H^r(\underline{x}; M) \simeq M/\underline{x}M$. By virtue of the spectral sequence it turns out that

$$E_\infty^{ij} = 0 \quad \text{for all } i \neq 0, \quad j \neq r.$$

Because of the subsequent stages of the spectral sequence

$$E_k^{i-k,r+k-1} \to E_k^{ir} \to E_k^{i+k,r-k+1}$$

and $E_k^{i-k,r+k-1} = E_k^{i+k,r-k+1} = 0$ for all $k \geq 2$ it yields that $E_\infty^{ir} \simeq H^i_{\mathfrak{m}}(M/\underline{x}M)$. By a similar consideration we obtain that $E_\infty^{0j} \simeq H^j(\underline{x}; M)$ for all $j \neq r$. Therefore it yields the following isomorphisms

$$H^i(C^\bullet) \simeq \begin{cases} H^i(\underline{x}; M) & \text{for } 0 \leq i < r, \\ H^{i-r}_{\mathfrak{m}}(M/\underline{x}M) & \text{for } r \leq i \leq d, \\ 0 & \text{otherwise.} \end{cases}$$

On the other hand there is the spectral sequence

$$'E_2^{ij} = H^j(K^\bullet(\underline{x}; K) \otimes_R H^i_{\mathfrak{m}}(M)) \Rightarrow {'E}^{i+j} = H^{i+j}(C^\bullet).$$

Because of $'E_2^{ij} = H^j(\underline{x}; H^i_{\mathfrak{m}}(M))$ it follows that $'E_2^{ij} = 0$ for all $j < 0$ and $j > r$. By the construction of the Koszul complex $'E_2^{ij}$ is a subquotient of the direct sum of $\binom{r}{j}$ copies of $H^i_{\mathfrak{m}}(M)$. Therefore $\mathfrak{a}_i(M)({}'E_2^{ij}) = 0$ for all $i, j \in \mathbb{N}$. Whence it implies that $\mathfrak{a}_i(M)({}'E_\infty^{ij}) = 0$ for all $i, j \in \mathbb{N}$. By view of the filtration of $H^{i+j}(C^\bullet)$ defined by ${}'E_\infty^{ij}$ it follows that

$$\begin{aligned} \mathfrak{a}_0(M) \cdot \ldots \cdot \mathfrak{a}_i(M) H^i(C^\bullet) &= 0 \quad \text{for } 0 \leq i < r \text{ and} \\ \mathfrak{a}_{i-r}(M) \cdot \ldots \cdot \mathfrak{a}_i(M) H^i(C^\bullet) &= 0 \quad \text{for } r \leq i \leq d. \end{aligned}$$

Hence, the above computation of $H^i(C^\bullet)$ proves the claim. □

For a filter regular sequence $\underline{x} = \{x_1, \ldots, x_r\}$ the proof of Theorem 3.3 provides that

$$\mathfrak{a}_0(M) \cdot \ldots \cdot \mathfrak{a}_i(M) H^i(\underline{x}_i M) = 0 \quad \text{for all } i < r.$$

Because of the finite length of $H^i(\underline{x}; M)$ for all $i < r$ this is a particular case of the results shown in [26]. Moreover the notion of M-filter regular sequences provides an interesting expression of the local cohomology modules of M.

LEMMA 3.4. *Let $\underline{x} = \{x_1, \ldots, x_r\}$ be M-filter regular. Put $\mathfrak{a} = (x_1, \ldots, x_r)R$. Then there are the following isomorphisms*

$$H^i_{\mathfrak{m}}(M) \simeq \begin{cases} H^i_{\mathfrak{a}}(M) & \text{for } 0 \leq i < r, \\ H^{i-r}_{\mathfrak{m}}(H^r_{\mathfrak{a}}(M)) & \text{for } r \leq i \leq d, \end{cases}$$

where $d = \dim_R M$.

PROOF. Consider the spectral sequence

$$E_2^{ij} = H^i_{\mathfrak{m}}(H^j_{\mathfrak{a}}(M)) \Rightarrow E^{i+j} = H^{i+j}_{\mathfrak{m}}(M).$$

By Proposition 2.5 we have that $\operatorname{Supp} H^j_{\mathfrak{a}}(M) \subseteq V(\mathfrak{m})$ for all $j < r$. Whence $E_2^{ij} = 0$ for all $i \neq 0$ and $j \neq r$. Furthermore, $E_2^{0j} = H^j_{\mathfrak{a}}(M)$ for $j \neq r$ and $E_2^{ir} = H^i_{\mathfrak{m}}(H^r_{\mathfrak{a}}(M))$. An argument similar to that of the proof given in Theorem 3.3 yields that

$$E_\infty^{0j} \simeq H^j_{\mathfrak{a}}(M) \quad \text{and} \quad E_\infty^{ir} \simeq H^i_{\mathfrak{m}}(H^r_{\mathfrak{a}}(M)).$$

Because of $E_\infty^{0j} = 0$ for $j > r$ the spectral sequence proves the claim. □

Let $\underline{x} = \{x_1, \ldots, x_r\}$ be a system of elements of R. For the following results put $\underline{x}^{(k)} = \{x_1^k, \ldots, x_r^k\}$ for an integer $k \in \mathbb{N}_+$.

COROLLARY 3.5. *Let $\underline{x} = \{x_1, \ldots, x_r\}$ be an M-filter regular sequence. The multiplication by $x_1 \cdots x_r$ induces a direct system $\{H^i_{\mathfrak{m}}(M/\underline{x}^{(k)}M)\}_{k \in \mathbb{N}_+}$, such that*

$$H^{i+r}_{\mathfrak{m}}(M) \simeq \varinjlim H^i_{\mathfrak{m}}(M/\underline{x}^{(k)}M)$$

for all $i \geq 0$.

PROOF. There is a direct system $\{M/\underline{x}^{(k)}M\}_{k\in IN_+}$ with homomorphisms induced by the multiplication by $x_1\cdots x_r$. By [10] there is an isomorphism

$$H^r_{\mathfrak{a}}(M) \simeq \varinjlim M/\underline{x}^{(k)}M.$$

Then the claim follows by Lemma 3.4 since the local cohomology commutes with direct limits. □

In order to produce an "upper" approximation of $\mathfrak{a}_i(M/\underline{x}M)$, $\underline{x} = \{x_1,\dots,x_r\}$ an M-filter regular sequence, a few preliminaries are necessary.

DEFINITION 3.6. Let $\underline{x} = \{x_1,\dots,x_r\}$ be an M-filter regular sequence. For $j = 0,1,\dots,r$ set

$$\mathfrak{a}_{ij}(\underline{x};M) = \bigcap_{k_1,\dots,k_j\geq 1} \mathfrak{a}_i(M/(x_1^{k_1},\dots,x_j^{k_j})M).$$

Furthermore define $\mathfrak{a}_i(\underline{x};M) = \bigcap_{j=0}^r \mathfrak{a}_{ij}(\underline{x};M)$.

The next results relates the cohomological annihilators of M to those of $M/\underline{x}M$. This extends results of [26], where a corresponding result for $H^0_{\mathfrak{m}}(M/\underline{x}M)$ is shown.

COROLLARY 3.7. *Let $\underline{x} = \{x_1,\dots,x_r\}$ be an M-filter regular sequence. Then*

$$\mathfrak{a}_i(M)\cdot\ldots\cdot\mathfrak{a}_{i+r}(M) \subseteq \mathfrak{a}_i(\underline{x};M) \subseteq \mathfrak{a}_i(M)\cap\ldots\cap\mathfrak{a}_{i+r}(M)$$

for all $0\leq i\leq d-r$. In particular, $\mathfrak{a}_i(\underline{x};M)$ and $\mathfrak{a}_i(M)\cap\ldots\cap\mathfrak{a}_{i+r}(M)$ have the same radical.

PROOF. By Theorem 3.3 it follows that $\mathfrak{a}_i(M)\cdot\ldots\cdot\mathfrak{a}_{i+j}(M) \subseteq \mathfrak{a}_{ij}(\underline{x};M)$ for $j = 0,1,\dots,r$. Recall that $\{x_1^{k_1},\dots,x_j^{k_j}\}$ forms an M-filter regular sequence. Whence the first inclusion is true. Moreover, by Corollary 3.5 it yields that

$$\mathfrak{a}_i(\underline{x};M) \subseteq \mathfrak{a}_{ij}(\underline{x};M) \subseteq \mathfrak{a}_{i+j}(M)$$

for all $j = 0,1,\dots,r$. This proves the second containment relation. □

4. (r,i)-Standard Sequences

Standard systems of parameters have been introduced by M. Brodmann, N. V. Trung and the second author. They play an important role in the theory of modules with finite cohomology (cf. [33], [28]). See Remark (i) in 4.4 for the precise definition. For our purposes we need to consider (r,i)-standard sequences, a generalization of standard system of parameters. We show sufficient conditions for an M-filter regular sequence to be an (r,i)-standard sequence. The main tool is a splitting result for local cohomology modules.

Let $x\in R$ denote an M-filter regular element. Then there is the following long exact cohomology sequence

$$0\to 0:_M x \to H^0_{\mathfrak{m}}(M) \xrightarrow{x} H^0_{\mathfrak{m}}(M) \to H^0_{\mathfrak{m}}(M/xM) \to H^1_{\mathfrak{m}}(M) \xrightarrow{x} H^1_{\mathfrak{m}}(M) \to \dots,$$

which is induced by multiplication by x (cf. [28]).

We begin with an easy observation which we will need in Section 6.

PROPOSITION 4.1. *Let $i \geq 0$ be an integer and let $\underline{x} = \{x_1, \dots, x_r\}$ be an M-filter regular sequence such that $x_n H^i_{\mathfrak{m}}(M/(x_1, \dots, x_{n-1})M) = 0$ for all $n = 1, \dots, r$. Then there is an embedding $H^i_{\mathfrak{m}}(M) \hookrightarrow H^i_{\mathfrak{m}}(M/\underline{x}M)$.*

PROOF. The exact sequence

$$\begin{aligned} H^i_{\mathfrak{m}}(M/(x_1, \dots, x_{n-1})M) &\xrightarrow{x_n} H^i_{\mathfrak{m}}(M/(x_1, \dots, x_{n-1})M) \\ &\rightarrow H^i_{\mathfrak{m}}(M/(x_1, \dots, x_n)M) \end{aligned}$$

and the assumption provides for all $n = 1, \dots, r$ an embedding

$$H^i_{\mathfrak{m}}(M/(x_1, \dots, x_{n-1})M) \hookrightarrow H^i_{\mathfrak{m}}(M/(x_1, \dots, x_n)M).$$

So the claim follows now by induction. □

By view of the definition of standard sequences this result suggests the following:

DEFINITION 4.2. Let $i \leq d - r$ denote an integer, where $d = \dim X$ and $1 \leq r \leq d$. An M-filter regular sequence $\underline{x} = \{x_1, \dots, x_r\} \subseteq \mathfrak{m}$ is said to be (r, i)-standard if and only if

$$x_{n+1} H^{i+j}_{\mathfrak{m}}(M/(x_1, \dots, x_n)M) = 0 \quad \text{for all } j \geq 0, \quad n \geq 0 \text{ with } j + n < r.$$

LEMMA 4.3. *An M-filter regular sequence $\underline{x} = \{x_1, \dots, x_r\}$ is (r, i)-standard if and only if*

$$\underline{x} H^{i+j}_{\mathfrak{m}}(M/(x_1, \dots, x_n)M) = 0 \quad \textit{for all } j \geq 0, \quad n \geq 0 \textit{ with } j + n < r.$$

PROOF. The "if-part" is trivial. For the other direction we induct on r. The case $r = 1$ is again trivial. Assume $r > 1$. Using the induction assumption it remains to show

$$x_r H^{i+j}_{\mathfrak{m}}(M/(x_1, \dots, x_n)M) = 0 \quad \text{if } j + n < r.$$

Now we induct on $r - n \geq 1$. If $n = r - 1$ it follows $j = 0$ and the assertion is clear from our assumption. Let $0 \leq n < r - 1$. Put $M' := M/(x_1, \dots, x_n)M$. By assumption we have for all j with $0 \leq j < r - (n + 1)$:

$$x_{n+1} H^{i+j}_{\mathfrak{m}}(M') = x_{n+1} H^{i+j+1}_{\mathfrak{m}}(M') = 0.$$

Hence the commutative diagram with exact rows and multiplication by x_r as vertical maps

$$\begin{array}{ccccccccc} 0 & \rightarrow & M'/(0 :_{M'} x_{n+1}) & \xrightarrow{x_{n+1}} & M' & \rightarrow & M'/x_{n+1}M & \rightarrow & 0 \\ & & \downarrow x_r & & \downarrow x_r & & \downarrow x_r & & \\ 0 & \rightarrow & M'/(0 :_{M'} x_{n+1}) & \xrightarrow{x_{n+1}} & M' & \rightarrow & M'/x_{n+1}M & \rightarrow & 0 \end{array}$$

implies the following commutative diagram with exact rows

$$\begin{array}{ccccccccc} 0 & \to & H^{i+j}_{\mathfrak{m}}(M') & \to & H^{i+j}_{\mathfrak{m}}(M'/x_{n+1}M') & \to & H^{i+j+1}_{\mathfrak{m}}(M') & \to & 0 \\ & & \downarrow x_r & & \downarrow x_r & & \downarrow x_r & & \\ 0 & \to & H^{i+j}_{\mathfrak{m}}(M') & \to & H^{i+j}_{\mathfrak{m}}(M'/x_{n+1}M') & \to & H^{i+j+1}_{\mathfrak{m}}(M') & \to & 0. \end{array}$$

By induction assumption on $r-n$ the middle vertical map is the zero map. Thus it follows $x_r H^{i+j}_{\mathfrak{m}}(M') = x_r H^{i+j+1}_{\mathfrak{m}}(M') = 0$ for all $j < r-(n+1)$. □

Remark 4.4. (i) The above lemma shows that $\underline{x} = \{x_1, \ldots, x_d\}$ is $(d,0)$-standard if and only if $\underline{x}$ is a standard sequence of parameters in the sense of M. Brodmann, N. V. Trung and the second author (cf. [33], Theorem 2.5).
(ii) If a module M possesses a standard system of parameters, then M has finite local cohomology. This is no longer true if we assume only the existence of an (r,i)-standard system where $r < d$. For example, let $(A, \mathfrak{n})$ be a local Buchsbaum ring of dimension 3 and depth 2. Then we consider the formal power series ring $R = A[[t]]$ over A with maximal ideal $\mathfrak{m}$. Then any A-regular subsystem of parameters $\{x_1, x_2\}$ is a $(2,2)$-standard sequence for R but $H^3_{\mathfrak{m}}(R)$ has infinite length.

We use the following notation. If m is a positive integer, then M^m denotes the direct sum of m copies of M. Our next result is a crucial point for the rest of the paper. It generalizes Theorem 3 (2) of [9] and Lemma 2.2 of [14]. In the proof we use an idea developed in step (a) of the proof of [30], Proposition I.2.1.

THEOREM 4.5. *Let $\underline{x} = \{x_1, \ldots, x_r\}$and $\underline{y} = \{y_1, \ldots, y_r\}$ be two systems of non-units of R. Suppose that $\{x_1y_1, \ldots, x_ry_r\}$ is an M-filter regular sequence and that*

$$\underline{x}H^{i+j}_{\mathfrak{m}}(M) = \underline{y}H^{i+j}_{\mathfrak{m}}(M) = 0 \quad \textit{for some } i < d-r \quad \textit{and for all } j = 0, \ldots, r.$$

Then

$$H^i_{\mathfrak{m}}(M/(x_1y_1, \ldots, x_ry_r)M) \simeq \bigoplus_{j=0}^{r} (H^{i+j}_{\mathfrak{m}}(M))^{\binom{r}{j}}.$$

PROOF. We use induction on r. Let $r = 1$. For short we put $x := x_1$, $y := y_1$. Consider the following commutative diagram with exact rows where the right vertical map is the canonical epimorphism:

$$\begin{array}{ccccccccc} 0 & \to & M/(0:_M xy) & \xrightarrow{xy} & M & \to & M/xyM & \to & 0 \\ & & \downarrow y & & \| & & \downarrow & & \\ 0 & \to & M/(0:_M x) & \longrightarrow & M & \to & M/xM & \to & 0. \end{array}$$

Due to Corollary 2.3 x is now M-filter regular. Hence, by the assumption this diagram implies the following commutative diagram with exact rows

$$\begin{array}{ccccccccc} 0 & \to & H^i_{\mathfrak{m}}(M) & \xrightarrow{\alpha} & H^i_{\mathfrak{m}}(M/xyM) & \to & H^{i+1}_{\mathfrak{m}}(M) & \to & 0 \\ & & \| & & \downarrow \gamma & & \downarrow y & & \\ 0 & \to & H^i_{\mathfrak{m}}(M) & \xrightarrow{\beta} & H^i_{\mathfrak{m}}(M/xM) & \xrightarrow{\delta} & H^{i+1}_{\mathfrak{m}}(M) & \to & 0. \end{array}$$

Since $yH^{i+1}_{\mathfrak{m}}(M) = 0$ by assumption the commutativity implies $\operatorname{im}\gamma \subset \ker\delta = \operatorname{im}\beta$. Thus we get a homomorphism $\epsilon := \beta^{-1}\circ\gamma$ whereby again by commutativity $\epsilon\circ\alpha = \operatorname{id}_{H^i_{\mathfrak{m}}(M)}$. Hence the top row in the above diagram splits and the assertion follows. Let $r > 1$. Then by the induction hypothesis we get

$$\begin{aligned} H^i_{\mathfrak{m}}(M/(x_1y_1,\dots,x_{r-1}y_{r-1})M) &\simeq \textstyle\bigoplus_{j=0}^{r-1}(H^{i+j}_{\mathfrak{m}}(M))^{\binom{r-1}{j}} \quad \text{and} \\ H^{i+1}_{\mathfrak{m}}(M/(x_1y_1,\dots,x_{r-1}y_{r-1})M) &\simeq \textstyle\bigoplus_{j=0}^{r-1}(H^{i+1+j}_{\mathfrak{m}}(M))^{\binom{r-1}{j}}. \end{aligned}$$

Since by assumption x_r and y_r annihilate the right-hand sides of the above isomorphisms and since x_ry_r is filter regular for $M/(x_1y_1,\dots,x_{r-1}y_{r-1})M$ we may apply the induction hypothesis and obtain

$$\begin{aligned} H^i_{\mathfrak{m}}(M/(x_1y_1,\dots,x_ry_r)M) \simeq\ & H^i_{\mathfrak{m}}(M/(x_1y_1,\dots,x_{r-1}y_{r-1})M) \ \oplus \\ & H^{i+1}_{\mathfrak{m}}(M/(x_1y_1,\dots,x_{r-1}y_{r-1})M). \end{aligned}$$

Then the above isomorphisms provide the assertion. □

We now state some consequences of the Theorem.

COROLLARY 4.6. *Let $\underline{x} = \{x_1,\dots,x_r\}$ and $\underline{y} = \{y_1,\dots,y_r\}$ be two systems of non-units of R such that $\{x_1y_1,\dots,x_ry_r\}$ is an M-filter regular sequence. Suppose that*

$$\underline{x}H^{i+j}_{\mathfrak{m}}(M) = \underline{y}H^{i+j}_{\mathfrak{m}}(M) = 0 \text{ for some } i \le d-r \text{and for all } j = 0,\dots,r-1.$$

Then $\{x_1y_1,\dots,x_{r-1}y_{r-1},x_r\}$ and $\{x_1y_1,\dots,x_{r-1}y_{r-1},y_r\}$ are (r,i)-standard sequences.

PROOF. Let $j+n<r$. By Theorem 4.5 we get

$$H^{i+j}_{\mathfrak{m}}(M/(x_1y_1,\dots,x_ny_n)M) \simeq \bigoplus_{l=0}^{n}(H^{i+j+l}_{\mathfrak{m}}(M))^{\binom{n}{l}}.$$

Since $\underline{x}$ and $\underline{y}$ annihilate the right-hand sides the assertion follows from Lemma 4.3. □

COROLLARY 4.7. *If $\underline{x} = \{x_1,\dots,x_r\}$ and $\underline{y} = \{y_1,\dots,y_r\}$ are (r,i)-standard sequences and if $\{x_1y_1,\dots,x_ry_r\}$ is an M-filter regular sequence, then*

$$\{x_1y_1,\dots,x_{r-1}y_{r-1},x_r\} \quad \text{and} \quad \{x_1y_1,\dots,x_{r-1}y_{r-1},y_r\}$$

are (r,i)-standard sequences.

PROOF. This follows immediately by Lemma 4.3 and Corollary 4.6. □

COROLLARY 4.8. *If $\underline{x} = \{x_1,\dots,x_r\} \subseteq \mathfrak{a}_i(M)\cap\dots\cap\mathfrak{a}_{i+r}(M)$ is an M-filter regular sequence, then the following is true:*

(a) *$H^i_{\mathfrak{m}}(M/\underline{x}^{(2)}M) \simeq H^i_{\mathfrak{m}}(M/(\underline{x}^{(n)}M)$ for all $n \ge 2$.*
(b) *$\mathfrak{a}_i(\underline{x}^{(2)};M) = \mathfrak{a}_i(M/\underline{x}^{(2)}M) = \mathfrak{a}_i(M)\cap\dots\cap\mathfrak{a}_{i+r}(M)$.*

PROOF. (a) is a consequence of Proposition 2.2 and Theorem 4.5. Therefore we get that $\mathfrak{a}_{ij}(\underline{x}^{(2)};M) = \mathfrak{a}_i(M)\cap\dots\cap\mathfrak{a}_{i+j}(M)$ and (b) follows. □

In the case that the cohomology modules $H^i_{\mathfrak{m}}(M), \ldots, H^{i+r}_{\mathfrak{m}}(M)$ have all finite length for some $i < d - r$, it is always possible to find $\underline{x}$ and $\underline{y}$ satisfying the assumptions of the above corollaries. Thus Corollary 4.8 (b) provides a lot of examples where we have equality on the right hand side of Corollary 3.7. The next result is immediately clear by view of Corollary 4.6.

PROPOSITION 4.9. *If M is a k-Buchsbaum module and if $\underline{x} = \{x_1, \ldots, x_r\} \subseteq \mathfrak{m}^k$ is a subsystem of a system of parameters of M, then $\{x_1^2, \ldots, x_{r-1}^2, x_r\}$ is an (r, i)-standard sequence for all i with $0 \le i \le d - r$.*

Note that L. T. Hoa and W. Vogel proved in [14], Proposition 2.5, that for a k-Buchsbaum module M **any** system of parameters $\underline{x} = \{x_1, \ldots, x_d\} \subseteq \mathfrak{m}^{2k}$ is a standard parameter system for M. Later on we will extend our last result to a characterization of k-Buchsbaum modules by standard parameter systems (cf. Proposition 5.3).

5. Modules With Finite Local Cohomology

In this Section we want to specialize the previous results to the case of finitely generated R-modules M such that $H^i_{\mathfrak{m}}(M)$, $i \neq \dim_R M$, are finitely generated R-modules. This kind of generalized Cohen-Macaulay modules were introduced in [27]. In the case R is a quotient of a Cohen-Macaulay ring this is equivalent to depth $M_{\mathfrak{p}} + \dim R/\mathfrak{p} = \dim M$ for all $\mathfrak{p} \in \operatorname{Supp} M \setminus \{\mathfrak{m}\}$, see [27]. So it follows that any system of parameters of M forms an M-filter regular sequence.

THEOREM 5.1. *Let M be a finitely generated R-module such that $H^j_{\mathfrak{m}}(M)$ is of finite length for all $j \neq \dim_R M$. Then*

$$\mathfrak{a}_i(\underline{x}; M) = \mathfrak{a}_{ir}(\underline{x}; M), \quad 0 \le i < d - r,$$

for any M-filter regular sequence $\underline{x} = \{x_1, \ldots, x_r\}$.

PROOF. Let $x \in \mathfrak{m}$ be an M-filter regular element. Then there is a short exact sequence

$$0 \to H^i_{\mathfrak{m}}(M)/xH^i_{\mathfrak{m}}(M) \to H^i_{\mathfrak{m}}(M/xM) \to 0 :_{H^{i+1}_{\mathfrak{m}}(M)} x \to 0$$

for all $i \in \mathbb{N}$. This is induced by the multiplication by x, see [28], (2.5). For a fixed M-filter regular sequence $\underline{x} = \{x_1, \ldots, x_r\}$ let us show

$$\mathfrak{a}_{i,j+1}(\underline{x}, M) \subseteq \mathfrak{a}_{ij}(\underline{x}, M)$$

for all $j < r$ and $i < d-r$. The previous short exact sequence implies an injection

$$0 \to H^i_{\mathfrak{m}}(M/(x_1^{k_1}, \ldots, x_j^{k_j})M) \otimes_R R/x_{j+1}^{k_{j+1}} R \to H^i_{\mathfrak{m}}(M/(x_1^{k_1}, \ldots, x_{j+1}^{k_{j+1}})M).$$

By definition of the cohomological annihilators it follows that

$$\mathfrak{a}_{i,j+1}(\underline{x}; M) H^i_{\mathfrak{m}}(M/(x_1^{k_1}, \ldots, x_j^{k_j})M) \subseteq \bigcap_{k_{j+1} \ge 1} x_{j+1}^{k_{j+1}} H^i_{\mathfrak{m}}(M/(x_1^{k_1}, \ldots, x_j^{k_j})M).$$

Because $H^i_{\mathfrak{m}}(M/(x_1^{k_1}, \ldots, x_j^{k_j})M)$ is a finitely generated R-module the Krull Intersection Theorem provides that the module on the right hand side vanishes. By the definition this means

$$\mathfrak{a}_{i,j+1}(\underline{x}; M) \subseteq \mathfrak{a}_i(M/(x_1^{k_1}, \ldots, x_j^{k_j})M) \quad \text{for all} \quad (k_1, \ldots, k_j) \in \mathbb{N}^j_+.$$

Therefore, it follows that $\mathfrak{a}_{i,j+1}(\underline{x}; M) \subseteq \mathfrak{a}_{ij}(\underline{x}; M)$ for all $0 \le j < r$. So

$$\mathfrak{a}_i(\underline{x}; M) = \bigcap_{j=0}^{r} \mathfrak{a}_{ij}(\underline{x}; M) = \mathfrak{a}_{ir}(\underline{x}; M),$$

which proves the claim. □

In terms of the integers $\lambda_i(M)$, see (3.2), now 5.1 implies that the annihilator of $H^i_{\mathfrak{m}}(M/\underline{x}M)$ grows linearly with respect to the annihilator of $H^i_{\mathfrak{m}}(M)$.

COROLLARY 5.2. *Let M be as in (5.1). Let $\underline{x} = \{x_1, \ldots, x_r\}$ denote an M-filter regular sequence. Then:*

(a) $\lambda_i(M) + \ldots + \lambda_{i+r}(M) \ge \lambda_i(M/\underline{x}M)$ *for* $0 \le i < d - r$.

(b) *For* $0 \le i < d-r$ *let* $k_i \ge \max\{\lambda_i(M), \ldots, \lambda_{i+r}(M)\}$ *denote an integer. Then*

$$\lambda_i(M/\underline{x}^{(2)}M) = \max\{\lambda_i(M), \ldots, \lambda_{i+r}(M)\}$$

for all $0 \le i < d - r$ *provided* $\underline{x} \subseteq \mathfrak{m}^{k_i}$.

(c) $\max\{\lambda_i(M), \ldots, \lambda_{i+r}(M)\} \le \max\{\lambda_i(M/\underline{x}^{(k)}M) : k \ge 1\} \le \lambda_i(M) + \ldots + \lambda_{i+r}(M)$.

PROOF. The inequality in (a) is a particular case of Theorem 3.3 and (b) follows by Corollary 4.8. Assertion (c) is now a consequence of (a), Proposition 2.2 and (b). □

Note that the above statement (c) is somewhat stronger than a combination of Corollary 3.7 and Theorem 5.1. As an application of the previous results we derive a characterization of k-Buchsbaum modules.

PROPOSITION 5.3. *Let M be a finitely generated R-module. Then M is a k-Buchsbaum module if and only if $\{x_1^2, \ldots, x_{d-1}^2, x_d\}$, $d = \dim_R M$, is a standard sequence for any system of parameters $\underline{x} = \{x_1, \ldots, x_d\} \subseteq \mathfrak{m}^k$ of M.*

PROOF. Assume that M is a k-Buchsbaum module. Then the claim follows by Proposition 4.9 and Remark 4.4 (i). For the converse first note that $H^i_{\mathfrak{m}}(M)$ is an R-module of finite length for all $i < d$. Then it follows that any parameter $x \in \mathfrak{m}^k$ annihilates $H^i_{\mathfrak{m}}(M)$ for all $i < d$. Indeed, we may find elements $x_1, \ldots, x_{d-1} \in \mathfrak{m}^k$ such that $\{x_1, \ldots, x_d\}, x_d = x$, forms a system of parameters of M. By the assumption $\{x_1^2, \ldots, x_{d-1}^2, x_d\}$ is a standard sequence. Hence Lemma 4.3 implies that $x_d H^i_{\mathfrak{m}}(M) = 0$. By Lemma 5.4 it follows that $\mathfrak{m}^k H^i_{\mathfrak{m}}(M) = 0$ for all $i \ne d$, i.e., M is a k-Buchsbaum module. □

In order to conclude the proof of (5.3) we need the following technical result.

LEMMA 5.4. *Let $\mathfrak{q}$ be an $\mathfrak{m}$-primary ideal of R. Let V denote an R-module of finite length with the property that $xV = 0$ for all $x \in \mathfrak{q}$ such that x is a parameter of M, a finitely generated R-module. Then $\mathfrak{q}V = 0$.*

PROOF. Let $x \in \mathfrak{q}$ be an arbitrary element. If x is a parameter of M, then $xV = 0$. Otherwise $x \in \mathfrak{p}_1, \dots, x \in \mathfrak{p}_s$ and $x \notin \mathfrak{p}_{s+1}, \dots, x \notin \mathfrak{p}_t$ for some $1 \leq s \leq t$, where $\{\mathfrak{p}_1, \dots, \mathfrak{p}_t\}$ denotes the set of highest dimensional associated prime ideals of M. Now choose $y' \notin \mathfrak{p}_1, \dots, y' \notin \mathfrak{p}_s$ and $y' \in \mathfrak{p}_{s+1}, \dots, y' \in \mathfrak{p}_t$. Take $k \gg 0$ such that $y := y'^k$ satisfies $yV = 0$ and $y \in \mathfrak{q}$. Then $x + y \in \mathfrak{q}$ is a parameter of M and $0 = (x+y)V = xV$, i.e., $\mathfrak{q}V = 0$ as required. □

The previous characterization of k-Buchsbaum modules suggests the question if it is suficient to require the standard sequence property of $\{x_{i_1}^2, \dots, x_{i_{d-1}}^2, x_{i_d}\}$ for a fixed system of parameters $\underline{x} = \{x_1, \dots, x_d\} \subseteq \mathfrak{m}^k$ and any permutation $(i_1, \dots, i_d)$ of $(1, \dots, d)$. This is not true by the following example. For a field K put

$$R = K[x, y, z]_{(x,y,z)}/(x^2) \cap (x^4, y, z).$$

Then R is not a 1-Buchsbaum ring because $xH_{\mathfrak{m}}^0(R) \neq 0$. On the other hand $\{y^2, z\}$ and $\{z^2, y\}$ form a standard system of parameters as easily seen.

Recall that in [14] also (k, s)-Buchsbaum modules, $k \in \mathbb{N}$ and $1 \leq s \leq \dim_R M$, are considered. A finitely generated R-module M is said to be a (k, s)-Buchsbaum module if $M/(x_1, \dots, x_{j-1})M$ is k-Buchsbaum for any subsystem $\{x_1, \dots, x_{j-1}\}$, $1 \leq j \leq s$, of any system of parameters $\{x_1, \dots, x_d\}$ of M. By Corollary 5.2 we immediately obtain a slight improvement of the main result Theorem 2.1 in [14].

PROPOSITION 5.5. *Suppose M is a finitely generated $R-$module such that $H_{\mathfrak{m}}^i(M)$ is finite for all $i \neq \dim_R M$. Let s denote an integer with $1 \leq s \leq d$. Put*

$$k = \max\{\lambda_i(M) + \ldots + \lambda_{i+s-1}(M) : 0 \leq i \leq d - s\}.$$

Then M is a (k, s)-Buchsbaum module.

6. Applications to Castelnuovo Bounds

In this Section we consider only graded objects. Hence we have to change our notation a little bit. Now $R = \oplus_{i \in \mathbb{N}} R_i$ is always a graded Noetherian ring such that $(R_0, \mathfrak{n})$ is a local ring and the R_0-algebra R is generated by the elements of R_1. We put $R_+ = \oplus_{i>0} R_i$. Moreover, M denotes always a $\mathbb{Z}$-graded Noetherian R-module of Krull-dimension d. Note that we may apply our results of the previous sections on the cohomological annihilators $\mathfrak{a}_i(M) := \operatorname{Ann}_R H_{R_+}^i(M)$.

For an arbitrary graded R-module N we define the end $e(N)$ of N by

$$e(N) := \sup\{j \in \mathbb{Z} : [N]_j \neq 0\}.$$

Thus $e(\{0\}) = -\infty$. Since the local cohomology modules $H_{R_+}^i(M)$ are graded R-modules we recall the following definition (cf. [20]).

DEFINITION 6.1. For any integer $s \geq 0$ we put

$$r_s(M) := \max\{i + e(H^i_{R_+}(M)) : i \geq s\}.$$

Then $\operatorname{reg}(M) := r_0(M) = r_{\operatorname{depth} M}(M)$ is said to be the Castelnuovo-Mumford regularity of M.

In order to derive our Castelnuovo bounds we need some preparatory results. In the following $\underline{x} = \{x_1, \dots, x_r\} \subseteq R_+$ will always denote a sequence of homogeneous elements of R.

LEMMA 6.2. *Let $\underline{x} = \{x_1, \dots, x_r\}$ be an M-filter regular sequence. Then*

$$e(H^{i+r}_{R_+}(M)) \leq e(H^i_{R_+}(M/\underline{x}M)) - \sum_{i=1}^{r} \deg(x_i)$$

for all $i \geq 0$.

PROOF. Let $x \in R$ be a homogeneous element of degree k which is M-filter regular. Then there is an exact sequence

$$H^i_{R_+}(M/xM) \to H^{i+1}_{R_+}(M)(-k) \to H^{i+1}_{R_+}(M)$$

for all $i \geq 0$. Since $e(H^{i+1}_{R_+}(M))$ is finite for all i we obtain

$$e(H^{i+1}_{R_+}(M)) \leq e(H^i_{R_+}(M/xM)) - k.$$

Hence the assertion follows by induction on r. □

LEMMA 6.3. *Let $i \leq d - r$ be an integer such that $\underline{x} = \{x_1, \dots, x_r\}$ is an (r, i)-standard sequence. Then we have a graded embedding $H^i_{R_+}(M) \hookrightarrow H^i_{R_+}(M/\underline{x}M)$ and*

$$e(H^{i+j}_{R_+}(M)) \leq e(H^i_{R_+}(M/\underline{x}M)) - \sum_{t=r-j+1}^{r} \deg(x_t)$$

for all $j = 0, \dots, r$.

PROOF. Analyzing the proof of Proposition 4.1 we see that the embedding given there is – in the graded situation – a homomorphism of degree zero. This proves our first assertion. Moreover it implies $e(H^i_{R_+}(M)) \leq e(H^i_{R_+}(M/\underline{x}M)$. Now the assumption allows the use of this result for $M/(x_1, \dots, x_{r-j})M$ where $1 \leq j \leq r$. Hence, by view of Lemma 6.2 we get

$$\begin{aligned} e(H^{i+j}_{R_+}(M)) &\leq e(H^{i+j}_{R_+}(M/(x_1, \dots, x_{r-j})M) \leq \\ &\leq e(H^i_{R_+}(M/\underline{x}M)) - \textstyle\sum_{t=r-j+1}^{r} \deg(x_t), \end{aligned}$$

which proves the statement. □

Now we are ready to prove our first bound of Castelnuovo type.

THEOREM 6.4. *Let $\underline{x} = \{x_1, \dots, x_r\}$ denote an M-filter regular sequence. Let $i \leq d - r$ be an integer such that $\underline{x} H^j_{R_+}(M) = 0$ for all j with $i \leq j < d$. Then there is the following bound*

$$r_i(M) \leq r_i(M/(x_1^2, \dots, x_{r-1}^2, x_r)M).$$

PROOF. At first Corollary 4.6 provides that $\{x_1^2, \dots, x_{r-1}^2, x_r\}$ is a (r, j)-standard sequence for all j with $i \leq j \leq d - r$. Then the above Lemma 6.3 applies and we obtain

$$\begin{aligned} e(H^j_{R_+}(M)) &\leq e(H^j_{R_+}(M/(x_1^2, \dots, x_{r-1}^2, x_r)M)) \quad \text{for all } i \leq j \leq d - r, \\ e(H^{d-r+j}_{R_+}(M)) &\leq e(H^{d-r}_{R_+}(M/(x_1^2, \dots, x_{r-1}^2, x_r)M)) - j \quad \text{for } j = 0, \dots, r, \end{aligned}$$

which finishes the proof. □

In the following we will show how this result applies in order to obtain Castelnuovo bounds for k-Buchsbaum modules. In order to do that we need the following result.

LEMMA 6.5. *Assume that M denotes an R-module of finite local cohomology. Let $\underline{x} = \{x_1, \dots, x_r\}$ be a subsystem of a homogeneous system of parameters for M. Then*

$$e(H^i_{R_+}(M/(x_1^{k_1}, \dots, x_r^{k_r})M) \leq e(H^i_{R_+}(M/\underline{x}M)) + \sum_{j=1}^{r} (k_j - 1) \deg(x_j).$$

for all positive integers $k_1, \dots, k_r$ and all $i \geq 0$

PROOF. We use induction on r. Let $r = 1$. For short put $x := x_1$ and $f := \deg x$. First we show the following Claim:

For all integers $n > 0$ and $i \geq 0$ there is an exact sequence

$$H^i_{R_+}(M/xM)(-nf) \to H^i_{R_+}(M/x^{n+1}M) \to H^i_{R_+}(M/x^n M).$$

To this end consider the epimorphism $\varphi\colon M/xM(-nf) \to x^n M/x^{n+1}M$ induced by the multiplication by x^n. We get $\ker \varphi = (0 :_M x^n + xM)/xM$. This is a module of finite length because x is M-filter regular. Hence the exact sequence

$$0 \to \ker \varphi \to M/xM(-nf) \to x^n M/x^{n+1}M \to 0$$

implies the exact sequence

$$(*) \qquad 0 \to \ker \varphi \to H^0_{R_+}(M/xM)(-nf) \to H^0_{R_+}(x^n M/x^{n+1}M) \to 0$$

and isomorphisms $H^i_{R_+}(M/xM)(-nf) \simeq H^i_{R_+}(x^n M/x^{n+1}M)$ for all $i > 0$. Using these isomorphisms the long exact cohomology sequence of

$$0 \to x^n M/x^{n+1}M \to M/x^{n+1}M \to M/x^n M \to 0$$

proves the claim if $i > 0$. Moreover, it provides an exact sequence

$$0 \to H^0_{R_+}(x^n M/x^{n+1}M) \to H^0_{R_+}(M/x^{n+1}M) \to H^0_{R_+}(M/x^n M).$$

Now the claim is shown to be true because this together with $(*)$ implies the statement in the case $i = 0$.

By the induction hypothesis on $n > 0$ the claim shows $e(H^i_{R_+}(M/x^nM)) \leq e(H^i_{R_+}(M/xM)) + (n-1)f$ proving the assertion if $r = 1$.

Let $r > 1$. By applying the induction hypothesis to $M' := M/x_1^{k_1}M$ we get

$$e(H^i_{R_+}(M/(x_1^{k_1}, \dots, x_r^{k_r})M)) = e(H^i_{R_+}(M'/(x_2^{k_2}, \dots, x_r^{k_r})M')) \leq$$
$$\leq e(H^i_{R_+}(M'/(x_2, \dots, x_r)M')) + \sum_{j=2}^{r}(k_j - 1)\deg x_j.$$

Since $\{x_2, \dots, x_r, x_1\}$ is a subsystem of a system of parameters for M we may apply the induction hypothesis to $M'' := M/(x_2, \dots, x_r)M$. Therefore

$$\begin{aligned} e(H^i_{R_+}(M'/(x_2, \dots, x_r)M')) &= e(H^i_{R_+}(M''/x_1^{k_1}M'')) \leq \\ &\leq e(H^i_{R_+}(M''/x_1M)) + (k_1 - 1)\deg(x_1) \\ &= e(H^i_{R_+}(M/(x_1, \dots, x_r)M)) + (k_1 - 1)\deg(x_1). \end{aligned}$$

Putting the above inequalities together the inductive step is complete. □

Remark 6.6. Let R be as above a graded algebra. We may assume that the field $R_0/\mathfrak{n}$ is infinite by tensoring R with $R_0[u]_{(u)}$ where u denotes an indeterminate. This follows because all the values which are involved are stable under this faithfully flat extension. Then there exists always an M-filter regular sequence $\{l_1, \dots, l_d\} \subseteq R_1$.

PROPOSITION 6.7. *Let M be k-Buchsbaum module, $k \geq 1$, and let $l = \{l_1, \dots, l_r\}$ be a subsystem of a system of parameters consisting of linear forms. Then*

$$r_j(M) \leq r_j(M/lM) + (2r-1)k - r.$$

for all $j \leq d - r$.

PROOF. Using Theorem 6.4 and Lemma 6.5 we get that

$$r_j(M) \leq r_j(M/(l_1^{2k}, \dots, l_{r-1}^{2k}, l_r^k)M) \leq r_j(M/lM) + (r-1)(2k-1) + k - 1,$$

which proves the claim. □

In the following we suppose that $R_0 = K$ is a field. As our first application we compare the regularity of R with the reduction exponent $r(\mathfrak{a})$ of a minimal reduction $\mathfrak{a}$ of R_+. Here $r(\mathfrak{a})$ is the least integer t such that $[R]_{t+1} = [\mathfrak{a}]_{t+1}$.

COROLLARY 6.8. *Let $k > 0$ be an integer. Suppose that R is a k-Buchsbaum graded K-algebra with $\dim R > 0$. Let $\mathfrak{a}$ be a minimal reduction of R_+. Then*

$$r(\mathfrak{a}) \leq \operatorname{reg}(R) \leq r(\mathfrak{a}) - d + (2d-1)k.$$

PROOF. By [34], Proposition 3.2, the first equality is true even without the assumption on k-Buchsbaumness. Moreover, due to [34], Lemma 3.1, there is a system of linear parameters $l = \{l_1, \dots, l_d\}$ generating $\mathfrak{a}$ minimally. As in the proof of Proposition 3.2 in [34] it follows that $r(\mathfrak{a}) = \operatorname{reg}(R/lR)$. Hence we may apply Proposition 6.7 with $j = 0$ and $r = d$. Now the conclusion follows. □

This result improves [14], Theorem 3.8. Now we will derive Castelnuovo bounds for schemes $X \subseteq \mathbb{P}_K^n$. We denote by $\mathcal{I}(X) \subseteq S = K[x_0, \dots, x_n]$ the defining ideal of X. Thus the graded K-algebra $R := S/\mathcal{I}(X)$ is the homogeneous coordinate ring of X. In the following we put $\operatorname{reg}(X) := \operatorname{reg}(\mathcal{I}(X)) = \operatorname{reg}(R) + 1$ and more generally $r_i(X) = r_{i+1}(\mathcal{I}(X)) = r_i(R) + 1$ for any integer $i \in \mathbb{Z}$. Recall that $\lceil a \rceil$ denotes the least integer $\geq a$ for $a \in \mathbb{R}$.

THEOREM 6.9. *Let $X \subseteq \mathbb{P}_K^n$ be a k-Buchsbaum scheme of dimension $d > 0$, where K is an algebraically closed field and $k > 0$. Then:*

(a) $\operatorname{reg}(X) \leq \deg(X) - d + (2d - 1)k$.

(b) *If X is in addition irreducible, reduced and nondegenerate, then:*

(i) $\operatorname{reg}(X) \leq \left\lceil \frac{\deg(X) - 1}{\operatorname{codim}(X)} \right\rceil - d + (2d - 1)k + 1$.

(ii) *If $d \geq 2$, then* $\operatorname{reg}(X) \leq \deg(X) - \operatorname{codim}(X) - d + (2d - 3)k + 2$.

PROOF. First we show (a). Let $l = \{l_1, \dots, l_d\}$ be a subsystem of a linear system of parameters and put $R' = R/lR$. Then R' is a 1-dimensional graded ring and $\operatorname{rank}_K[H^1_{R_+}(R')]_0 = \deg(\mathcal{I}(X) + l \cdot S) - 1 = \deg(X) - 1$. By virtue of [21], Lemma 4.3, it follows that

$$r_1(R') = 1 + e(H^1_{R_+}(R')) \leq \deg(X) - 1.$$

Because of $\operatorname{reg}(X) = 1 + r_1(R)$ we get the first assertion (a) by Proposition 6.7. In order to prove (b) we use the above notation. If we choose the linear forms $l_1, \dots, l_d$ generically, then due to [1] $R'' := R'/H^0_{R_+}(R')$ is the homogeneous coordinate ring of a set of $\deg(X)$ points in linear semi-uniform position. Now we have

$$1 + e(H^1_{R_+}(R')) = 1 + e(H^1_{R_+}(R'')) \leq \left\lceil \frac{\deg(X) - 1}{\operatorname{codim}(X)} \right\rceil$$

(cf. [1] and [22]). Then Proposition 6.7 shows (i). For proving (ii) we put $R^* := R/(l_1, \dots, l_{d-1})/H^0_{R_+}(R/(l_1, \dots, l_{d-1})R)$. Then R^* is the coordinate ring of a curve. Hence by [11] there is the following bound

$$r_1(R/(l_1, \dots, l_{d-1})R) = r_1(R^*) \leq \deg(X) - \operatorname{codim}(X).$$

Therefore the assertion follows as above. □

COROLLARY 6.10. *Let $X \subseteq \mathbb{P}_K^n$ be a 1-Buchsbaum, irreducible, reduced and nondegenerate subvariety over the algebraically closed field K. Then we have*

$$\operatorname{reg}(X) \leq \deg(X) - \operatorname{codim}(X) + 1$$

provided $\deg(X) \geq n + \dim(X) - 1$ *or* $\dim(X) \leq 3$.

Proof. If $\operatorname{codim}(X) = 1$, then X is a hypersurface and it is well-known that $\operatorname{reg}(X) = \deg(X)$. Thus we may assume $\operatorname{codim}(X) \geq 2$. If $\deg(X) \geq n + \dim(X) - 1$, then

$$\deg(X) - \operatorname{codim}(X) + 1 \geq \left\lceil \frac{\deg(X) - 1}{\operatorname{codim}(X)} \right\rceil + \dim(X).$$

To this end note that if $\deg(X) \geq n + \dim(X) - 1$, then

$$\deg(X) \geq n + \frac{\dim(X) - 1}{\operatorname{codim}(X) - 1} \text{ and } \deg(X) + 1 - n \geq \left\lceil \frac{\deg(X) - 1}{\operatorname{codim}(X)} \right\rceil$$

by a few nasty but elementary calculations. Recall that $n = \dim(X) + \operatorname{codim}(X)$. Thus, in this case the assertion follows by Theorem 6.9 (b) (i).

If $\dim(X) = 1$ the assertion is true by [11]. If $\dim(X) = 2$ the result follows by Theorem 6.9 (b) (ii). Suppose $\dim(X) = 3$. Let $H = V(l)$ be a general hyperplane of $\mathbb{P}^n_K$. Then $\operatorname{reg}(X) \leq \operatorname{reg}(X \cap H)$ by Theorem 6.4. But $X \cap H$ is a surface which is 2-Buchsbaum by Corollary 5.2. Hence Theorem 4.10 of [5] gives the desired bound. □

Remark 6.11. (i) The Corollary in the introduction is covered by Theorem 6.9 with $k = 1$ and by Corollary 6.10. Note that we do not require any assumption on the characteristic of the field K.
(ii) Under the assumptions of Corollary 6.10 we have always $\deg(X) \geq n - \dim(X) + 1$. Hence our above result shows the Eisenbud conjecture in [6] for a new rather large class of subvarieties.
(iii) In Theorem 6.9 suppose in addition that char $(K) = 0$. Then the ring R'' in the proof of Theorem 6.9 is the coordinate ring of a set of points in uniform position. This allows to improve the bounds on $e(H^1_{R_+}(R''))$, thus the bounds on $\operatorname{reg}(X)$ too (cf. [23] and [14], Theorem 3.4).
(iv) If R is a Cohen-Macaulay ring, i.e. $k = 0$, then the following bound corresponding to 6.8 holds: $\operatorname{reg} R \leq r(\mathfrak{a})$. In the case $k = 0$ in 6.9, i.e. X is arithmetically Cohen-Macaulay it follows that $\operatorname{reg}(X) \leq \deg(X)$, (cf. [24]), instead of 6.9 (a) resp. $\operatorname{reg}(X) \leq \left\lceil \frac{\deg(X)-1}{\operatorname{codim}(X)} \right\rceil$, (cf. [23], Corollary 15), instead of 6.9 (b).
(v) When the present paper was finished it came to our attention that L. T. Hoa and C. Miyazaki, see [15], have proved similar but slightly weaker Castelnuovo bounds as those of Theorem 6.9 with a completely different method.

Example 6.12. (i) Let $R = K[x,y,z,w]/\mathfrak{a}$, $\mathfrak{a} = (x,y) \cap (z,w) \cap (x, y^2, z, w^2)$, be a graded K-algebra. Then $H^0_{R_+}(R) \simeq K(-2)$ and $H^1_{R_+}(R) \simeq K$ (cf. [30], Lemma I.2.14). In particular R is a 1-Buchsbaum ring with $\operatorname{reg}(R) = 2$. Let $\{l_1, l_2\}$ be a linear system of parameters. Then

$$H^0_{R_+}(R/(l_1,l_2)R) \simeq R/(l_1,l_2)R \simeq S/(\mathfrak{a} + (l_1,l_2)S),$$

where $S = K[x, y, z, w]$. Let l_1 and l_2 be chosen generically. Then

$$[S]_2 = [(xz, xw, yz, yw, l_1, l_2)S]_2 \subseteq [\mathfrak{a} + (l_1, l_2)S]_2,$$

and therefore $[H^0_{R+}(R/(l_1, l_2)R))]_2 = 0$. Hence we have

$$\operatorname{reg}(R/(l_1, l_2)R) = e(H^0_{R+}(R/(l_1, l_2)R) = 1 = \operatorname{reg} R - 1.$$

Since R is a 1-Buchsbaum ring it shows that the bound in Proposition 6.7 is sharp in this case. Moreover, by Proposition 4.9 it follows that $\{l_1^2, l_2\}$ is a standard system of parameters for R. The example shows that this result is also optimal in the following sense: A general chosen $\{l_1, l_2\}$ can not be a standard system of parameters. Otherwise Lemma 6.3 provides an embedding

$$K(-2) \simeq H^0_{R+}(R) \hookrightarrow H^0_{R+}(R/(l_1, l_2)R)$$

which is impossible because of $[H^0_{R+}(R/(l_1, l_2)R)]_2 = 0$. Since $\{l_1, l_2\}$ is in general not a standard system of parameters R/l_1R can not be a 1-Buchsbaum ring. But Corollary 5.2 states that R/l_1R is a 2-Buchsbaum ring which is optimal in this case. Note that for a certain special choice, e.g. $l_1 = x - z, l_2 = y - w$, the system $\{l_1, l_2\}$ is a standard system of parameters.
(ii) Let $R' = S/\mathfrak{b}\mathfrak{m}$, where $\mathfrak{b} = (x, y) \cap (z, w)$ and $\mathfrak{m} = (x, y, z, w)$ are ideals of $S = K[x, y, z, w]$. It is easy to see that R' is a 1-Buchsbaum ring. On one hand

$$\operatorname{rank}_K[H^0_{R'_+}(R')]_2 = \operatorname{rank}_K[\mathfrak{b}/\mathfrak{b}\mathfrak{m}]_2 = 4.$$

On the other hand it follows that

$$\operatorname{rank}_K[H^0_{R'_+}(R'/(l_1, l_2)R')]_2 = \operatorname{rank}_K[S/(\mathfrak{b}\mathfrak{m} + (l_1, l_2)S)]_2 = 3$$

for any linear system of parameters $\{l_1, l_2\}$. Note that $\mathfrak{b}\mathfrak{m}$ does not contain any quadric. Hence $H^0_{R'_+}(R')$ can not be embedded into $H^0_{R'_+}(R'/(l_1, l_2)R'))$. By Lemma 6.3 the system $\{l_1, l_2\}$ is not a standard system of parameters, while this is true for $\{l_1^2, l_2\}$, see Proposition 4.9.

Acknowledgement. The authors are grateful to the referee for his helpful comments.

References

1. E. Ballico, *On singular curves in positive characteristic*, Math. Nachr. **141** (1989), 267–273.
2. D. Bayer, M. Stillman, *On the complexity of computing syzygies*, J. Symbolic Comput. **6** (1988), 135–147.
3. M. Brodmann, *A priori bounds of Castelnuovo type for cohomological Hilbert functions*, Comment. Math. Helv. **65** (1990), 478–518.
4. M. Brodmann, U. Nagel, *Bounding cohomological Hilbert functions by hyperplane sections*, Preprint, Universities of Paderborn and Zürich, 1992.
5. M. Brodmann, W. Vogel, *Bounds for the cohomology and the Castelnuovo regularity of certain surfaces*, Preprint, Universities of Halle and Zürich, 1992.
6. D. Eisenbud, S. Gôto, *Linear free resolutions and minimal multiplicity*, J. Algebra **88** (1984), 89–133.

7. S. Gôto, *A note on quasi-Buchsbaum rings*, Proc. Amer. Math. Soc. **90** (1984), 511–516.
8. S. Gôto, K. Watanabe, *On graded rings, I*, J. Math. Soc. Japan **30** (1978), 179–213.
9. S. Gôto, K. Yamagishi, *The theory of unconditioned strong d-sequences and modules of finite local cohomology*, Preprint (unpublished).
10. A. Grothendieck, *"Local cohomology"*, Lecture Notes in Math. **41**, Springer, 1967.
11. L. Gruson, R. Lazarsfeld, C. Peskine, *On a theorem of Castelnuovo, and the equations defining space curves*, Invent. Math. **72** (1983), 491–506.
12. M. Herrmann, S. Ikeda, U. Orbanz, *"Equimultiplicity and Blowing Up"*, Springer, 1988.
13. L. T. Hoa, R. M. Miró-Roig, W. Vogel, *On numerical invariants of locally Cohen-Macaulay schemes in $\mathbb{P}^n$*, Hiroshima Math. J. (to appear).
14. L. T. Hoa, W. Vogel, *Castelnuovo-Mumford regularity and hyperplane sections*, J. Algebra (to appear).
15. L. T. Hoa, C. Miyazaki, *Bounds on Castelnuovo-Mumford regularity for generalized Cohen-Macaulay graded rings*, Preprint, Institut d'Estudis Catalans (Barcelona).
16. I. Kaplansky, *"Commutative rings"*, Boston 1970.
17. J. McCleary, *"User's Guide to Spectral Sequences"*, Publish or Perish, Delaware, 1985.
18. J. Migliore, R. M. Miró-Roig, *On k-Buchsbaum curves*, Commun. Algebra **18** (8) (1990), 2403–2422.
19. D. Mumford, *"Lectures on curves on an algebraic surface"*, Ann. of Math. Studies **59**, Princeton Univ. Press, 1966.
20. U. Nagel, *Castelnuovo's regularity and Hilbert functions*, Compositio Math. **76** (1990), 265–275.
21. U. Nagel, *On bounds for cohomological Hilbert functions*, J. Algebra **150** (1992), 231–244.
22. U. Nagel, *Über Gradschranken für Syzygien und kohomologische Hilbertfunktionen*, Thesis, Paderborn, 1990.
23. U. Nagel, W. Vogel, *Bounds for Castelnuovo's regularity and the genus of projective varieties, In*: Topics in Algebra. Banach Center Publications, Vol. 26, Part 2, 163–183, PWN-Polish Scientific Publishers, Warsaw, 1990.
24. A. Ooishi, *Castelnuovo's regularity of graded rings and modules*, Hiroshima Math. J. **12** (1982), 627–644.
25. P. Roberts, *Two applications of dualizing complexes over local rings*, Ann. Sci. Ec. Norm. Sup. (4) **9** (1976), 103–106.
26. P. Schenzel, *"Dualisierende Komplexe in der lokalen Algebra und Buchsbaum-Ringe"*, Lecture Notes in Math. **907**, Springer, 1982.
27. P. Schenzel, *Einige Anwendungen der lokalen Dualität und verallgemeinerte Cohen-Macaulay-Moduln*, Math. Nachr. **69** (1975), 227–242.
28. P. Schenzel, *Standard systems of parameters and their blowing-up rings*, J. Reine Angew. Math. **344** (1983), 201–220.
29. P. Schenzel, N. V. Trung, N. T. Cuong, *Verallgemeinerte Cohen-Macaulay-Moduln*, Math. Nachr. **85** (1978), 57–73.
30. J. Stückrad, W. Vogel, *"Buchsbaum Rings and Applications"*, VEB Deutscher Verlag der Wissenschaften, Berlin, 1986.
31. J. Stückrad, W. Vogel, *Castelnuovo bounds for locally Cohen-Macaulay schemes*, Math. Nachr. **136** (1988), 307–320.
32. J. Stückrad, W. Vogel, *Castelnuovo bounds for certain subvarieties in $\mathbb{P}^n$*, Math. Ann. **276** (1987), 341–352.
33. N. V. Trung, *Towards a theory of generalized Cohen-Macaulay modules*, Nagoya Math. J. **102** (1986), 1–49.
34. N. V. Trung, *Reduction exponent and degree bound for the defining equations of graded rings*, Proc. Amer. Math. Soc. **102** (1987), 229– 236.

Fachbereich 17 – Mathematik und Inform., Universität-Gesamthochschule Paderborn, D — 33 095 Paderborn, Germany

E-mail address: uwen@uni-paderborn.de schenzel@uni-paderborn.de

Contemporary Mathematics
Volume 159, 1994

LOCAL-GLOBAL PRINCIPLE FOR ANNIHILATION OF LOCAL COHOMOLOGY

K. N. RAGHAVAN

ABSTRACT. Let R be a Noetherian ring, let M be a finitely generated R-module, and let I, J be ideals of R. Consider the following question, which will be referred to as the local-global principle for annihilation of local cohomology: if, for every prime ideal $\mathbf{p}$ of R, there exists an integer $k(\mathbf{p})$, depending on $\mathbf{p}$, such that $J^{k(\mathbf{p})}$ kills the local cohomology module $H^j_{I_\mathbf{p}}(M_\mathbf{p})$ for every integer j less than a fixed integer s, then does there exist an integer k such that $J^k H^j_I(M) = 0$ for every $j < s$? We answer this question in the affirmative in two cases: (1) R is a homomorphic image of a Noetherian regular ring; (2) for $s = 2$ with no condition on R.

Let R be a Noetherian ring, let M be a finitely generated R-module, and let I and J be ideals of R. The local cohomology modules are not always finitely generated; the local-global principle is non-trivial only for those values of s for which not all local cohomology modules with indices below s are finitely generated. Actually, even if all local cohomology modules with indices below s have finitely many associated primes, the local-global principle clearly has an affirmative answer, but it is an open problem whether every local cohomology module has finitely many associated primes (see, however, [4]). Of course, one could ask the local-global principle for a fixed integer instead of for every $j < s$ (this question remains open to the best of the author's knowledge), but the following theorem of Faltings provides some hope that the problem as stated may be easier to handle, and indeed the first thing that we do in this note is to deduce the local-global principle as an easy corollary of another theorem of Faltings.

Theorem. ([2, Lemma 3], [3, §2, pp.48–49]) *For an ideal I of a Noetherian ring R, a finitely generated R-module M, and a positive integer s, the following conditions are equivalent:*

(1) $H^j_I(M)$ *is finitely generated* $\forall\ j < s$.
(2) $H^j_{I_\mathbf{p}}(M_\mathbf{p})$ *is finitely generated* $\forall\ j < s$ *and* $\forall\ \mathbf{p} \in \operatorname{Spec}(R)$.
(3) $\exists$ *an integer k such that* $I^k H^j_I(M) = 0\ \forall\ j < s$.

We first point out how, if R is a homomorphic image of a (Noetherian) regular ring, the local-global principle follows easily from a theorem of Faltings [2, Satz 1].

1991 *Mathematics Subject Classification.* Primary 13D45.
This paper is in final form and no version of it will be submitted for publication elsewhere.

Definition. Let $D(J)$ denote the set of prime ideals of R that do not contain J. Define

$$s(I,J,M) = \min_{\mathbf{p}\in D(J)}\{\operatorname{depth}(M_{\mathbf{p}}) + \operatorname{ht}(\frac{I+\mathbf{p}}{\mathbf{p}})\}$$

Depth of the zero module and height of the unit ideal are both ∞ by convention. Also let $s(I,J,M) = \infty$ if $D(J)$ is empty. Given a prime ideal $\mathbf{p}$ of R, we say that $s(I,J,M)$ is *attained* at $\mathbf{p}$ if $\mathbf{p}$ belongs to $D(J)$ and $s(I,J,M) = \operatorname{depth}(M_{\mathbf{p}}) + \operatorname{ht}(I+\mathbf{p}/\mathbf{p})$.

Theorem. [2, Satz 1] *If R is a homomorphic image of a Noetherian regular ring, then there is an integer k such that $J^k H^j_I(M) = 0$ for $j < s(I,J,M)$. Conversely, if there exist k, s such that $J^k H^j_I(M) = 0$ for every $j < s$, then $s \le s(I,J,M)$.*

Corollary. *The local-global principle holds for a homomorphic image of a Noetherian regular ring.*

Proof. By hypothesis, we know that for every prime ideal $\mathbf{p}$ of R, there exists an integer $k(\mathbf{p})$, depending on $\mathbf{p}$, such that $J^{k(\mathbf{p})}$ kills the local cohomology module $H^j_{I_{\mathbf{p}}}(M_{\mathbf{p}})$ for every integer j less than a fixed integer s. We claim that $s \le s(I,J,M)$. Combining this claim with the theorem just above immediately yields the desired conclusion.

To prove the claim, there is nothing to be done if $s(I,J,M) = \infty$. Otherwise, let $\mathbf{p}$ be a prime ideal at which $s(I,J,M)$ is attained, and let $\mathbf{q}$ be a prime ideal that contains $I + \mathbf{p}$. Then $s(I,J,M) = s(I_{\mathbf{q}}, J_{\mathbf{q}}, M_{\mathbf{q}})$. Since, $J^{k(\mathbf{q})} H^j_{I_{\mathbf{q}}}(M_{\mathbf{q}})$ for every $j < s$, it follows from the previous theorem applied to the local ring $R_{\mathbf{q}}$ that $s \le s(I_{\mathbf{q}}, J_{\mathbf{q}}, M_{\mathbf{q}}) = s(I,J,M)$. This finishes the proof of the claim. $\square$

We now prove the local-global principle for $s = 2$ without assuming any condition on the ring R.

Theorem. *Let R be a Noetherian ring, let M be a finitely generated R-module, and let I, J be ideals of R. If $J_{\mathbf{p}} \subseteq \sqrt{\operatorname{ann} H^1_{I_{\mathbf{p}}}(M_{\mathbf{p}})}$ $\forall$ $\mathbf{p} \in \operatorname{Spec}(R)$, and $J \subseteq \sqrt{\operatorname{ann} H^0_I(M)}$, then $J \subseteq \sqrt{\operatorname{ann} H^1_I(M)}$. If, in addition, $H^0_J(M) = H^0_I(M) = 0$, then $H^1_I(M)$ is finitely generated.*

Proof. We first show that the first statement follows from the second. Let $L = H^0_J(M)$ and $N = M/L$. Notice that $J \subseteq \sqrt{\operatorname{ann} L}$, and hence $J \subseteq \sqrt{\operatorname{ann} H^j_I(L)}$ $\forall$ j. We also have $H^0_J(N) = 0$. From the long exact sequence on local cohomology induced by the short exact sequence $0 \to L \to M \to N \to 0$, it follows that $J \subseteq \sqrt{\operatorname{ann} H^0_I(N)}$ and that $J_{\mathbf{p}} \subseteq \sqrt{\operatorname{ann} H^1_{I_{\mathbf{p}}}(N_{\mathbf{p}})}$ $\forall$ $\mathbf{p} \in \operatorname{Spec}(R)$. We claim that $H^0_I(N) = 0$. To prove the claim, let x be an element of N that belongs to $H^0_I(N)$. Lift x to an element of M, and denote this also by x. We have $I^n x \in L = H^0_J(M) \Rightarrow J^m I^n x = 0 \Rightarrow J^m x \in H^0_I(M) \Rightarrow$ (since $J \subseteq \sqrt{\operatorname{ann} H^0_I(M)}$) $J^p x = 0 \Rightarrow x \in L$. Thus the image of x in N is zero, which proves the claim. Thus N satisfies the hypothesis of the second statement. By the conclusion of the second statement, $H^1_I(N)$ is finitely generated. This along with the hypothesis that $J_{\mathbf{p}} \subseteq \sqrt{\operatorname{ann} H^1_{I_{\mathbf{p}}}(M_{\mathbf{p}})}$ implies that $J \subseteq \sqrt{\operatorname{ann} H^1_I(N)}$. Using the long exact sequence on local cohomology induced by $0 \to L \to M \to N \to 0$ once again, we conclude that $J \subseteq \sqrt{\operatorname{ann} H_I^1(M)}$.

To prove the second statement, we claim that $H_I^1(M)$ has finitely many associated primes. To prove the claim, let x be an element of J that is regular on M. It is enough to show that $\operatorname{Ass} H_I^1(M) \subseteq \bigcup_k \operatorname{Ass} H_I^0(M/x^kM)$, for $\operatorname{Ass} H_I^0(M/x^kM)$ is contained in $\operatorname{Ass}(M/x^kM)$ since $H_I^0(M/x^kM)$ is a submodule of M/x^kM, and by Brodmann's theorem [1], $\bigcup_k \operatorname{Ass}(M/x^kM)$ is finite. Consider the long exact sequence on local cohomology induced by the short exact sequence $0 \to M \xrightarrow{x^k} M \to M/x^kM \to 0$:

$$0 \to H_I^0(M/x^kM) \to H_I^1(M) \xrightarrow{x^k} H_I^1(M) \to \cdots$$

Let $\mathbf{p} \in \operatorname{Ass} H_I^1(M)$. Then $\mathbf{p} = (0 :_R y)$ for some $y \in H_I^1(M)$. Since there exists an integer k such that $y \in \operatorname{ann}(x^k)$, it follows that y is in the image of the injective map $0 \to H_I^0(M/x^kM) \to H_I^1(M)$ for such a k. Thus $\mathbf{p} \in \operatorname{Ass} H_I^0(M/x^kM)$, and the claim is proved.

Let k be an integer so large that $J_{\mathbf{p}}^k H_{I_{\mathbf{p}}}^1(M_{\mathbf{p}}) = 0 \; \forall \, \mathbf{p} \in \operatorname{Ass} H_I^1(M)$. Observe that for a module Q and an ideal $\mathbf{b}$ over a Noetherian ring A, we have $\operatorname{Ass}(Q/(0 : \mathbf{b})) \subseteq \operatorname{Ass}(Q)$: let $a_1, \ldots, a_n$ be a finite generating set for $\mathbf{b}$; if $\mathbf{p} \in \operatorname{Ass}(Q/(0 :_Q \mathbf{b}))$, then there exists $q \in Q$ such that $\mathbf{p} = ((0 :_Q \mathbf{b}) :_R q) = (0 :_R \mathbf{b}q) = \bigcap_{i=1}^{n} (0 : a_i q) \Rightarrow \mathbf{p} = (0 : a_i q)$ for some i. It follows from this observation that $\operatorname{Ass}(H_I^1(M)/(0 : J^k)) \subseteq \operatorname{Ass} H_I^1(M)$. But $(H_I^1(M)/(0 : J^k))_{\mathbf{p}} = 0 \; \forall \, \mathbf{p} \in \operatorname{Ass} H_I^1(M)$. Thus $J^k H_I^1(M) = 0$. Now let x be an element of J that is regular on M. Since $x^k H_I^1(M) = 0$, it follows from the long exact sequence on local cohomology induced by $0 \to M \xrightarrow{x^k} M \to M/x^kM \to 0$ that $H_I^1(M)$ is a homomorphic image of $H_I^0(M/x^kM)$ and hence finitely generated. □

References

1. M. Brodmann, *Asymptotic stability of* $\operatorname{Ass}(M/I^nM)$, Proc. Amer. Math. Soc. **74** *no. 1* (1979), 16–18.
2. G. Faltings, *Über die Annulatoren lokaler Kohomologiegruppen*, Arch. Math. **30** (1978), 473–476.
3. ———, *Der Endlichkeitssatz der lokalen kohomologie*, Math. Ann. **255** (1981), 45–56.
4. C. Huneke and R. Sharp, *Bass numbers of local cohomology modules of a regular local ring of positive characteristic*, Trans. Amer. Math. Soc. (to appear).

Department of Mathematics, Michigan State University, East Lansing, MI 48824–1027, USA

E-mail address: raghavan@math.msu.edu

Contemporary Mathematics
Volume 159, 1994

Multiplicities and Chern Classes

PAUL C. ROBERTS

ABSTRACT. This paper describes several connections between the theory of Chern classes of matrices over Noetherian rings and the theory of multiplicities of ideals. These include a new notion of mutiplicities for modules of finite length and stronger versions of the concept of reduction of an ideal. In addition, we show a relation with the Monomial Conjecture and work out a class of examples in detail.

The aim of this paper is to present some results concerning Chern classes defined by matrices which are analogous to well-known properties of multiplicities of primary ideals.

In the first section, we present a brief description of the Chern classes and Chern character defined by a matrix with support at the maximal ideal of a local ring. We first give a somewhat non-standard definition of the multiplicity of an ideal (but one which is clearly equivalent to the usual definition), and the more general definition of Chern classes of matrices is then seen to be a natural generalization.

The second section presents a new definition of multiplicities of modules of finite length, which we compare to the Buchsbaum-Rim multiplicity. The third is devoted to a discussion of a generalization of the concept of reduction of an ideal or module. Classical results of Rees and others show that there is a close connection between reductions and multiplicities, and we investigate similar connections in the context of Chern classes. In addition, we prove a connection between reductions (in our sense) of ideals defined from parameter ideals and the homological conjectures, in particular the Monomial Conjecture.

In the last section we give a complete description of what the Chern characters mean in a very specific case, that of a module which is a quotient of monomial ideals in dimension two. We also discuss some open questions concerning the positivity of local Chern characters of matrices in general.

1991 *Mathematics Subject Classification.* Primary 13H15; Secondary 13D25, 14C35, 19L10.
This research was supported in part by NSF Grant #DMS-9204297
This paper is in final form and no version of it will be submitted for publication elsewhere.

1. The Construction of Chern Classes defined by a Matrix

In this section we describe the algebraic construction of the Chern classes and Chern character defined by a matrix with support at the maximal ideal of a local ring. We describe how these are computed; for more details, as well as proofs that they are well-defined, we refer to Roberts [14]. For a more general description see Fulton ([3], Chapter 18); this construction is an algebraic version of the construction of Baum, Fulton, MacPherson [1]. The case of the multiplicity of an m-primary ideal is the case in which the matrix is a row matrix. We first discuss this special case, leading to a somewhat non-standard version of the definition which can be extended to the case of a more general matrix.

We first recall the definition of the multiplicity of an m-primary ideal of a local ring (we refer to Matsumura [8], section 14 for more details on this subject). Let A be a commutative Noetherian local domain of dimension d, and let I be an ideal such that A/I has finite length. There is then a polynomial $P(n)$ of degree d such that $P(n) = \text{length}(A/I^n)$ for large n. The *multiplicity* of the ideal is defined to be $d!$ times the leading coefficient of this polynomial.

We next present a slightly different version of this definition. Let I be an m-primary ideal as above, and let $R(I) = \oplus_n I^n$ be the Rees ring of I. Let G be the associated graded ring of I, so that $G = R(I) \otimes_A A/I$ and $G_n = I^n/I^{n+1}$ for all n. Let $i_1, \ldots i_k$ be a set of generators of I. We can then map a polynomial ring in k generators over A/I, which we denote $(A/I)[T_1, \ldots T_k]$, onto G by sending T_j to i_j in degree 1. We thus have a map of graded rings, and in this way we can consider G as defining a closed subscheme of dimension $d-1$ of projective space of dimension $k-1$ over A/I. In this formulation, the multiplicity of I is simply the degree of this subscheme of projective space.

The degree of a graded ring G can be defined in several equivalent ways. The usual algebraic definition is to let the degree be $(d-1)!$ times the leading coefficient of the Hilbert polynomial of G, where d is the Krull dimension of G. Using this definition of degree, it is clear that this definition of multiplicity agrees with that given above. From the geometric point of view, the degree of a subscheme of projective space can be considered to be the the number of points obtained by intersecting the subscheme with $d-1$ generic hyperplanes. The connection between these definitions comes from the following considerations: if x is any element of degree one in G such that $\dim(G/xG) = \dim(G) - 1$, then, letting G_x denote the set of elements annihilated by x, multiplication by x induces a short exact sequence

$$0 \to G_x \to G \to G \to G/xG \to 0.$$

Since multiplication by x shifts degrees by 1, we obtain that $P(n) - P(n-1)$ is a difference of the Hilbert polynomials of G/xG and G_x, which have lower dimension. In the case in which x is a non-zero-divisor, this is simply the Hilbert polynomial of G/xG. Since the leading coefficient of the polynomial given by $P(n) - P(n-1)$ is the degree of $P(n)$ times the leading coefficient of $P(n)$, this argument shows that the degree of G/xG is equal to the degree of G. Hence, if we can continue to find non-zero-divisors, we may continue, and eventually get down to dimension zero, where the scheme is a finite set of points and the degree is the number of these points. Since dividing by an element of degree one is the algebraic version of intersecting with a hyperplane, this shows that the two definitions agree. Thus if G is Cohen-Macaulay, the degree of G is simply the length of $G/(x_1, \ldots x_d)$, where $(x_1, \ldots x_d)$ is a system of parameters of degree one.

If G is not Cohen-Macaulay, the answer is somewhat more complicated, but it can still be viewed as the number of points obtained by intersecting generic hyperplanes. To define this intersection rigorously, it is necessary to define the intersection of a hyperplane on cycles and the cycle associated to a module as in the first chapters of Fulton [3]. One can then show that if x is a zero-divisor on G, the intersection of the class of the cycle defined by G with the hyperplane defined by x is the difference of the classes defined by G/xG and by G_x as in the above exact sequence. Note that this difference is simply the alternating sum of the homology modules of the Koszul complex on the one element x tensored with G. If we then have a system of parameters $(x_1, \ldots x_d)$, the degree can be computed as the alternating sum of lengths of the homology modules of the Koszul complex on $(x_1, \ldots x_d)$. The fact that this gives the correct answer for the intersection of d hyperplanes may be proven using induction and the standard exact sequences of Koszul complexes of the form

$$0 \to K_\bullet(x_1, \ldots, x_{k-1}) \to K_\bullet(x_1, \ldots, x_k) \to K_\bullet(x_1, \ldots, x_{k-1})[-1] \to 0.$$

We present one further variation on this definition before generalizing it. One may replace the ring A and ideal I by a polynomial ring $A[s]$ in one variable and the ideal $(I, s) = J$; it is easy to see that J has the same multiplicity as I. Next, take the Rees ring $R(J) = \oplus_n J^n$ of J. Instead of tensoring with A/J to obtain the asociated graded ring, we tensor with $A[s]/sA[s]$. The graded ring thus obtained will not have components of finite length; the component of degree n will in fact be isomorphic to

$$(s^n A + s^{n-1} I + \ldots + I^n)/(s^{n+1} A + s^n I + \ldots + sI^n) \cong A/I \oplus I/I^2 \oplus \ldots \oplus I^{n-1}/I^n \oplus I^n.$$

While the components do not have finite length, it is still possible to define the degree of this graded ring. In fact, only the last summand in the above decomposition does not have finite length, and if we divide this ring by any m-primary ideal K of A which annihilates the other components (such as I, for example), we will have a graded ring with components of finite length and can compute its degree. This degree does not depend on the choice of K, since the results will differ by a polynomial of degree at most $d - 1$. Thus we may define the multiplicity as the degree of the graded ring $R(J) \otimes_{A[s]} (A[s]/sA[s])$, which has dimension $d + 1$ and defines a projective scheme of dimension d.

This last description is almost (but not quite) the definition of the multiplicity of an m-primary ideal as a special case of the definition of the Chern character of a matrix. We continue this example and give the precise description later in this section.

We now generalize this definition to define the Chern classes of a matrix M with support at the maximal ideal. We let A be a local domain as above, and let M be a matrix with entries in A. Let r be the rank of M; since A is a domain, the rank is unambiguously defined. We say that M has support at the maximal ideal if the ideal of r by r minors is m-primary.

Let $A[s]$ be a polynomial ring over A in one variable. Let M be an m by n matrix with support at the maximal ideal, and let $\tilde{M} = [M^t \;\; sI]$, where sI is s times an n by n identity matrix, that is, a matrix with s in each diagonal position and 0 everywhere else, and where M^t denotes the transpose of M. Then $\tilde{M}$ is an n by $m+n$ matrix with coefficients in $A[s]$. Let J be the ideal generated by all maximal minors of $\tilde{M}$. Take the Rees ring $R(J)$ of J, and let Q be the graded ring obtained by tensoring $R(J)$ with $A[s]/sA[s]$. As before, there is a part not of finite length which can be taken care of by tensoring with A/I, where I is the ideal of maximal minors of M (or any other m-primary ideal which annihilates the rest of $R(J) \otimes A[s]/sA[s]$).

Now let $i_1, \dots i_k$ be positive integers with $i_1 + \dots + i_k = d$. For each j between 1 and k, choose a generic $n+i_j-1$ by n matrix consisting of linear combinations of columns of $\tilde{M}$. Next, divide by the ideal of Q generated by the n by n minors of all these matrices, considered as elements of Q of degree one. If Q is Cohen Macaulay, the result we are after is the degree of the resulting quotient. If Q is not Cohen-Macaulay, the computation is more complicated, but it can be computed as an alternating sum of degrees of homology modules as in the case of ordinary multiplicities. We denote this number by $c_{i_1}c_{i_2}\cdots c_{i_k}$.

The justification for this procedure and the proof that it is well-defined can be seen by identifying this construction with an intersection with standard Chern classes in a Grassmannian; see Fulton [3] and Roberts [14]). Just as the graded ring G defined in the definition of the multiplicity of an ideal defines a subscheme of projective space, the graded ring Q, which is generated by the maximal minors of a matrix, defines a subscheme of a Grassmannian. In this interpretation, the number $c_{i_1}c_{i_2}\cdots c_{i_k}$ defined above is the intersection of this subscheme with the Chern classes $c_{i_1}, c_{i_2}, \dots c_{i_k}$ of the canonical quotient bundle of the Grassmannian. In the case of projective space, the only significant Chern class is c_1, and in this notation the degree which we computed above is $c_1c_1\dots c_1 = c_1^d$.

We next define the local Chern character defined by the matrix M. We consider $c_1, \dots c_d$ as elementary symmetric polynomials in $x_1, \dots x_d$. Write

$$\frac{1}{d!}(x_1^d + \dots + x_d^d)$$

in terms of elementary symmetric polynomials. Then the Chern character of M, $\mathrm{ch}_d(M)$, is this polynomial in $c_1, \dots c_d$ with each term evaluated by the procedure above. For large d this expression is complicated, but in small cases it is not so diifficult. For example, if $d = 2$, we have

$$\mathrm{ch}_2(M) = \frac{1}{2}(x_1^2 + x_2^2) = \frac{1}{2}(x_1 + x_2)^2 - x_1 x_2 = \frac{c_1^2}{2} - c_2.$$

In this case $\frac{c_1^2}{2}$ is one half the degree of the graded ring Q, which can be shown to be the multiplicity of the ideal of r by r minors of M. The term $-c_2$ is more mysterious; in general (if Q is Cohen-Macaulay) it is the degree of the quotient of Q by the minors of a generic n by $n+1$ submatrix of $\tilde{M}$. We give an example in the next section where c_2 can be interpreted as a mixed multiplicity.

We now return to the case of the multiplicity of an ideal. Let I be an m-primary ideal with generators $i_1, \ldots, i_k$. These generators can be represented by either a row matrix or a column matrix. We show that if they are represented as a row matrix, which we denote M, the general definition of the Chern character $\mathrm{ch}_d(M)$ will give the multiplicity of I divided by $d!$, while if it is represented by a column matrix M^t, the answer will be $(-1)^{d+1}$ times this number. (This is a special case of the relation between the Chern character of a matrix and that of its transpose; we shall prove it directly in this case.)

The matrix $\tilde{M}$ as in the general construction will be

$$\begin{pmatrix} i_1 & s & & \\ \vdots & & \ddots & \\ i_k & & & s \end{pmatrix},$$

while the matrix $\tilde{M}^t$ will be

$$(i_1 \quad \ldots \quad i_k \quad s).$$

The ideals J in the construction of the graded ring are $(s^k, s^{k-1}I)$ and (s, I) respectively. We note that the Rees rings and the graded rings Q defined by these ideals are isomorphic, so they are both isomorphic to the graded ring defined previously. In the previous description, we showed that the multiplicity of I is the degree of this graded ring, which in terms of Chern classes is c_1^d. Now in the first case, the matrix $\tilde{M}$ is a k by $k+1$ matrix. We recall that c_2 is defined by intersecting with the ideal of minors of a generic k by $k+1$ matrix whose columns are linear combinations of columns of $\tilde{M}$; in this case, since $\tilde{M}$ is itself k by $k+1$, this ideal will include all the minors of $\tilde{M}$. These minors generate all of Q_1, so the contribution of any term containing c_2, and similarly c_i for any $i > 2$, will be zero. Thus $\mathrm{ch}_d(M) = \frac{c_1^d}{d!}$ = the multiplicity of I divided by $d!$ as claimed.

In the second case, $\tilde{M}^t$ is a row matrix, and c_i is obtained by intersecting with the ideal J_i generated by the minors of a generic 1 by i matrix with entries in the ideal generated by the entries of $\tilde{M}^t$. Thus the ideal J_i is generated i generic elements of Q_1, so that we have the equality $c_i = c_1^i$ for each i. It can be shown that substituting c_i with c_1^i in the general formula for ch_d gives $(-1)^{d+1}\frac{c_1^d}{d!}$, which gives the required equality in this case. Thus up to sign, the multiplicity of an m-primary ideal can be obtained as the Chern character defined by either a row or the column matrix whose entries generate the ideal.

We conclude with a summary of some of the basic properties of local Chern characters.

If M can be written in block form

$$M = \begin{pmatrix} M_1 & 0 \\ 0 & M_2 \end{pmatrix},$$

Then $\mathrm{ch}_d(M) = \mathrm{ch}_d(M_1) + \mathrm{ch}_d(M_2)$. This additivity property is perhaps the main reason for considering Chern characters as opposed to the simpler Chern classes.

If M_1 and M_2 are two matrices such that the ideal of minors J_1 of $\tilde{M}_1$ is naturally contained in the ideal of minors J_2 of $\tilde{M}_2$, and J_2 is integral over J_1, then the Chern classes defined by the two matrices are equal. This statement is a little vague, since it is not clear when one ideal of minors is "naturally" contained in another. One specific instance, which will be developed in section three, is where one of the matrices is diagonal as in the previous paragraph, and the other is triangular with the same matrices on the diagonal.

A simple but important property is that the Chern classes are independent of change of basis, so that two matrices which represent the same map of free modules with respect to different bases have the same Chern classes.

Finally, we recall a special case of the local Riemann-Roch formula. Let A be a regular local ring of dimension d, and let N be a module of finite length. Let $F_\bullet$ be a free resolution of N, and for each i between 1 and d, let M_i be a matrix defining the map from F_i to F_{i-1}. Then

$$\mathrm{length}(N) = \sum_{i=1}^{d} (-1)^{i+1} \mathrm{ch}_d(M_i).$$

This connection between the length of a module and the Chern characters of the matrices in its free resolution is one of their most important properties. In section 4 we work out an example on modules defined by monomials where this formula has a geometric interpretation.

2. Multiplicities of Modules of finite length

In this section we discuss a definition of the multiplicity of a module of finite length which arises naturally in the context of local Chern characters and discuss its relation to the Buchsbaum-Rim multiplicity. This latter extension of the idea of the multiplicity of an ideal to the case of more general matrices and modules is due to Buchsbaum and Rim [2], and is given another description in Kirby [6]. Recently Kirby and Rees [7] have given very general definitions of multiplicities of this type for modules over graded rings.

The multiplicity we define here for a module N of finite length is the dth Chern character of the matrix which defines a presentation of N.

DEFINITION. Let N be a module of finite length over an integral domain A of dimension d, and let

$$F_1 \to F_0 \to N$$

be a presentation of N. We define the Chern multiplicity of N to be $d!$ times $\mathrm{ch}_d(M)$, where M is a matrix defining the map from F_1 to F_0.

As outlined in the previous section, if $N = A/I$, the Chern multiplicity of A/I is the ordinary multiplicity of I.

We next compare the Chern multiplicity to the Buchsbaum-Rim multiplicity. We recall the definition of Buchsbaum-Rim multiplicity as given by Rees. Let N be a module of finite length over a local ring A, which, as usual, we assume is an integral domain. Let F be a free module of rank r mapping onto N, and let K be the kernel. Let S be the symmetric algebra on F, which is a polynomial algebra over A on r variables, and let T be the subalgebra of S generated by K. The Buchsbaum-Rim multiplicity is then defined to be the n^D coefficient of the polynomial which gives the length of S_n/T_n for large n, multiplied by $D!$, where $D = \dim(A) + r - 1$.

It is clear that this definition also coincides with the usual definition for m-primary ideals. However, for non-cyclic modules the two concepts are quite different. Note that, since local Chern characters are additive on direct sums of matrices, the Chern multiplicity is additive on direct sums of modules (although it is not additive on short exact sequences in general).

EXAMPLE 2.1. As a first example we compute the Buchsbaum-Rim multiplicity of a direct sum of t copies of the residue field k. Here $r = t$, and $K = mF$, so that $S_n/T_n \cong (A/m^n)^g$, where g is the number of monomials of degree n in t variables. The number of monomials is a polynomial of degree $t-1$ with leading coefficient $1/(t-1)!$; that giving the length of A/m^n is of degree $d = \dim(A)$ with leading coefficient $e/d!$, where e is the multiplicity of the ideal m in A. Thus the Buchsbaum-Rim multiplicity in this case is $e(d+t-1)!/d!(t-1)!$. The Chern multiplicity is te, by additivity.

EXAMPLE 2.2. Let I and J be m-primary ideals in a two dimensional local domain. We compute the Buchsbaum-Rim multiplicity of $A/I \oplus A/J$. In this case the submodule is $I \oplus J$, and its nth power is $I^n \oplus I^{n-1}J \oplus \ldots \oplus J^n$. We recall (see for example Teissier [15]) that there is a polynomial in two variables of degree two $P(r,s) = \frac{e(I)}{2}r^2 + e(I,J)rs + \frac{e(J)}{2}s^2+$ lower degree terms, whose value at (r,s) is the length of A/I^rJ^s for large r and s. Here $e(I)$ and $e(J)$ are the multiplicities of I and J respectively, and $e(I,J)$ is called the mixed multiplicity of I and J. Putting this in the above formula, we obtain that the length of $A/I^n \oplus A/I^{n-1}J \oplus \ldots \oplus A/J^n$ is $P(n,0) + P(n-1,1) + \ldots + P(0,n)$. Working this out shows that the Buchsbaum-Rim multiplicity is $\frac{1}{6}(e(I) + e(I,J) + e(J))$.

In this case the local Chern character can also be easily computed in terms of mixed multiplicities. Since local Chern characters are additive on direct sums, we have that $\mathrm{ch}_2(A/I \oplus A/J) = \frac{1}{2}(e(I) + e(J))$, and the Chern multiplicity is $(e(I) + e(J))$. Now if M is the matrix whose cokernel is $A/I \oplus A/J$, we have that the ideal of 2 by 2 minors is IJ. Since $\dfrac{c_1^2}{2}$ is the multiplicity of this ideal, we have that $\dfrac{c_1^2}{2} = e(I)/2 + e(I,J) + e(J)/2$. Hence in this case $c_2 = e(I,J)$ is the mixed multiplicity of the two ideals.

While the additivity property of the Chern multiplicity seems natural, a property which is not obvious is positivity. Since the formula for the local Chern character has both positive and negative terms, it is not clear that the Chern multiplicity of a non-zero module must always be positive. It can be shown that in dimension two positivity follows from a basic inequality of Teissier [16], proven in general by Rees and Sharp [12], but it is not known whether it holds in higher dimension.

3. Reductions of ideals and modules

One of the main results in the theory of multiplicities is the theorem of Rees that (under suitable conditions) the multiplicity of an m-primary ideal I is the same as that of an ideal J contained in I if and only if J is a reduction of I (see Rees [10]). We recall that J is said to be a reduction of I if every element $x \in I$ satisfies a polynomial $x^n + a_1x^{n-1} + \ldots + a_n = 0$ with $a_i \in J^i$ for all i. In this situation I is also said to be integral over J. In this section we define, for each integer k between 1 and the dimension of A, the notion of "k-reduction". This definition is related to the Chern characters defined by the matrices in free resolutions of A/I and A/J. A 1-reduction is the same as a usual reduction, and as k increases, the condition for J to be a k reduction of I becomes stronger.

We make this defintion not only for ideals, but also for submodules of a free module with quotient of finite length. The concept of reduction has been generalized to this case by Rees [11], and the corresponding theorem has been proven using the Buchsbaum-Rim multiplicity.

Let $K \subseteq L$ be submodules of a free module F_0 such that F_0/K has finite length. Let $F_\bullet$ be a free resolution of F_0/K. We may extend this resolution to one of L by adding new generators to the resolution $F_\bullet$ at each stage. That is, we begin by defining a map from $F_1 \oplus G_1$ to F_0 by mapping the generators of the free module G_1 onto elements of L which generate L/K. Continuing in this way, we obtain a resolution $H_\bullet$ of F_0/L such that $H_i = F_i \oplus G_i$ for each i. This resolution will not in general be minimal, but it can be chosen so that the quotient complex $G_\bullet$ is minimal. The complex $G_\bullet$ will be a free resolution of L/K with degree shifted by one; in fact, our construction is simply the mapping cone of the map from a resolution of L/K to a resolution of F_0/K induced by the inclusion $L/K \subseteq F_0/K$. The matrices which define the maps

$H_i = F_i \oplus G_i \to H_{i-1} = F_{i-1} \oplus G_{i-1}$ will be in triangular form

$$\begin{pmatrix} A_i & B_i \\ 0 & C_i \end{pmatrix}$$

where A_i, B_i, and C_i are submatrices of appropriate sizes.

Let M_i denote the ith matrix in this resolution, and let N_i denote the matrix $\begin{pmatrix} A_i & 0 \\ 0 & C_i \end{pmatrix}$. Let $J(M_i)$ and $J(N_i)$ be the ideals of the extended matrices $\tilde{M}_i$ and $\tilde{N}_i$ as described in section 1. It is clear that every minor of $\tilde{N}_i$ is also a minor of $\tilde{M}_i$, so we have $J(M_i) \supseteq J(N_i)$.

DEFINITION. Let $K \subseteq L \subseteq F_0$ be submodules of a free module as above. We say that K is a k-reduction of L if for each $i = 1, \ldots k$ the ideal $J(N_i)$ is a reduction of $J(M_i)$.

THEOREM 3.1. *Let $K \subseteq L$ be submodules of F_0 such that F_0/K has finite length. Then K is a 1-reduction of L if and only if K is a reduction of L.*

PROOF. Suppose that K is a reduction of L. The matrix M_1 in the resolution of F_0/L has the form $(A_1 \;\; B_1)$ where the columns of A_1 generate K and those of B_1 are added to complete a set of generators of L. By Rees [11], K is a reduction of L if and only if for every map to a discrete valuation ring, the image of the columns of B_1 are in the module generated by the columns of A_1. If this holds, then the images of minors are in the ideal generated by minors of A_1. This implies that the ideal of t by t minors of M_1 is integral over the ideal of t by t minors of A_1 for each t, and this implies that $J(M_1)$ is integral over $J(A_1)$, which is the condition to be a 1-reduction.

Conversely, suppose that $J(M_1)$ is integral over $J(A_1)$. If we map A to a discrete valuation ring, this condition will still hold. Hence the Chern characters defined by the matrices M_1 and A_1 over the discrete valuation ring will be equal. However, in the case of a discrete valuation ring the Chern character is simply the length of the quotient, M_1 must define the same submodule as A_1. Hence K is a reduction of L.

One of the motivations for attempting to generalize the notion of reduction is the recent introduction of the concept of tight closure of Hochster and Huneke (see for instance [5]). They have used this idea to give new proofs of several of the homological conjectures. While these proofs still rest on the use of iterations of the Frobenius map as used by Peskine and Szpiro in their proof of the Intersection Theorem [9], there is some hope that a general definition of a closure operator of this type could lead to proofs of some of the questions, such as the Monomial Conjecture of Hochster [4], which are still open in mixed characteristic. In one direction, it has been shown that the integrality of certain module extensions is sufficient to imply this conjecture (see Roberts [13]); these extensions are contained in the tight closure in positive characteristic. If one wishes to restrict attention to ideals, a basic object of study is the relation between an ideal J generated by part of a system of parameters and its extension I, which consists of all elements annihilated modulo J by the next element in a system of parameters. Of course, if A is Cohen-Macaulay, then $I = J$, and Hochster and Huneke have shown that I is always contained in the tight closure of J in positive characteristic. We show below that in dimension 3, if J is a 2-reduction of I, then the Monomial Conjecture must hold. However, we first show that in the case of finite projective dimension, a module is a k-reduction of its tight closure for all k and give an example to show that the converse does not hold.

THEOREM 3.2. *Let $K \subseteq L$ be submodules of a free module as above, and assume that K has finite projective dimension. If L is contained in the tight closure of K, then K is a k-reduction of L for all k.*

PROOF. In this case, we may assume that the resolution $F_\bullet$ of F_0/K is finite. Hence the complex $F_\bullet$ remains exact when tensored any number of times with the Frobenius map. We use the notation above, but assume that we have tensored with some iteration of the Frobenius map. Let c be a non-zero element of A such that c times any column in M_1 is in the submodule generated by the columns of A_1. We replace each of the upper triangular matrices in the resolution of F_0/L by the matrix obtained by replacing B_i in the upper corner with cB_i. Since the map from H_1 to F_0 in this new complex can be factored through $F_1 \to F_0$, and since $F_\bullet$ is exact, we can find a retraction of $H_\bullet$ onto $F_\bullet$. Thus the ideals of minors of the new matrices (those with B_i replaced by cB_i) are contained in the ideals of minors of the corresponding diagonal matrix. Since there is a fixed integer m (depending on the ranks of the matrices in the resolution) such that c^m times the ideals of minors of the original matrices are contained in the ideals of minors of the new matrices, we deduce that the ideals of minors of the matrices in the resolution of F_0/L are integral over (and in fact in the tight closure of) the ideals of minors of the corresponding diagonal matrices. Hence L is k-integral over K for all k.

While the converse of Theorem 3.2 is not true in general (we give an example below), if both K and L have finite projective dimension, then it can be shown that if K is a k-reduction of L for all k, then the Chern multiplicity of the module $\mathrm{Ext}^d(L/K, A)$ must be zero. Thus, for example, in a regular local ring, in any extension of ideals with quotient of length one, it is impossible for the smaller module to be a k reduction of the larger one for all k.

EXAMPLE 3.1. This example is due to Hochster, who used it to show that the regular closure of an ideal can be larger than its tight closure. Let $A = k[x, y, z]/(x^3 + y^3 + z^3)$. Then Hochster has shown that x is in the regular closure but not the tight closure of (y, z). The resolution of (y, z) is simply the Koszul complex. The resolution of (x, y, z) has first kernel defined by the matrix:

$$\begin{pmatrix} -y & -z & x^2 & 0 \\ x & 0 & y^2 & -z \\ 0 & x & z^2 & y \end{pmatrix}.$$

The corresponding "diagonal" matrix is:

$$\begin{pmatrix} -y & -z & x^2 & 0 \\ 0 & 0 & 0 & -z \\ 0 & 0 & 0 & y \end{pmatrix}.$$

Since x is integral over (y, z), it is clear that all minors are integral over those of the product, and hence x is 2-integral over (y, z).

EXAMPLE 3.2. Let $A = k[x, y, z]$ and $J = (x^n, y^n, z^n)$, and let I be an ideal generated by J and one other monomial. It is known that a monomial is integral over J, and hence J is a 1-reduction of I, if and only if its degree is at least n. Furthermore, it can be shown (with a little computation) that I is 2-integral over J if and only if the degree of the new monomial is at least $2n$. It follows from the above results on modules of finite projective dimension (and can be verified directly in this case) that I is 3-integral over J only if $I = J$. It would be interesting to know if these results extend to higher dimensions.

We conclude this section by proving the connection between k-reductions and the Monomial Conjecture referred to above. Let $R = \mathbb{Z}_p[[x, y]]$, and let A be a finite extension of R of depth 2. Let I_n be the ideal (x^n, y^n), and let $\overline{I^n}$ be the set of elements x of A such that $p^k x \in I_n$ for some k.

THEOREM 3.3. *If I_n is a 2-reduction of $\overline{I^n}$ for all n, then the Monomial Conjecture holds for A.*

PROOF. If the Monomial Conjecture does not hold, then there is an integer n and elements a, b, and c of A such that

$$p^{n-1}x^{n-1}y^{n-1} = ax^n + by^n + cp^n.$$

Rewriting this equation, we obtain that

$$p^{n-1}(x^{n-1}y^{n-1} - cp) = ax^n + by^n,$$

or that $x^{n-1}y^{n-1}$ is congruent modulo p to an element of $\overline{I^n}$.

We next resolve I_n (by a Koszul complex) and extend it to a resolution of $\overline{I^n}$ as described above. In so doing, we take x^n and y^n as generators for I_n and an element of $\overline{I^n}$ congruent to $x^{n-1}y^{n-1}$ modulo p for the next generator of $\overline{I^n}$. The first column of the next map in the resolution, which we denote M, will have entries $-y^n, x^n, 0, \dots 0$. The matrix M has the form

$$\begin{pmatrix} -y^n & B \\ x^n & \\ 0 & C \end{pmatrix}.$$

By choosing a minimal set of new generators, we may assume that C has entries in the maximal ideal of A. By hypothesis, I_n is a 2-reduction of $\overline{I^n}$, which implies that the ideal 2 by 2 of minors of M is integral over (x^n, y^n) times the ideal generated by the entries of C. Now reduce modulo p. Since p is not a zero-divisor modulo $\overline{I^n}$, the entire resolution is still exact, and it is clear that the integrality condition will continue to hold. The first three generators of $\overline{I^n}$ modulo p are $x^n, y^n, x^{n-1}y^{n-1}$. Thus, since the complex is still exact, some linear combination of its columns must produce the matrix :

$$\begin{pmatrix} y^{n-1} & 0 \\ 0 & x^{n-1} \\ -x & -y \end{pmatrix}.$$

Two minors of this submatrix are x^n and y^n. Hence our hypotheses imply that x^n and y^n are integral over $m(x^n, y^n)$, which is impossible. Thus the Monomial Conjecture must hold.

4. Monomial modules in two variables

In this section we discuss a simple but interesting example, the case of a monomial module over a polynomial ring $A = k[x, y]$ in two variables. By "monomial module" we mean a module N which is a quotient I/J, where I and J are ideals of A generated by monomials. We assume that I and J are m-primary, so that N has finite length. Since I and J are generated by monomials, they must each contain a power of x and a power of y.

If $N = I/J$ is a monomial module, then the set S of monomials which are in I but not in J forms a basis for N, and the number of such monomials is the length of N. We may consider the set of monomials as representing lattice points in the plane, which we take to be in the first quadrant, with powers of x along the usual x-axis and powers of y along the y-axis. If we take the union of all unit squares whose lower left corner is in S we obtain a region R in the plane whose area is the length of N.

Let M_1 and M_2 be the matrices in a minimal free resolution of N. We recall that by the local Riemann-Roch formula the length of N, which is the area of the region R, is the sum of $\mathrm{ch}_2(M_1)$ and -$\mathrm{ch}_2(M_2)$. We will give an expression for these numbers which is described geometrically in terms of the region R. More precisely, $\mathrm{ch}_2(M_1)$ can be computed as follows: let L be the shortest broken line between the upper left corner and the lower right corner of R which lies entirely in R. Then $\mathrm{ch}_2(M_1)$ is the area of the part of R which lies below this line.

The following diagram illustrates this situation in the case where $I = (x^5, y^2)$ and $J = (x^6, x^2y^3, y^5)$.

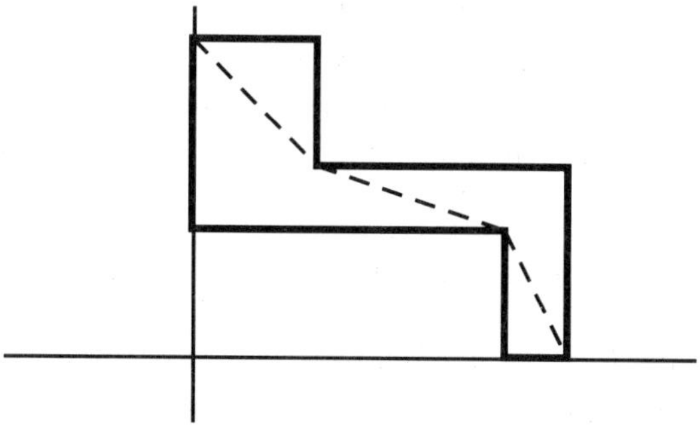

The heavy line is the boundary of the region R and the dotted line is the line L.

We next give a more precise definition of the region under consideration. Let I be generated by the monomials $x^{i_0}, x^{i_1}y^{j_1}, \dots x^{i_{r-1}}y^{j_{r-1}}, y^{j_r}$. Let J be generated by $x^{m_0}, x^{m_1}y^{n_1}, \dots x^{m_{s-1}}y^{n_{s-1}}, y^{n_s}$. By eliminating unnecessary generators we may assume that the sequence $\{i_k\}$ is strictly decreasing and that the sequence $\{j_k\}$ strictly increasing, and similarly for the generators of J. We then let $a_0 = m_0 - i_0$ and $a_k = i_{k-1} - i_k$ for $k \geq 1$. Similarly, let $b_k = j_{k+1} - j_k$ for $k \leq r$, and let $b_r = n_s - j_r$. We denote the generators of the modules N which are the images of those of I listed above by $e_0, \dots, e_r$.

Consider the following lattice points: $(m_0, 0), (i_0, j_0), \dots (i_r, j_r), (0, n_s)$. We refer to these as *lower points.* Similarly, we refer to the points $(m_0, , 0), (m_1, n_1)$, $\dots, (m_{s-1}, n_{s-1}), (0, n_s)$ as *upper points*; note that $(m_0, 0)$ and $(0, n_s)$ are both lower and upper.

The line L separating the two regions of R as described above can be defined more precisely as follows: begin with the point $(0, n_s)$. Consider the slopes between $(0, n_s)$ and the upper and lower points. If the slopes to the upper points are all at least as large as the slopes to the lower points, simply connect $(0, n_s)$ to $(m_0, 0)$ Otherwise, take the point with lowest x-coordinate which violates this condition; if it is an upper point and has x coordinate $\tilde{x}$, connect $(0, n_s)$ to the lower point with largest x-coordinate less than $\tilde{x}$. If it is a lower point, proceed similarly with lower and upper reversed. We then continue this process from the point which we just connected to $(0, n_s)$, and proceed until we reach $(m_0, 0)$. We denote this broken line L. It is not hard to see that L lies entirely within R and is the shortest such broken line. In particular, if we construct this line similarly beginning at $(m_0, 0)$ we obtain the same result.

As mentioned above, we will show that $\mathrm{ch}_2(M_1)$ is the area of the part of R which lies below L. The proof is by reduction to the case in which $I = A$, where this result is known. We first note that the local Riemann-Roch theorem implies that this result is true if and only if -$\mathrm{ch}_2(M_2)$ is the area of the part of R which is above L. Now taking the dual $\mathrm{Hom}_k(N, k)$ takes N to another monomial module in which the two regions are reversed and the matrices in the resolution are interchanged and transposed. It thus follows that the theorem is true for N if and only if it is true for the dual of N.

While this theorem is known for the case in which N is cyclic, it is difficult to find a reference, so we present a brief summary of the proof. In this case $\mathrm{ch}_2(M_1)$ is one half times the multiplicity of the ideal J, so we have

$$\mathrm{ch}_2(M_1) = \lim_{n\to\infty} \frac{\mathrm{length}(A/J^n)}{n^2}.$$

In addition, the line L is the boundary of the convex hull of the region consisting of monomials in the ideal J.

PROPOSITION 4.1. *If J is a $\mathfrak{m}$-primary ideal generated by monomials, then the multiplicity of J is the area outside the convex hull of the region consisting of monomials in J.*

PROOF. We note first that the area outside J itself is the length of A/J. The nth power of J is the ideal generated by the nth powers of the generators of J together with products of distinct generators of J, which fill up lattice points on lines between the nth powers. Dividing by n^2 can be interpreted geometrically by contracting in each direction by n, which moves the nth powers back to the original generators and the points between them to points between them with rational coordinates with denominators divisible by n. Taking the limit, we fill up the convex hull of the region of monomials in J, so the limit is the area of the complement of the convex hull as required.

THEOREM 4.2. *Let the monomial module $N = I/J$, the region R, and the broken line L be as above. Then $\mathrm{ch}_2(M_1)$ is the area of the part of R which is below L.*

PROOF.. We prove this theorem by two reductions. First, we prove it in the case in which L is a straight line connecting $(m_0, 0)$ and $(0, n_s)$ by reduction to the cyclic case outlined above. Second, we use induction on the number of straight line segments in L to deduce the general case. In both cases the result follows from a lemma which states that if we add a new generator to the relations on N, so that we add another column to the matrix M_1 to get a matrix M_1', the position of the new generator relative to the line L implies that the ideals of t by t minors of M_1' are integral over the ideals of t by t minors of M_1 for all t. Thus the local Chern character remains the same; in the constructions we use the area below L also remains the same, allowing us to conclude the result by induction.

The first step is to describe the matrix M_1. If J is generated by two elements x^{m_0}, y^{n_1}, then M_1 is an $r+1$ by $r+2$ matrix. Using the above notation for the generators of I and J, M_1 is the matrix

$$\begin{pmatrix} x^{a_0} & y^{b_0} & 0 & \dots & 0 \\ 0 & x^{a_1} & y^{b_1} & \dots & 0 \\ 0 & \dots & \dots & \dots & 0 \\ 0 & \dots & \dots & x^{a_r} & y^{b_r} \end{pmatrix}.$$

If there are more generators for J, they contribute extra columns to the matrix M_1, each of which may be chosen to contain one non-zero entry. If J is generated by two elements, then its dual is cyclic, so the result is true in this case. We prove the general case in which L is a straight line by induction on the number of generators of J.

Let M_1' be obtained from M_1 by adding a new generator of J which lies above the line L. Let the column defining the new generator have non-zero entry $x^g y^h$ in the kth row, so that the new relation is a multiple of e_k. We first consider a t by t minor whose columns consist of the new one and $t-1$ adjacent columns of the left half of the matrix including at least one of the columns with non-zero entry in row k. To be non-zero, the rows of the minor must also be adjacent and contain the kth row. In this case the minor is a monomial of the form $y^{b_{k-i}} \dots y^{b_{k-1}} x^g y^h x^{a_{k+1}} \dots x^{a_{k-i+t-1}}$. We claim that this element is integral over the ideal generated by the two minors obtained by taking the same set of rows and the same columns from the original matrix, adding first the next column to the left and then the one to the right. These two minors are $x^{a_{k-i}} \dots x^{a_{k-i+t-1}}$ and $y^{b_{k-i}} \dots y^{b_{k-i+t-1}}$ respectively. This can be seen most easily as follows: let the new generator be the element $x^c y^d$ as an element of J; divide $x^c y^d$ by the new minor to obtain a monomial $x^e y^f$, and multiply this monomial by each of the two minors of the original matrix listed above. These points must be lower points of R, so they lie below L. On the other hand, the new minor lies above L, and hence above the line connecting these products. This implies that the new minor is integral over the ideal generated by the old ones, as required.

The following diagram should clarify this argument:

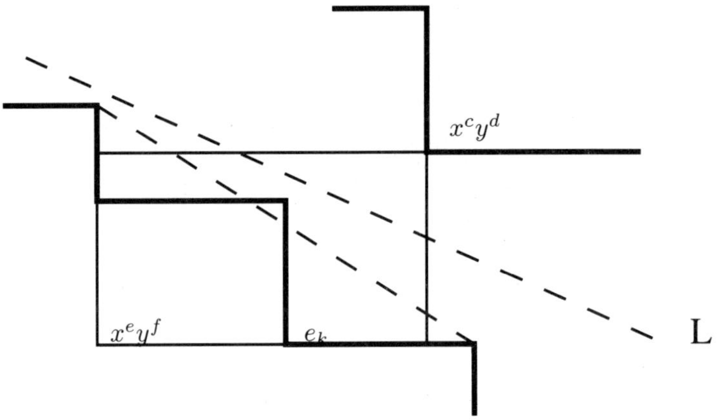

In this diagram, the heavy lines represent the upper and lower boundaries of R. Here $t = 2$, and the minor under consideration is of the form

$$\begin{vmatrix} x^{a_{r-1}} & 0 \\ y^{b_r} & x^g y^h \end{vmatrix}.$$

One dotted line is L, and the other is the line connecting the two lower points as in the argument above. The rectangle surrounded by mostly thin lines has sides of length h and $g+a_{r-1}$, and it corresponds to the above minor. The distances from the lower left corner to the two lower points of the argument (those connected by the dotted line) are $a_{r-1} + a_r$ and $b_{r-1} + b_r$, which correspnd to the two minors of the original matrix. Thus the diagram shows that $x^{g+a_{r-1}}y^h$ is integral over the ideal generated by $x^{a_{r-1}+a_r}$ and $y^{b_{r-1}+b_r}$.

The general case where the line L is straight reduces to the case just proven. Consider the largest set of adjacent columns containing those with non-zero entries in the kth row in the minor under consideration. The entire minor is then a product of this subminor with a product of monomials from the other rows. In the above argument we showed that this subminor is integral over two minors obtained by adding a column to the right and a column to the left. Neither of these columns are in the original minor, and the minor obtained by the product with the other rows are thus minors of the original matrix. Hence the minor is integral over the ideal generated by these two minors, and hence over the ideal of minors of the original matrix.

We next consider the case in which the line L has at least one bend. By replacing the module N by its dual if necessary, we may assume that it bends down at a lower point of the region R. We claim that if we add the monomial at this point to J, the minors of the new matrix are integral over the minors of the previous matrix. The new module is now decomposable, so that we may assume that the theorem holds for each piece by induction; since the region below L is the same in the new module, this proves the theorem.

Again we give a diagram to illustrate this argument:

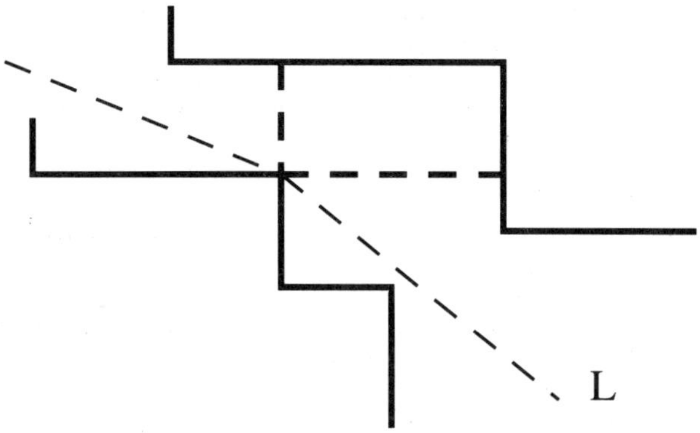

The solid heavy lines in this diagram represent the upper and lower boundaries of the region R as before. The lower left corner of the heavy dotted line is the new generator of J, so that adding this generator extends J and splits R into two separate regions. Note that the sum of the parts of the areas of these regions below L is the same as the original area below L. Thus it suffices to show that the new generator of J produces minors in M_1 which are integral over the original ones. The proof is similar to the previous case, so we do not give all the details, but point out the places in which the argument is different.

As in the previous case, we first consider the case where the columns of the minor consist of the new column plus a set of adjacent columns in the original matrix, and the rows are a set of adjacent rows. The argument is exactly the same until the diagram passes a point where L bends upward. Such a point corresponds to a generator of J. At this point, instead of adding the next column of the matrix, we add the column corresponding to this generator of J and the argument works as before.

The reduction to the case of adjacent rows and columns works as in the previous case.

It would be interesting to know if a similar geometric description exists for the local Chern characters in a resolution of a monomial module in higher dimensions.

Another question is the positivity of the contributions of the different matrices in a resolution, say of a module of finite length over a regular local ring. In high dimensions, it is not even known if the first of these (the Chern multiplicity defined in section 2) is positive. While it seems reasonable to conjecture that the multiplicity is positive, the higher ones are not well understood, even in the case of ideals defined by monomials.

5. References

References

1. P. Baum, W. Fulton, and R. MacPherson, *Riemann-Roch for singular varieties*, Publ. Math. IHES **45** (1975), 101–145.
2. D. A. Buchsbaum and D. S. Rim, *A generalized Koszul complex II. Depth and multiplicity*, Trans. Amer. Math. Soc **111** (1964), 197–224.
3. W. Fulton, *Intersection Theory*, Springer-Verlag, Berlin, Heidelberg, and New York, 1984.
4. M. Hochster, *Canonical elements in local cohomology modules and the direct summand conjecture*, J. of Algebra **84** (1983), 503–553.
5. M. Hochster and C. Huneke, *Tight closure, invariant theory, and the Briançon-Skoda Theorem*, J. of the Amer. Math. Soc. **1** (1990), 31–116.
6. D. Kirby, *On the Buchsbaum-Rim multiplicity associated with a matrix*, J. London Math. Soc. **(2) 32** (1985), 57-61.
7. D. Kirby and D. Rees, *Multiplicities in graded rings 1: The general theory* (to appear).
8. H. Matsumura, *Commutative ring theory*, Cambridge University Press, Cambridge, 1986.
9. C. Peskine and L. Szpiro, *Dimension projective finie et cohomologie locale*, Publ. Math. IHES **42** (1973), 47–119.
10. D. Rees, *Transforms of local rings and a theorem on multiplicities of ideals*, Math. Proc. Camb. Philos. Soc. **57** (1961), 8–17.
11. D. Rees, *Reduction of modules*, Math. Proc. Camb. Philos. Soc. **101** (1987), 431–449.
12. D. Rees and R. Sharp, *On a theorem of B. Teissier on multiplicities of ideals in local rings*, J. London Math. Soc. **13** (1978), 449–463.
13. P. Roberts, *An infinitely generated symbolic blow-up in a power series ring and a new counterexample to Hilbert's fourteenth problem*, J. of Algebra **132** (1990), 461–473.
14. P. Roberts, *Chern classes of matrices over Noetherian rings* (to appear).
15. B. Teissier, *Cycles évanescents, sections planes, et condition de Whitney*, Singularités à Cargèse 1972, Astérisque 7–8, 1973, pp. 285–362.
16. B. Teissier, *Sur une inégalité à la Minkowski pour les multiplicités*, Ann. of Math. **106** (1977), 38–44.

Department of Mathematics, University of Utah, Salt Lake City, UT 84112

E-mail address: roberts@math.utah.edu

Contemporary Mathematics
Volume **159**, 1994

A Computation of Local Cohomology

PAUL C. ROBERTS

ABSTRACT. It is shown that a certain local cohomology module, whose vanishing would imply the Monomial Conjecture, does not vanish for rings of characteristic zero.

Several years ago Mel Hochster showed that some of the homological conjectures, including the Monomial Conjecture, would follow if it could be shown that a certain local cohomology module vanished (Hochster [1]). In positive characteristic, this module is always zero, but it was not known whether or not it vanished in other cases. More recently, this example has come up again in Hochster's theory of solid closure [2]. We show here that in characteristic zero this module does not vanish. We would like to thank Mel Hochster for a careful reading of a preliminary version of this paper and for suggesting several improvements.

We first describe the local cohomology module which is under consideration. Let $R = \mathbb{Z}[a, b, c, x, y, z]$, let $\eta = x^2y^2z^2 - ax^3 - by^3 - cz^3$, and let $A = R/\eta$. This ring arises in a particular case of the Monomial Conjecture, which states that $x^2y^2z^2$ cannot be in the ideal generated by x^3, y^3, z^3 if x, y, z are a system of parameters of a local ring. Equivalently, the conjecture states that there is no homomorphism from A into a local ring under which x, y, z go to a system of parameters. It can be shown (we refer to Hochster [1] for details) that the vanishing of the third local cohomology group of A with support in the ideal (x, y, z) implies that such a map cannot exist. We will show here that this local cohomology module is non-zero, and in fact that it is non-zero when $\mathbb{Z}$ is replaced by any field of characteristic zero. We denote this module H^3.

The module H^3 can be constructed as the direct limit over n of the modules $H_n = A/(x^n, y^n, z^n)$, where the map from H_n to H_{n+1} is multiplication by xyz. We will show that the image of the generator of H_2 is not zero in the limit, or, in

1991 *Mathematics Subject Classification.* Primary 54C40, 14E20; Secondary 46E25, 20C20.
The author was supported in part by a grant from the National Science Foundation
This paper is in final form and no version of it will be submitted for publication elsewhere.

other words, that $x^{n-2}y^{n-2}z^{n-2}$ is not in the ideal of R generated by x^n, y^n, z^n, and η for any n.

Fix n, and let $B = R/(x^n, y^n, z^n)$. Another way of expressing the conclusion that we wish to prove is to say that $x^{n-2}y^{n-2}z^{n-2}$ is not a multiple of η in B. If it were, there would be an element ϕ of B with $\phi\eta = x^{n-2}y^{n-2}z^{n-2}$. The proof that such a ϕ cannot exist is a fairly complicated computation using two gradings on B to break ϕ into more manageable pieces. For the rest of this paper we assume that ϕ is an element of B with $\phi\eta = x^{n-2}y^{n-2}z^{n-2}$.

First, we define a $\mathbb{Z}^3$ (multi-)grading on B letting x, y, z, a, b, c have degrees $(1,0,0), (0,1,0), (0,0,1), (-1,2,2), (2,-1,2), (2,2,-1)$ respectively. In this grading, η is homogeneous of degree $(2,2,2)$ and $x^{n-2}y^{n-2}z^{n-2}$ is homogeneous of degree $(n-2, n-2, n-2)$. Hence if $x^{n-2}y^{n-2}z^{n-2} = \phi\eta$, we may assume that ϕ is homogeneous of degree $(n-4, n-4, n-4)$.

We next write $\phi = \phi_0 + \ldots + \phi_r$ in homogeneous components using the simple grading in which a, b, c have degree 1 and x, y, z have degree 0. Similarly, we write $\eta = \eta_0 - \eta_1$, where $\eta_0 = x^2y^2z^2$ and $\eta_1 = ax^3 + by^3 + cz^3$; note that η_0 and η_1 are the parts of η of degrees 0 and 1 respectively under this grading. Since ϕ is homogeneous of degree $(n-4, n-4, n-4)$ in the $\mathbb{Z}^3$-grading, each ϕ_m will be homogeneous of degree $(n-4, n-4, n-4)$ as well.

We next work out what these homogeneity conditions imply as to which monomials can occur in ϕ_m. Let $a^ib^jc^kx^ry^sz^t$ be a monomial with non-zero coefficient in ϕ_m. Its degree in the $\mathbb{Z}^3$-grading can be computed to be $(r - i + 2j + 2k, s + 2i - j + 2k, t + 2i + 2j - k)$, which must equal $(n-4, n-4, n-4)$, since ϕ_m is assumed to have this degree. Subtracting $(2m, 2m, 2m)$ from both sides of this equality, using the fact that $i + j + k = m$, we get

$$(r - 3i, s - 3j, t - 3k) = (n - 4 - 2m, n - 4 - 2m, n - 4 - 2m).$$

Thus, if we factor out $(xyz)^{n-4-2m}$, in the remaining monomial the exponent of x will be three times that of a, and similarly for y and b and for z and c. In other words, we may write $\phi_m = (xyz)^{n-4-2m}\psi_m(ax^3, by^3, cz^3)$, where ψ_m is homogeneous of degree m.

We have shown that if ϕ is a polynomial in $B = \mathbb{Z}[a, b, c, x, y, z]/(x^n, y^n, z^n)$ with $\phi\eta = x^{n-2}y^{n-2}z^{n-2}$, then ϕ can be written in the form

$$x^{n-4}y^{n-4}z^{n-4}\psi_0(ax^3, by^3, cz^3) + x^{n-6}y^{n-6}z^{n-6}\psi_1(ax^3, by^3, cz^3) + \ldots$$

where each ψ_m is homogeneous of degree m. We wish to show that a polynomial of this form cannot be multiplied by $\eta = x^2y^2z^2 - (ax^3 + by^3 + cz^3)$ to give $x^{n-2}y^{n-2}z^{n-2}$ in B. Note that if we allowed ϕ to be a power series and have negative exponents, we could solve the equation $\phi\eta = x^{n-2}y^{n-2}z^{n-2}$ simply by letting $\psi_m = (ax^3 + by^3 + cz^3)^m$ for each m. We will show that any solution to this equation must agree with this power series solution up to a multiple of a type of polynomial which we call "extremal". No power of $ax^3 + by^3 + cz^3$ can be written in this form in characteristic zero, so a polynomial solution to the equation cannot exist.

To simplify notation, let $s = ax^3, t = by^3$, and $u = cz^3$. We call a monomial $s^i t^j u^k$ *extremal* if $3i, 3j$, or $3k$ is at least equal to $2m + 2$, where $m = i + j + k$ is the degree of the monomial. We call a polynomial $f(s, t, u)$ extremal if each monomial in f with non-zero coefficient is extremal. We note that if the monomials of degree m are arranged in a triangle in the usual way, the extremal ones are close to the corners of the triangle. The theorem we are proving will result from three Lemmas on extremal polynomials.

LEMMA 1. *Let $\rho(s, t, u)$ be a homogeneous polynomial of degree m. If the product $(xyz)^{n-2-2m}\rho(ax^3, by^3, cz^3)$ is in the ideal generated by x^n, y^n, and z^n, then ρ is extremal.*

PROOF. Let $s^i t^j u^k$ be a monomial with non-zero coefficient in ρ; we then have $i + j + k = m$. The hypothesis implies that $(xyz)^{n-2-2m}(ax^3)^i(by^3)^j(cz^3)^k$ is in the ideal (x^n, y^n, z^n), which means that the exponent of x, y, or z is at least n. Suppose that the exponent of x is at least n. This means that $n-2-2m+3i \geq n$, or that $3i \geq 2m + 2$, which is the required inequality. Arguing similarly for y and z, we see that $s^i t^j u^k$, and hence ρ, is extremal.

LEMMA 2. *For each m, we have*

$$\psi_m(s, t, u) = (s + t + u)^m + \sum_{k=2}^{m}(s + t + u)^{m-k}\mu_k(s, t, u)$$

where each $\mu_k(s, t, u)$ is an extremal homogeneous polynomial of degree k.

PROOF. We remark that the sum begins with $k = 2$ because there are no non-zero extremal polynomials of degree less than 2.

The proof is by induction on m. For $m = 0$, we have $x^{n-2}y^{n-2}z^{n-2} = \eta_0\phi_0 = x^2y^2z^2((xyz)^{n-4}\psi_0)$, so that $\psi_0 = 1$ as required.

We now let $m > 0$. Since $x^{n-2}y^{n-2}z^{n-2}$ has degree zero in a, b, c, the higher degree terms must cancel, and we have the equation $\eta_1\phi_{m-1} = \eta_0\phi_m$, or

$$(ax^3 + by^3 + cz^3)(xyz)^{n-4-2m+2}\psi_{m-1} = x^2y^2z^2(xyz)^{n-4-2m}\psi_m.$$

Factoring out $(xyz)^{n-2-2m}$, we obtain

$$(xyz)^{n-2-2m}(\psi_m - (ax^3 + by^3 + cz^3)\psi_{m-1}) = 0.$$

By Lemma 1, this equation implies that $\psi_m(s, t, u) - (s + t + u)\psi_{m-1}(s, t, u)$ is extremal. Let $\mu_m(s, t, u) = \psi_m(s, t, u) - (s+t+u)\psi_{m-1}(s, t, u)$. By induction, we may write ψ_{m-1} in the required form. Let

$$\psi_{m-1}(s, t, u) = (s + t + u)^{m-1} + \sum_{k=2}^{m-1}(s + t + u)^{m-1-k}\mu_k(s, t, u)$$

with μ_k extremal for $k = 2, \dots m-1$. From the definition of μ_m we now have

$$\psi_m(s,t,u) = (s+t+u)\psi_{m-1}(s,t,u) + \mu_m(s,t,u) =$$

$$= (s+t+u)((s+t+u)^{m-1} + \sum_{k=2}^{m-1}(s+t+u)^{m-1-k}\mu_k(s,t,u)) + \mu_m(s,t,u) =$$

$$= (s+t+u)^m + \sum_{k=2}^{m}(s+t+u)^{m-k}\mu_k(s,t,u),$$

and, as shown above, μ_m is extremal. Thus the Lemma follows.

Up to this point we have not used the fact that the coefficient ring has characteristic zero. This comes in the next Lemma:

LEMMA 3. *Let $\mu(s,t,u)$ be a homogeneous polynomial. If $(s+t+u)\mu(s,t,u)$ is extremal, then so is $\mu(s,t,u)$.*

PROOF. Let $\mu(s,t,u)$ be a homogeneous polynomial of degree m, and assume that $(s+t+u)\mu(s,t,u)$ is extremal. Write μ in powers of s; $\mu(s,t,u) = s^m\mu_0(t,u) + \dots + \mu_m(t,u)$. Let g be the smallest integer greater than or equal to $\frac{2m+2}{3}$, and let h be the smallest integer greater than or equal to $\frac{2(m+1)+2}{3}$. Then a monomial $s^it^ju^k$ of degree m (resp. m+1) is extremal if and only if i, j, or k is at least g (resp. h). Let $k = g-1$. Then $s^k\mu_{m-k}(t,u)$ and the successive terms are those which are not extremal due to the power of s, although they will be extremal if the powers of t or u which occur are high enough.

We next use the hypothesis that $(s+t+u)\mu(s,t,u)$ is extremal. For each $i = m-k, \dots m-1$ this gives the equation

$$\mu_i(t,u)(t+u) + \mu_{i+1}(t,u) \equiv 0 \quad \text{modulo } (t^h, u^h).$$

From these equations we deduce that

$$(**) \qquad \mu_{m-k}(t,u)(t+u)^r \equiv \mu_{m-k+r} \quad \text{modulo } (t^h, u^h).$$

for each $r = 0, \dots k$ and, using again that $(s+t+u)\mu(s,t,u)$ is extremal, we get that

$$(***) \qquad \mu_k(t,u)(t+u)^{k+1} \equiv 0 \mod (t^h, u^h).$$

From equation (**) it suffices to show that $\mu_{m-k} = 0$ to show that μ is extremal. We next show that equation (***) implies that $\mu_{m-k} = 0$.

The idea of this part of the proof is as follows: if d is the degree of μ_{m-k}, then there are $d+1$ terms in μ_{m-k}, and if $d+1$ consecutive terms of $\mu_{m-k}(t+u)^n$ are zero, then $\mu_{m-k} = 0$. The degree of μ_{m-k} is $m-k = m-g+1$. The number of terms of degree $m+1$ in t and u modulo (t^h, u^h) is $m+2-2(m+1-h+1) = -m+2h-2$. Thus we must show that $-m+2h-2 \geq m-g+2$, or $2h+g \geq 2m+4$. This is straightforward, dividing into three cases depending on the value of m

modulo 3; for example, if $m \equiv 1 \bmod 3$, we have $g = \dfrac{2m+4}{3}$ and $h = \dfrac{2m+4}{3}$, so $2h+g = \dfrac{6m+12}{3} = 2m+4$. Similarly, one obtains $2h+g = 2m+4$ if $m \equiv 2$ and $2m+5$ if $m \equiv 0$.

The matrix defining a sequence of $d+1$ consecutive coefficients of $\mu_{m-k}(t+u)^n$ in terms of the coefficients of μ_{m-k} is a matrix of binomial coefficients of the form

$$\begin{pmatrix} \binom{n}{p} & \binom{n}{p+1} & \cdots & \binom{n}{p+d} \\ \binom{n}{p-1} & \binom{n}{p} & \cdots & \binom{n}{n+d-1} \\ \vdots & \vdots & & \vdots \\ \binom{n}{p-d} & \binom{n}{p-d+1} & \cdots & \binom{n}{p} \end{pmatrix}.$$

This matrix is evaluated in Muir [3], p. 682, where it is shown to be equal to

$$\frac{\binom{n}{p}\binom{n+1}{p}\cdots\binom{n+d}{p}}{\binom{p}{p}\binom{p+1}{p}\cdots\binom{p+d}{p}}.$$

For completeness, we outline this computation here. The computation is a sequence of row and column operations, using the standard relations between binomial coefficients, in particular the relation

$$\binom{m}{n} = \binom{m-1}{n} + \binom{m-1}{n-1}.$$

We first add the second row from the bottom to the bottom row, which changes the last row to

$$\binom{n+1}{p-d+1}\binom{n+1}{p-d+2}\cdots\binom{n+1}{p+1}.$$

We next add the third from last row to the second from last, then the second from last to the last. We then repeat this procedure starting with the fourth from the last, and so on until we carry it out starting with the first row. The result of these $\dfrac{d(d+1)}{2}$ operations is the matrix

$$\begin{pmatrix} \binom{n}{p} & \binom{n}{p+1} & \cdots & \binom{n}{p+d} \\ \binom{n+1}{p} & \binom{n+1}{p+1} & \cdots & \binom{n+1}{p+d} \\ \vdots & \vdots & & \vdots \\ \binom{n+d}{p} & \binom{n+d}{p+1} & \cdots & \binom{n+d}{p+d} \end{pmatrix}.$$

Denote the determinant of this matrix $\Delta(n,p,d)$. We next use the identity $\binom{m}{k} = \dfrac{m}{k}\binom{m-1}{k-1}$ on each of the entries of this matrix. This gives a factor of $n+k$ for the $(k-1)^{st}$ row and $\dfrac{1}{p+k}$ for the $(k-1)^{st}$ column. Taking out all these factors, we have the formula

$$\Delta(n,p,d) = \frac{n(n+1)\dots(n+d)}{p(p+1)\dots(p+d)}\Delta(n-1,p-1,d).$$

Using this relation p times, we obtain

$$\Delta(n,p,d) = \frac{\binom{n}{p}\binom{n+1}{p}\cdots\binom{n+d}{p}}{\binom{p}{p}\binom{p+1}{p}\cdots\binom{p+d}{p}}\Delta(n-p,0,d).$$

Thus it remains to show that $\Delta(n-p,0,d) = 1$. This can be seen be the following operations: subtract the next to the last row from the last, then the third from the last from the second from last, and so on. This transforms the matrix into a matrix with 1 in the upper left corner and zeros below it, and with the d by d submatrix in the lower right the matrix whose determinant is $\Delta(n-p,0,d-1)$. Hence this statement can be proven by induction.

We are now ready to prove the main result.

THEOREM. *The local cohomology module $H^3_{(x,y,z)}(A)$ is not zero.*

PROOF. Suppose that $x^{n-2}y^{n-2}z^{n-2}$ were equal to $\eta\phi$, and let $\psi = \psi_0 + \ldots + \psi_r$ as defined above. From Lemma 2, we may write

$$(*) \qquad \psi_r(s,t,u) = (s+t+u)^r + \sum_{k=2}^{r}(s+t+u)^{r-k}\mu_k(s,t,u)$$

with μ_k extremal. Since r is the degree of ϕ in a,b,c, and since $x^{n-2}y^{n-2}z^{n-2}$ has degree zero, we must have $\eta_1\phi_r = 0$. This implies that

$$(ax^3+by^3+cz^3)(xyz)^{m-4-2r}\psi_r(ax^3,by^3,cz^3) = 0$$

in $B = R/(x^n,y^n,z^n)$. Thus by Lemma 1, $(s+t+u)\psi_r(s,t,u)$ is extremal. Writing ψ as in equation (*), and multiplying by $s+t+u$, we may thus write

$$(s+t+u)^{r+1} = -\sum_{k=2}^{r+1}(s+t+u)^{r+1-k}\mu_k(s,t,u)$$

with μ_{r+1} extremal. Now it follows from this equation that μ_{r+1} is divisible by $s+t+u$, say $\mu_{r+1} = (s+t+u)\mu'_r(s,t,u)$. By Lemma 3, μ' is also extremal. Thus this term can be added to $(s+t+u)\mu_r$ and the entire expression may be divided by $s+t+u$, giving an expression with the same properties but with r replaced by $r-1$. This can be continued until we obtain that $s+t+u$ is extremal, which is a contradiction. Thus an element ϕ with $x^{n-2}y^{n-2}z^{n-2} = \eta\phi$ cannot exist, so $H^3_{(x,y,z)}(A) \neq 0$.

REFERENCES

1. M. Hochster, *Canonical elements in local cohomology modules and the direct summand conjecture*, J. of Algebra **84** (1983), 503–553.
2. M. Hochster, *Solid closure*, Proceedings of this conference.
3. T. Muir, *A treatise on the theory of determinants*, revised by W. H. Metzler, Albany, New York, 1930.

DEPARTMENT OF MATHEMATICS, UNIVERSITY OF UTAH, SALT LAKE CITY, UT 84112

E-mail address: roberts@math.utah.edu

Contemporary Mathematics
Volume 159, 1994

Algebra Structures for Graded Free Resolutions

HEMA SRINIVASAN

ABSTRACT. We give some sufficient conditions for the existence of an algebra structure on the graded free resolutions of homogeneous cyclic modules. In particular, if R is a positively graded ring and I is a homogeneous ideal, then the minimal resolution of R/I has an algebra structure provided the twists in the minimal homogeneous resolution of R/I satisfy certain inequalities. We give examples to show that these conditions are tight, but not necessary. We apply these conditions to ideals in the linkage class of a complete intersection (LICCI ideals). We show that for a LICCI Gorenstein ideal I generated by homogeneous elements of the same degree if twists in the minimal resolution $\boldsymbol{F}$ of R/I do note exceed that of the corresponding Koszul complex, then $\boldsymbol{F}$ has an algebra structure.

Let R be a commutative noetherian ring and I an ideal of R with an R–free resolution $\boldsymbol{F}$ of R/I. If $\boldsymbol{F}$ admits a multiplication giving it a structure of an associative, commutative differential graded R–algebra then it is called an algebra resolution. Every cyclic module R/I does have an algebra resolution, called the Tate resolution [T].

When R is a local ring, one can ask whether the minimal resolution has such an algebra structure. When I is generated by a regular sequence, then the minimal resolution of R/I is the Koszul complex which has a natural multiplication making it an algebra resolution. When this happens, R/I is said to have a minimal algebra resolution. In [B–E] Buchsbaum and Eisenbud began the study of minimal algebra resolutions where they showed that there is always a multiplication on the minimal resolution making it a commutative differential graded algebra which is at least homotopy associative. However, not all cyclic modules possess minimal algebra resolutions [A]. In fact, the vanishing of Avramov obstructions are necessary for the existence of an algebra structure on the minimal resolutions [A].

1991 *Mathematics Subject Classification.* 13D25; 13C10; 18G10.

Partially supported by NSF.
The paper is in final form, and no version of it will be submitted for publication elsewhere.

In this paper, we consider graded free resolutions over a polynomial ring R. If m is the irrelevant maximal ideal of R, then we call a graded free resolution $\boldsymbol{F}$ with boundary map $\partial : \boldsymbol{F} \to \boldsymbol{F}$ minimal if $\partial(\boldsymbol{F})$ is contained in $m\boldsymbol{F}$. We give some sufficient conditions for the existence of an algebra structure on the graded free resolutions of homogeneous cyclic modules [Theorem 2.1]. We also give some examples to illustrate that these sufficient conditions are tight but by no means necessary.

In section 3, we apply these conditions to the ideals in the linkage class of a complete intersection or LICCI ideals. It is not known whether cyclic modules defined by LICCI ideals possess minimal algebra resolutions. Some special cases with small linking numbers are known to possess this property [AKM]. We show that for LICCI Gorenstein ideals I generated by homogeneous elements of the same degree, the minimal resolution of R/I has an algebra structure as long as the twists in resolution do not exceed the ones in the associated Koszul complex.

1. Definitions and Notations

A free resolution

$$\boldsymbol{F}: \quad \cdots \longrightarrow F_n \xrightarrow{\partial_n} F_{n-1} \longrightarrow \cdots \xrightarrow{\partial_2} F_1 \xrightarrow{\partial_1} F_0 = R$$

of a cyclic module R/I is said to have an algebra structure or is called an algebra resolution if there is a multiplication $\boldsymbol{F} \bigotimes \boldsymbol{F} \to \boldsymbol{F}$ on $\boldsymbol{F}$ which makes it an associative commutative differential graded algebra over R. If R is local with maximal ideal m, then $\boldsymbol{F}$ is called minimal if $\partial(\boldsymbol{F}) \subset m\boldsymbol{F}$.

Suppose R is a graded ring and M is a graded R module. Then M_i denotes the i'th graded piece of M. Let M and N be graded R modules. An R module homomorphism $f : M \to N$ is said to be of degree d if $f(M_i) \subset N_{i+d}$ for every integer i. For any integer d, $M(d)$ is the twisted graded module whose i'th graded piece is M_{i+d}. Thus if $M = \bigoplus_{i \geq 0} M_i$ and d is any positive integer, then $M(-d)$ has no elements in degree less than d. Clearly, if $f : M \to N$ is of degree d, then $f : M(-d) \to N$ will be a degree zero homomorphism.

Suppose $R = k[x_1, x_2, \ldots, x_r]$ is a polynomial ring over a field k with the irrelevant maximal ideal m. Let I be a homogeneous ideal of R. Then the minimal homogeneous graded resolution of R/I is the resolution of the form

$$0 \longrightarrow \sum_{i=1}^{b_n} R(-d_{n_i}) \xrightarrow{\partial_n} \cdots \longrightarrow \sum_{i=1}^{b_2} R(-d_{2_i}) \xrightarrow{\partial_2} \sum_{i=1}^{b_1} R(-d_{1_i}) \xrightarrow{\partial_1} R \longrightarrow R/I$$

where the shifts on the free summands are such that the boundary maps ∂_t's are all degree zero homomorphisms and b_t is the t'th betti number of R/I.

An R-module M is said to have a pure resolution of type $d_1, d_2, \ldots, d_n$ if the minimal homogeneous resolution of M is of the form

$$0 \longrightarrow R^{b_n}(-d_n) \xrightarrow{\partial_n} \cdots \longrightarrow R^{b_2}(-d_2) \xrightarrow{\partial_2} R^{b_1}(-d_1) \xrightarrow{\partial_1} R \longrightarrow R/I$$

2. Algebra Structure on the Resolutions

Throughout this section, $R = k[x_1, x_2, \ldots, x_r]$ over a field k. Let $I \subset R$ be a homogeneous ideal with the minimal graded resolution $\boldsymbol{F}$

(2.1)

$$0 \longrightarrow \sum_{i=1}^{b_n} R(-d_{ni}) \xrightarrow{\partial_n} \cdots \longrightarrow \sum_{i=1}^{b_2} R(-d_{2i}) \xrightarrow{\partial_2} \sum_{i=1}^{b_1} R(-d_{1i}) \xrightarrow{\partial_1} R \longrightarrow R/I$$

Let $F_t = \sum_{i=1}^{b_t} R(-d_{t_i})$ and $\boldsymbol{F} = \sum_{t=0}^{n} F_t$. $(\boldsymbol{F}, \partial)$ is a differential graded R module in the natural way. Since each F_t is also a graded R module, there is a second grading on $\boldsymbol{F}$ induced by the grading on R. For an element $a \in (\boldsymbol{F}_t)_i \subseteq \boldsymbol{F}$ we say that a is of degree i in F_t or a is of bidegree (t, i). Thus 1 in $R(-d_{t_i})$ has bidegree (t, d_{t_i}).

Suppose that $\boldsymbol{F}$ does have an algebra structure. An algebra structure on $\boldsymbol{F}$ is apriori a graded multiplication on the differential graded R module $\boldsymbol{F}$. This graded multiplication actually respects the second grading on $\boldsymbol{F}$ induced by the grading on R. This is simply because the multiplication on $\boldsymbol{F}$ defining an algebra structure must be a map of complexes $\boldsymbol{F} \bigotimes \boldsymbol{F} \to \boldsymbol{F}$. This key point is used to arrive at sufficient conditions for the existence of an algebra structure on $\boldsymbol{F}$ and is the essential new idea in the proof of Theorem 2.1 below.

THEOREM 2.1. *Let $R = k[x_1, x_2, \ldots, x_r]$ and $I \subset R$ be a homogeneous ideal with the minimal graded resolution $\boldsymbol{F}$*

$$0 \longrightarrow \sum_{i=1}^{b_n} R(-d_{n_i}) \xrightarrow{\partial_n} \cdots \longrightarrow \sum_{i=1}^{b_2} R(-d_{2_i}) \xrightarrow{\partial_2} \sum_{i=1}^{b_1} R(-d_{1_i}) \xrightarrow{\partial_1} R \longrightarrow R/I$$

Suppose that $d_{(a+b+c+1)_i} > d_{a_j} + d_{b_k} + d_{c_l}$ for all a, b, c, i, j, k, l. Then $\boldsymbol{F}$ has an algebra structure.

PROOF. We will construct a graded commutative multiplication $*$ on $\boldsymbol{F}$ which is a map of complexes $\boldsymbol{F} \bigotimes \boldsymbol{F} \xrightarrow{*} \boldsymbol{F}$ where each homomorphism $\sum_{i+j=t} F_i \bigotimes F_j \to F_t$ is of degree zero. That is, we will define, by induction on $t = r + s$, an R-module homomorphism, $F_r \bigotimes F_s \xrightarrow{*} F_{r+s}$, such that for all

$$a \in (F_r)_i,\ b \in (F_s)_j$$
$$\text{(D)} \quad a * b \in (F_{r+s})_{i+j}$$
$$\text{(C)} \quad a * b = (-1)^{rs} b * a$$
$$\text{(L)} \quad \partial_{r+s}(a * b) = \partial_r(a) * b + (-1)^r a * \partial_s(b).$$

Define $R \times F. \xrightarrow{*} F.$ and $F. \times R \xrightarrow{*} F.$ as the module multiplication. This defines $*$ for $t = 1$. Assume we have $*$ for all $r + s < t$.

Now, let $r+s=t$. $\partial(a)*b$ and $a*\partial(b)$ are defined by induction. Consider the map

$$l : F_r \bigotimes F_s \to F_{t-1}$$

given by $l(a*b)=\partial_r(a)*b+(-1)^r a*\partial(b)$. If a is of bidegree (r,i) and b is of bidegree (s,j), then $l(a*b)$ is of bidegree $(r+s-1,i+j)=(t-1,i+j)$. Now $\partial_{t-1}(l(a*b))=0$. So $\exists\phi : F_r\bigotimes F_s \to F_t$ such that $\partial_t\circ\phi=l$ and ϕ satisfies (D) and (L). Now, if $r=s=$ odd, then $l(a*b)=-l(b*a)$ and if $r=s=$ even, $l(a*b)=l(b*a)$. So if $r=s=$ odd, l factors through $\overset{2}{\wedge}F_r$ and if $r=s=$ even, l factors through D_2F_r. Let $\bar{l}$ denote the map $\overset{2}{\wedge}F_r\to F_{2r-1}$ when r is odd; $D_2F_r\to F_{2r-1}$ when r is even and $F_r\bigotimes F_s\to F_{r+s-1}$ when $r<s$. Again, we have, $\partial_{t-1}\circ\bar{l}=0$. Let ϕ be the corresponding map which satisfies $\partial_t\circ\phi=\bar{l}$. If $r>s$, then define $F_r\bigotimes F_s\xrightarrow{\phi}F_t$ by $\phi(a\bigotimes b)=(-1)^{rs}\phi(b\bigotimes a)$. This ϕ satisfies (C), (L) and (D) and is the required $*$.

Alternatively, to be more concrete, let us choose a basis

$$\{e_{t_i}\,|\,1\le i\le b_t, 0\le t\le n\}$$

for $\boldsymbol{F}$ so that each e_{t_i} is a generator of $R(-d_{ti})$. Thus e_{t_i} has bidegree (t,d_{t_i}). Now, to define $\sum_{r+s=t}F_r\bigotimes F_s\xrightarrow{*}F_t$, it suffices to define $*$ on the basis elements. i.e. We must specify $e_{r_i}*e_{s_j}$ for all $r+s=t$.

Order the basis elements

$$\{e_{t_i}\,|\,1\le i\le b_t, 0\le t\le n\} \text{ by } e_{r_i}\le e_{s_j} \text{ if } r<s \text{ or if } r=s \text{ and } i\le j.$$

We will specify $e_{r_i}*e_{s_j}$ for $e_{r_i}\le e_{s_j}$ and define $e_{r_i}*e_{s_j}=(-1)^{rs}e_{s_j}*e_{r_i}$ if $e_{r_i}>e_{s_j}$. Now $l(e_{r_i}*e_{s_j})=\partial_r(e_{r_i})*e_{s_j}+(-1)^n e_{r_i}*\partial_s(e_{s_j})$ is defined by induction and is a cycle in F_{t-1} of degree $d_{r_i}+d_{s_j}$.

If $e_{r_i}=e_{s_j}$, and $r=s=$ odd define $e_{r_i}*e_{s_j}=0$. For all other $e_{r_i}\le e_{s_j}$, define $e_{r_i}*e_{s_j}$ to be any element in F_t of degree $d_{r_i}+d_{s_j}$ such that

$$\partial_t(e_{r_i}*e_{s_j})=l(e_{r_i}*e_{s_j}).$$

This defines $r+s=t$ and completes the definition of $*$ by induction.

To complete the proof, we must show that $*$ is associative. Clearly it suffices to check this on basis elements. Since $*$ satisfies (D) i.e. respects both the gradings on $\boldsymbol{F}$, $e_{a_j}*(e_{b_k}*e_{c_l})-(e_{a_j}*e_{b_k})*e_{c_l}$ is homogeneous of bidegree $(a+b+c, d_{a_j}+d_{b_k}+d_{c_l})$. Also, $\partial_{a+b+c}(e_{a_j}*(e_{b_k}*e_{c_l}))=\partial((e_{a_j}*e_{b_k})*e_{c_l})$ by (L). So $e_{a_j}*(e_{b_k}*e_{c_l})-(e_{a_j}*e_{b_k})*e_{c_l}$ is a cycle in F_{a+b+c}. Thus, there is an element ξ in $F_{a+b+c+1}$ of degree $d_{a_j}+d_{b_k}+d_{c_l}$ such that

$$\partial_{a+b+c+1}(\xi)=e_{a_j}*(e_{b_k}*e_{c_l})-(e_{a_j}*e_{b_k})*l_{c_l}.$$

But $\min_i d_{a+b+c+1,i}>d_{a_j}+d_{b_k}+d_{c_l}$ for all a,b,c,i,j,k,l. So there is no such a ξ in $F_{a+b+c+1}$ other than 0. Thus

$$e_{a_j}*(e_{b_k}*e_{c_l})-(e_{a_j}*e_{b_k})*e_{c_l}=0 \text{ for all } a,b,c,j,k,l.$$

Hence $*$ is associative. This proves that $\boldsymbol{F}$ has an algebra structure. $\square$

Remark 2.2. In the above proof, the function $\phi : \sum_{r+s=t} F_r \bigotimes F_s \rightarrow F_{r+s}$ satisfying $\partial_{r+s} \circ \phi = l$ satisfies both (D) and (L). Thus ϕ is a map of complexes of $\boldsymbol{F} \bigotimes \boldsymbol{F} \rightarrow \boldsymbol{F}$ which respects both the gradings on $\boldsymbol{F}$. We just showed that such a multiplication given by ϕ has to be associative if the twists in $\boldsymbol{F}$ satisfy the inequalities of the theorem. This multiplication ϕ can easily be modified to satisfy graded commutatity as shown in [B–E].

COROLLARY 2.3. *Suppose R/I has a pure resolution $\boldsymbol{F}$ of type $d_1, \ldots, d_n$ such that $d_t > (t-1)d_1$ for all $4 \leq t \leq n$; then $\boldsymbol{F}$ has an algebra structure.*

PROOF. Since the resolutions is pure;

$$d_a = \min_i d_{a_i} \leq ad_1.$$

So $d_a + d_b + d_c \leq (a+b+c)d_1 \leq d_{a+b+c+1}$ by hypothesis. By Theorem 2.1, $\boldsymbol{F}$ has an algebra structure. $\square$

COROLLARY 2.4. *Suppose that I is generated by homogeneous forms of same degree d_1 and the minimal graded resolution $\boldsymbol{F}$ of R/I is of the form (2.1). If*

$$(t-1)d_1 < \min_i d_{t_i} \leq \max_i d_{t_i} \leq td_1,$$

for all $1 < t \leq n$, then $\boldsymbol{F}$ has an algebra structure.

PROOF. $d_{a_j} + d_{b_k} + d_{c_l} \leq (a+b+c)d_1 < \min_i d_{(a+b+c+1)i}$ by hypothesis. By Theorem 2.1 $\boldsymbol{F}$ has an algebra structure. $\square$

The inequalities in Theorem 2.1 are sufficient but are by no means necessary.

Example 2.5. Let $X = (x_{ij})$ be a generic $m \times n$ matrix x with $n \geq m \geq 2$. Let $R = k[x_{ij}]$ with k a field of characteristic zero rationals and I be the ideal of maximal minors of X. Then the minimal resolution of R/I is the Eagon–Northcott complexes [E–N], $\boldsymbol{E}$. $\boldsymbol{E}$ is a pure resolution of the following form.

$$0 \longrightarrow R^{b_{(n-m+1)}}(-dn) \longrightarrow \cdots \longrightarrow R^{b_t}(-(m+t-1)) \longrightarrow R^{b_1}(-m) \longrightarrow R.$$

So

$$d_{a+b+c+1} - (d_a + d_b + d_c) = (m + (a+b+c+1) - 1) - (3m + a + b + c - 3) = 3 - 2m < 0.$$

Thus E fails to satisfy any of the inequalities in Theorem 2.1. Nevertheless, E does have an algebra structure [S1].

On the other hand, the inequalities of Theorem 2.1 are also tight.

Example 2.6. Let $X = (x_{ij})$ be a 6×6 generic alternating matrix x with $R = [x_{ij}]$ where k is a field of characteristic $\neq 2$. Let I be the ideal of 4×4 praffians of X. Then R/I has the minimal resolution which looks like

$$0 \longrightarrow R(-9) \longrightarrow R^{b_1}(-7) \longrightarrow R^{b_2}(-6) \longrightarrow \begin{matrix} R^a(-4) \\ \bigoplus \\ R^6(-5) \end{matrix} \longrightarrow R^{b_2}(-3) \longrightarrow R^{b_1}(-2) \longrightarrow R.$$

Here $d_4 = 6 = 3(d_1)$ and hence $d_4 \not> 3d_1$. Also, $d_5 = 7 = 2+2+3$ and $d_6 = 9 = 5+2+2$.

Thus, for this resolution, $\min_i d_{(a+b+c+1)_i}$ actually equals $d_{a_j} + d_{b_k} + d_{c_l}$ for every $a+b+c+1$. We know from [S2] that this resolution does not admit an algebra structure.

Example 2.7. Let $R = k[x_{ij}, l \leq i < j \leq 5, y_i, 1 \leq i \subseteq 5]$ with $\operatorname{char} k \neq 2$. Let X be the 5×5 generic alternating matrix x and Y be a generic 1×5 matrix. Let $I = Pf_4(X) + I_1(YX)$ be the ideal generated by the 4×4 praffians of X together with the entries of the matrix YX. The minimal resolution $\mathbb{P}$ of R/I does not admit an algebra structure [S3]. This resolution $\mathbb{P}$ looks like

$$0 \longrightarrow R(-8) \longrightarrow R^{10}(-6) \longrightarrow R^{16}(-5) \longrightarrow R^{16}(-3) \longrightarrow R^{10}(-2) \longrightarrow R$$

$\mathbb{P}$ satisfies $d_5 = 8 > 3+2+2$. But $d_4 = 6 = 3d_1$. So $\mathbb{P}$ fails to satisfy only one inequality of Theorem 2.1 (which becomes an equality for $\mathbb{P}$). This shows that these inequalities can be very tight.

We remark that a koszul complex associated to sequence of homogeneous elements of same degree will satisfy the conditions.

The following example (Example 5.14 of [H–U–2]) is an interesting case where the inequalities of Theorem 2.1 are satisfied.

Example 2.8. Let $R = k[x_1, \dots, x_{12}]$ be a polynomial ring in 12 indeterminates over a field k. Let I be the homogeneous ideal of R generated by $(x_8x_{10} - x_9x_{11}, x_7x_{12} - x_9x_{10}, x_7x_{11} - x_8x_{10}, x_4x_{10} + x_5x_{11} + x_6x_{12}, x_1x_{10} + x_2x_{11} + x_3x_{12}, x_1x_7 + x_2x_8 + x_3x_9, x_3x_6 + x_4x_7 + x_5x_8)$. By [H–U–2], the minimal resolution of R/I is of the form

$$\boldsymbol{F} : 0 \longrightarrow R(-9) \longrightarrow R^7(-7) \longrightarrow \begin{matrix} R^{11}(-5) \\ \bigoplus \\ R^5(-6) \end{matrix} \longrightarrow \begin{matrix} R^5(-3) \\ \bigoplus \\ R^{11}(-4) \end{matrix} \longrightarrow R^7(-2) \longrightarrow R \longrightarrow 0.$$

We have, $d_5 > 2+2+4$, $d_4 > 3(2)$. So by Theorem 2.1, $\boldsymbol{F}$ has an algebra structure.

Also, as a corollary to the proof of Theorem 2.1, we get

COROLLARY 2.9. *In the notations of Theorem* 2.1, *if*

$$\min_i d_{(a+b+1)_i} > d_{a_j} + d_{b_k}$$

for all a, b, j, k, *then there is a unique multiplication on* $\boldsymbol{F}$ *making it a commutative differential graded algebra which is homotopy associative.*

PROOF. The map ϕ in the proof of 2.1 is determined up to boundary. Thus $e_{t_i} * e_{s_j}$ is an element of bidegree $(t+s, d_{t_i} + d_{s_j})$ which is determined up to a cycle in F_{t+s}. Any cycle in F_{t+s} lives in degree at least $\min_i d_{(t+s+1)_i}$, which exceeds $d_{t_i} + d_{s_j}$ by hypothesis. So, $e_{t_i} * e_{s_j}$ is uniquely determined. □

3. Some Applications to LICCI Ideals

Let R be a regular local ring. Two proper ideals I and J of R are said to be linked if there is a regular sequence α contained in $I \cap J$ such that $I = (\alpha : J)$ and $J = (\alpha : I)$. We write $I \sim J$ if I is linked to J. An ideal I is called LICCI if there is a sequence of ideals $I_0, \dots, I_n$ such that $I = I_0 \sim I_1 \sim I_2 \sim \cdots \sim I_n$ and I_n is generated by a regular sequence. LICCI is short for linkage class of a complete intersection.

Suppose $S = k[x_1, \dots, x_n]$, a polynomial ring over a field k with $\deg x_i > 0$. Let $m = (x_1, \cdots, x_r)$ be the irrelevant maximal ideal. Let I be a homogeneous ideal of S. $\hat{S} = k[[x_1, \dots, x_r]]$ and S_m are both local rings with maximal ideal $m\hat{S}$ and mS_m respectively.

The minimal graded resolution $\boldsymbol{F}$ of S/I is the graded resolution $(\boldsymbol{F}, \partial)$ of S/I with $\partial(\boldsymbol{F}) \subseteq m\boldsymbol{F}$. $\boldsymbol{F} \bigotimes_S \hat{S}$ and $\boldsymbol{F}_m$ are respectively the minimal resolutions of $\hat{S}/I\hat{S}$ and S_m/IS_m. If $\boldsymbol{F}$ has an algebra structure, then the minimal resolutions of $\hat{S}/I\hat{S}$ and S_m/IS_m also have algebra structures.

I is called LICCI if IS_m is. From now on, $S = k[x_1, \dots, x_n]$ and I is a homogeneous LICCI ideal with minimal homogeneous resolution $\boldsymbol{F}$ as in (2.1). Then we have the following theorem of Huneke and Ulrich.

THEOREM [H–U–1] 3.1. *Let I be a LICCI homogeneous ideal of grade n. Let the minimal graded free resolution $\boldsymbol{F}$ of R/I be*

$$0 \longrightarrow \sum_{i=1}^{b_n} R(-d_{n_i}) \xrightarrow{\partial_n} \cdots \xrightarrow{\partial_2} \sum_{i=1}^{b_2} R(-d_{2_i}) \xrightarrow{\partial_1} \sum_{i=1}^{b_1} R(-d_{1_i}) \longrightarrow R.$$

Then $\max_i d_{n_i} > (n-1)\min_i d_{1_i}$.

When I is Gorenstein $b_n = 1$ and hence $\max_i d_{n_i} = d_n = \min_i d_{n_i}$. So, for a Gorenstein LICCI ideal generated by elements of same degree d, we must have $d_n = \min_i d_{n_i} > (n-1)\max_i d_{1_i} = (n-1)d$. This is an important tool in Theorem 3.2.

THEOREM 3.2. *Suppose* $R = S_m$ *or* $\hat{S}$ *and* I *is a Gorenstein LICCI ideal of* R *generated by homogeneous elements of same degree* d. *Suppose* $\max_i d_{t_i} \leq td$ *for all* t. *Then the minimal resolution of* R/I *has an algebra structure.*

PROOF. Since R/I is Gorenstein of grade, say, n, the minimal resolution $\boldsymbol{F}$ of R/I satisfies,

$$\mathrm{Hom}(\boldsymbol{F}, R) \cong \boldsymbol{F} \text{ and } F_n = R(-d_n).$$

Thus $\boldsymbol{F}$ looks like

$$0 \longrightarrow R(-d_n) \xrightarrow{\partial_n} R^{b_1}(-d_n + d_1) \xrightarrow{\partial_{n-1}} \cdots \longrightarrow \sum_{i=1}^{b_t} R(-d_n + d_{(n-t)_i}) \xrightarrow{\partial_t} \cdots$$
$$\cdots \xrightarrow{\partial_2} R^{b_1}(-d_1) \xrightarrow{\partial_1} R.$$

Thus

$$F_t = \sum_{i=1}^{b_t} R(-d_n + d_{(n-t)i})$$

and

$$\{d_{t_i} \mid 1 \leq i \leq b_t\} = \{d_n - d_{(n-t)_i} \mid 1 \leq i \leq b_t\},$$

for all t.

So

$$\begin{aligned} \min_i d_{t_i} &= d_n - \max_i (d_{(n-t)_i}) \\ &\geq d_n - (n-t)d_1, \end{aligned}$$

by hypothesis. Using Theorem 3.1, we get

$$\begin{aligned} \min_i d_{t_i} &\geq d_n - (n-t)d_1 \\ &> (n-1)d_1 - (n-t)d_1 \\ &= (t-1)d_1 \end{aligned}$$

By Corollary 2.4, $\boldsymbol{F}$ has an algebra structure. □

REFERENCES

[A] L. L. Avramov: *Obstructions to the existence of Algebra Structures on Resolutions*, American J. Math. **103** (1984), 1–31.

[A–K–M] L. L. Avramov, A. Kustin and M. Miller: *Poincaré Series of Modules over local rings of small embedding codepth or small linking number*, J. Algebra **118** (1988), 162–204.

[B–E] D. Buchsbaum and D. Eisenbud: *Algebra structure for finite free resolutions and some structure theorems for ideals of codimension three*, Amer. J. Math. **99** (1977), 447–485.

[E–N] J. Eagon and D. Northcott: *Ideals defined by matrices and a certain complex associated to them*, Proc. Roy. Soc. London Ser. A **269** (1962), 188–204.

[H–K] J. Herzog and M. Kühl: *One the Betti numbers of finite pure and linear resolutions*, Comm. in Algebra **12** (1984), 1627–1646.

[H–U–1] C. Huneke and B. Ulrich: *Structure of Algebraic linkage*, Annals of Math. **126** (1987), 277–334.

[H–U–2] C. Huneke and B. Ulrich: *Divisor Class groups and Deformation*, American J. Math. **107** (1985), 1265–1303.

[S1] H. Srinivasan: *Algebra Structures on some canonical resolutions*, J. Algebra **122** (1) (1989), 150–187.

[S2] H. Srinivasan: *Vanishing of Avramov obstructions is not sufficient to ensure the existence of minimal algebra resolutions*, J. Algebra **146** (2) (1992), 251–266.

[S3] H. Srinivasan: *A grade five Gorenstein cyclic module with no minimal algebra resolution*, preprint.

[T] J. Tate: *Homology in Noetherian rings and local rings*, Illinois J. Math. **1** (1957), 14–27.

[G] T. Gulliksen: *A Proof of the existence of Minimal R–algebra resolutions*, Acta. Math. **120** (1968), 53–57.

Department of Mathematics, University of Missouri, Columbia, Missouri 65211
E-mail address: mathhs@mizzou1.missouri.edu

Contemporary Mathematics
Volume 159, 1994

Primary Decompositions of Powers of Ideals

IRENA SWANSON

ABSTRACT. Let R be a Noetherian ring and I an ideal. We prove that there exists an integer k such that for all $n \geq 1$ there exists an irredundant primary decomposition $I^n = q_1 \cap \cdots \cap q_l$ such that $\sqrt{q_i}^{nk} \subseteq q_i$ whenever $\operatorname{ht}(q_i/I) \leq 1$. In particular, if R is a local ring with maximal ideal m and I is a prime ideal of dimension 1, then $m^{kn} I^{(n)} \subseteq I^n$, where $I^{(n)}$ denotes the n'th symbolic power of I.

We study some asymptotic properties of primary decompositions of powers of ideals in a Noetherian ring. In particular, we consider the following question:

Let (R, m) be a regular local ring and P a prime ideal of dimension 1. Then for some $c_n \in \mathbb{N}$, $m^{c_n} P^{(n)} \subseteq P^n$. How does this c_n depend on n? (cf. [**2**])

The main theorem 1 says that c_n is bounded linearly, i.e. there exists an integer k such that $m^{nk} P^{(n)} \subseteq P^n$ for all $n \geq 0$.

Note that if $P^n = P^{(n)} \cap J_n$ is a primary decomposition of P^n with $m \subseteq \sqrt{J_n}$ and $m^{c_n} \subseteq J_n$, then $m^{c_n} P^{(n)} \subseteq P^n$. So we tackle the question via selected irreducible primary components of powers of ideals. Hence we consider more generally: if I is an ideal in a Noetherian ring and $P \in \cup_{n=1}^{\infty} \operatorname{Ass}(R/I^n)$, does there exist an irredundant primary decomposition of $I^n = q_{n1} \cap \cdots \cap q_{nk_n}$ such that if $\sqrt{q_{ni}} = P$, then the least integer c_n for which $P^{c_n} \subseteq q_{ni}$ is bounded linearly with respect to n?

We would also like to know whether there are good primary decompositions of ideals of the form $(x_1^q, \ldots, x_n^q)$, q ranging over powers of the characteristic of R. A positive answer to this question would solve the open question whether tight closure commutes with localization, at least for ideals generated by elements $x_1, \ldots, x_n$ for which $\cup_q \operatorname{Ass}(R/(x_1^q, \ldots, x_n^q))$ is a finite set. There may be infinitely many primes associated to such ideals if q is allowed to vary over all positive integers. An example of Hochster is the following: let $R = \mathbb{Z}[X, Y]$, a polynomial ring in two variables over the ring of integers. Let $x_1 = X$, $x_2 = Y$,

1991 *Mathematics Subject Classification.* Primary 13H99.
This paper is in final form and no version of it will be submitted for publication elsewhere.

and $x_3 = X + Y$. It is easy to see that for each prime integer p, (p, X, Y) is a prime ideal of R associated to (x_1^p, x_2^p, x_3^p). No such examples are known for local rings or for rings containing fields, so there is some hope for this method in proving that tight closure commutes with localization.

In this paper we give partial answers to the three stated lines of inquiry. The main theorem is the following:

THEOREM 1. *Let R be a Noetherian ring and I an ideal. Then there exists an integer k such that for all $n \geq 1$ there exists an irredundant primary decomposition $I^n = q_1 \cap \cdots \cap q_l$ such that $\sqrt{q_i}^{nk} \subseteq q_i$ whenever $ht(q_i/I) \leq 1$.*

Note that the theorem holds trivially if all of the associated primes of I are minimal over I since the minimal primary components are uniquely determined. Namely, if k is a positive integer such that P^k is contained in the P-primary component $IR_P \cap R$ of I for every minimal prime P over I (there are only finitely many of them as R is Noetherian), then this k works also for all higher powers of I. For let n be a positive integer. Then $P^{kn} \subseteq (IR_P \cap R)^n \subseteq I^n R_P \cap R$, which is the P-primary component of I^n.

If P is not minimal over I, P-primary component of I is not uniquely determined. In fact, it is not true that $\sqrt{q_i}^{kn} \subseteq q_i$ for *every* irreducible primary decomposition $I^n = q_1 \cap \cdots \cap q_l$. We show this in an example:

Let $R = k[X, Y]$, a polynomial ring in two variables X and Y over a field k. Let $I = (X^2, XY)$. It is easy to verify that $I = (X) \cap (X^2, XY, Y^m)$ for all positive integers m. Each one of these decompositions is an irredundant primary decomposition, but for any integer k there exists an integer m such that $(X, Y)^k \not\subseteq (X^2, XY, Y^m)$. So the theorem can only hold for *some* primary decompositions.

Before we prove the theorem we state two results needed in the proof:

THEOREM 2. *(Ratliff* [**5**]*) Let I be an ideal in a Noetherian ring R. Then $\cup_{n \geq 1} Ass(R/I^n)$ is a finite set.*

However, in general $\mathrm{Ass}(R/I^n) \neq \mathrm{Ass}(R/I)$. Brodmann showed in [**1**] that for large n, $\mathrm{Ass}(R/I^n)$ stabilizes.

THEOREM 3. *(Katz-McAdam [4, (1.5)]) Let R be a Noetherian ring. For any ideal I there exists an integer l such that $I^n : J^{nl} = I^n : J^{nl+1}$ for all ideals J and all $n \geq 0$.*

Actually, Katz and McAdam prove existence of such an integer l depending on I and J, but their argument can be easily extended to show that there is an l independent of J:

Proof: Let $S = R[It, t^{-1}]$, where t is an indeterminate over R. It is clear that $I^n :_R J = (t^{-n}S :_S J) \cap R$ for all n and all ideals J of R. Thus it suffices to show that there exists an integer l such that $t^{-n}S :_S J^{nl} = t^{-n}S :_S J^{nl+1}$ for all

n and all ideals J of S. So by replacing R by S we may assume without loss of generality that I is a principal ideal generated by a regular element a.

Now let $(a) = q_1 \cap \cdots \cap q_s$ be a primary decomposition of $I = (a)$. Let l be such that $(\sqrt{q_i})^l \subseteq q_i$ for all $i = 1, \ldots, s$.

We prove by induction on n that this l works for all ideals J. First assume that $n = 1$. If $J \subseteq \sqrt{q_i}$ then by the choice of l we have $J^l \subseteq q_i$. If, however, $J \not\subseteq \sqrt{q_i}$, then $q_i : J^l = q_i : J^{l+1} = q_i$ as q_i is primary. Thus $I : J^l = (q_1 : J^l) \cap \cdots \cap (q_s : J^l) = (q_1 : J^{l+1}) \cap \cdots \cap (q_s : J^{l+1}) = I : J^{l+1}$, so we are done.

Now let $n \geq 1$. It is enough to show that $I^n : J^{nl+1} \subseteq I^n : J^{nl}$. Let x be an element of $I^n : J^{nl+1}$. Then $x \in I^{n-1} : J^{nl+1}$ and by induction assumption x lies in $I^{n-1} : J^{(n-1)l}$. Let y be an element of $J^{(n-1)l}$ and z an element of J^{l+1}. Then xy lies in I^{n-1}, so $xy = ba^{n-1}$ for some b. As yz is in J^{ln+1}, we get that xyz lies in I^n, so $xyz = ca^n$ for some c. These two equations say that $ca^n = bza^{n-1}$, hence that $ca = bz$. Since z is an arbitrary element of J^{l+1}, this says that $b \in (a) : J^{l+1}$. So by induction assumption for $n = 1$ we get that $b \in (a) : J^l$. Now let w be an element of J^l. Then $xyw = ba^{n-1}w \in (a^n)$. This holds for arbitrary $y \in J^{(n-1)l}$ and arbitrary $w \in J^l$, so $x \in I^n : J^{nl}$. □

Author's original motivation for studying asymptotic properties of primary decompositions of powers of an ideal was the question whether tight closure commutes with localization. Part of this problem is determining asymptotic properties of primary components of ideals of the form $(x_1^{n_1}, \ldots, x_s^{n_s})$. If the x_i form a regular sequence, a similar argument as above, together with an argument from [**3**, Theorem 4.5], gives:

PROPOSITION 4. *Let R be a Noetherian ring and $x_1, \ldots, x_s$ a regular sequence in R. Let l be as in Theorem 3 for $I = (x_1, \ldots, x_s)$. Then for any $n_1, \ldots, n_s \geq 1$ and any ideal J,*

$$(x_1^{n_1}, \ldots, x_s^{n_s}) : J^{lN} = (x_1^{n_1}, \ldots, x_s^{n_s}) : J^{lN+1},$$

where $N = (\sum n_i) - s + 1$. □

Now we are ready to prove Theorem 1:

Proof: By the remark immediately after the statement of the theorem there exists an integer k' which works for all minimal associated primes. By Theorem 2 there are only finitely many prime ideals $P_1, \ldots, P_t$ which are associated to some power of I and are of height 1 over I. By prime avoidance we can choose an element b contained in each one of these primes P_i but not contained in any minimal prime over I. Let l be as in Theorem 3.

Claim 1: $I^n = (I^n : b^{ln}) \cap (I^n + (b^{ln}))$.

Proof of claim 1: If $a + rb^{ln} \in I^n : b^{ln}$ for some $a \in I^n$, then $ab^{ln} + rb^{2ln} \in I^n$. Since $a \in I^n$ then $rb^{2ln} \in I^n$, so $r \in I^n : b^{2ln} = I^n : b^{ln}$. Hence $rb^{ln} \in I^n$. The other inclusion is easy.

Claim 2: $\mathrm{Ass}\,(R/(I^n : b^{ln})) = \{P \in \mathrm{Ass}\,(R/I^n) | b \notin P\}$.

Proof of claim 2: If $P \in \operatorname{Ass}(R/(I^n : b^{ln}))$, then $P = (I^n : b^{ln}) : c$ for some $c \in R$. Then $P = I^n : b^{ln}c$, so $P \in \operatorname{Ass}(R/I^n)$. If $b \in P = I^n : cb^{ln}$, then $c \in I^n : b^{ln+1} = I^n : b^{ln}$, so $P = I^n : cb^{ln} = R$, which is impossible. Conversely, if $P \in \operatorname{Ass}(R/I^n)$ and b is not in P, then $P = I^n : c$ for some $c \in R$. As b is not in P, $P = P : b^{ln} = I^n : cb^{ln}$, so $P \in \operatorname{Ass}(R/(I^n : b^{ln}))$. This proves the claim.

It follows that $\{P \in \operatorname{Ass}(R/I^n) | b \in P\} \subseteq \operatorname{Ass}(R/I^n + (b^{ln}))$.

By the choice of b, each P_i is minimal over $I + (b^l)$. Let k'' be such that $P_i^{k''} \subseteq (I + (b^l))R_{P_i} \cap R$ for all i. Then

$$\begin{aligned} P_i^{2nk''} &\subseteq (I + (b^l))^{2n} R_{P_i} \cap R \\ &\subseteq (I^n + (b^{ln}))R_{P_i} \cap R \end{aligned}$$

which is the P_i-primary component of $I^n + (b^{ln})$ and by the claims also a (possibly redundant) P_i-primary component of I^n.

Finally, set $k = \max\{k', 2k''\}$. □

The question remains whether there are good primary decompositions with similarly bounded properties for primary components of height greater than 1 over the chosen ideal.

Mark Johnson observed that the argument above shows that if (R, m) is a regular local ring and P is a prime ideal of dimension 1, then $k \geq 2\left((ed!)^{1/d} - d + 1\right)$, where $d = \dim(R)$ and $e = e(R/P)$.

Proof: (Due to Johnson) We use notation from above. As $m^{k''} \subseteq P + (b^l)$,

$$\binom{d + k'' - 1}{d} = \lambda\left(R/m^{k''}\right) \geq \lambda\left(R/(P + (b^l))\right).$$

But R/P is one-dimensional Cohen-Macaulay, so

$$\lambda\left(R/(P + (b^l))\right) = e_m\left(R/(P + (b^l))\right) \geq e_m\left(R/P\right).$$

Thus $\binom{d+k''-1}{d} \geq e$ and

$$(d + k'' - 1)^d \geq d! \binom{d + k'' - 1}{d} \geq ed!,$$

from which we get

$$k = \max\{1, 2k''\} \geq 2\left((ed!)^{1/d} - d + 1\right). \quad \square$$

Note that Theorem 1 may hold for smaller k, it is only the k from the proof which has this lower bound.

Acknowledgment. I thank Craig Huneke and Mark Johnson for the conversations regarding this paper.

References

1. M. Brodmann. Asymptotic stability of $\mathrm{Ass}(M/I^n M)$. *Proc. Amer. Math. Soc.*, **74** (1979), 16–18.
2. J. Herzog. A homological approach to symbolic powers. In *Commutative Algebra, Proc. of a Workshop held in Salvador, Brazil, 1988*, Lecture Notes in Mathematics 1430, pp. 32–46. Springer-Verlag, Berlin, 1990.
3. M. Hochster and C. Huneke. F-regularity, test elements, and smooth base change. Preprint.
4. D. Katz and S. McAdam. Two asymptotic functions. *Comm. in Alg.*, **17** (1989), 1069–1091.
5. L. J. Ratliff, Jr. On prime divisors of I^n, n large. *Michigan Math. J.*, **23** (1976), 337–352.

Department of Mathematics, University of Michigan, Ann Arbor, Michigan 48109-1003, USA

E-mail address: iswanson@math.lsa.umich.edu

Contemporary Mathematics
Volume **159**, 1994

Artin-Nagata Properties and Reductions of Ideals

BERND ULRICH

Dedicated to Professor Ernst Kunz on his sixtieth birthday

ABSTRACT. For an ideal I in a local Cohen-Macaulay ring and an integer s, we introduce the Artin-Nagata property AN_s, by which we mean that for every $i \leq s$ and every i-residual intersection J of I, R/J is a Cohen-Macaulay ring. We describe and characterize these conditions in terms of the depths of powers and symbolic powers of I. As an application, we extend results by Huckaba, Huneke, and Vasconcelos ([10], [11], [27]), who have studied Rees algebras of certain ideals by passing to their minimal reductions.

Introduction

In this paper we investigate the Cohen-Macaulay property of residual intersections. To recall the definition, let I be an ideal of height g in a local Cohen-Macaulay ring R, and let $s \geq g$ be an integer; an *s-residual intersection* of I is a proper R-ideal J such that $J = \mathfrak{a} : I$ for some R-ideal $\mathfrak{a} \subset I$ with $ht\ J \geq s \geq \mu(\mathfrak{a})$ (μ denoting minimal number of generators) (Definition 1.1). This notion, essentially introduced by Artin and Nagata ([2]), generalizes the concept of linkage to the case where the two "linked" ideals I and J need not have the same height. Unlike for linkage however, it is by no means clear when residual intersections are Cohen-Macaulay. This issue has been addressed in a series of results (e.g. [15], [9], [18]), which require depth conditions on *all* Koszul homology modules of I, either the "strong Cohen-Macaulayness" or the weaker "sliding depth condition". Here one says that $I = (a_1, \dots, a_n)$ is *strongly Cohen-Macaulay* if all Koszul homology modules $H_j = H_j(a_1, \dots, a_n; R)$ are Cohen-Macaulay ([15]), and that I satisfies *sliding depth* if depth $H_j \geq \dim R - n + j$ for every j ([9])

1991 *Mathematics Subject Classification.* Primary 13H10; Secondary 13C15, 13A30, 13C40.
Supported by the NSF.
This paper is in final form and no version of it will be submitted for publication elsewhere.

(both definitions are independent of the chosen generating set $a_1, \ldots, a_n$). Considering recent papers by Huckaba, Huneke, and Vasconcelos ([10], [11], [27]) that highlight the interplay between residual intersections and reductions of ideals, it seems natural to ask whether the above depth conditions on the Koszul homology can be weakened if we restrict ourselves to considering residual intersections only up to a given height. To formalize this, we say that I has the *Artin-Nagata property* AN_s if for every $g \le i \le s$ and every i-residual intersection J of I, R/J is a Cohen-Macaulay ring (Definition 1.2). Also recall that I satisfies G_s if $\mu(I_p) \le \dim R_p$ for every $p \in V(I)$ with $\dim R_p \le s-1$, and that I is G_∞ in case it satisfies G_s for every s ([2]). Now assume R to be Gorenstein with infinite residue class field. By linkage theory ([22]), I has property AN_g if $H_0 \cong R/I$ is Cohen-Macaulay, and by [20], I satisfies AN_{g+1}, in case I is G_{g+1} and H_0 as well as H_1 are Cohen-Macaulay. The latter result has prompted the question of whether AN_s holds, if I satisfies G_s and H_j are Cohen-Macaulay in the range $0 \le j \le s-g$ ([20]).

We are going to obtain an affirmative answer to this question (Remark 2.10) as a special case of the following stronger result (Theorem 2.9): If I is G_s and depth $R/I^j \ge \dim R - g - j + 1$ for $1 \le j \le s-g+1$, then I satisfies AN_s, and for every $g \le i \le s$ and every i-residual intersection J of I, the canonical module $\omega_{R/J}$ has an "expected form". We also prove the converse for unmixed ideals of positive height satisfying G_{s+1} (Theorem 2.15). Although depth assumptions on the powers of a G_∞-ideal are in general weaker than the Cohen-Macaulay property of Koszul homology modules, one can nevertheless show that the strong Cohen-Macaulayness is forced once we require sufficiently many powers to have the expected depth (Corollary 2.13).

We obtain similar results when replacing powers by symbolic powers. Here the *j-th symbolic power* $I^{(j)}$ of an arbitrary ideal I is the intersection over all isolated primary components of the ordinary power I^j. Now assume that for every minimal $p \in V(I)$, $\mu(I_p) = g \ge 1$, and that for every non-minimal $p \in V(I)$ with $\dim R_p \le s$, the analytic spread $\ell(I_p)$ is less than $\dim R_p$; then we prove that I satisfies AN_s if and only if depth $R/I^{(j)} \ge \dim R - g - j + 1$ for $1 \le j \le s-g+1$ (Theorem 3.3). As a consequence, one can deduce the sliding depth property of an ideal from depth assumptions on suitably many symbolic powers (Corollary 3.4). We also obtain a characterization of the Artin-Nagata property AN_{g+1} in terms of depth $\Omega(R/I)$, where Ω denotes the module of Kähler differentials (Corollary 3.5).

The above results can be applied to studying ideals via their reductions. Let I be an ideal in a Noetherian local ring (R, m, k); then $\mathcal{R}(I)$ denotes the Rees algebra $\oplus_{j \ge 0} I^j$, and $\ell(I)$ stands for the *analytic spread* of I, which is defined to be $\dim \mathcal{R}(I) \otimes_R k$. An R-ideal $K \subset I$ is a *reduction* of I if $I^{r+1} = KI^r$ for some integer $r \ge 0$; the least such integer is called the *reduction number* of I with respect to K and is denoted by $r_K(I)$. A reduction K of I is *minimal* if K is minimal among the reduction of I. If k is infinite then every reduction K of

I contains a minimal reduction, and furthermore $\mu(K) \geq \ell(I)$, where equality holds if and only if K is minimal ([21]).

Now in [10], [11], and [27], Huckaba, Huneke, and Vasconcelos use reductions to study Rees algebras of ideals having "too many" generators. Most of their main results can be summarized as follows: Let R be a local Cohen-Macaulay ring with infinite residue class field, let I be an R-ideal with grade $I = g \geq 2$ and $\ell(I) = s$, and assume that there exists a minimal reduction K of I such that $r_{K_p}(I_p) \leq 1$ for every $p \in V(I)$ with $\dim R_p = s$. Further suppose that either (i) ([10], [11]) $s = g+1$, depth $R/I \geq \dim R - s$, and $\mu(I_p) = g$ for every minimal $p \in V(I)$; or (ii) ([10], [11]) $s = g+2$, R is Gorenstein, R/I is Cohen-Macaulay, and $\mu(I_p) = g$ for every $p \in V(I)$ with $\dim R_p \leq g+1$; or (iii) ([27]) I satisfies G_s and sliding depth. With these assumptions, Huckaba, Huneke, and Vasconcelos prove that $r_K(I) \leq 1$, that $\mathcal{R}(I)$ is Cohen-Macaulay, and that (with some mild additional conditions) $I^{(j)} = I^j$ for every $j \geq 1$ if and only if $\ell(I_p) \leq \dim R_p - 1$ for every non-minimal $p \in V(I)$ with $\dim R_p \leq s$. We are going to extend these results to ideals with depth $R/I \geq \dim R - s$, satisfying G_s and AN_{s-2} instead of the stronger conditions (i), (ii), or (iii) (Theorem 4.1). More concrete statements are obtained in conjunction with earlier results of this paper (Theorem 4.9, Corollaries 4.10 and 4.11). In showing Theorem 4.1 we also give new proofs of the above cases (i) and (ii) (cf. also [27]), but we do not add anything to the proof of (iii). On the contrary, we rely heavily on a crucial theorem due to Vasconcelos, that provides the transfer of Cohen-Macaulayness from $\mathcal{R}(K)$ to $\mathcal{R}(I)$ ([27], cf. Theorem 4.8).

1. Artin-Nagata Properties

In this section we introduce the conditions AN_s and prove several basic facts about them.

DEFINITION 1.1. ([18, 1.1]) Let R be a local Cohen-Macaulay ring, let I be an R-ideal of grade g, let J be a proper R-ideal, and let $s \geq g$ be an integer.

(a) J is called an *s-residual intersection* of I if there exists an R-ideal $\mathfrak{a} \subset I$, such that $J = \mathfrak{a} : I$ and $ht\ J \geq s \geq \mu(\mathfrak{a})$.
(b) J is called a *geometric s-residual intersection* of I, if J is an s-residual intersection of I and if in addition $ht\ I + J \geq s+1$.

When saying "$\mathfrak{a} : I$ is an s-residual intersection", we always mean that $\mathfrak{a}$ is an ideal as in Definition 1.1.

DEFINITION 1.2. Let R be a local Cohen-Macaulay ring, let I be an R-ideal of grade g, and let s be an integer.

(a) We say that I satisfies AN_s^- if for every $g \leq i \leq s$ and every geometric i-residual intersection J of I, R/J is Cohen-Macaulay.
(b) We say that I satisfies AN_s if for every $g \leq i \leq s$ and every i-residual intersection J of I, R/J is Cohen-Macaulay.

Notice that AN_s always holds for $s \leq g-1$. Before we can investigate the above conditions, we need several lemmas about the local generation of ideals.

LEMMA 1.3. *Let R be a Noetherian ring, let M be a finitely generated R-module, consider (not necessarily distinct) prime ideals $p_1, \ldots, p_n$ of R, and submodules $N_1, \ldots, N_n$ of M. Assume that either*

(i) *(R, m, k) is local with k infinite, or*
(ii) *$p_1, \ldots, p_n$ are pairwise distinct (which can always be achieved if $N_1 = \cdots = N_n$).*

Then there exists $x \in M$ such that for every $1 \leq i \leq n$, $\mu((M/N_i + (x))_{p_i}) = \max\{0, \mu((M/N_i)_{p_i}) - 1\}$.

PROOF. We may assume that $(M/N_i)_{p_i} \neq 0$ for every $1 \leq i \leq n$.

First consider the case where (i) holds. Let M_i denote the preimage of $(N_i)_{p_i} + p_i M_{p_i}$ in M. By Nakayama's Lemma, $M_{p_i} \neq (N_i)_{p_i} + p_i M_{p_i}$, and therefore $M \neq M_i$, which, again by Nakayama's Lemma, gives that $M \neq M_i + mM$. But then, $M \neq \bigcup_{i=1}^n (M_i + mM)$, since, being a vector space over the infinite field k, $M \otimes_R k$ cannot be a finite union of proper subspaces. In particular, $M \setminus \bigcup_{i=1}^n M_i \neq \emptyset$. Any element x in this set has the desired properties.

Next, assume that (ii) holds. We proceed by induction on n, the assertion being clear for $n = 1$. So let $n \geq 2$, and arrange $p_1, \ldots, p_n$ in such a way that $p_i \not\subset p_n$ for $1 \leq i \leq n-1$; then there exists $a \in (\bigcap_{i=1}^{n-1} p_i) \setminus p_n$. By induction hypothesis, we have elements u and v in M such that $\mu((M/N_i + (u))_{p_i}) < \mu((M/N_i)_{p_i})$ for $1 \leq i \leq n-1$ and $\mu((M/N_n + (v))_{p_n}) < \mu((M/N_n)_{p_n})$. Now if in M_{p_n}, $\frac{u}{1} \notin (N_n)_{p_n} + p_n M_{p_n}$, then take $x = u$. Otherwise, set $x = u + av$. ∎

LEMMA 1.4. *Let R be a Noetherian local ring with infinite residue class field, let $\mathcal{P}$ be a finite (possibly empty) subset of Spec(R), let s be an integer, let $\mathfrak{a} \subset I$ be R-ideals with ht $\mathfrak{a} : I \geq s$ (possibly $\mathfrak{a} = I$ in which case ht $\mathfrak{a} : I = \infty$), and assume that I satisfies G_s.*

Then there exist elements $a_1, \ldots, a_s$ in $\mathfrak{a}$ such that for every $0 \leq i \leq s$, and every subset $\{\nu_1, \ldots, \nu_i\}$ of $\{1, \ldots, s\}$,

(i) *$\mu((\mathfrak{a}/(a_1, \ldots, a_s))_p) = \max\{0, \mu(\mathfrak{a}_p) - s\}$ whenever $p \in \mathcal{P}$;*
(ii) *ht $(a_{\nu_1}, \ldots, a_{\nu_i}) : I \geq i$;*
(iii) *ht $I + (a_{\nu_1}, \ldots, a_{\nu_i}) : I \geq i+1$ whenever $0 \leq i \leq s-1$;*
(iv) *ht $I + (a_1, \ldots, a_s) : I \geq s+1$, provided that ht $I + \mathfrak{a} : I \geq s+1$ and I satisfies G_{s+1}.*

PROOF. By induction on k, $0 \leq k \leq s$, we are going to construct elements $a_1, \ldots, a_k$ in $\mathfrak{a}$ such that for every $0 \leq i \leq k$, and every subset $\{\nu_1, \ldots, \nu_i\} \subset \{1, \ldots, k\}$,

(i′) $\mu((\mathfrak{a}/(a_1, \ldots, a_k))_p) = \max\{0, \mu(\mathfrak{a}_p) - k\}$ whenever $p \in \mathcal{P}$;
(ii′) $(\mathfrak{a}/(a_{\nu_1}, \ldots, a_{\nu_i}))_p = 0$ whenever $p \in \text{Spec}(R)$ and $\dim R_p \leq i-1$;

(iii′) $\mu((\mathfrak{a}/(a_{\nu_1},\dots,a_{\nu_i}))_p) \leq \dim R_p - i$ whenever $p \in V(I)$ and $i \leq \dim R_p \leq s-1$;

(iv′) $\mu((\mathfrak{a}/(a_{\nu_1},\dots,a_{\nu_i}))_p) \leq \dim R_p - i$ whenever $p \in V(I)$ and $\dim R_p = s$, provided that $ht\ I + \mathfrak{a} : I \geq s+1$ and I satisfies G_{s+1}.

It is clear that $a_1, \dots, a_s$ will have the desired properties (i) through (iv). This follows because $I_p = \mathfrak{a}_p$ for every $p \in \operatorname{Spec}(R)$ with $\dim R_p \leq s-1$ and, in the situation of (iv), for every $p \in V(I)$ with $\dim R_p = s$.

Notice that $\mathfrak{a}$ satisfies G_s and that with the assumptions of (iv′), $\mu(\mathfrak{a}_p) \leq \dim R_p$ for $p \in V(I)$ with $\dim R_p = s$. Therefore (i′) through (iv′) certainly hold if $k = 0$. So let $1 \leq k \leq s$, and assume that $a_1, \dots, a_{k-1}$ have been constructed. In order to obtain a_k, we wish to apply Lemma 1.3 to the module $M = \mathfrak{a}$ and the finite family $\mathcal{M}$ of all submodules of the form $N = (a_{\nu_1}, \dots, a_{\nu_i})$, where $0 \leq i \leq k-1$ and $\{\nu_1, \dots, \nu_i\} \subset \{1, \dots, k-1\}$. For $0 \leq j \leq s$ consider the j-th Fitting ideals of the R-modules $\mathfrak{a}/N$, $F_j = F_j(\mathfrak{a}/N)$, which define the loci $V(F_j) = \{p \in \operatorname{Spec}(R) \mid \mu((\mathfrak{a}/N)_p) > j\}$. Now let $\mathcal{Q}$ be the finite subset of $\operatorname{Spec}(R)$ consisting of all prime ideals in the set $\mathcal{P}$, of all minimal prime ideals in $V(F_0)$, and of all minimal prime ideals in $V(I + F_j)$, for $0 \leq j \leq s$ and every $N \in \mathcal{M}$. By Lemma 1.3, there exists an element $x = a_k \in \mathfrak{a}$ such that $\mu((\mathfrak{a}/N + (a_k))_p) = \max\{0, \mu((\mathfrak{a}/N)_p) - 1\}$ for every $p \in \mathcal{Q}$ and every $N \in \mathcal{M}$.

Now $a_1, \dots, a_k$ certainly satisfy (i′), because $\mathcal{P} \subset \mathcal{Q}$, because $(a_1, \dots, a_{k-1}) \in \mathcal{M}$, and because $a_1, \dots, a_{k-1}$ satisfy (i′). As to (ii′), (iii′), and (iv′), it suffices to check these conditions for factor modules of the form $\mathfrak{a}/(a_{\nu_1}, \dots, a_{\nu_i}, a_k) = \mathfrak{a}/N + (a_k)$ with $N \in \mathcal{M}$, since we already know that $a_1, \dots, a_{k-1}$ satisfy (ii′), (iii′), and (iv′). Again we write $F_j = F_j(\mathfrak{a}/N)$.

In order to verify (ii′), let $p \in \operatorname{Spec}(R)$ with $\dim R_p \leq i$. If $p \notin V(F_0)$, then $(\mathfrak{a}/N)_p = 0$ and we are done. Thus we may assume that $F_0 \subset p$. Furthermore, by (ii′) applied to $\mathfrak{a}/N$, $ht\ F_0 \geq i$, so that $\dim R_p = i$ and p is minimal in $V(F_0)$. Hence $p \in \mathcal{Q}$, and from our choice of a_k we conclude that $\mu((\mathfrak{a}/N + (a_k))_p) = \max\{0, \mu((\mathfrak{a}/N)_p) - 1\}$. On the other hand, if $p \notin V(\mathfrak{a})$, then $\mu((\mathfrak{a}/N)_p) \leq 1$; whereas if $p \in V(\mathfrak{a})$, then $p \in V(I)$, and hence by (iii′) applied to $\mathfrak{a}/N$, $\mu((\mathfrak{a}/N)_p) = 0$. In either case, $\mu((\mathfrak{a}/N + (a_k))_p) = 0$, which shows (ii′).

To show (iii′) and (iv′), let $p \in V(I)$ with $i + 1 \leq \dim R_p \leq s$. Write $j = \dim R_p - i - 1$, and notice that $0 \leq j \leq s - i - 1$. If $p \notin V(I + F_j)$, then $\mu((\mathfrak{a}/N)_p) \leq j$, and we are done. So we may assume that $I + F_j \subset p$. On the other hand, by (iii′) applied to $\mathfrak{a}/N$, $ht(I + F_j) \geq j + i + 1$ for $0 \leq j \leq s - i - 1$. Therefore $j + i + 1 = \dim R_p \geq ht(I + F_j) \geq j + i + 1$, which shows that p is minimal in $V(I + F_j)$. Thus $p \in \mathcal{Q}$; now our choice of a_k implies that $\mu((\mathfrak{a}/N + (a_k))_p) = \max\{0, \mu((\mathfrak{a}/N)_p) - 1\}$. On the other hand, if $\dim R_p \leq s-1$ then by (iii′) applied to $\mathfrak{a}/N$, $\mu((\mathfrak{a}/N)_p) \leq \dim R_p - i$; and if the assumptions of (iv′) are satisfied then $\mu((\mathfrak{a}/N)_p) \leq \dim R_p - i$ as well. Thus in either case $\mu((\mathfrak{a}/N + (a_k))_p) \leq \max\{0, (\dim R_p - i) - 1\} = \dim R_p - i - 1$. ∎

LEMMA 1.5. *Let R be a Northerian ring, let $\mathcal{P}$ be a finite (possibly empty) subset of Spec(R), let s be an integer, let $\mathfrak{a} \subset I$ be R-ideals with ht $\mathfrak{a} : I \geq s$ (possibly $\mathfrak{a} = I$ in which case ht $\mathfrak{a} : I = \infty$), and assume that I satisfies G_s.*

Then there exist elements $a_1, \ldots, a_s$ in $\mathfrak{a}$ such that

(i) $\mu((\mathfrak{a}/(a_1, \ldots, a_s))_p) = \max\{0, \mu(\mathfrak{a}_p) - s\}$ *whenever* $p \in \mathcal{P}$;
(ii) *ht* $(a_1, \ldots, a_i) : I \geq i$ *whenever* $0 \leq i \leq s$;
(iii) *ht* $I + (a_1, \ldots, a_i) : I \geq i + 1$ *whenever* $0 \leq i \leq s - 1$;
(iv) *ht* $I + (a_1, \ldots, a_s) : I \geq s + 1$, *provided that ht* $I + \mathfrak{a} : I \geq s + 1$ *and* I *satisfies* G_{s+1}.

PROOF. By induction on k, $0 \leq k \leq s$, one constructs suitable elements $a_1, \ldots, a_k$ like in the proof of Lemma 1.4. However now, in order to obtain a_k, $1 \leq k \leq s$, it suffices to apply Lemma 1.3 to the family of submodules consisting of the single module $N = (a_1, \ldots, a_{k-1})$. Thus assumption (ii) of Lemma 1.3 is satisfied, and our assertions follow as in the previous proof. ■

COROLLARY 1.6. *Let (R, m, k) be a Noetherian local ring, let s be an integer, let $\mathfrak{a} \subset I$ be R-ideals with ht $\mathfrak{a} : I \geq s \geq \mu(\mathfrak{a})$, and assume that I satisfies G_s.*

(a) *(cf. also [2, 2.3]) There exists a generating sequence $a_1, \ldots, a_s$ of $\mathfrak{a}$ such that, with $\mathfrak{a}_i = (a_1, \ldots, a_i)$ and $J_i = \mathfrak{a}_i : I$, ht $J_i \geq i$ if $0 \leq i \leq s$ and ht $I + J_i \geq i + 1$ if $0 \leq i \leq s - 1$.*
(b) *If k is infinite then there exists a generating sequence $a_1, \ldots, a_s$ of $\mathfrak{a}$ such that for every subset $\{\nu_1, \ldots, \nu_i\} \subset \{1, \ldots, s\}$, ht $(a_{\nu_1}, \ldots, a_{\nu_i}) : I \geq i$ if $0 \leq i \leq s$, and ht $I + (a_1, \ldots, a_{\nu_i}) : I \geq i + 1$ if $0 \leq i \leq s - 1$.*

PROOF. Our assertions follow by applying Lemmas 1.5 and 1.4 with $\mathcal{P} = \{m\}$. ■

The next proposition contains some basic facts about Artin-Nagata properties; it also shows that in studying s-residual intersections, it often suffices to assume AN_{s-1} rather than AN_s. Several aspects of this result have been observed in other contexts (cf. [15, the proof of 3.1]).

PROPOSITION 1.7. *Let R be a local Cohen-Macaulay ring of dimension d, let I be an R-ideal of grade g satisfying G_s for some $s \geq g$, let $J = \mathfrak{a} : I$ be an s-residual intersection, and let $J_i = \mathfrak{a}_i : I$ be ideals as in Corollary 1.6 (a). Further assume that for every $g \leq i \leq s - 1$, R/J_i is Cohen-Macaulay (which is automatically the case if I satisfies AN^-_{s-1}). Now let $0 \leq i \leq s$, and let "$\,'\,$" denote images in $R' = R/J_i$. Then:*

(a) *J_i is unmixed of height i;*
(b) *depth $R/\mathfrak{a}_i = d - i$;*
(c) *$\mathfrak{a}_i = I \cap J_i$ if $0 \leq i \leq s - 1$, or if $i = s$ and J is a geometric s-residual intersection;*
(d) *ht $I' = 1$ if $g - 1 \leq i \leq s - 1$, or if $i = s$ and J is a geometric s-residual intersection;*

(e) $R'/I' \cong R/I + J_i$ *is Cohen-Macaulay, if* $0 \le i \le s-1$ *and* R/I *is Cohen-Macaulay;*
(f) $J'_{i+1} = (a'_{i+1}) : I'$ *and* a'_{i+1} *is regular on* R', *if* $0 \le i \le s-1$;
(g) $J_i = \mathfrak{a}_i : (a_{i+1})$ *if* $0 \le i \le s-1$.

PROOF. After adjoining a variable to R and $\mathfrak{a}$, and localizing, we may suppose that $g \ge 1$. We proceed by induction on i, $0 \le i \le s$, the cases $0 \le i \le g-1$ being trivial. Now assume that our assertions hold for i, $g-1 \le i \le s-1$, and let "$-$" denote images in $\bar{R} = R/J_{i+1}$.

Since $R' = R/J_i$ is Cohen-Macaulay, (f) together with (a) implies that J_{i+1} is unmixed of height $i+1$, which verifies assertion (a) for J_{i+1}. Next, from (g), we obtain an exact sequence

$$0 \longrightarrow R/J_i \longrightarrow R/\mathfrak{a}_i \longrightarrow R/\mathfrak{a}_{i+1} \longrightarrow 0,$$

where by (a) and (b), depth R/J_i = depth $R/\mathfrak{a}_i$ $= d-i$. Therefore depth $R/\mathfrak{a}_{i+1} \ge d-i-1$. On the other hand, $IJ_{i+1} \subset \mathfrak{a}_{i+1}$ with $I \not\subset \mathfrak{a}_{i+1}$ and *ht* $J_{i+1} = i+1$. Hence depth $R/\mathfrak{a}_{i+1} = d-i-1$, which is part (b) for $i+1$.

Now assume that $i+1 \le s-1$, or that $i+1 = s$ and J is a geometric s-residual intersection; then in either case *ht* $I + J_{i+1} \ge i+2$ (cf. Corollary 1.6 (a)). On the other hand, by (b) applied to $R/\mathfrak{a}_{i+1}$, $\dim R_p \le i+1$ for every $p \in Ass\ R/\mathfrak{a}_{i+1}$. Since furthermore $\mathfrak{a}_{i+1} \subset I \cap J_{i+1}$ and $J_{i+1} = \mathfrak{a}_{i+1} : I$, it follows that $\mathfrak{a}_{i+1} = I \cap J_{i+1}$. This proves part (c) for $i+1$. With the same assumptions, and using (d) for i, and (a) for i and $i+1$, we conclude that *ht* $I' + J'_{i+1} \ge 2 > 1 =$ *ht* $J'_{i+1} =$ *ht* I', and therefore $\sqrt{I'} \ne \sqrt{(a'_{i+1})} \ne \sqrt{J'_{i+1}}$. Since on the other hand, $I'J'_{i+1} \subset (a'_{i+1})$, the Abhyankar-Hartshorne connectedness lemma ([1, 4.9], [6, 2.1]) implies that grade $I' + J'_{i+1}/(a'_{i+1}) \le 1$. Therefore $2 \ge$ grade $I' + J'_{i+1} =$ *ht* $I' + J'_{i+1} \ge 2$, which gives *ht* $I' + J'_{i+1} = 2$ and *ht* $\bar{I} = 1$. Thus we have shown (d) for $i+1$.

Next assume that $i+1 \le s-1$ and that R/I is Cohen-Macaulay. From (a), (b), and (c) for $i+1$, we have an exact sequence

$$0 \longrightarrow R/\mathfrak{a}_{i+1} \longrightarrow R/I \oplus R/J_{i+1} \longrightarrow R/I + J_{i+1} \longrightarrow 0,$$

where depth $R/\mathfrak{a}_{i+1}$ = depth $R/J_{i+1} = d-i-1$ and depth $R/I = d-g \ge d-i-1$. Therefore depth $R/I + J_{i+1} \ge d-i-2$. On the other hand, by (a) and (d) for $i+1$, $\dim R/I + J_{i+1} = \dim \bar{R}/\bar{I} = (d-i-1)-1 = d-i-2$. This proves (e) for $\bar{R}/\bar{I}$.

Next we show (f) for $i+1 \le s-1$. Since $\mathfrak{a}_{i+1} = I \cap J_{i+1}$, it follows that

$$\begin{aligned} J_{i+2} &= (\mathfrak{a}_{i+1}, a_{i+2}) : I = (I \cap J_{i+1}, a_{i+2}) : I = (J_{i+1}, a_{i+2}) : I \\ &= (J_{i+1}, a_{i+2}) : (J_{i+1}, I) \quad \text{(cf. also [15, the proof of 3.1]).} \end{aligned}$$

Hence $\overline{J_{i+2}} = (\overline{a_{i+2}}) : \bar{I}$. Furthermore $\overline{a_{i+2}}$ has to be $\bar{R}$-regular, because J_{i+1} is unmixed and *ht* $J_{i+1} = i+1 < i+2 \le$ *ht* J_{i+2} as well as *ht* $J_{i+1} = i+1 < i+2 \le$ *ht* $I + J_{i+1}$.

Finally, to prove (g) for $i+1 \leq s-1$, notice that $J_{i+1} \subset \mathfrak{a}_{i+1} : (a_{i+2})$. On the other hand, $(\mathfrak{a}_{i+1} : (a_{i+2}))a_{i+2} \subset \mathfrak{a}_{i+1} \subset J_{i+1}$, where a_{i+2} is regular on $\bar{R} = R/J_{i+1}$ by the previous part. Therefore $\mathfrak{a}_{i+1} : (a_{i+2}) = J_{i+1}$. ■

The notion of a d-sequence has been introduced by Huneke ([12]); it can be described in the following way: a sequence of elements $a_1, \ldots, a_n$ in a Noetherian local ring is a *d-sequence* if $a_1, \ldots, a_n$ minimally generate an ideal I and if $[(a_1, \ldots, a_i) : (a_{i+1})] \cap I = (a_1, \ldots, a_i)$ for $0 \leq i \leq n-1$. A d-sequence is called *unconditioned* in case every permutation of the elements forms a d-sequence. Our next two results are refinements of [15, 5.2] and [9, 3.4].

COROLLARY 1.8. *Let R be a local Cohen-Macaulay ring of dimension d, let I be an R-ideal with grade $I = g$ and $\mu(I) = n$, assume that I satisfies G_n, and use the notation of Corollary 1.6 for $\mathfrak{a} = I$ and $s = n$.*

(a) *Assume that for every $g \leq i \leq n-2$, R/J_i is Cohen-Macaulay; then:*

  (i) *the sequence $a_1, \ldots, a_n$ of Corollary 1.6 (a) is a d-sequence generating I;*

  (ii) *I satisfies sliding depth if depth $R/I \geq d-n$.*

(b) *Assume that I satisfies AN^-_{n-2}; then the sequence $a_1, \ldots, a_n$ of Corollary 1.6 (b) is an unconditioned d-sequence generating I;*

(c) *Assume that for every $g \leq i \leq n-1$, R/J_i is Cohen-Macaulay; then I satisfies sliding depth.*

PROOF. We may assume that $n \geq g+1$.

To prove (a), notice that by Corollary 1.6 (a), $J_{n-1} = \mathfrak{a}_{n-1} : I$ is a geometric $n-1$ residual intersection of I. Thus we may apply Proposition 1.7 with $s = n-1$. Parts (g) and (c) of this proposition imply that $a_1, \ldots, a_n$ from a d-sequence. But then $a_1, \ldots, a_n$ are a proper sequence ([12, Proposition 2.1] and [5]). Furthermore, by Proposition 1.7 (b), depth $R/\mathfrak{a}_i = d-i$ for $0 \leq i \leq n-1$; and if we assume that depth $R/I \geq d-n$, then depth $R/(a_1, \ldots, a_i) \geq d-i$ for every $0 \leq i \leq n$. In this situation, [9, 3.7] implies that the ideal I generated by the proper sequence $a_1, \ldots, a_n$ satisfies sliding depth.

The proof of (b) is analogous to the proof of part (a.i); and in order to show (c), recall that by Proposition 1.7 (g), $\mathfrak{a}_{i+1}/\mathfrak{a}_i \cong R/J_i$ for $0 \leq i \leq n-2$. Since moreover $\mathfrak{a}_n/\mathfrak{a}_{n-1} \cong R/\mathfrak{a}_{n-1} : \mathfrak{a}_n = R/\mathfrak{a}_{n-1} : I = R/J_{n-1}$, we have that $\mathfrak{a}_{i+1}/\mathfrak{a}_i \cong R/J_i$ are Cohen-Macaulay for every $0 \leq i \leq n-1$. Again, using the fact that $a_1, \ldots, a_n$ form a proper sequence, we conclude from [9, 3.7] that $I = (a_1, \ldots, a_n)$ satisfies sliding depth. ■

COROLLARY 1.9. *Let R be a local Cohen-Macaulay ring of dimension d, let I be an R-ideal satisfying G_{d-1} and AN^-_{d-2}.*

Then I satisfies AN^-_s for every s. If in addition I is G_∞, then I satisfies sliding depth.

PROOF. Our first claim follows from Proposition 1.7 (a) for $i = s = d - 1$, and the second assertion is a consequence of Corollary 1.8 (c). ∎

We now need to clarify how Artin-Nagata properties behave under localization.

LEMMA 1.10. *Let R be a local Cohen-Macaulay ring, let I be an R-ideal, consider $p \in V(I)$, and let s be an integer.*

(a) *If I satisfies G_{s+1} and AN_s^-, then I_p satisfies AN_s^-.*
(b) *If I satisfies G_s and AN_s, then I_p satisfies AN_s.*

PROOF. We prove (b) and (a) simultaneously. So assume that I satisfies G_s and AN_s (or G_{s+1} and AN_s^-, respectively). It suffices to consider an s-residual intersection (or a geometric s-residual intersection, respectively) in R_p, $J = \mathfrak{a} : I_p$, and to show that R_p/J is Cohen-Macaulay. We are done once we have found an s-residual intersection (or a geometric s-residual intersection, respectively) in R, $\mathfrak{c} : I$, with $\mathfrak{c}_p = \mathfrak{a}$.

To this end consider the natural map $\varphi : R \to R_p$, and define $\mathfrak{b} = I \cap \varphi^{-1}(\mathfrak{a})$; then $\mathfrak{b}_p = \mathfrak{a}$ and $I/\mathfrak{b}$ embeds into $I_p/\mathfrak{a}$. Therefore $\mathfrak{b} :_R I = \mathrm{ann}_R I/\mathfrak{b} = \varphi^{-1}(\mathrm{ann}_{R_p} I_p/\mathfrak{a}) = \varphi^{-1}(J)$; hence every minimal prime ideal of $\mathfrak{b} :_R I$ has height at least s and is contained in p. It follows that $ht\ \mathfrak{b} :_R I \geq s$ and, if J is a geometric s-residual intersection, that $ht\ I + \mathfrak{b} :_R I \geq s + 1$. Now applying Lemma 1.5 with $\mathcal{P} = \{p\}$ we conclude that there exist elements $b_1, \ldots, b_s$ in $\mathfrak{b}$, such that for $\mathfrak{c} = (b_1, \ldots, b_s)$ the following conditions hold: $\mu((\mathfrak{b}/\mathfrak{c})_p) = \max\{0, \mu(\mathfrak{b}_p) - s\} = \max\{0, \mu(\mathfrak{a}) - s\} = 0$, which gives $\mathfrak{c}_p = \mathfrak{b}_p = \mathfrak{a}$; $\mathfrak{c} : I$ is an s-residual intersection of I if I satisfies G_s; $\mathfrak{c} : I$ is a geometric s-residual intersection of I, if I satisfies G_{s+1} and if J is a geometric s-residual intersection of I_p. ∎

Our next result is a generalization of [25, 4.1]. Recall that an ideal I is said to be of *linear type* if the natural epimorphism from the symmetric algebra $S(I) = \oplus_{j \geq 0} S_j(I)$ to the Rees algebra $\mathcal{R}(I)$ is an isomorphism.

PROPOSITION 1.11. *Let (R, m, k) be a local Cohen-Macaulay ring, let I be an R-ideal, let s be an integer, and assume that k is infinite in case $s \geq \dim R$. Further suppose that I satisfies AN_{s-2}^-, that for every minimal $p \in V(I)$ with $\dim R_p \leq s$, I_p is a complete intersection, and that for every non-minimal $p \in V(I)$ with $\dim R_p \leq s$, either $\mu(I_p) \leq \dim R_p$ or $\ell(I_p) \leq \dim R_p - 1$.*

Then for every $p \in V(I)$ with $\dim R_p \leq s$, I_p satisfies sliding depth and is of linear type; in particular, I satisfies G_{s+1}.

PROOF. We induct on s, the assertion being trivial for $s << 0$. So let $p \in V(I)$ with $\dim R_p = s$. By our induction hypothesis I satisfies G_s, and hence by Lemma 1.10 (a), we do not change our assumptions when localizing at p to assume that $p = m$ and that k is infinite. Again by induction hypothesis, I_q is of linear type for every $q \in \operatorname{Spec}(R) \setminus \{m\}$. Now by Corollary 1.9 and [7, 2.3 and 2.5], we will be done once we have shown that $\mu(I) \leq \dim R = s$.

To this end, we may assume that $\ell(I) \leq \dim R - 1 = s - 1$. So let $\mathfrak{a} = (a_1, \dots, a_{s-1})$ be a reduction of I and suppose that $\mathfrak{a} \neq I$. By the above, $I_q = \mathfrak{a}_q$ for every $q \in \operatorname{Spec}(R) \setminus \{m\}$, hence $ht\ \mathfrak{a} : I = s$, which shows that $\mathfrak{a} : I$ is an $(s-1)$-residual intersection of I. On the other hand, since I satisfying G_{s-1} and AN_{s-2}^-, we conclude from Proposition 1.7 (a) that $ht\ \mathfrak{a} : I = s - 1$. This contradiction shows that $I = \mathfrak{a} = (a_1, \dots, a_{s-1})$. ∎

We conclude this section with a remark and a lemma that will be needed later.

REMARK 1.12. Let R be a local Cohen-Macaulay ring, let s be an integer, let $\mathfrak{a} \subset I$ be R-ideals with $ht\ \mathfrak{a} : I \geq s + 1$, and assume that I satisfies G_s.

If I satisfies AN_s or AN_s^-, then $\mathfrak{a}$ satisfies AN_s or AN_s^-, respectively.

PROOF. It suffices to consider an s-residual intersection (or a geometric s-residual intersection) $\mathfrak{b} : \mathfrak{a}$ of $\mathfrak{a}$. Since $I_p = \mathfrak{a}_p$ for every $p \in \operatorname{Spec}(R)$ with $\dim R_p \leq s$, it follows that $\mathfrak{b} : I$ is an s-residual intersection of I (or a geometric s-residual intersection of I, respectively). But then by Proposition 1.7 (a), for every associated prime p of $\mathfrak{b} : I$, $\dim R_p = s$, and therefore $(\mathfrak{b} : I)_p = (\mathfrak{b} : \mathfrak{a})_p$. Since on the other hand, $\mathfrak{b} : I \subset \mathfrak{b} : \mathfrak{a}$, we conclude that $\mathfrak{b} : I = \mathfrak{b} : \mathfrak{a}$. Thus $R/\mathfrak{b} : \mathfrak{a} = R/\mathfrak{b} : I$ is Cohen-Macaulay. ∎

Recall that a pair $(\tilde{R}, \tilde{I})$ of a Noetherian local ring $\tilde{R}$ and an $\tilde{R}$-ideal $\tilde{I}$ is said to be a *deformation* of a pair (R, I), if $R \cong \tilde{R}/(\mathbf{x})$ and $I = \tilde{I}R$ for some sequence $\mathbf{x} = x_1, \dots, x_n$ in $\tilde{R}$ that is regular on $\tilde{R}$ and on $\tilde{R}/\tilde{I}$.

LEMMA 1.13. *Let R be a local Cohen-Macaulay ring, let s be an integer, let I be an R-ideal, let $(\tilde{R}, \tilde{I})$ be a deformation of (R, I), and assume that $\tilde{I}$ satisfies G_s.*

If $\tilde{I}$ satisfies AN_s^-, then I satisfies AN_s^-.

PROOF. It suffices to consider a geometric s-residual intersection $J = \mathfrak{a} : I$ of I. Let $\pi : \tilde{R} \longrightarrow R$ be the projection map, and pick an $\tilde{R}$-ideal $\tilde{\mathfrak{a}}$ with $\tilde{\mathfrak{a}} \subset \tilde{I}$, $\mu(\tilde{\mathfrak{a}}) \leq s$, and $\pi(\tilde{\mathfrak{a}}) = \mathfrak{a}$. Define $\tilde{J} = \tilde{\mathfrak{a}} : \tilde{I}$. Since $\ker \pi$ is generated by a sequence that is regular on $\tilde{R}/\tilde{I}$, we may use [18, 4.1] to conclude that $\sqrt{\pi(\tilde{J})} = \sqrt{J}$. Therefore $ht\ \tilde{J} \geq ht\ \pi(\tilde{J}) = ht\ J \geq s$, and likewise $ht\ \tilde{I} + \tilde{J} \geq ht\ \pi(\tilde{I} + \tilde{J}) = ht\ I + J \geq s + 1$. Hence $\tilde{J}$ is a geometric s-residual intersection of $\tilde{I}$. Now our assumption on $\tilde{I}$ and Proposition 1.7 (a) imply that $\tilde{R}/\tilde{J}$ is Cohen-Macaulay with $ht\ \tilde{J} = s$. The Cohen-Macaulayness of R/J then follows from [18, 4.2 (ii)], because J is a geometric s-residual intersection. ∎

2. Depth conditions on powers

In this section we relate the Artin-Nagata property to the depths of ideal powers. To do so, we first prove two lemmas that describe the Cohen-Macaulayness of certain colon ideals in terms of canonical modules.

LEMMA 2.1. *Let R be a local Cohen-Macaulay ring with a canonical module $\omega = \omega_R$, write $-^\vee = Hom_R(-, \omega)$, let I be an R-ideal, let α be an R-regular element contained in I, let $J = (\alpha) : I \neq R$, and write $\bar{R} = R/J$. Then:*

(a) $\omega_{\bar{R}} \cong (I\omega)^{\vee\vee}/\alpha\omega$;

(b) *$\bar{R}$ is Cohen-Macaulay if and only if $(I\omega)^{\vee\vee}$ is a Cohen-Macaulay R-module.*

PROOF. There are natural isomorphisms

$$(\alpha) : I \cong \mathrm{Hom}_R(I, (\alpha)) \cong \mathrm{Hom}_R(I, R)\alpha \cong \mathrm{Hom}_R(I, \mathrm{Hom}_R(\omega, \omega))\alpha$$
$$\cong \mathrm{Hom}_R(I \otimes_R \omega, \omega)\alpha \cong \mathrm{Hom}_R(I\omega, \omega)\alpha,$$

which give an identification

$$J \cong (I\omega)^\vee \alpha. \tag{2.2}$$

Now applying $-^\vee$ to the exact sequence

$$0 \longrightarrow J \longrightarrow R \longrightarrow \bar{R} \longrightarrow 0,$$

one obtains an exact sequence

$$0 \longrightarrow \omega \longrightarrow J^\vee \longrightarrow \omega_{\bar{R}} \longrightarrow 0,$$

which, when combined with (2.2), implies (a). Part (b) is also an immediate consequence of (2.2), because $\bar{R}$ is Cohen-Macaulay if and only if $J \cong (I\omega)^\vee$ is a Cohen-Macaulay R-module, which in turn is equivalent to the Cohen-Macaulayness of $(I\omega)^{\vee\vee}$. ■

LEMMA 2.3. *Let R be a Cohen-Macaulay ring with a canonical module $\omega = \omega_R$, let I be an R-ideal of grade g, let $\underline{\alpha} = \alpha_1, \dots, \alpha_g$ be an R-regular sequence contained in I, let $J = (\underline{\alpha}) : I \neq R$, and write $\bar{R} = R/J$. Consider the conditions:*

(a) *$\omega/I\omega$ is a Cohen-Macaulay R/I-module;*

(b) *$\bar{R}$ is a Cohen-Macaulay ring with $\omega_{\bar{R}} \cong I\omega/(\underline{\alpha})\omega$.*

Then (a) implies (b), and the converse holds if for every $p \in V(I + J)$ with $\dim R_p = g + 1$, I_p is unmixed of grade g, and either

(i) *I_p is strongly Cohen-Macaulay, or*

(ii) *$ht(I + J)_p \geq g + 1$ and $\bar{R}_p$ is generically Gorenstein.*

PROOF. After adjoining a variable to R and $\underline{\alpha}$, and localizing, we may assume that $g \geq 1$, and after factoring out $(\alpha_1, \dots, \alpha_{g-1})$, we may even suppose that $g = 1$ (the strong Cohen-Macaulayness is preserved under these modifications, as has been shown in [14, 1.6]). But then $\omega/I\omega$ is a Cohen-Macaulay R/I-module if and only if $I\omega$ is a Cohen-Macaulay R-module. Now Lemma 2.1 gives that (a) implies (b).

Conversely assume that (b) holds. By Lemma 2.1, $(I\omega)^{\vee\vee}$ is a Cohen-Macaulay R-module, and therefore $I\omega$ is a Cohen-Macaulay R-module if and only if the natural embedding $\iota : I\omega \longrightarrow (I\omega)^{\vee\vee}$ is surjective. Now by part (b) and Lemma 2.1, ι induces a commutative diagram with exact rows,

$$\begin{array}{ccccccccc} 0 & \longrightarrow & \alpha\omega & \longrightarrow & I\omega & \longrightarrow & \omega_{\bar{R}} & \longrightarrow & 0 \\ & & \| & & \downarrow \iota & & \downarrow \varepsilon & & \\ 0 & \longrightarrow & \alpha\omega & \longrightarrow & (I\omega)^{\vee\vee} & \longrightarrow & \omega_{\bar{R}} & \longrightarrow & 0 \end{array}, \tag{2.4}$$

where ι is surjective if and only if ε has this property. But $\varepsilon \in \operatorname{Hom}_{\bar{R}}(\omega_{\bar{R}}, \omega_{\bar{R}}) \cong \bar{R}$, hence $\varepsilon = \mu_b$ for some $b \in \bar{R}$, and therefore $\dim \operatorname{coker} \varepsilon \geq \dim \bar{R} - 1$ in case ε is not an epimorphism. Thus it suffices to show that $\iota \otimes_R R_p$ is surjective for every $p \in V(J)$ with $\dim R/p \geq \dim \bar{R} - 1$, or equivalently, with $\dim R_p \leq g + 1$. Since $\iota \otimes_R R_p$ is obviously an isomorphism for $p \notin V(I)$, it suffices to consider the case where $p \in V(I + J)$. But then, localizing at p we may assume that I is unmixed of grade one, that $\dim R \leq g + 1 = 2$, and that conditions (i) or (ii) are satisfied. We will prove that $I\omega$ is a Cohen-Macaulay R-module, for which we may suppose that $\dim R = 2$.

Now let $x \in R$ with α, x forming an R-regular sequence, and let "$'$" denote images in $R' = R/(x)$. We need to prove that depth $I\omega \otimes_R R' \geq 1$. To do so, we will establish the equality $J' = (\alpha') : I'$. Once this is done, then $\bar{R} \otimes_R R' \cong R'/(\alpha') : I'$, and hence $\omega_{\bar{R}} \otimes_R R' \cong \omega_S$, where $S = R'/(\alpha') : I'$. Therefore, tensoring the first row of (2.4) with $\otimes_R R'$ we obtain an exact sequence

$$0 \longrightarrow \alpha'\omega_{R'} \longrightarrow I\omega \otimes_R R' \longrightarrow \omega_S \longrightarrow 0. \tag{2.5}$$

On the other hand, $I'\omega_{R'}$ being a Cohen-Macaulay R'-module, part (b) applied to S yields an exact sequence

$$0 \longrightarrow \alpha'\omega_{R'} \longrightarrow I'\omega_{R'} \longrightarrow \omega_S \longrightarrow 0. \tag{2.6}$$

Now, comparing (2.5) and (2.6), one can see that the natural epimorphism $\pi : I\omega \otimes_R R' \twoheadrightarrow I'\omega_{R'}$ induces an epimorphism on ω_S. Since every surjective endormorphism of ω_S is bijective, it follows that π has to be an isomorphism. Therefore depth $I\omega \otimes_R R' =$ depth $I'\omega_{R'} \geq 1$.

It remains to show that $J' = (\alpha') : I'$. To do so, we may factor out (α) and (α') ([14, 1.6]); then I is unmixed of grade zero satisfying conditions (i) or (ii), $\dim R = 1$, and x is regular on R and on R/I. We have to prove that $J' = 0' : I'$.

If I is strongly Cohen-Macaulay, then this equality has been established in [17, the proof of Proposition 1.1]. Thus we may assume that grade $I + J = 1$ and that $\bar{R}$ is generically Gorenstein. Let H be the image of the natural map $\operatorname{Hom}_R(\omega, R) \otimes_R \omega \longrightarrow R$. For every $p \in Ass_R \bar{R}$, $R_p \cong \bar{R}_p$ is Gorenstein and therefore $\omega \otimes_R R_p$ is a free R_p-module, which implies that $H \not\subset p$. In fact, H cannot be contained in the union over all associated primes of $\bar{R}$, and hence there exists $\varphi \in \operatorname{Hom}_R(\omega, R)$ such that $\varphi(\omega)$ contains an $\bar{R}$-regular element. Now consider the $\bar{R}$-ideal $\bar{\omega} = \varphi(\omega) + J/J$; then grade $\bar{I}\bar{\omega} = 1$. On the other hand, since $I\omega \cong \omega_{\bar{R}}$ and $\bar{R}$ is generically Gorenstein, it follows that $I\omega$ is a torsion free $\bar{R}$-module of rank one. Therefore the natural epimorphism $I\omega \twoheadrightarrow \bar{I}\bar{\omega}$ is injective, because source and target are torsion free $\bar{R}$-modules of the same rank. In summary, $\bar{I}\bar{\omega} \cong \omega_{\bar{R}}$, which forces the natural embedding $\bar{R} \hookrightarrow \mathrm{Hom}_{\bar{R}}(\bar{I}\bar{\omega}, \bar{I}\bar{\omega})$ to be an isomorphism. Now, writing K for the total ring of quotients of $\bar{R}$, we have a commutative diagram of natural injective homomorphisms,

$$\begin{array}{ccccc} & & \mathrm{Hom}_{\bar{R}}(\bar{I}\bar{\omega}, \bar{I}\bar{\omega}) & \stackrel{\sim}{\longleftarrow} & \bar{I}\bar{\omega} :_K \bar{I}\bar{\omega} \\ & \swarrow & & & \\ \bar{R} & & & & \uparrow \\ & \searrow & & & \\ & & \mathrm{Hom}_{\bar{R}}(\bar{I}, \bar{I}) & \stackrel{\sim}{\longrightarrow} & \bar{I} :_K \bar{I} \end{array},$$

which implies that $\mathrm{Hom}_{\bar{R}}(\bar{I}, \bar{I})/\bar{R} = 0$. On the other hand, by [18, 4.3], $\mathrm{Ext}^1_R(R/I, R) \cong \mathrm{Hom}_{\bar{R}}(\bar{I}, \bar{I})/\bar{R}$, and therefore $\mathrm{Ext}^1_R(R/I, R) = 0$. Thus, when applying $\mathrm{Hom}_R(R/I, -)$ to the exact sequence

$$0 \longrightarrow R \stackrel{x}{\longrightarrow} R \longrightarrow R' \longrightarrow 0,$$

one sees that the natural map $J = 0 :_R I \longrightarrow 0' :_{R'} I'$ is surjective. This proves the asserted equality $J' = 0' : I'$. ∎

Our next lemma provides the crucial link between the depths of residual intersections and the depths of ideal powers.

LEMMA 2.7. *Let R be a local Cohen-Macaulay ring of dimension d, let I be an R-ideal of grade $g \geq 1$, let $s \geq g$ and $0 \leq r \leq s-g+1$, assume that I satisfies G_s and AN^-_{s-1}, and that depth $R/I^j \geq d-g-j+1$ whenever $1 \leq j \leq r$. Let $J = \mathfrak{a} : I$ with $\mathfrak{a} = (a_1, \ldots, a_s)$ be an s-residual intersection, for $0 \leq i \leq s$ let $J_i = \mathfrak{a}_i : I$ with $\mathfrak{a}_i = (a_1, \ldots, a_i)$ be ideals as defined in Corollary 1.6 (a), and for $1 \leq i \leq s$ and $2 \leq j$ consider the complexes*

$$C_{ij}. : 0 \longrightarrow I^{j-1}/\mathfrak{a}_{i-1}I^{j-2} \longrightarrow I^j/\mathfrak{a}_{i-1}I^{j-1} \longrightarrow I^j/\mathfrak{a}_i I^{j-1} \longrightarrow 0,$$

where the first map is induced by multiplication with a_i. Then:

(a) *The complexes $C_{ij}.$ are exact whenever $2 \leq j \leq r+1$ and $1 \leq i \leq s$;*
(b) *depth $I^j/\mathfrak{a}_i I^{j-1} \geq \min\{d-i, d-g-j+2\}$ whenever $1 \leq j \leq r$ and $0 \leq i \leq s$;*
(c) *$I^j/\mathfrak{a}_i I^{j-1} \cong I^j + J_i/J_i$ whenever $1 \leq j \leq r$ and $0 \leq i \leq s-1$, and, if J is a geometric $s-$residual intersection, whenever $1 \leq j \leq r+1$ and $0 \leq i \leq s$.*

PROOF. We proceed by induction on i, $0 \leq i \leq s$. If $i = 0$, then (a) is vacuous, (b) follows from our assumptions, and (c) is trivially satisfied. So let $i \geq 1$ and assume that our assertions have been shown for $i-1$.

In proving part (a) it suffices to show that the first map of $C_{ij\cdot}$ is injective. However, there is a commutative diagram,

$$\begin{array}{ccc} I^{j-1}/\mathfrak{a}_{i-1}I^{j-2} & \xrightarrow{\varphi} & I^j/\mathfrak{a}_{i-1}I^{j-1} \\ \downarrow \pi & \# & \downarrow \\ I^{j-1}+J_{i-1}/J_{i-1} & \xrightarrow{\psi} & I^j+J_{i-1}/J_{i-1} \end{array},$$

where φ and ψ are induced by multiplication with a_i. By Proposition 1.7 (f), a_i is regular on R/J_{i-1} and therefore ψ is injective. On the other hand, π is an isomorphism by part (c) for $i-1$. Thus φ is injective and $C_{ij\cdot}$ is exact.

To prove (b) we first consider the case $j=1$; then depth $I^j/\mathfrak{a}_i I^{j-1} =$ depth $I/\mathfrak{a}_i \geq \min\{$ depth $R/\mathfrak{a}_i$, depth $R/I+1\} = \min\{d-i, d-g+1\}$, where the last equality follows from Proposition 1.7(b). Thus we may assume that $2 \leq j \leq r$. But then the exactness of $C_{ij\cdot}$ as well as part (b) for $i-1$ imply that depth $I^j/\mathfrak{a}_i I^{j-1} \geq \min\{d-i, d-g-j+2\}$.

Finally we prove (c). First notice that by Proposition 1.7 (c), $\mathfrak{a}_i = I \cap J_i$. In particular, (c) holds for $j=1$. Thus we may assume that $j \geq 2$. Furthermore, Proposition 1.7 (a) shows that $\mu(\mathfrak{a})$ cannot be less than s. Therefore by Proposition 1.7 (g) and (c), $a_1, \ldots, a_s$ form a d-sequence in R, and hence the images of $a_{i+1}, \ldots, a_s$ form a d-sequence in $R/\mathfrak{a}_i$ ([12, p. 252]). Then it follows from [12, Theorem 2.1] that $\mathfrak{a}_i\mathfrak{a}^{j-1} = \mathfrak{a}_i \cap \mathfrak{a}^j$. On the other hand by the above, $\mathfrak{a}_i \cap \mathfrak{a}^j = I \cap J_i \cap \mathfrak{a}^j = J_i \cap \mathfrak{a}^j$. Putting this together, we conclude that

$$\mathfrak{a}^j/\mathfrak{a}_i\mathfrak{a}^{j-1} \cong \mathfrak{a}^j + J_i/J_i. \tag{2.8}$$

Now consider the natural epimorphism $\pi : I^j/\mathfrak{a}_i I^{j-1} \twoheadrightarrow I^j + J_i/J_i$, and let $p \in \operatorname{Spec}(R)$. If $I \not\subset p$, then $(\mathfrak{a}_i)_p = (J_i)_p$, and π_p is an isomorphism. If either $\dim R_p \leq s-1$, or else, $\dim R_p \leq s$ and $I \subset p$ and J is a geometric s-residual intersection, then $I_p = \mathfrak{a}_p$, and (2.8) implies that π_p is an isomorphism. Thus we may assume that $\dim R_p \geq s$, or else, that $\dim R_p \geq s+1$ if $j = r+1$ or $i = s$. In either case, $\dim R_p - i \geq 1$, and if $j = r+1$ then $\dim R_p \geq s+1$. We are going to show that $\operatorname{depth}(I^j/\mathfrak{a}_i I^{j-1})_p \geq 1$, which will imply that π is an isomorphism. First consider the case where $2 \leq j \leq r$; then by the inequality of part (b) and its local nature,

$$\begin{aligned} \operatorname{depth}(I^j/\mathfrak{a}_i I^{j-1})_p &\geq \min\{\dim R_p - i, \dim R_p - g - j + 2\} \\ &\geq \min\{1, s-g-r+2\} = 1. \end{aligned}$$

Next assume that $j = r+1$, in which case $\dim R_p \geq s+1$ and J is a geometric s-residual intersection; then again by (b),

$$\begin{aligned} \operatorname{depth}(I^r/\mathfrak{a}_{i-1}I^{r-1})_p &\geq \min\{\dim R_p - i + 1, \dim R_p - g - r + 2\} \\ &\geq \min\{2, s+1-g-r+2\} = 2. \end{aligned}$$

Furthermore, by part (c) for $i-1$ and by Proposition 1.7 (a), $\operatorname{depth}(I^{r+1}/\mathfrak{a}_{i-1}I^r)_p \geq 1$; and by (a), the complex $(\mathcal{C}_{i,r+1}.)_p$ is exact. It follows that $\operatorname{depth}(I^{r+1}/\mathfrak{a}_iI^r)_p \geq 1$. ■

We are now ready to prove our first main result.

THEOREM 2.9. *Let R be a local Gorenstein ring of dimension d, let I be an R-ideal of grade g, let s be an integer, assume that I is generically a complete intersection, and that for every non-minimal $p \in V(I)$ with $\dim R_p \leq s-1$, either $\mu(I_p) \leq \dim R_p$ or $\ell(I_p) \leq \dim R_p - 1$. Further suppose that depth $R/I^j \geq d-g-j+1$ for $1 \leq j \leq s-g+1$. Then:*

(a) *I satisfies G_s and AN_s;*

(b) *for every $g \leq i \leq s$ and every i-residual intersection $J = \mathfrak{a} : I$ of I, $\omega_{R/J} \cong I^{i-g+1}/\mathfrak{a}I^{i-g}$, where $\omega_{R/J} \cong S_{i-g+1}(I/\mathfrak{a})$ in case $I^{i-g+1} \cong S_{i-g+1}(I)$ with $g \geq 1$, and $\omega_{R/J} \cong I^{i-g+1} + J/J$ in case J is a geometric i-residual intersection.*

PROOF. After adjoining a variable and localizing, we may assume that $g \geq 1$ and that R has infinite residue class field. We will proceed by induction on $s \geq g$. For $s = g$ our assertions are well known from linkage theory ([22]). Thus we may assume that $s \geq g+1$ and that (a) and (b) hold for $s-1$. In particular, I satisfies AN_{s-1}, and hence by Proposition 1.11, I satisfies G_s as well. Thus we may apply Lemma 2.7 with $r = s-g+1$.

Let $J = \mathfrak{a} : I$ be an s-residual intersection, use the notation of Lemma 2.7, and let "$'$" denote images in $R' = R/J_{s-1}$. Notice that J_{s-1} is a geometric $(s-1)$-residual intersection of I. Now by our induction hypothesis, R' is Cohen-Macaulay with $\omega_{R'} \cong I^{s-g}/\mathfrak{a}_{s-1}I^{s-g-1} \cong I'^{s-g}$, and by Proposition 1.7, $J' = (a'_s) : I'$ where a'_s is R'-regular and $ht\ I' = 1$. Furthermore, parts (b) and (c) of Lemma 2.7 imply that depth $I'^{s-g} \geq d-s+1$ and depth $I'^{s-g+1} \geq d-s+1$. Therefore depth $\omega_{R'}/I'\omega_{R'} =$ depth $I'^{s-g}/I'^{s-g+1} \geq d-s$, which forces $\omega_{R'}/I'\omega_{R'}$ to be a Cohen-Macaulay module over R'/I'. But then by Lemma 2.3, $R'/J' \cong R/J$ is Cohen-Macaulay with $\omega_{R/J} \cong I'\omega_{R'}/a'_s\omega_{R'} \cong I'^{s-g+1}/a'_sI'^{s-g}$. On the other hand, parts (a) and (c) of Lemma 2.7 yield the exact sequence

$$\mathcal{C}_{s,s-g+1}. : 0 \longrightarrow I'^{s-g} \xrightarrow{a'_s} I'^{s-g+1} \longrightarrow I^{s-g+1}/\mathfrak{a}I^{s-g} \longrightarrow 0,$$

which implies that $I'^{s-g+1}/a'_sI'^{s-g} \cong I^{s-g+1}/\mathfrak{a}I^{s-g}$. Therefore $\omega_{R/J} \cong I^{s-g+1}/\mathfrak{a}I^{s-g}$.

Now if $I^{s-g+1} \cong S_{s-g+1}(I)$ then trivially $\omega_{R/J} \cong S_{s-g+1}(I/\mathfrak{a})$, whereas if J is a geometric residual intersection then $\omega_{R/J} \cong I^{s-g+1} + J/J$ by Lemma 2.7 (c). ■

REMARK 2.10. In the setting of Theorem 2.9, the assumption that depth $R/I^j \geq d-g-j+1$ for $1 \leq j \leq s-g+1$, is automatically satisfied if the Koszul homology modules H_j on some generating sequence of I are Cohen-Macaulay modules for $0 \leq j \leq s-g$.

PROOF. We may assume that R has infinite residue class field. Again, we proceed by induction on $s \geq g$, the case $s = g$ being trivial. So let $s \geq g+1$ and suppose that our assertion holds for $s-1$. Then by Theorem 2.9, I satisfies AN_{s-1}, and hence by Proposition 1.11, I is G_s. Let $a_1, \ldots, a_n$ be a generating sequence of I giving $H_j = H_j(a_1, \ldots, a_n; R)$, let $F = R^n$, and let

$$\mathcal{M}_{j\cdot} : 0 \longrightarrow H_j \otimes_R S_0(F) \longrightarrow \cdots \longrightarrow H_{j-i} \otimes_R S_i(F) \longrightarrow \cdots \longrightarrow H_0 \otimes_R S_j(F)$$

be the graded pieces of the $\mathcal{M}$-complex on $a_1, \ldots, a_n$ (cf. [7]). Since by the Cohen-Macaulay assumption on H_j, depth $H_j \geq d-g$ for $0 \leq j \leq s-g$ ([17, 1.5]), and since I satisfies G_s, one can use the Acyclicity Lemma to conclude that the complexes $\mathcal{M}_{j\cdot}$ are acyclic for $0 \leq j \leq s-g$ (cf. also [7, the proof of 2.5]). Moreover, the acyclicity of these complexes implies $H_0(\mathcal{M}_{j\cdot}) \cong I^j/I^{j+1}$ for $0 \leq j \leq s-g$ ([7, the proof of 2.3]). Now again using that depth $H_j \geq d-g$ for $0 \leq j \leq s-g$, one obtains that depth $R/I^{j+1} \geq d-g-j$ for $0 \leq j \leq s-g$. ∎

EXAMPLE 2.11. Let R be a local Gorenstein ring, and let I be a perfect R-ideal with grade $I = 3$ and $r(R/I) = 2$; then by [26, 2.10], $H_1(I)$ is Cohen-Macaulay. Hence if I is generically a complete intersection, then I satisfies AN_4 (Remark 2.10 and Theorem 2.9; cf. also [20, 6.1]).

There are natural instances however, where the assumptions of Theorem 2.9 are satisfied, despite some bad behavior of the Koszul homology modules. I am grateful to Mark Johnson for pointing out to me the following example of an ideal whose second and third powers are Cohen-Macaulay:

EXAMPLE 2.12. Let k be a regular local ring with infinite residue class field, let X be an alternating 5 by 5 matrix of variables, let Y be a 5 by 1 matrix of variables, let $R = k[X, Y]$ (possibly localized at the irrelevant maximal ideal), and let $I = Pf_4(X) + I_1(XY)$ be the ideal generated by the 4 by 4 Pfaffians of X and the entries of the product matrix XY.

This example has played an important role in other contexts (e.g. [24]). The factor ring R/I is the associated graded ring of the ideal $Pf_4(X)$ in $k[X]$. Therefore R/I is a Gorenstein domain ([14, 1.18 and 2.2]) satisfying Serre's condition R_4 ([13, Proposition 2.1]). Furthermore, grade $I = 5$, $\mu(I) = 10$, and $\ell(I) \leq 9$. Thus I satisfies G_∞ and $\ell(I_p) \leq \dim R_p - 1$ for every non-minimal $p \in V(I)$. In addition, R/I^j are Cohen-Macaulay for $1 \leq j \leq 3$ (M. Johnson, using *Macaulay*).

Now Theorem 2.9 implies that I satisfies AN_7. On the other hand, I is not even syzygetic (i.e., $S_2(I) \ncong I^2$, or equivalently, $H_1(I)$ is not torsion free over R/I; cf. [23, 1.2]) (M. Johnson). In particular, Corollary 1.8 (a) implies that I does not satisfy AN_8^-.

Contrasting to this and similar examples, one can show that depth assumptions on sufficiently many powers of I force the Koszul homology modules to be Cohen-Macaulay:

COROLLARY 2.13. *Let R be a local Gorenstein ring of dimension d, let I be an R-ideal with grade $I = g$ and $\mu(I) = n$, assume that I is generically a complete intersection, and that for every non-minimal $p \in V(I)$, either $\mu(I_p) \leq \dim R_p$ or $\ell(I_p) \leq \dim R_p - 1$. Further suppose that R/I is Cohen-Macaulay, and that depth $R/I^j \geq d - g - j + 1$ for $1 \leq j \leq n - g - 1$.*

Then I satisfies G_∞ and is strongly Cohen-Macaulay. Moreover, depth $R/I^j \geq \max\{d - n, d - g - j + 1\}$ for $j \geq 1$, where equality holds if $j \geq n - g + 1$ and $\operatorname{projdim}_R I < \infty$.

PROOF. We may assume that R has infinite residue class field. By Theorem 2.9, I satisfies AN_{n-2}. Therefore I is G_n (Proposition 1.11), and hence can be generated by a d-sequence $a_1, \ldots, a_n$ (Corollary 1.8 (a.i)). Let H_j be the j-th Koszul homology on $a_1, \ldots, a_n$, and write $F = R^n$. Since $a_1, \ldots, a_n$ form a d-sequence, one knows from [8, 4.1] and [7, 2.3] that the approximation complex $\mathcal{M}.$ on $a_1, \ldots, a_n$ is acyclic with $H_0(\mathcal{M}.) \cong gr_I(R)$. Now the graded pieces of $\mathcal{M}.$ give rise to exact sequences

$$\begin{aligned} 0 \longrightarrow H_j \otimes_R S_0(F) \longrightarrow \cdots \longrightarrow H_{j-i} \otimes_R S_i(F) \longrightarrow \cdots \longrightarrow H_0 \otimes_R S_j(F) \\ \longrightarrow I^j/I^{j+1} \longrightarrow 0, \end{aligned}$$

from which we deduce, by induction on j, that H_j are Cohen-Macaulay R/I-modules for $0 \leq j \leq n - g - 2$. By duality ([16, 2.13 and 2.22]), this suffices to conclude that all Koszul homology modules of I are Cohen-Macaulay.

Now since we know that I is strongly Cohen-Macaulay, the remaining assertions about the depth of R/I^j follow from Remark 2.10 and its proof and from [19, 2.7]. ∎

By using a deformation argument, one can relax the conditions on the local number of generators or the local analytic spread that were imposed in Theorem 2.9. This works, because residual intersections specialize whenever their canonical module has the "expected form" ([18]). Our next result is a generalization of ([18, 5.1]).

THEOREM 2.14. *Let R be a local Gorenstein ring of dimension d, let I be an R-ideal of grade g, and let s be an integer. Assume that (R, I) has a deformation $(\tilde{R}, \tilde{I})$ such that $\tilde{I}$ is generically a complete intersection, and that for every non-minimal $p \in V(\tilde{I})$ with $\dim \tilde{R}_p \leq s$, either $\mu(\tilde{I}_p) \leq \dim \tilde{R}_p$ or $\ell(\tilde{I}_p) \leq \dim \tilde{R}_p - 1$. Further suppose that depth $\tilde{R}/\tilde{I}^j \geq \dim \tilde{R} - g - j + 1$ for $1 \leq j \leq s - g + 1$. Then:*

(a) *I satisfies AN_s;*
(b) *for every $g \leq i \leq s + 1$ and every i-residual intersection $J = \mathfrak{a} : I$ of I,*

 (i) J is unmixed of height i,
 (ii) depth $R/\mathfrak{a} = d - i$,

(iii) $\mathfrak{a} = I \cap J$ *in case* J *is a geometric residual intersection;*

(c) *for every* $g \le i \le s$ *and every* i*-residual intersection* $J = \mathfrak{a} : I$ *of* I, $\omega_{R/J} \cong S_{i-g+1}(I/\mathfrak{a})$ *in case* $I^{i-g+1} \cong S_{i-g+1}(I)$ *with* $g \ge 1$, *and* $\omega_{R/J} \cong I^{i-g+1} + J/J$ *in case* J *is a geometric* i*-residual intersection.*

PROOF. After adjoining variables and localizing, we may assume that $g \ge 1$ and that the residue class field is infinite. By Theorem 2.9 (a), $\tilde{I}$ satisfies AN_s, and then by Proposition 1.11, $\tilde{I}$ is G_{s+1}. Now let $g \le i \le s+1$, and let $J = \mathfrak{a} : I$ be an i-residual intersection of I. We consider a generic i-residual intersection $RI(i; \tilde{I}) = \tilde{\mathfrak{a}} : \tilde{I}\tilde{R}[X]$ of $\tilde{I}$ in some polynomial extension $\tilde{R}[X]$ (for the definition of generic residual intersections, cf. [18, 3.1]). Now there exists $q \in \operatorname{Spec}(\tilde{R}[X])$ and an epimorphism $\pi : S = \tilde{R}[X]_q \longrightarrow R$, such that $\pi(\tilde{I}S) = I$, $\pi(\tilde{\mathfrak{a}}S) = \mathfrak{a}$, and $\ker \pi$ is generated by a sequence that is regular on S as well as $S/\tilde{I}S$ ([18, the proof of 5.1]). Write $\tilde{J} = RI(i; \tilde{I})S$. Since $\tilde{I}$ satisfies G_{s+1}, $\tilde{J} = \tilde{\mathfrak{a}}S : \tilde{I}S$ is an i-residual intersection of $\tilde{I}S$ ([18, the proof of 3.2]), and therefore, since $\tilde{I}$ is G_{s+1} and AN_s, depth $S/\tilde{\mathfrak{a}}S = \dim S - i$ (Proposition 1.7 (b)). But then by [18, 4.2 (i)], the S-regular sequence generating $\ker \pi$ forms a regular sequence on $S/\tilde{\mathfrak{a}}S$. Hence depth $R/\mathfrak{a} = d - i$, which suffices to imply part (b) for $J = \mathfrak{a} : I$.

Now assume that $g \le i \le s$; then $\tilde{J}$ is a geometric i-residual intersection of $\tilde{I}S$ ([18, 3.2]). Furthermore, by Proposition 1.7 (a) and Theorem 2.9, *ht* $\tilde{J} = i$ and $S/\tilde{J}$ is Cohen-Macaulay with $\omega_{S/\tilde{J}} \cong (\tilde{I}S)^{i-g+1} + \tilde{J}/\tilde{J}$. In this situation, [18, 4.7] implies that $R/J \cong S/\tilde{J} \otimes_S R$ and that the S-regular sequence generating $\ker \pi$ forms a regular sequence on $S/\tilde{J}$. Therefore R/J is Cohen-Macaulay, which proves part (a). Moreover, $\omega_{R/J} \cong \omega_{S/\tilde{J}} \otimes_S R$.

Now if $I^{i-g+1} \cong S_{i-g+1}(I)$, then $(\tilde{I}S)^{i-g+1} \cong S_{i-g+1}(\tilde{I}S)$, and therefore by Theorem 2.9 (b), $\omega_{S/\tilde{J}} \cong S_{i-g+1}(\tilde{I}S/\tilde{\mathfrak{a}}S)$, which yields $\omega_{R/J} \cong S_{i-g+1}(\tilde{I}S/\tilde{\mathfrak{a}}S) \otimes_S R \cong S_{i-g+1}(I/\mathfrak{a})$. On the other hand, if J is a geometric residual intersection, then the natural epimorphism $\omega_{R/J} \cong [(\tilde{I}S)^{i-g+1} + \tilde{J}/\tilde{J}] \otimes_S R \twoheadrightarrow I^{i-g+1} + J/J$ is an isomorphism because $\operatorname{grade}_I \omega_{R/J} \ge 1$. This shows part (c). ■

We are now going to prove a converse of Theorem 2.9.

THEOREM 2.15. *Let* R *be a local Gorenstein ring of dimension* d, *let* I *be an unmixed* R*-ideal of grade* $g \ge 1$, *let* s *be an integer, and assume that the residue class field of* R *is infinite in case* $s \ge d$. *Further suppose that* I *is generically a complete intersection, and that for every nonminimal* $p \in V(I)$ *with* $\dim R_p \le s$, *either* $\mu(I_p) \le \dim R_p$ *or* $\ell(I_p) \le \dim R_p - 1$. *The following are equivalent:*

(a) *depth* $R/I^j \ge d - g - j + 1$ *for* $1 \le j \le s - g + 1$;

(b) I *satisfies* AN_s^-, *and for every* $g \le i \le s$ *and every geometric* i*-residual intersection* J *of* I, $\omega_{R/J} \cong I^{i-g+1} + J/J$.

PROOF. In the light of Theorem 2.9 it suffices to prove that (b) implies (a), which we are going to do by induction on $s \ge g$. So assume that (b) holds;

then by Proposition 1.11, I satisfies G_{s+1}. Now let $s = g$; in this case R/I is Cohen-Macaulay since I is unmixed, generically a complete intersection, and AN_g^- ([22, 1.3]). Thus we may assume that $s \geq g+1$ and that by induction hypothesis, $\dim R/I^j \geq d-g-j+1$ for $1 \leq j \leq s-g$. Therefore Lemma 2.7 applies with $r = s-g$. We may further suppose that $\mu(I) \geq s+1$, because otherwise our assertions follow from Corollary 2.13. We need to prove that depth $R/I^{s-g+1} \geq d-s$.

Since $\mu(I) \geq s+1$ and I satisfies G_{s+1}, we may apply Lemma 1.5 with $\mathfrak{a} = I$ to obtain elements $a_1, \ldots, a_s$ in I, such that $J_i = (a_1, \ldots, a_i) : I$ are geometric i-residual intersections for $g \leq i \leq s$. Let "$'$" and "$^-$" denote images in $R' = R/J_{s-1}$ and $\bar{R} = R/J_s$, respectively. By our assumption (b) and Proposition 1.7, R' is Cohen-Macaulay with $\dim R' = d-s+1$, $ht\ I' = 1$, R'/I' is Cohen-Macaulay, $J'_s = (a'_s) : I'$ with a'_s an R'-regular element, and $ht\ I' + J'_s \geq 2$. Furthermore, if $p \in Ass_R \bar{R}$, then again by Proposition 1.7 (a), $\dim R_p = s$ and therefore $J_p = (a_1, \ldots, a_s)_p$ is a complete intersection, which shows that $\bar{R}$ is generically Gorenstein. Thus we are in the setting of Lemma 2.3. By our assumption, $\omega_{R'} \cong I'^{s-g}$ and $\bar{R}$ is Cohen-Macaulay with $\omega_{\bar{R}} \cong \bar{I}^{s-g+1}$. Moreover, from parts (a) and (c) of Lemma 2.7 we have an exact sequence

$$\mathcal{C}_{s,s-g+1\cdot} : 0 \longrightarrow I'^{s-g} \xrightarrow{a'_s} I'^{s-g+1} \longrightarrow \bar{I}^{s-g+1} \longrightarrow 0,$$

which implies that $\omega_{\bar{R}} \cong I'\omega_{R'}/a'_s\omega_{R'}$. In this situation, we may use Lemma 2.3 to conclude that $\omega_{R'}/I'\omega_{R'}$ is a Cohen-Macaulay R'/I'-module. Therefore, since $\dim \omega_{R'}/I'\omega_{R'} = \dim R'/I' = d-s$, one obtains depth $I'\omega_{R'} \geq d-s+1$, and hence by Lemma 2.7 (c),

$$\text{(2.16)} \qquad \operatorname{depth} I^{s-g+1}/\mathfrak{a}_{s-1}I^{s-g} \geq d-s+1.$$

On the other hand, Lemma 2.7 (b) implies that

$$\text{(2.17)} \qquad \operatorname{depth} I^{s-g}/\mathfrak{a}_i I^{s-g-1} \geq d-s+2, \text{ for } 0 \leq i \leq s-2.$$

Furthermore, by Lemma 2.7 (a), the complexes $\mathcal{C}_{i,s-g+1\cdot}$ are exact for $1 \leq i \leq s-1$. Now feeding (2.16) and (2.17) into these complexes, and using decreasing induction on i, $1 \leq i \leq s-1$, we conclude that depth $I^{s-g+1} \geq d-s+1$. Therefore depth $R/I^{s-g+1} \geq d-s$. ∎

REMARK 2.18. Let R be a local Gorenstein ring of dimension d, let I be an unmixed R-ideal of grade $g \geq 1$, let s be an integer, and assume that for $g \leq i \leq s$ there exists a sequence of geometric i-residual intersections $J_i = (a_1, \ldots, a_i) : I$ with R/J_i being Cohen-Macaulay and $\omega_{R/J_i} \cong I^{i-g+1} + J_i/J_i$; then depth $R/I^j \geq d-g-j+1$ for $1 \leq j \leq s-g+1$.

PROOF. The assertion follows from the proof of Theorem 2.15. ∎

3. Depth conditions on symbolic powers

In this section we characterize the Artin-Nagata properties in terms of depths of symbolic powers. We first prove a lemma that is the analogue of Lemma 2.7.

LEMMA 3.1. *Let R be a local Cohen-Macaulay ring of dimension d, let I be an R-ideal of grade $g \geq 1$, let $s \geq g$, and assume that the residue class field of R is infinite in case $s \geq d$. Further suppose that for every minimal $p \in V(I)$, $\mu(I_p) = g$, that for every non-minimal $p \in V(I)$ with $\dim R_p \leq s$, $\ell(I_p) \leq \dim R_p - 1$, and that I satisfies AN_{s-1}^- (and hence G_{s+1}, cf. Proposition 1.11). Also assume that depth $R/I^{(j)} \geq d-g-j+1$ whenever $1 \leq j \leq s-g$. Let $J = \mathfrak{a} : I$ with $\mathfrak{a} = (a_1, \dots, a_s)$ be an s-residual intersection, for $0 \leq i \leq s$ let $J_i = \mathfrak{a}_i : I$ with $\mathfrak{a}_i = (a_1, \dots, a_i)$ be ideals as defined in Corollary 1.6 (a), and for $1 \leq i \leq s$ and $2 \leq j$ consider the complexes*

$$\mathcal{D}_{ij\cdot} : 0 \longrightarrow I^{(j-1)}/\mathfrak{a}_{i-1}I^{(j-2)} \longrightarrow I^{(j)}/\mathfrak{a}_{i-1}I^{(j-1)} \longrightarrow I^{(j)}/\mathfrak{a}_i I^{(j-1)} \longrightarrow 0,$$

where the first map is induced by multiplication with a_i. Then:

(a) *The complexes $\mathcal{D}_{ij\cdot}$ are exact whenever $2 \leq j \leq s-g+2$ and $1 \leq i \leq s$;*
(b) *depth $I^{(j)}/\mathfrak{a}_i I^{(j-1)} \geq \min\{d-i, d-g-j+2\}$ whenever $1 \leq j \leq s-g$ and $0 \leq i \leq s$;*
(c) *depth$(I^{(j)}/\mathfrak{a}_i I^{(j-1)})_p \geq k$ for $1 \leq k \leq 2$ and every $p \in$ Spec(R) with $\dim R_p \geq \max\{i+k, g+j-2+k\}$, whenever $1 \leq j \leq s-g+1$ and $0 \leq i \leq s$;*
(d) *$I^{(j)}/\mathfrak{a}_i I^{(j-1)} \cong I^{(j)} + J_i/J_i$ whenever $1 \leq j \leq s-g+1$ and $0 \leq i \leq s-1$;*
(e) *$I^{(j)}/\mathfrak{a}_i I^{(j-1)} \cong (I^j + J_i/J_i)^{\vee\vee}$ with $-^\vee = Hom_{R/J_i}(-, \omega_{R/J_i})$, whenever $1 \leq j \leq s-g+1$ and $g+j-2 \leq i \leq s-1$.*

PROOF. As in the proof of Lemma 2.7, inducting on i, we may assume that $i \geq 1$ and that our assertions hold for $i-1$.

Now the exactness of $\mathcal{D}_{ij\cdot}$ follows from Proposition 1.7 (f) and from part (d), applied to $i-1$, $j-1$. As to (b), first consider the case $j = 1$; then by Proposition 1.7 (b) and since $s-g \geq 1$, depth $I^{(j)}/\mathfrak{a}_i I^{(j-1)} =$ depth $I^{(1)}/\mathfrak{a}_i \geq \min\{\text{depth } R/\mathfrak{a}_i, \text{depth } R/I^{(1)}+1\} \geq \min\{d-i, d-g+1\}$. On the other hand, if $2 \leq j \leq s-g$, then (b) follows from the exactness of $\mathcal{D}_{ij\cdot}$ and from part (b), applied to $i-1, j-1$ and $i-1, j$.

To prove (c) for i, first assume that $j = 1$. In this case $\dim R_p \geq \max\{i+k, g-1+k\}$ and hence as above, depth$(I^{(j)}/\mathfrak{a}_i I^{(j-1)})_p =$ depth$(I^{(1)}/\mathfrak{a}_i)_p \geq \min\{\text{depth}(R/\mathfrak{a}_i)_p, \text{depth}(R/I^{(1)})_p + 1\} \geq \min\{(i+k)-i, (g-1+k)-g+1\} = k$. On the other hand, if $2 \leq j \leq s-g+1$, then by part (b), applied to $i-1, j-1$, depth$(I^{(j-1)}/\mathfrak{a}_{i-1}I^{(j-2)})_p \geq k+1$, whereas by part (c), applied to $i-1, j$, depth$(I^{(j)}/\mathfrak{a}_{i-1}I^{(j-1)})_p \geq k$. Now the exactness of $\mathcal{D}_{ij\cdot}$ implies that depth$(I^{(j)}/\mathfrak{a}_i I^{(j-1)})_p \geq k$.

Before proving parts (d) and (e), we first show that for every $j \geq 1$,

$$(3.2) \qquad (I^{(j)})_p = (I^j)_p \quad \text{whenever} \quad p \in \text{Spec}(R) \text{ with } \dim R_p \leq s.$$

It suffices to see this for $p \in V(I)$. However by Proposition 1.11, I_p is G_∞ and satisfies sliding depth. In this situation, our assumption on the local analytic spread of I implies (3.2) ([8, 6.1] and [4, 3.3]).

In order to prove part (d), let $1 \le j \le s-g+1$, $1 \le i \le s-1$. Consider the natural epimorphism $\pi : I^{(j)}/\mathfrak{a}_i I^{(j-1)} \twoheadrightarrow I^{(j)} + J_i/J_i$, and let $p \in \operatorname{Spec}(R)$. If $\dim R_p \le s-1$, then by (3.2), $(I^{(j)})_p = (I^j)_p = (\mathfrak{a}^j)_p$, and it follows from (2.8) that π_p is an isomorphism (for (2.8) to hold, one only needed that I satisfies G_s and AN^-_{s-1}). If on the other hand, $\dim R_p \ge s$, then $\dim R_p \ge s \ge \max\{i+1, g+j-2+1\}$, and hence by part (c), $\operatorname{depth}(I^{(j)}/\mathfrak{a}_i I^{(j-1)})_p \ge 1$. This shows that π is an isomorphism.

Finally, to prove (e), assume $1 \le j \le s-g+1$ and $g+j-2 \le i \le s-1$. Let "$'$" denote images in $R' = R/J_i$, and recall that R' is a Cohen-Macaulay ring with $\dim R' = d-i$ (Proposition 1.7 (a)). By part (d) it suffices to prove that $(I^j)'^{\vee\vee} \cong (I^{(j)})'$. To this end consider the natural embedding $\iota : (I^j)' \hookrightarrow (I^{(j)})'$, and let $p \in V(J_i)$ with $\dim R'_p \le 1$; then $\dim R_p \le i+1 \le s$, which by (3.2) forces ι_p to be an isomorphism. Therefore $\iota^{\vee\vee}$ is an isomorphism, and we are reduced to proving that $(I^{(j)})' \cong (I^{(j)})'^{\vee\vee}$, or equivalently, that $\operatorname{depth}(I^{(j)})'_p \ge 2$ for every $p \in V(J_i)$ with $\dim R'_p \ge 2$. However, for every such p, $\dim R_p \ge i+2 = \max\{i+2, g+j-2+2\}$, and hence by parts (c) and (d), $\operatorname{depth}(I^{(j)})'_p \ge 2$. ■

We are now able to prove the main result of this section.

THEOREM 3.3. *Let R be a local Gorenstein ring of dimension d, let I be an R-ideal of grade $g \ge 1$, let s be an integer, and assume that the residue class field of R is infinite in case $s \ge d$. Further suppose that for every minimal $p \in V(I)$, $\mu(I_p) = g$, and that for every non-minimal $p \in V(I)$ with $\dim R_p \le s$, $\ell(I_p) \le \dim R_p - 1$. The following are equivalent:*

(a) *depth $R/I^{(j)} \ge d-g-j+1$ for $1 \le j \le s-g+1$;*
(b) *I satisfies AN^-_s;*
(c) *I satisfies AN_s.*

PROOF. We are going to prove our assertion by induction on $s \ge g$, the case $s = g$ being well known ([22, 1.3]). So let $s \ge g+1$, and assume that by induction hypothesis, depth $R/I^{(j)} \ge d-g-j+1$ for $1 \le j \le s-g$ and I satisfies AN_{s-1}. In particular, I is G_{s+1} (Proposition 1.11), and we may assume that $d \ge s+1$. Thus, applying Lemma 1.5 with $\mathfrak{a} = mI$, where m is the maximal ideal of R, one sees that I admits a geometric s-residual intersection. Now choose such a geometric s-residual intersection $J = \mathfrak{a} : I$ (in proving that (b) implies (a)), or consider an arbitrary s-residual intersection $J = \mathfrak{a} : I$ (in proving that (a) implies (c)). We need to show that depth $R/I^{(s-g+1)} \ge d-s$ if and only if R/J is Cohen-Macaulay. To this end, we may use Lemma 3.1 and the notation introduced there.

First, we prove by induction on i, $0 \le i \le s-1$, that $\omega_{R/J_i} \cong (I^{i-g+1} + J_i/J_i)^{\vee\vee}$, where $(I^k + J_i/J_i) = R/J_i$ for $k \le 0$. This is obvious for

$0 \le i \le g-1$, because then $\omega_{R/J_i} \cong R/J_i \cong (I^{i-g+1}+J_i/J_i)^{\vee\vee}$. If $i = g$, then by linkage theory, $\omega_{R/J_g} \cong I^{(1)}/\mathfrak{a}_g$, and by Lemma 3.1 (e), $I^{(1)}/\mathfrak{a}_g \cong (I+J_g/J_g)^{\vee\vee}$. So assume $g+1 \le i \le s-1$, and let "$'$" denote images in $R' = R/J_{i-1}$. From our induction hypothesis, we know that $\omega_{R'} \cong (I')^{i-g\vee\vee}$, where now $-^\vee = \operatorname{Hom}_{R'}(-, \omega_{R'})$. Since R' is Cohen-Macaulay and $J_i' = (a_i') : I'$ with a_i' an R'-regular element (Proposition 1.7 (f)), it follows from Lemma 2.1 (a) that $\omega_{R/J_i} \cong (I'\omega_{R'})^{\vee\vee}/a_i'\omega_{R'} \cong (I')^{i-g+1\vee\vee}/a_i'(I')^{i-g\vee\vee}$. On the other hand, since $i-g+1 \ge 2$, we may use parts (a) and (e) of Lemma 3.1 to obtain an exact sequence

$$\mathcal{D}_{i,i-g+1}. : 0 \longrightarrow (I')^{i-g\vee\vee} \xrightarrow{a_i'} (I')^{i-g+1\vee\vee} \longrightarrow I^{(i-g+1)}/\mathfrak{a}_i I^{(i-g)} \longrightarrow 0,$$

which identifies $\omega_{R/J_i} \cong I^{(i-g+1)}/\mathfrak{a}_i I^{(i-g)}$. Once more by Lemma 3.1 (e), $\omega_{R/J_i} \cong (I^{i-g+1} + J_i/J_i)^{\vee\vee}$, where now $-^\vee = \operatorname{Hom}_{R/J_i}(-, \omega_{R/J_i})$.

Now let "$'$" denote images in $R' = R/J_{s-1}$. We know that $(I'\omega_{R'})^{\vee\vee} \cong (I')^{s-g+1\vee\vee}$. Thus by Proposition 1.7 (f) and Lemma 2.1 (b), R/J is Cohen-Macaulay if and only if $(I')^{s-g+1\vee\vee}$ is a Cohen-Macaulay R'-module. By Lemma 3.1 (e), $(I')^{s-g+1\vee\vee} \cong I^{(s-g+1)}/\mathfrak{a}_{s-1} I^{(s-g)}$, so we have to prove that depth $I^{(s-g+1)}/\mathfrak{a}_{s-1} I^{(s-g)} \ge d-s+1$ if and only if depth $I^{(s-g+1)} \ge d-s+1$. However, parts (a) and (b) of Lemma 3.1 yield exact sequences for $1 \le i \le s-1$,

$$\mathcal{D}_{i,s-g+1}. : 0 \longrightarrow I^{(s-g)}/\mathfrak{a}_{i-1} I^{(s-g-1)} \longrightarrow I^{(s-g+1)}/\mathfrak{a}_{i-1} I^{(s-g)} \longrightarrow I^{(s-g+1)}/\mathfrak{a}_i I^{(s-g)} \longrightarrow 0$$

with depth $I^{(s-g)}/\mathfrak{a}_{i-1} I^{(s-g-1)} \ge \min\{d-i+1, d-g-(s-g)+2\} = d-s+2$. Now it follows by decreasing and increasing induction on i that depth $I^{(s-g+1)}/\mathfrak{a}_{s-1} I^{(s-g)} \ge d-s+1$ if and only if depth $I^{(s-g+1)} \ge d-s+1$. ∎

As in Remark 2.18, one can replace condition (b) of Theorem 3.3 by the assumption that for $g \le i \le s$ there exists a sequence of geometric i-residual intersections $J_i = (a_1, \ldots, a_i) : I$ with R/J_i being Cohen-Macaulay.

COROLLARY 3.4. *Let R be a local Gorenstein ring of dimension d, let I be an R-ideal with grade $I = g$ and $\mu(I) = n$, assume that for every minimal $\mathfrak{p} \in V(I)$, $\mu(I_\mathfrak{p}) = g$, that for every non-minimal $\mathfrak{p} \in V(I)$ with $\dim R_\mathfrak{p} \le n-2$, $\ell(I_\mathfrak{p}) \le \dim R_\mathfrak{p} - 1$, and that for every non-minimal $\mathfrak{p} \in V(I)$ with $\dim R_\mathfrak{p} = n-1$, either $\mu(I_\mathfrak{p}) \le \dim R_\mathfrak{p}$ or $\ell(I_\mathfrak{p}) \le \dim R_\mathfrak{p} - 1$. Further suppose that depth $R/I \ge d-n$, and that depth $R/I^{(j)} \ge d-g-j+1$ for $1 \le j \le n-g-1$.*

Then I satisfies G_∞ and sliding depth.

PROOF. After adjoining variables and localizing, we may assume that $g \ge 1$ ([9, 3.5]) and that the residue class field of R is infinite. Now by Theorem 3.3, I satisfies AN_{n-2}, and hence by Proposition 1.11, I is G_n. But then our assertion follows from Corollary 1.8 (a.ii). ∎

Generalizing [20, 6.1], we can now relate the property AN_{g+1} to the depth of the module of differentials. Recall that for a factor ring $A = R/I$ of a regular local ring R, the *deviation* $d(A)$ of A is defined to be the difference $\mu(I) - \operatorname{grade} I$.

COROLLARY 3.5. *Let R be a regular local ring essentially of finite type over a perfect field k, let I be an R-ideal of grade g, assume that $A = R/I$ is Cohen-Macaulay, reduced, and a complete intersection in codimension one.*

(a) *I satisfies AN_{g+1} (or AN_{g+1}^-) if and only if depth $\Omega_k(A) \geq \dim A - 2$.*
(b) *If I satisfies sliding depth then depth $\Omega_k(A) \geq \dim A - 2$.*
(c) *If $d(A) \leq 3$ and $d(A_p) \leq 2$ for every $p \in Spec(A)$ with $\dim A_p \leq 2$, then I satisfies sliding depth if and only if depth $\Omega_k(A) \geq \dim A - 2$.*

PROOF. Consider the exact sequence

$$0 \longrightarrow I/I^{(2)} \longrightarrow \Omega_k(R) \otimes_R A \cong \oplus A \longrightarrow \Omega_k(A) \longrightarrow 0,$$

and recall that by [9, 3.3], I satisifes AN_{g+1}^- in case I has the sliding depth property. Now our assertions follow from Theorem 3.3 and Corollary 3.4. ■

EXAMPLE 3.6. Let k be a perfect field, let $R = k[X]_{(X)}$, where X is either a symmetric 3 by 3 matrix of variables, or else a 2 by 4 matrix of variables, let $I = I_2(X)$ be the R-ideal generated by the 2 by 2 minors of X, and write $A = R/I$. In either case, we may apply Corollary 3.5 (c). In the first case, however, I satisfies sliding depth ([9, 2.3]), whereas in the second case, I does not have this property, corresponding to the fact that depth $\Omega_k(A) = 1 = \dim A - 2$ or depth $\Omega_k(A) = 2 = \dim A - 3$ ([28, 3.4]), respectively.

4. Rees algebras and reductions

In this section we apply our earlier results to study the Rees algebra of an R-ideal I by passing to a reduction K. Our main result here is Theorem 4.1, which extends work by Huckaba and Huneke ([10], [11]), and by Vasconcelos ([27]). Parts of our proof were inspired by these earlier papers, but our main strategy is different in that we aim at reducing to the case where $\mu(K) = \dim R$. Once this is done, we know that K satisfies sliding depth, and we can invoke a result by Vasconcelos to pass from $\mathcal{R}(K)$ back to $\mathcal{R}(I)$ ([27]). Strictly speaking, we need a modified version of Vasconcelos's theorem (Theorem 4.8), but its proof is literally the same as in [27] and can safely be omitted.

THEOREM 4.1. *Let R be a local Cohen-Macaulay ring of dimension d with infinite residue class field, let I be an R-ideal with grade $I \geq 2$, let s be an integer, let K be a reduction of I with $\mu(K) \leq s$, assume that depth $R/I \geq d - s$, that for every minimal $p \in V(I)$, I_p is a complete intersection, that for every non-minimal $p \in V(I)$ with $\dim R_p \leq s - 1$, either $\mu(I_p) \leq \dim R_p$ or $\ell(I_p) \leq \dim R_p - 1$, and that for every $p \in V(I)$ with $\dim R_p = s$, either $r_{K_p}(I_p) \leq 1$ or*

$\mu(I_p) \leq \dim R_p$ or $\ell(I_p) \leq \dim R_p - 1$. Further suppose that I satisfies AN^-_{s-2}. Then:

(a) *$r_K(I) \leq 1$;*
(b) *$\mathcal{R}(I)$ is Cohen-Macaulay;*
(c) *$I^{(j)} = I^j$ for every $j \geq 1$ if and only if $\ell(I_p) \leq \dim R_p - 1$ for every non-minimal $p \in V(I)$ with $\dim R_p \leq s$.*

PROOF. Part (c) is a well known consequence of (b) (e.g. [4, 3.3] via [13, Proposition 1.1]), hence it suffices to prove (a) and (b).

Proposition 1.11 implies that I satisfies G_s, that for every $p \in V(I)$ with $\dim R_p \leq s-1$, I_p satisfies sliding depth and is of linear type, and that for every $p \in V(I)$ with $\dim R_p = s$, $r_{K_p}(I_p) \leq 1$. In particular, if $d \leq s-1$, then I satisfies G_∞ and sliding depth, and our assertions follows from [8, 6.1]. Thus we may from now on assume that $d \geq s$. Furthermore, $ht\ K : I \geq s$, and therefore by Remark 1.12, K satisfies AN^-_{s-2} and G_∞. If $\mu(K) \leq s-1$, then $K = I$ by Proposition 1.7 (a), and K satisfies sliding depth by Corollary 1.8 (c). Once more [8, 6.1] would imply all our claims. Hence we may assume that $\mu(K) = s$. Applying Corollary 1.6 (b) to the ideal K, we find a minimal generating sequence $a_1, \dots, a_s$ of K such that for every subset $\{\nu_1, \dots, \nu_i\} \subset \{1, \dots, s\}$, $ht\ (a_{\nu_1}, \dots, a_{\nu_i}) : K \geq i$ and $ht\ K + (a_{\nu_1}, \dots, a_{\nu_i}) : K \geq i+1$. Now write $\mathfrak{b} = (a_1)$ and $\mathfrak{a} = (a_2, \dots, a_s)$. (The idea of splitting up the generating set of a reduction stems from [10], [11]). Then $a_1, \dots, a_s$ form a d-sequence by Corollary 1.8 (a.i), and a_1 is an R-regular element since grade K = grade $I \geq 1$. Furthermore, since $ht\ \mathfrak{a} : K \geq s-1$, $\mathfrak{a}$ satisfies G_∞ and sliding depth by Remark 1.12 and Corollary 1.8 (c).

Let $H_i = H_i(a_2, \dots, a_s)$ be the i-th Koszul homology of $a_2, \dots, a_s$, let $S = R[T_2, \dots, T_s]$ be a polynomial ring, and consider the approximation complex on $a_2, \dots, a_s$,

$$\mathcal{M}. : 0 \longrightarrow H_{s-2} \otimes_R S(-s+2) \longrightarrow \cdots \longrightarrow H_i \otimes_R S(-i) \longrightarrow \cdots \longrightarrow H_0 \otimes_R S.$$

Since $\mathfrak{a}$ satisfies G_∞ and sliding depth, the complex $\mathcal{M}.$ is acyclic with $H_0(\mathcal{M}.) \cong gr_{\mathfrak{a}}(R)$ ([7, 2.5]). In particular,

$$\text{depth } \mathfrak{a}^j \geq d - s + 2, \quad \text{for } j \geq 0. \tag{4.2}$$

Now let $E = \oplus E_j$ denote the first homology of the truncated complex $(\mathcal{M}_i \mid i \geq 1)$; then $\text{depth}_R E_j \geq d-s+2$. On the other hand, there is an exact sequence

$$E \xrightarrow{\iota} \mathcal{M}_0 \otimes_R R/I \cong S \otimes_R R/I \longrightarrow H_0(\mathcal{M}.) \otimes_R R/I \cong gr_{\mathfrak{a}}(R) \otimes_R R/I \longrightarrow 0.$$

To see that ι is injective, we consider $p \in Ass_R E_j$. By the above depth estimate, $\dim R_p \leq s-2$, thus $I_p = K_p = \mathfrak{a}_p$ and hence ι_p is injective because the complex $\mathcal{M}. \otimes_R R_p$ is acyclic. Therefore ι is injective, which together with (4.2) implies that

$$\text{depth } \mathfrak{a}^j I \geq d - s + 1, \quad \text{for } j \geq 0. \tag{4.3}$$

(Cf. [27, the proof of 4.13] for similar arguments.)

Next we claim that

$$\mathfrak{b} \cap \mathfrak{a}^j I = \mathfrak{a}^j \mathfrak{b}, \quad \text{for } j \geq 0. \tag{4.4}$$

Since $\mathfrak{a}^j\mathfrak{b} \cong \mathfrak{a}^j$ and because of (4.2), we know that $\dim R_p \leq s-1$ for every $p \in Ass_R(R/\mathfrak{a}^j\mathfrak{b})$. Thus it suffices to prove (4.4) locally at $p \in \operatorname{Spec}(R)$ where $\dim R_p \leq s-1$. We may further assume that p contains $\mathfrak{b}+\mathfrak{a} = K$. Thus by our choice of $a_1, \ldots, a_s$, $\mathfrak{a}_p = K_p$. Since on the other hand, $K_p = I_p$, we are now reduced to showing the inclusion $\mathfrak{b}_p \cap \mathfrak{a}_p^{j+1} \subset \mathfrak{a}_p^j \mathfrak{b}_p$. However, $a_1, \ldots, a_s$ forming a d-sequence, it follows from [12, Theorem 2.1] that $\mathfrak{b} \cap \mathfrak{a}^{j+1} \subset \mathfrak{a}^j\mathfrak{b}$, for $j \geq 0$ (cf. also the proof of (2.8)). This proves (4.4).

From (4.4) we obtain an exact sequence

$$0 \longrightarrow \mathfrak{a}^j\mathfrak{b} \cong \mathfrak{a}^j \longrightarrow \mathfrak{b} \oplus \mathfrak{a}^j I \longrightarrow \mathfrak{b} + \mathfrak{a}^j I \longrightarrow 0,$$

which together with (4.2) and (4.3) yields the depth estimates

$$\operatorname{depth} R/\mathfrak{b} + \mathfrak{a}^j I \geq d - s, \quad \text{for } j \geq 0. \tag{4.5}$$

We are now going to show that

$$\mathfrak{b} \cap I^2 = \mathfrak{b} I. \tag{4.6}$$

Since $\mathfrak{b}I \cong I$ and depth $I \geq d-s+1$, it suffices to prove (4.6) locally at $p \in V(I)$ where $\dim R_p \leq s$. However, for every such p, $r_{K_p}(I_p) \leq 1$. Therefore $I_p^2 = (KI)_p$. On the other hand, $\mathfrak{b} \cap KI = \mathfrak{b} \cap (\mathfrak{b}+\mathfrak{a})I = \mathfrak{b}I + \mathfrak{b} \cap \mathfrak{a}I$, where $\mathfrak{b} \cap \mathfrak{a}I = \mathfrak{a}\mathfrak{b}$ by (4.4). Hence $(\mathfrak{b} \cap I^2)_p = (\mathfrak{b} \cap KI)_p = (\mathfrak{b}I + \mathfrak{a}\mathfrak{b})_p = (\mathfrak{b}I)_p$, which gives (4.6).

We are now ready to prove part (a). If $p \in Ass_R(R/\mathfrak{b}+\mathfrak{a}I)$, then by (4.5), $\dim R_p \leq s$, and therefore $(I^2)_p = (KI)_p \subset (\mathfrak{b}+\mathfrak{a}I)_p$. From this we conclude that $I^2 \subset \mathfrak{b}+\mathfrak{a}I$. Hence $I^2 = (\mathfrak{b}+\mathfrak{a}I) \cap I^2 = \mathfrak{b} \cap I^2 + \mathfrak{a}I$, where $\mathfrak{b} \cap I^2 = \mathfrak{b}I$ by (4.6). It follows that $I^2 = \mathfrak{b}I + \mathfrak{a}I = KI$, which is the assertion of (a).

In order to prove part (b), we claim that

$$\mathfrak{b} \cap I^{j+1} = \mathfrak{b} I^j, \quad \text{for } j \geq 0. \tag{4.7}$$

The assertion being trivial for $j = 0$, we may assume that $j \geq 1$. But then by part (a) and (4.4),

$$\begin{aligned} \mathfrak{b} \cap I^{j+1} &= \mathfrak{b} \cap K^j I = \mathfrak{b} \cap (\mathfrak{b}+\mathfrak{a})^j I = \mathfrak{b}I^j + \mathfrak{b} \cap \mathfrak{a}^j I \\ &= \mathfrak{b}I^j + \mathfrak{a}^j\mathfrak{b} = \mathfrak{b}I^j, \end{aligned}$$

which gives (4.7).

Now let "$-$" denote images in $\bar{R} = R/\mathfrak{b}$, and let "$'$" denote leading forms in $gr_I(R)$. It follows from (4.7) that $gr_I(R)/(a_1') \cong gr_{\bar{I}}(\bar{R})$ and that a_1' is a regular element on $gr_I(R)$. On the other hand, by part (a) and (4.5), depth $\bar{R}/\bar{I}^j = \operatorname{depth} R/\mathfrak{b} + I^j \geq d-s$ for every $j \geq 1$, and hence there exists an R-regular sequence $x_1, \ldots, x_{d-s}$ that is regular on $gr_{\bar{I}}(\bar{R}) \cong gr_I(R)/(a_1')$ ([3]). Now $a_1', x_1', \ldots, x_{d-s}'$ form a regular sequence of homogeneous elements on $gr_I(R)$, and thus $x'_1, \ldots, x'_{d-s}, a'_1$ are a regular sequence as well. Therefore, with "$*$" denoting images in $R^* = R/(x_1, \ldots, x_{d-s})$, we have that $\mathcal{R}(I) \otimes_R R^* \cong \mathcal{R}(I^*)$ and that $x_1, \ldots, x_{d-s}$ form a regular sequence on $\mathcal{R}(I)$. Thus it suffices to show the Cohen-Macaulayness of $\mathcal{R}(I^*)$.

The regular sequence $x_1, \ldots, x_{d-s}$ can be chosen in such a way that, in addition, I^* still satisfies G_s and $ht(K : I)^* \geq s$. Thus $ht\ K^* : I^* \geq s$ and $K^* = (a_1, \ldots, a_s)^*$ is G_∞. By Lemma 1.13, I^* satisfies AN^-_{s-2}, and hence by Remark 1.12, K^* has the same property AN^-_{s-2}. Since now on the other hand $\dim R^* = s$, we may use Corollary 1.9 to conclude that the ideal K^* satisfies sliding depth. Furthermore, we still have $r_{K^*}(I^*) \leq 1$. In this situation the Cohen-Macaulayness of $\mathcal{R}(I^*)$ follows from Theorem 4.8, which is a slight modification of a result by Vasconcelos. ∎

THEOREM 4.8. *([27, 5.1 and its proof]) Let R be a local Cohen-Macaulay ring of dimension d, let I be an R-ideal with grade $I \geq 2$, let K be a reduction of I with $\mu(K) \leq s$, assume that depth $R/I \geq d - s$, that $ht\ K : I \geq s$, that $r_K(I) \leq 1$, and that K satisfies G_s and sliding depth.*

Then $\mathcal{R}(I)$ is Cohen-Macaulay.

By combining Theorem 4.1 with earlier results of this paper, we obtain the following theorem:

THEOREM 4.9. *Let R be a local Cohen-Macaulay ring of dimension d, let I be an R-ideal of grade g, let s be an integer, and let K be a reduction of I with $\mu(K) \leq s$. Assume that depth $R/I \geq d - s$, that for every minimal $p \in V(I)$, I_p is a complete intersection, that for every non-minimal $p \in V(I)$ with $\dim R_p \leq s - 1$, either $\mu(I_p) \leq \dim R_p$ or $\ell(I_p) \leq \dim R_p - 1$, and that for every $p \in V(I)$ with $\dim R_p = s$, either $r_{K_p}(I_p) \leq 1$ or $\mu(I_p) \leq \dim R_p$ or $\ell(I_p) \leq \dim R_p - 1$. Further suppose that one of the following conditions hold:*

(i) *([10], [11], [27]) $s \leq g + 1$;*
(ii) *([10, [11]) $s = g + 2$, R is Gorenstein, and $R/I^{(1)}$ is Cohen-Macaulay;*
(iii) *([27]) I satisfies sliding depth;*
(iv) *R is Gorenstein, and depth $R/I^j \geq d - g - j + 1$ for $1 \leq j \leq s - g - 1$;*
(v) *R is Gorenstein, depth $R/I^{(j)} \geq d - g - j + 1$ for $1 \leq j \leq s - g - 1$, and $\ell(I_p) \leq \dim R_p - 1$ for every non-minimal $p \in V(I)$ with $\dim R_p \leq s - 2$.*

Then:

(a) *$r_K(I) \leq 1$;*
(b) *$gr_I(R)$ is Cohen-Macaulay;*
(c) *$\mathcal{R}(I)$ is Cohen-Macaulay if $g \geq 2$;*
(d) *$I^{(j)} = I^j$ for every $j \geq 1$ if and only if $\ell(I_p) \leq \dim R_p - 1$ for every non-minimal $p \in V(I)$ with $\dim R_p \leq s$.*

PROOF. After adjoining a variable to R, and localizing, we may assume that R has infinite residue class field, and after adjoining variables to R and I, and localizing, we may assume that $g \geq 2$ in (a), (b), (d). Also notice that part (b) is a consequence of (c) ([13, Proposition 1.1]). Now our assertions will follow from Theorem 4.2 once we have shown that I satisfies AN_{s-2}^{-}. We may assume that $s-2 \geq g$, which finishes (i). Condition (ii) is a special case of (v) (or is taken care of by [22, 1.3]), and for (iii), I satisfies AN_{s-1}^{-} by [25, 4.1] and [9, 3.3]. Finally, for I as in (iv) and (v) respectively, the property AN_{s-2} follows from Theorem 2.9 or Theorem 3.3, respectively. ∎

Our next corollary is a special case of Theorem 4.9 (v).

COROLLARY 4.10. *Let k be a perfect field, let R be a regular local ring essentially of finite type over k with infinite residue class field, let I be an R-ideal with grade $I = g$ and $\ell(I) \leq g+3$, assume that $A = R/I$ is Cohen-Macaulay, reduced, and a complete intersection in codimension three, and that depth $\Omega_k(A) \geq \dim A - 2$.*

Then $r_K(I) \leq 1$ for every minimal reduction K of I, $\mathcal{R}(I)$ is Cohen-Macaulay, and $I^{(j)} = I^j$ for every $j \geq 1$.

COROLLARY 4.11. *Let R be a local Gorenstein ring of dimension d with infinite residue class field, let I be an R-ideal with grade $I = g$ and $\ell(I) \neq \mu(I)$, and let s be an integer with $\ell(I) \leq s \leq \mu(I)$. Assume that for every minimal $p \in V(I)$, I_p is a complete intersection, and that for every non-minimal $p \in V(I)$ with $\dim R_p \leq s$, either $\mu(I_p) \leq \dim R_p$ or $\ell(I_p) \leq \dim R_p - 1$. Further suppose that depth $R/I \geq d-s$ and that depth $R/I^j \geq d-g-j+1$ for $1 \leq j \leq s-g-1$.*

Then $\ell(I) = s$, $r_K(I) = 1$ for every minimal reduction K of I, I satisfies the assertions of Theorem 4.9 (b), (c), (d), and depth $R/I^j = d-s$ for $j \geq s-g$.

PROOF. As before, we may assume that $g \geq 1$. Let K be any minimal reduction of I; then K and I satisfy the assumptions of Theorem 4.9 (iv), and hence the assertions of this theorem hold. In particular, $gr_I(R)$ is Cohen-Macaulay. Also notice that $r_K(I) = 1$ since $K \neq I$.

As before, Theorem 2.9 and Proposition 1.11 imply that I satisfies G_{s+1} and that $ht\ K : I \geq s+1$. In particular, $K : I$ is an s-residual intersection of I. Now suppose that depth $R/I^{s-g} \geq d-s+1$. Then by Theorem 2.9, I would satisfy AN_{s-1}, and hence by Proposition 1.7 (a), $ht\ K : I = s$. This contradiction shows that depth $R/I^{s-g} \leq d-s \leq d-\ell(I)$. Since $gr_I(R)$ is Cohen-Macaulay, [4, 3.3] then implies that depth $R/I^j = d-\ell(I)$ whenever $j \geq s-g$. Hence $s = \ell(I)$, and depth $R/I^j = d-s$ for $j \geq s-g$. ∎

EXAMPLE 4.12. Corollary 4.11 applies to the ideal I of Example 2.12 with $s = 9$. In particular notice that $gr_I(R)$ is a Gorenstein domain, depth $R/I^j = 10$ for $1 \leq j \leq 3$, and depth $R/I^j = 6$ for $j \geq 4$.

REFERENCES

1. S. Abhyankar, Concepts of order and rank on a complex space, and a condition for normality, Math. Ann **141** (1969), 171-192.

2. M. Artin and M. Nagata, Residual intersections in Cohen-Macaulay rings, J. Math. Kyoto Univ. **12** (1972), 307-323.
3. M. Brodmann, Asymptotic stability of $Ass(M/I^n M)$, Proc. Amer. Math. Soc. **74** (1979), 16-18.
4. D. Eisenbud and C. Huneke, Cohen-Macaulay Rees algebras and their specializations, J. Algebra **81** (1983), 202-224.
5. M. Fiorentini, On relative regular sequences, J. Algebra **18** (1971), 384-389.
6. R. Hartshorne, Complete intersections and connectedness, Amer. J. Math. **84** (1962), 497-508.
7. J. Herzog, A. Simis, and W.V. Vasconcelos, Approximation complexes of blowing-up rings, J. Algebra **74** (1982), 466-493.
8. J. Herzog, A. Simis, and W.V. Vasconcelos, Approximation complexes of blowing-up rings, II, J. Algebra **82** (1983), 53-83.
9. J. Herzog, W.V. Vasconcelos, and R. Villarreal, Ideals with sliding depth, Nagoya Math. J. **99** (1985), 159-172.
10. S. Huckaba and C. Huneke, Powers of ideals having small analytic deviation, Amer. J. Math. **114** (1992), 367-403.
11. S. Huckaba and C. Huneke, Rees algebras of ideals having small analytic deviation, Trans. Amer. Math. Soc. (to appear).
12. C. Huneke, The theory of *d*-sequences and powers of ideals, Adv. Math. **46** (1982), 249-279.
13. C. Huneke, On the associated graded ring of an ideal, Illinois J. Math. **26** (1982), 121-137.
14. C. Huneke, Linkage and Koszul homology of ideals, Amer. J. Math. **104** (1982), 1043-1062.
15. C. Huneke, Strongly Cohen-Macaulay schemes and residual intersections, Trans. Amer. Math. Soc. **277** (1983), 739-763.
16. C. Huneke, Numerical invariants of liaison classes, Invent. Math. **75** (1984), 301-325.
17. C. Huneke, The Koszul homology of an ideal, Adv. Math. **56** (1985), 295-318.
18. C. Huneke and B. Ulrich, Residual intersections, J. Reine Angew. Math. **390** (1988), 1-20.
19. C. Huneke and B. Ulrich, Powers of licci ideals, in *Commutative Algebra, Proceedings of a Microprogram held June 15-July 2, 1987*, eds.: M. Hochster, C. Huneke, J. Sally, Springer-Verlag, New York, 1988, 339-346.
20. A. Kustin, M. Miller, and B. Ulrich, Generating a residual intersection, J. Algebra **146** (1992), 335-384.
21. D.G. Northcott and D. Rees, Reductions of ideals in local rings, Proc. Comb. Phil. Soc. **50** (1954), 145-158.
22. C. Peskine and L. Szpiro, Liaison des variétés algébriques, Invent. Math. **26** (1974), 271-302.
23. A. Simis and W. Vasconcelos, The syzygies of the conormal module, Amer. J. Math. **103** (1981), 203-224.
24. H. Srinivasan, A grade five cyclic Gorenstein module with no minimal algebra resolutions, preprint.
25. B. Ulrich and W. Vasconcelos, The equations of Rees algebras of ideals with linear presentation, Math. Z. (to appear).
26. W. Vasconcelos, Koszul homology and the structure of low codimension Cohen-Macaulay ideals, Trans. Amer. Math. Soc. **301** (1987), 591-613.
27. W. Vasconcelos, Hilbert functions, analytic spread, and Koszul homology, in this volume.
28. U. Vetter, The depth of the module of differentials of a generic determinantal singularity, Comm. Algebra **11** (1983), 1701-1724.

Department of Mathematics, Michigan State University, East Lansing, Michigan 48824

E-mail address: 21144bfu@msu.bitnet

Contemporary Mathematics
Volume **159**, 1994

Hilbert Functions, Analytic Spread, and Koszul Homology

WOLMER V. VASCONCELOS

ABSTRACT. Let R be a Noetherian ring and let $R[It]$ be the blowup ring of an ideal I. We introduce devices that permit the arithmetical study of some of these algebras by techniques that were previously restricted to ideals generated by d-sequences. They rely on being able to pass to a minimal reduction of I properties of the Koszul homology of I. There are applications to the behaviour of the coefficients of the Hilbert polynomial of I (in the case of primary ideals), to restrictions on the analytic spread of I, to the number of generators of certain prime ideals, and to the Cohen–Macaulayness of $R[It]$.

1. Introduction

The Rees algebra of an ideal I of a commutative ring R is the subalgebra of the ring of polynomials $R[t]$

$$R[It] = R \oplus It \oplus \cdots \oplus I^n t^n \oplus \cdots .$$

It gathers many properties of the variety defined by I, particularly of analytic interest, and gives an algebraic realization for the blowup of Spec(R) along the subvariety $V(I)$. Its arithmetical properties, such as depth and normality, have a telling effect on the cohomology groups of the blowup. In another instance, if I is an ideal of dimension zero, the behaviour of its Hilbert functions is faithfully mirrored in the algebraic properties of the associated graded ring

$$\mathrm{gr}_I(R) = \bigoplus_{n=0}^{\infty} I^n/I^{n+1} = R[It] \otimes R/I.$$

1991 *Mathematics Subject Classification.* Primary 13H10; Secondary 13C40, 13D40.
Key words and phrases. Approximation complex, Cohen–Macaulay ring, Hilbert function, Koszul homology, reduction, Rees algebra.
The author was partially supported by the NSF.
"This paper is in final form and no version of it will be submitted for publication elsewhere."

The arithmetical study of these algebras can proceed more systematically when the ideal is generated by a d–sequence, a notion introduced by Huneke ([**18**]), which is a strong form of analytic independence and a far–reaching generalization of regular sequences. A first property of these sequences is that the canonical mapping from the symmetric algebra of I onto its Rees algebra

$$S(I) \longrightarrow R[It],$$

is an isomorphism—a situation summed up by saying I is of *linear type*. There are also acyclic complexes—the *approximation complexes* of [**11**]—that mediate between the Koszul homology of I and depth properties of $R[It]$.

A drawback in this approach lies on the limitation that, locally, the number of generators of I should not exceed the dimension of R, $\nu(I) \leq \dim R$. To help span the gap it is appropriate to view the Rees algebra of a 'general' ideal as an extension of the Rees algebra of an ideal of linear type. The vehicle for this theory is the notion of reduction of an ideal, an object first isolated by Northcott and Rees ([**28**]), and which plays a premier role in the study of Rees algebras. An ideal $J \subset I$ is a *reduction* of I if $JI^r = I^{r+1}$ for some integer r; the least such integer, $r_J(I)$, is the *reduction number* of I with respect to J. Phrased otherwise, the extension

$$R[Jt] \hookrightarrow R[It]$$

is a finite morphism of graded algebras, with the reduction number giving the maximal degree of a set of homogeneous, minimal, generating set.

Among the reductions of an ideal, one plays a distinguished role. If $(R, \mathfrak{m})$ is a local ring, a reduction J of I is *minimal* if J is a minimal ideal among the reductions of I. In this case, it is related to another invariant of I. The *special fiber* of $R[It]$ is the ring $R[It] \otimes (R/\mathbf{m})$; its Krull dimension is the *analytic spread* of I and shall be denoted $\ell(I)$. When I contains regular elements (which shall be assumed throughout), $\dim R[It] = \dim R + 1$, and $\ell(I) \leq \dim R$. If $R/\mathfrak{m}$ is an infinite field, then $\ell(I)$ is simply the Krull dimension of a (graded) Noether normalization of $R[It] \otimes (R/\mathfrak{m})$. From this, it will follow that $\ell(I) = \nu(J)$, for any minimal reduction J of I. It may be still be the case that the reduction number is not the same for all minimal reductions. Nevertheless, $J \subset I$ is a minimal reduction if and only if J is a reduction of I generated by analytically independent elements.

The *deviation* of I is the non–negative integer $\nu(I) -$ height I. Huckaba and Huneke [**16**] have defined the *analytic deviation* of I as $\ell(I) -$ height I; for ideals of linear type $\nu(I) = \ell(I)$. The ideals of analytic deviation zero are called equimultiple. There is still room for another definition: the difference $\nu(I) - \ell(I)$ will be called the *second deviation* of the ideal I, the terminology arising because it is the difference between the other two deviations. An useful guideline here is that ideals with low deviations provide a more direct access to the algebraic property of the blowup.

There are two elements in this study. First, we introduce the *Sally module* and the *mixed graded algebra* associated to a reduction of an ideal. They form a backdrop in which to make comparisons of the Rees algebra of an ideal and the Rees algebra of one of its reductions. It tends to work better when the deviations and the reduction numbers are low.

The other technical novelty is an analysis of the Koszul homology of reductions. It extends the applicability of the approximation complexes in the theory of Rees algebras (cf. [**11**]) to a much larger class of ideals. They will be used to estimate analytic spreads and exhibit the Cohen–Macaulayness of certain Rees algebras. The devices that are introduced here accomplish some of the aims stated at the outset by transfering certain properties from an ideal I to one of its minimal reductions J, and then studying the algebra structure of $R[It]$ in terms of its module structure over $R[Jt]$.

This paper is organized as follows. In section 2 we introduce the Sally module of a reduction, and discuss some of its general dimension–theoretic properties. Its vanishing is shown to be related to the Cohen–Macaulayness of the associated graded ring. In section 3 we show how certain results in the recent literature on the vanishing of Hilbert coefficients and the depth of graded rings have a very direct explanation. The next two sections contain our main results. To formulate the most representative result we recall the definitions of two technical requirements on an ideal I (cf. [**12**], [**14**] and [**20**]).

DEFINITION 1.1. *Let I be an ideal of a local ring $(R, \mathfrak{m})$ of dimension d. Let $\mathbb{K}$ be the Koszul complex on the set $\{a_1, \dots, a_n\}$ of generators of I. Denote by $H_i(\mathbb{K})$ the homology modules of $\mathbb{K}$. I satisfies* sliding depth *if*

$$\textit{depth } H_i(\mathbb{K}) \geq d - n + i, \ \forall i.$$

This notion works as a mechanism to control Cohen–Macaulay residual intersections as introduced by Artin and Nagata ([**1**]; see [**21**] and [**24**]). There is a more stringent condition on the depths of the Koszul homology modules:

DEFINITION 1.2. *The ideal I is* strongly Cohen–Macaulay *if the modules $H_i(\mathbb{K})$ are Cohen–Macaulay.*

These conditions are independent of the chosen generating set and are stable under localization. They play a fundamental role in the theory of residual intersection of Huneke and Ulrich ([**24**]). (These sources contain detailed listings of classes of ideals with sliding depth.) They are used in the approximation complexes (see [**12**]) to produce Rees algebras with good depth properties once the following condition is realized:

DEFINITION 1.3. *Let R be a Noetherian ring and let I be an ideal of R. I is said to satisfy the condition $\mathcal{F}_1$ if for the prime ideals $I \subset \wp$, $\nu(I_\wp) \leq$ height $\wp$.*

When restricted to prime ideals of codimension less than s, this condition was called G_s in **[1]**. It is naturally expressed by the heights of the Fitting ideals of I.

Now we are able to state our main result:

Theorem 5.1: *Let R be a local Cohen–Macaulay ring of dimension d, let I be an ideal of codimension at least two, let J be a reduction of I with $\nu(J) = s \leq d$, and assume that I satisfies $\mathcal{F}_1$, locally in codimension $\leq s-1$. Suppose that for each prime ideal $\wp \supset I$, depth $R_\wp/I_\wp \geq \dim R_\wp - \nu(J_\wp)$. If I has sliding depth and $r_J(I) = 1$, then $R[It]$ is a Cohen–Macaulay algebra.*

The condition on the local depth of R/I is realized if I is Cohen–Macaulay or more generally depth $R/I \geq \dim R/I - 1$. It is reminiscent of Serre's conditions S_k. Some attention is given to how the hypotheses, such as the equality $r_J(I) = 1$, are to be tested.

It will be a consequence of its proof that, if a Cohen–Macaulay prime ideal $\wp$ of a regular local ring R satisfies sliding depth and is such that all the powers $\wp^m$ are $\wp$–primary then $\wp$ can be generated by $\dim R - 1$ elements. The techniques are also used to give alternate proofs to several known results.

We have benefited from the interest of several colleagues, especially Sam Huckaba, Craig Huneke and Judy Sally. We are particularly grateful to Bernd Ulrich, whose intervention enriched it considerably, in particular the formulation of Theorem 4.11 and its elegant proof are due to him.

2. The Sally Module of a Reduction

Let R be a Noetherian ring, and let I be an ideal with a reduction J. Consider the exact sequence of finitely generated $R[Jt]$–modules:

$$(1) \quad 0 \to I \cdot R[Jt] \longrightarrow I \cdot R[It] \longrightarrow S_J(I) = \bigoplus_{n=0}^{\infty} I^{n+1}t^n/IJ^nt^n \to 0.$$

DEFINITION 2.1. *The group $S_J(I)$, as an $R[Jt]$–module, is the* Sally module *of I with respect to J.*

A motivation for this definition is the work of Sally, particularly in **[31]**, **[32]**, **[33]**, and **[34]**, where there is an extensive analysis of the structure of the components of $S_J(I)$. One could also define higher versions of it, $S_J^{(s)}(I)$, by using $I^s \cdot R[Jt]$ instead, but we shall not have the occasion here to use them.

To be useful, this sequence requires information about $I \cdot R[Jt]$—which is readily available in many cases—and more fine properties of $S_J(I)$. Its main feature is the relationship it bears with $R[Jt]$, a ring that often has many well–known properties.

First we shall develop some general properties of the Sally module, and promptly make use of them. Let us begin by pointing out a simple but critical property of $S_J(I)$:

PROPOSITION 2.2. *Let* $(R, \mathfrak{m})$ *be a Cohen–Macaulay local ring of dimension* d, *let* I *be a* $\mathfrak{m}$*-primary ideal and let* J *be a minimal reduction of* I. *If* $S_J(I) \neq 0$ *its associated prime ideals have codimension* 1, *in particular its Krull dimension as an* $R[Jt]$*-module is* d.

PROOF. may assume $d \geq 1$. For each prime ideal P of $A = R[Jt]$, of codimension at least two, it suffices to use the homology sequence of the functor $\mathrm{Hom}_A(A/P, \cdot)$ on the sequence (2.1), where we take into account that $I \cdot R[Jt]$ is a maximal Cohen–Macaulay module. □

2.1. From Rees algebras to associated graded rings. From now on when speaking of a Rees algebra $R[It]$ it is to be understood that I contains a regular element. An elementary but important relationship between Rees algebras and associated graded is the following result of [**19**]:

THEOREM 2.3. *Let* R *be a Cohen–Macaulay ring and let* I *be an ideal containing regular elements such that* $R[It]$ *is Cohen–Macaulay. Then* $\mathrm{gr}_I(R)$ *is Cohen–Macaulay.*

Because we are going to refer to them repeatedly, we display the two exact sequences that occur in its proof. They relate depth data in R, $R[It]$, and $\mathrm{gr}_I(R)$:

$$0 \to I \cdot R[It] \longrightarrow R[It] \longrightarrow \mathrm{gr}_I(R) \to 0 \tag{2}$$

$$0 \to It \cdot R[It] \longrightarrow R[It] \longrightarrow R \to 0. \tag{3}$$

There is no converse to Theorem 2.3. A significant case where it holds is for ideals generated by d–sequences ([**12**, Corollary 6.4]). An open question asks whether the converse holds for prime ideals of regular local rings. A more specific problem is whether the Rees algebra of a four generated prime ideal of a three–dimensional regular local ring is always Cohen–Macaulay.

2.2. Reduction number one. Several authors have studied, under various guises, what amounts to the vanishing of some $S_J(I)$. In addition to the aforementioned research of Sally on $\mathfrak{m}$–primary ideals, recent work by Huckaba and Huneke ([**16**], [**17**]) has established the vanishing of the Sally module for large classes of ideals of positive dimension.

We want to see this module as a vehicle for bookeeping information on algebraic properties of the associated graded ring of an ideal. This occurs as in:

PROPOSITION 2.4. *Let $(R, \mathfrak{m})$ be a Cohen–Macaulay local ring of dimension d. Suppose I is a Cohen–Macaulay ideal and $J \subset I$ is generated by a regular sequence of s elements. If $I^2 = JI$ then $\mathrm{gr}_I(R)$ is a Cohen–Macaulay ring.*

PROOF. In this case $S_J(I) = 0$, while the exact sequence

$$0 \to I \cdot R[Jt] \longrightarrow R[Jt] \longrightarrow R/I[T_1, \ldots, T_s] \to 0,$$

means that $I \cdot R[Jt]$ is a Cohen Macaulay module of dimension $\dim R/I + s + 1 = d + 1$. Plugging this in equation (3), and then in (2), yields the assertion. □

COROLLARY 2.5. ([**37**, Proposition 3.1]) *Let $(R, \mathfrak{m})$ be a Cohen–Macaulay local ring of dimension d and let I be a $\mathfrak{m}$-primary ideal. Suppose J is a reduction of I generated by a regular sequence and $I^2 = JI$. Then $\mathrm{gr}_I(R)$ is a Cohen–Macaulay ring.*

2.3. Reduction number two. If the reduction number is 2, the structure of $S_J(I)$ depends on the structure of the module I^2/JI. Note that $S_J(I)$ is then a module over $\mathrm{gr}_J(R)$.

PROPOSITION 2.6. *Let $(R, \mathfrak{m})$ be a Cohen–Macaulay local ring of dimension d and let I be a $\mathfrak{m}$-primary ideal. Suppose J is a reduction of I generated by a regular sequence and $I^3 = JI^2$.*

(a) *If $I^2/JI = R/\mathfrak{m}$ then* depth $\mathrm{gr}_I(R) \geq d - 1$.

(b) *If $I^2/JI = (R/\mathfrak{m})^2$ then* depth $\mathrm{gr}_I(R) \geq d - 2$.

PROOF. (a) The assumptions imply that $S_J(I)$ is a cyclic $R[Jt]$-module, annihilated by $\mathfrak{m}R[Jt]$. But in the sequence of definition of $S_J(I)$, with $I \cdot R[Jt]$ Cohen–Macaulay, no associated prime of $S_J(I)$ can have height greater than 1. This shows that $S_J(I) \simeq R/\mathfrak{m}[T_1, \ldots, T_d]$. From this we again go to the other two sequences to obtain the desired depth estimate.

(b) The module $S_J(I)$ is isomorphic to two copies of the ring of polynomials $B = R/\mathfrak{m}[T_1, \ldots, T_d]$, or to one its ideals generated by two elements. In first case, one proceeds as in (a). If $S_J(I)$ is an ideal of B, it has depth $d - 1$. This means that $I \cdot R[It]$ has also depth $d - 1$ from which depth $\mathrm{gr}_I(R) \geq d - 2$ follows. □

How to tell the cases in (b) from one another? It may take place as follows. Suppose $I^3 = JI^2$ and $I^2 = (a_1, \ldots, a_s, JI) \subset J$, and $\mathfrak{m}I^2 \subset JI$. It is clear that $S_J(I)$ will be isomorphic to an ideal of $R/\mathfrak{m}[T_1, \ldots, T_d]$ generated by indeterminates and therefore its depth is available. Of course the analysis becomes harder if $S_J(I)$ is not a module over B.

The Cohen–Macaulayness of the Rees algebra $R[It]$ is more delicate. For $\mathfrak{m}$-primary ideals there is a general condition due to Goto–Shimoda [**6**, Theorem 3.1] (see also [**8**, Theorem 4.8], [**15**], [**25**]).

THEOREM 2.7. *The Rees algebra $R[It]$ is Cohen–Macaulay if and only if the associated graded ring $\mathrm{gr}_I(R)$ is Cohen–Macaulay and the reduction number of I is at most $d-1$.*

2.4. Prime ideals. There are other partial converses to Theorem 2.3 but which require detailed knowledge of the canonical module.

We recall the notion of the *a–invariant* of a positively graded algebra $A = A_0 \oplus A_1 \oplus \cdots$, where we assume A_0 is a local ring (see [**7**]). Suppose $\dim A = d$ and let

$$N = \mathfrak{m} \oplus A_1 \oplus A_2 \oplus \cdots$$

be the maximal graded ideal of A. The local cohomology module $H_N^d(A)$ is a graded, Artinian module. The *a–invariant* of A is defined to be

$$a(A) = \sup\{j \mid H_N^d(A)_j \neq 0\}.$$

If A has a canonical module ω_A, by the theorem of local duality ([**9**]), $a(A)$ is the negative of the infimum degree of the non–vanishing components of ω_A.

Let I be an unmixed ideal of codimension g. Using the terminology of [**13**], we shall say that the canonical module of $S = R[It]$ has the expected form if $\omega_S = \omega_R(a, at)^s$, for some integer s. If I is generically a complete intersection $s = g-2$.

PROPOSITION 2.8. *Let R be a Gorenstein local ring and let I be an ideal whose Rees algebra $R[It]$ has the expected canonical module. If $\mathrm{gr}_I(R)$ is Cohen–Macaulay then $R[It]$ is Cohen–Macaulay.*

PROOF. Let $N = (\mathfrak{m}, It)$ be the irrelevant maximal ideal of $S = R[It]$. We are going to apply local cohomology relative to N in both sequences (2) and (3). We set $L = It \cdot R[It]$ and note that $I \cdot R[It] = L[+1]$. Let $\dim R = d$, set $G = \mathrm{gr}_I(R)$, and consider the respective two long cohomology exact sequences:

$$H_N^{r-1}(R)_j \longrightarrow H_N^r(L)_j \longrightarrow H_N^r(S)_j \longrightarrow H_N^r(R)_j \tag{4}$$

$$H_N^{r-1}(G)_j \longrightarrow H_N^r(L)_{j+1} \xrightarrow{\varphi} H_N^r(S)_j \longrightarrow H_N^r(G)_j. \tag{5}$$

They show that, in each dimension $r < d$ and for each degree j, we have

$$\begin{cases} H_N^r(L)_j \simeq H_N^r(S)_j \\ H_N^r(L)_{j+1} \hookrightarrow H_N^r(S)_j \end{cases}$$

Since $H_N^r(L)_j = 0$ for $j \gg 0$ (even for $r \leq d$), these relations imply that $H_N^r(S) = 0$ and therefore depth $S \geq d$.

To complete the proof, it suffices to show that in equation (5) the mapping

$$\varphi : H_N^d(L)_{j+1} \longrightarrow H_N^d(S)_j$$

is an isomorphism. To this end, let A be a Cohen–Macaulay ring of dimension $d+1$ and canonical module Ω which is a presentation of S, $S = A/J$. Dualizing the sequence (2) with respect to Ω, we obtain, as in [**13**, proof of Corollary 2.5], that

$$\varphi^* : \mathrm{Ext}^1_A(S, \Omega) \simeq \mathrm{Ext}^1_A(L[+1], \Omega),$$

which is an isomorphism by local duality ([**9**, Theorem 6.3]), implying our assertion. □

COROLLARY 2.9. *Let R be a regular local ring and let P be a prime ideal whose associated graded ring $\mathrm{gr}_P(R)$ is a Cohen–Macaulay domain. Then $R[Pt]$ is a Cohen–Macaulay ring.*

PROOF. The canonical module of $R[Pt]$ has the expected form according to [**13**]. □

3. Hilbert Functions

There will be no attempt here at dealing with the finer properties of the Hilbert function of ideals. We will however indicate the possibilities by proving some known inequalities on the coefficients of the Hilbert polynomial of a primary ideal.

PROPOSITION 3.1. *Let $(R, \mathfrak{m})$ be a Cohen–Macaulay local ring of dimension d, with infinite residue field, and let I be an $\mathfrak{m}$-primary ideal. Denote by $H_I(n) = \lambda(R/I^n)$ the Hilbert function of I, and let*

$$(6)\quad e_0\binom{n+d-1}{d} - e_1\binom{n+d-2}{d-1} + \cdots + (-1)^{d-1}e_{d-1}\binom{n}{1} + (-1)^d e_d$$

be its Hilbert polynomial. Suppose J is a minimal reduction of I and let $S = S_J(I)$ be the corresponding Sally module. Then for $n \gg 0$

$$(7)\quad H_I(n) = e_0\binom{n+d-1}{d} + (\lambda(R/I) - e_0)\binom{n+d-2}{d-1} - \lambda(S_{n-1}).$$

PROOF. It is a straightforward calculation once one takes into account that $e_0 = \lambda(R/J)$. □

COROLLARY 3.2. *If the Sally module $S_J(I) \neq 0$ then the function $\lambda(S_n)$ has the growth of a polynomial of degree $d-1$.*

PROOF. According to Proposition 2.2, if $S_J(I) \neq 0$ then it is a module whose Krull dimension is d. □

COROLLARY 3.3. *Let $s_0, s_1, \ldots, s_{d-1}$ be the coefficients of the Hilbert polynomial of $S_J(I)$. Then*

$$(8)\quad \begin{aligned} e_1 &= e_0 - \lambda(R/I) + s_0 \\ e_{i+1} &= s_i \ \text{for}\ i \geq 1. \end{aligned}$$

COROLLARY 3.4. *The following hold:*

(a) $\lambda(R/I) \geq e_0 - e_1$ ([**27**]).

(b) *In case of equality, then* $S_J(I) = 0$ ([**22**, Theorem 2.1], [**29**]).

PROOF. (a) This is clear since e_1 will obtained by adding to $e_0 - \lambda(R/I)$ the contribution from the leading term in the Hilbert polynomial of $S_J(I)$.

(b) For equality to occur there must be no contribuition from $S_J(I)$, meaning that $S_J(I)$ is a module of Krull dimension $\leq d-1$. From Proposition 2.2 this implies $S_J(I) = 0$. □

PROPOSITION 3.5. *Let $(R, \mathfrak{m})$ be a Cohen–Macaulay local ring of dimension d, with infinite residue field, and let I be an $\mathfrak{m}$-primary ideal with a minimal reduction J. Let e_0 and e_1 be the first two coefficients of the Hilbert polynomial of I. If*

$$e_1 = e_0 - \lambda(R/I) + 1, \tag{9}$$

then $S_J(I)$ is isomorphic to an ideal of $B = R/\mathfrak{m}[T_1, \ldots, T_d]$.

PROOF. The assumption is that the Hilbert polynomial of $S_J(I)$ has the form

$$\binom{d+n-1}{d-1} - e_2\binom{d+n-2}{d-2} + \cdots .$$

From Proposition 2.2 it follows that $\mathfrak{m}R[Jt]$ is the only associated prime of $S_J(I)$. Let $0 \neq a \in S_J(I)_1$ be an element annihilated by $\mathfrak{m}$. There exists a form $f \in R[Jt] \setminus \mathfrak{m}R[Jt]$ such that

$$f \cdot S_J(I) \subset R[Jt]a.$$

Because f is a regular element on $S_J(I)$, the equation implies that $S_J(I)$ is annihilated by $\mathfrak{m}$. □

The claim can be rephrased as follows. There exists an ideal $L \subset B$ and a form $f \in B$ such that $f \cdot S_J(I) = L \cdot a$. There are several conditions on the associated graded ring $\mathrm{gr}_I(R)$ that impact on the ideal L. We are going to present one discovered by Sally ([**34**, Theorem 1.4]).

THEOREM 3.6. *Let $(R, \mathfrak{m})$ be a Cohen–Macaulay local ring of dimension d, with infinite residue field, and let I be an $\mathfrak{m}$-primary ideal. Let e_0, e_1 and e_2 be the first three coefficients of the Hilbert polynomial of I. If $e_1 = e_0 - \lambda(R/I) + 1$ and $e_2 \neq 0$, then $L \simeq B$. In particular I has reduction number 2.*

PROOF. In the equality $f \cdot S_J(I) = L \cdot a$ we take f of least degree. Note that if f is a constant then L must be equal to B. Suppose L has a gcd $= \Delta \neq 1$, $L = \Delta L_0$. There is then an equation $fa = \Delta ca$ that shows $\Delta | f$.

Suppose f has degree $h \geq 1$. From Proposition 3.5, the Hilbert polynomial of $f \cdot S$ is then

$$\begin{aligned} H_{fS}(n) = H_S(n-h) &= \binom{d+n-h-1}{d-1} - e_2\binom{d+n-h-2}{d-2} + \text{lower terms} \\ &= \binom{d+n-1}{d-1} - (e_2+h)\binom{d+n-2}{d-2} + \text{lower terms,} \end{aligned}$$

while the Hilbert polynomial of La has the form

$$H_{La}(n) = H_L(n-1) = \binom{d+n-2}{d-1} - b\binom{d+n-4}{d-3} + \text{lower terms.}$$

This equation arises from the fact that the Hilbert polynomial of L is the difference between the Hilbert polynomials of B and of B/L, and B/L has dimension at most $d-2$.

A comparison of coefficients shows that $h + e_2 = 1$. By [**26**] $e_2 \geq 0$, while $h \geq 1$. Thus $e_2 = 0$, which is a contradiction. □

COROLLARY 3.7. *Suppose, in addition to the hypotheses above, that* $R[It]$ *is integrally closed. If* $e_1 = e_0 - \lambda(R/I) + 1$ *then* $L = B$.

PROOF. The hypothesis on $R[It]$ says that it satisfies the condition S_2 of Serre, so that from the exact sequences (2) and (3) it follows that $I \cdot R[It]$ will inherit the condition. It is now easy to see from the definition of $S_J(I)$ that it has the S_2 property as a B–module. This says that L is a reflexive ideal and it is therefore free. □

REMARK 3.8. *If* R *is a ring that satisfies the condition* S_2 *of Serre and* I *is an ideal of height at least two, then* $R[It]$ *has the condition* S_2 *precisely when* $\mathrm{gr}_I(R)$ *has no embedded prime (see* [**3**, Theorem 1.5]*).*

The proof of the theorem shows:

COROLLARY 3.9. *If* $e_1 = e_0 - \lambda(R/I) + 1$ *and* $L \neq B$, *then* depth $\mathrm{gr}_I(R) \geq$ depth B/L.

The technique also suggests that one may approach the computation of Hilbert functions by using more general reductions as long as their Rees algebras are nice to handle. Elsewhere we shall give a fuller structure for Sally modules.

4. Koszul Homology and the Mixed Graded Algebra

Let R be a Cohen–Macaulay local ring of infinite residue field, with canonical module ω_R, and suppose I is an unmixed ideal which is generically a complete intersection. We examine some homological properties of the reductions of I. It will permit estimates of analytic spreads of several classes of ideals and of depths of associated graded rings. Unless said otherwise, we assume that I is not generated by analytically independent elements.

4.1. The top Koszul homology of a reduction. We begin to derive properties of the Koszul homology of a reduction and the constraints they place on the analytic spread.

PROPOSITION 4.1. *Let J be a reduction of I. Then I and J have the same top Koszul homology module, and the rings R/I and R/J have the same canonical module.*

PROOF. Suppose height $I = g$, and consider the sequence

$$0 \to I/J \longrightarrow R/J \longrightarrow R/I \to 0.$$

I/J is a module annihilated by a power of I, whose support contains no minimal prime of I; therefore it has dimension at most $d - g - 1$, $d = \dim R$. In the cohomology sequence

$$\mathrm{Ext}_R^{g-1}(I/J, \omega_R) \to \mathrm{Ext}_R^g(R/I, \omega_R) \to \mathrm{Ext}_R^g(R/J, \omega_R) \to \mathrm{Ext}_R^g(I/J, \omega_R)$$

the modules at the ends vanish. The other claim has a similar proof. □

The next result reveals several obstructions for a Cohen–Macaulay ideal to have analytic deviation 1. It is the strand of an idea traced to [**35**, Proposition 4].

THEOREM 4.2. *Let R be a Gorenstein local ring of dimension d, and infinite residue field. Let I be an ideal of codimension g that is generically a complete intersection and has analytic spread $\ell(I) = g + 1$. Let J be a minimal reduction of I.*

(a) *If I is Cohen–Macaulay then* depth $R/J = d - g - 1$.
(b) *If R/I has the condition S_2 of Serre then J has no associated prime ideal of codimension $g + 2$.*

PROOF. (a) Let $J = (a_1, \dots, a_{g+1})$, and consider the acyclic portion of the Koszul complex $\mathbb{K}$ built on the a_i's:

$$0 \to K_{g+1} \longrightarrow K_g \longrightarrow \cdots \longrightarrow K_2 \longrightarrow B_1 \to 0. \tag{10}$$

From Proposition 4.1 we have that $H_1(\mathbb{K})$, the canonical module of R/J, is isomorphic to the canonical module of R/I, and therefore is Cohen–Macaulay. If Z_1 denotes of 1–cycles of $\mathbb{K}$, the assertion is that depth $Z_1 = d - g + 1$. Let

$$0 \to B_1 \longrightarrow Z_1 \longrightarrow H_1(\mathbb{K}) \to 0 \tag{11}$$

be the sequence that defines $H_1(\mathbb{K})$. Since B_1 has projective dimension $g - 1$, it has depth $d - g + 1$ and thus the depth of Z_1 is at least $d - g$ since $H_1(\mathbb{K})$ is Cohen–Macaulay of dimension $d - g$. Into the cohomology exact sequence

$$\mathrm{Ext}_R^{g-1}(B_1, R) \longrightarrow \mathrm{Ext}_R^g(H_1(\mathbb{K}), R) \longrightarrow \mathrm{Ext}_R^g(Z_1, R) \longrightarrow \mathrm{Ext}_R^g(B_1, R)$$

we feed

$$\begin{cases} \operatorname{Ext}_R^{g-1}(B_1, R) = R/J \\ \operatorname{Ext}_R^{g}(B_1, R) = 0 \\ \operatorname{Ext}_R^{g}(H_1(\mathbb{K}), R) = R/I \end{cases}$$

the first two from the Koszul complex, and the last from the local duality theorem since R/I is Cohen–Macaulay:

$$R/J \longrightarrow \operatorname{Ext}_R^g(H_1(\mathbb{K}), R) = R/I \longrightarrow \operatorname{Ext}_R^g(Z_1, R) \to 0.$$

We claim that $\operatorname{Ext}_R^g(Z_1, R)$ vanishes. By induction on the dimension of R, we may assume that this module has finite length, and $\dim R/I \geq 2$. If the module is different from zero, from the sequence it must have the form $R/(I, a)$ for some a. By Krull's theorem this is not possible.

Finally, to argue that the equality holds, note that if depth $R/J = d - g$, I and J would necessarily have the same associated prime ideals, and therefore would be equal since they agree at such primes.

(b) The equality $\operatorname{Ext}_R^g(H_1(\mathbb{K}), R) = R/I$ still holds and therefore one still gets that $\operatorname{Ext}_R^g(Z_1, R) = 0$. This means that $\operatorname{Ext}_R^{g+2}(R/J, R) = 0$, which is a strong form of the assertion. □

In the following corollaries, I is still assumed to be generically a complete intersection but not necessarily generated by analytically independent elements.

COROLLARY 4.3. *Let I be a Cohen–Macaulay ideal of codimension g that is of linear type in codimension $\leq g+1$. Then $\ell(I) \geq g+2$.*

COROLLARY 4.4. *Let I be a Cohen–Macaulay prime ideal all of whose powers I^m are unmixed. Then $g+2 \leq \ell(I) \leq \dim R - 1$. In particular, if $\dim R/I = 2$ no such prime exists, and if $\dim R/I = 3$ then $\ell(I) = g+2$.*

PROOF. The upper bound is given by [**4**], while the equality $\ell(I) = g+1$ is ruled out by [**5**]. □

The following is a special case of [**17**, Theorem 2.1]:

COROLLARY 4.5. *Let I be a Cohen–Macaulay ideal of analytic deviation* 1 *and reduction number* 1. *Then $\operatorname{gr}_I(R)$ is Cohen–Macaulay.*

PROOF. Let J be a minimal reduction of I, of reduction number 1. The previous result saying that J is an ideal with sliding depth, we have by [**2**],[**23**, Theorem 3.1], or [**12**], that $R[Jt]$ is Cohen–Macaulay. On the other hand, in the approximation complex of J

(12)$\to H_1(\mathbb{K}) \otimes B[-1] \longrightarrow R/J \otimes B \longrightarrow S(J/J^2) = R[Jt]/J \cdot R[Jt] \to 0,$

where $B = R[T_1, \dots T_{g+1}]$, we have that it splits as a complex of R–modules at the minimal primes of I. Since $H_1(\mathbb{K})$ is already a R/I–module of rank 1, tensoring the sequence with R/I we obtain an exact sequence of R/I–modules

$$0 \to H_1(\mathbb{K}) \otimes B[-1] \longrightarrow R/I \otimes B \longrightarrow S(J/J^2) \otimes R/I \to 0, \tag{13}$$

where $S(J/J^2) \otimes R/I = R[Jt]/I \cdot R[J]t$ and from which it is easy to see that $I \cdot R[Jt]$ is a maximal Cohen–Macaulay module. We can then argue as in Proposition 2.4 since by hypothesis $I^2 = JI$. □

4.2. The mixed graded algebra and reduction number one. Motivated by the previous proof, we introduce the following algebra.

DEFINITION 4.6. *Let I be an ideal and J be one of its reductions. The* mixed graded algebra *of I relative to J is the ring $gr_J(R) \otimes R/I$.*

To show its usefulness, let us give another treatment of the first of several interesting facts in [**16**]. Although not appreciably shorter, this proof will display features that can be used under other hypotheses. For this purpose, we also place ourselves in their context which does not assume that R is a Gorenstein ring.

THEOREM 4.7. ([**16**, Theorem 2.2]) *Let R be a Cohen–Macaulay local ring and I an ideal of analytic deviation one and height $g \geq 1$. Let $r \geq 1$ be an integer and assume that the minimal primes of R/I have the same height, and the associated primes of R/I have height at most $g + r$. Assume also that I is a complete intersection at each of its minimal primes and there exists a minimal reduction J of I such that $r_J(I_\wp) \leq 1$ for every prime ideal $\wp \supset I$ with height $\wp/I \leq r$. Then $r_J(I) = 1$.*

PROOF. Without loss of generality we may assume that the residue field of R is infinite. Let $J = (a_1, \dots, a_g, c)$ be a reduction of I where we may assume that $J_0 = (a_1, \dots, a_g)$ generates I at each of its minimal primes and c is a regular element. Consider the complex defining the conormal module of J

$$0 \to H_1(J) \longrightarrow (R/J)^{g+1} \longrightarrow J/J^2 \to 0, \tag{14}$$

where $H_1(J)$ is the 1–dimensional Koszul homology module on the $g+1$ generators of J.

From the exact sequence (11), it follows that $H_1(J)$ is an R/J–module that satisfies the condition S_2 of Serre. In particular its minimal primes are all of codimension g. As a consequence the equality $I \cdot H_1(J) = 0$ results by localization. This means that if we tensor the sequence (14) with R/I we obtain the complex

$$0 \to H_1(J) \longrightarrow (R/I)^{g+1} \longrightarrow J/JI \to 0, \tag{15}$$

whose exactness follows from the acyclicity lemma. In turn, this implies that the associated prime ideals of J/JI have codimension at most $g+r$. Thus to show the equality $I^2 = JI$ it suffices to prove $I^2 \subset J$, as it leads to an embedding

$$I^2/JI \hookrightarrow J/JI,$$

and the vanishing of I^2/JI would follow from the assumptions.

The point is to show $I^2 \subset J_0 + cI$, as the height of the associated primes of the latter can be estimated. Indeed, the exact sequence induced by addition of ideals

$$0 \to J_0 \cap cI \longrightarrow J_0 \oplus cI \longrightarrow J_0 + cI \to 0,$$

and the elementary equality $J_0 \cap cI = cJ_0$ ([**16**, Remark 2.1(iii)]), we observe that the associated primes of $J_0 + cI$ have codimension at most $g+r$; the inclusion $I^2 \subset J_0 + cI$ can then be checked at such primes and the remaining assumption takes care of the rest. □

REMARK 4.8. *Another known result, but in analytic deviation two* [**16**, Theorem 3.1], *also yields to the same analysis.*

We shall use the higher degree components of the sequence (13) to give one of the main applications of the previous result. For an unmixed ideal I, of codimension g, we recall that its nth symbolic power is the intersection of the primary components of I^n of codimension g.

COROLLARY 4.9. *(*[**16**, Theorem 2.5]*) Let R be a Cohen–Macaulay local ring and I an analytic deviation one unmixed ideal of R. Assume height $I = g \geq 1$, I is generically a complete intersection. Then the following are equivalent:*

(a) $I^n = I^{(n)}$ *for every n;*

(b) $\ell(I_\wp) <$ *height $\wp$ for every $\wp \supset I$ of codimension $g+1$.*

PROOF. (a) $\Rightarrow$ (b): Follows from Burch's formula [**4**].

(b) $\Rightarrow$ (a): Let J be a reduction of I generated by $g+1$ elements. By Theorem 4.7, $I^2 = JI$. As in [**16**], we argue by induction on n, that is, assume that the equality $I^{(i)} = I^i$ holds for $i \leq n$ and that $I^{(n+1)}$ and I^{n+1} have the same primary components in the punctured spectrum of R. This implies that $I^{(n+1)} \subset I^n$, and we want to argue that $I^{(n+1)} \subset J^n$. To this end, we seek to embed $I^{(n+1)}$ into a subideal of J^n whose associated primes do not include the maximal ideal $\mathfrak{m}$. Using the notation of the previous proof and ([**16**, Remark 2.1(iii)]) we have that $\mathfrak{m}$ is not an associated prime of $J_0^n + cI^n$. This means that the inclusion

$$I^{(n+1)} \subset J_0^n + cI^n \subset J^n$$

can be checked on the punctured spectrum only. In the current case this chore is taken care by the hypotheses.

Finally, we may assume that

$$I^{(n+1)}/I^{n+1} \subset J^n/IJ^n.$$

From the degree n component of the approximation complex (13), $\mathfrak{m}$ is not an associated prime of J^n/IJ^n, which shows that $I^{(n+1)}/I^{n+1}$ must vanish. □

4.3. Cohen–Macaulay type. The following computation determines the Cohen–Macaulay type of the algebra $R[Jt]$.

THEOREM 4.10. *Let I be a Cohen–Macaulay ideal as in Theorem 4.2, and let J be a minimal reduction. Then $R[Jt]$ is a Cohen–Macaulay ring of type $\nu(I)-2$.*

PROOF. We have only to prove the statement about the Cohen–Macaulay type. Denote $S = R[Jt]$ and ω_S its canonical module, and recall the formula ([**12**, Theorem 6.5]):

$$\omega_S/S_+\omega_S = \bigoplus_{i=1}^{g} \operatorname{Ext}_R^{g-i}(Z_i, R),$$

where Z_i are the cycles of the Koszul complex on a minimal set of generators of J. A simple calculation shows:

$$\omega_S/S_+\omega_S \quad = \quad R \oplus (R/J)^{g-2} \oplus \operatorname{Ext}_R^{g+1}(R/J, R). \tag{16}$$

We must now produce the module $\operatorname{Ext}_R^{g+1}(R/J, R)$, for which we use an exact sequence in the proof of [**35**, Proposition 4]. The ideal J has obviously I for the intersection of the isolated primary components; a remainder of the intersection of the codimension $g+1$ components can be obtained as follows (cf. [**35**]). Let $J = (a_1, \ldots, a_g, a_{g+1})$ and assume that $J_0 = (a_1, \ldots, a_g)$ generates J at each minimal prime of I. Let $K = J_0 \colon I$; then a_{g+1} is regular on R/K and $J = I \cap L$, $L = (K, a_{g+1})$. There results the exact sequence of natural maps

$$0 \to R/J \longrightarrow R/I \oplus R/L \longrightarrow R/(I+L) \to 0.$$

Note that $(I, L) = (I, K)$, and that the latter is a Gorenstein ideal of codimension $g+1$ ([**30**, Remarque 1.4]). Consider now the long exact sequence of Ext's:

$$\begin{gathered}\operatorname{Ext}_R^g(R/(I+L), R) \to \operatorname{Ext}_R^g(R/I, R) \oplus \operatorname{Ext}_R^g(R/L, R) \to \operatorname{Ext}_R^g(R/J, R) \to \\ R/(I+L) \to \operatorname{Ext}_R^{g+1}(R/I, R) \oplus \operatorname{Ext}_R^{g+1}(R/L, R) \to \operatorname{Ext}_R^{g+1}(R/J, R) \to 0.\end{gathered}$$

From the proof of Proposition 4.1, the sequence above yields the short exact sequence

$$0 \to R/(I+L) \xrightarrow{\phi} \operatorname{Ext}_R^{g+1}(R/L, R) = I/(J_0, Ia_{g+1}) \to \operatorname{Ext}_R^{g+1}(R/J, R) \to 0.$$

As the class of a_{g+1} lies in the image of ϕ, $\operatorname{Ext}_R^{g+1}(R/J, R) = I/J$. Since J is generated by a subset of minimal generators of I, the Cohen–Macaulay type of $R[Jt]$ from equation (16) is given by the count

$$1 + (g-2) + \nu(I) - (g+1) = \nu(I) - 2,$$

as claimed. □

4.4. Koszul homology of reductions. The next result shows that a great deal of syzygetic information on an ideal is captured by some of its reductions.

THEOREM 4.11. *Let R be a local Cohen–Macaulay ring of dimension d, let I be an ideal, let J be a reduction of I with $\nu(J) = s \leq d$, and assume that I satisfies $\mathcal{F}_1$, locally in codimension $\leq s-1$. If I has sliding depth, then J has sliding depth, and in particular J is of linear type and $R[Jt]$ is Cohen–Macaulay.*

PROOF. We may assume that $g = \text{height } J < s \leq d$. Further notice that $I_\wp = J_\wp$ for all $\wp \in \text{Spec}(R)$ with $\dim R_\wp \leq s-1$.

At this juncture, it is no restriction to assume that the residue field of R is infinite. Now let $a_1, \ldots, a_s$ be a generating sequence of J, and for $g \leq i \leq s-1$ write $L_i = (a_1, \ldots, a_i)$ and $K_i = L_i\colon J$. Since J satisfies $\mathcal{F}_1$, we may choose $a_1, \ldots, a_s$ in such a way that $J_\wp = (L_i)_\wp$ for all $\wp \in \text{Spec}(R)$ with $\dim R_\wp \leq i-1$ and all $\wp \in V(J)$ with $\dim R_\wp \leq i$. In other words, height $K_i \geq i$ and height$(J+K_i) \geq i+1$. We claim that for $g \leq i \leq s-1$, R/K_i is Cohen–Macaulay of dimension $d-i$ and $J \cap K_i = L_i$. Once this is shown, the theorem will follow from [**14**, Theorem 3.4.$(b) \Rightarrow (a)$ and its proof].

To prove the claim notice that since $i \leq s-1$, we have $I_\wp = J_\wp = (L_i)_\wp$ for all $\wp \in \text{Spec}(R)$ with $\dim R_\wp \leq i-1$ and all $\wp \in V(I)$ with $\dim R_\wp \leq i$. Thus height $L_i\colon I \geq i$ and height$(I + (L_i\colon I)) \geq i+1$. However, $\nu(I_\wp) \leq \dim R_\wp$ for all $\wp \in V(I)$ with $\dim R_\wp \leq i \leq s-1$, and I is assumed to have sliding depth. In this situation, [**14**, Theorem 3.3] implies that $R/L_i\colon I$ is a Cohen–Macaulay ideal of dimension $d-i$ and $I \cap (L_i\colon I) = L_i$.

Now it suffices to prove that $L_i\colon I = K_i$. The inclusion $L_i\colon I \subset L_i\colon J = K_i$ being trivial, we only need to show the asserted equality at every associated prime $\wp$ of $L_i\colon I$. Since the latter ideal is Cohen–Macaulay of height i, we know that $\dim R_\wp = i \leq s-1$. Thus $I_\wp = J_\wp$, and $(L_i\colon I)_\wp = (K_i)_\wp$.

Finally, that $R[Jt]$ is Cohen–Macaulay follows from [**12**, Theorem 6.2]. □

COROLLARY 4.12. *Let I and J be as in* Theorem 4.11. *Let $\mathbb{K}$ be the Koszul complex on a minimal set of generators of J. Then I annihilates $H_i(\mathbb{K})$ for $i > 0$.*

PROOF. Since $I \cdot H_i(\mathbb{K}) \hookrightarrow H_i(\mathbb{K})$, and the latter is a module of depth $\geq d-s+i$, it suffices to check the prime ideals $\wp$ of codimension $s-i \leq s-1$. But in this range we have that $I_\wp = J_\wp$. □

This result provides for a broad generalization of Corollary 4.5:

COROLLARY 4.13. *Let R, I and J be as above. Suppose also that* depth $R/I \geq$ $\dim R - s$. *Then $I \cdot R[Jt]$ is a maximal Cohen–Macaulay module. If the reduction number $r_J(I) = 1$ then* $\text{gr}_I(R)$ *is Cohen–Macaulay.*

PROOF. As in the proof of Corollary 4.5 we use the approximation complex $\mathcal{M}(J)$ of J:

$$0 \to H_{s-g}(J) \otimes B[s-g] \to \cdots \to H_1(J) \otimes B[-1] \to H_0 \otimes B \to S(J/J^2) \to 0,$$

where each $H_i(J)$, $i > 0$, is actually an R/I–module. Tensoring the approximation complex of J with R/I gives rise to a complex very similar $\mathcal{M}(J)$; furthermore by the acyclicity lemma it is an exact complex, and as in the earlier argument, we conclude that $\mathrm{gr}_J(R) \otimes R/I$ is Cohen–Macaulay. □

COROLLARY 4.14. *Let R be a local Cohen–Macaulay ring of dimension d, with infinite residue field, and let I be an ideal with sliding depth. If I is of linear type in codimension h then $\ell(I) \geq \inf\{h+1, \nu(I)\}$.*

5. Cohen–Macaulay Rees Algebras

In this section the main strands of this paper, the Sally modules and the Koszul homology of reductions, come together to study the Cohen–Macaulayness of the Rees algebra of the ideal. The emphasis will still be on reduction number 1.

5.1. Main result. We shall now dig deeper into the properties of the approximation complexes to lift the assertion of Corollary 4.13 to the Rees algebra of the ideal.

THEOREM 5.1. *Let I and J be ideals as in* Theorem 4.11. *Assume also that depth $R/I \geq \dim R - s$. If I has codimension at least two and $r_J(I) = 1$ then $R[It]$ is a Cohen–Macaulay algebra.*

PROOF. It will be necessary to re-assemble some elements that appeared earlier. First, from Theorem 4.11, $R[Jt]$ is a Cohen–Macaulay ring, while by Corollary 4.13 $\mathrm{gr}_J(R) \otimes R/I$ is also Cohen–Macaulay. This implies that $I \cdot R[Jt] = I \cdot R[It]$ is a maximal Cohen–Macaulay module.

To show that $R[It]$ is Cohen–Macaulay, we are going to prove the vanishing of certain local cohomology modules. We may assume that $(R, \mathfrak{m})$ is a local ring and write $N = (R[It]_+, \mathfrak{m})$ for the maximal graded ideal of $R[It]$.

Let $\dim R = d$. Write also $L = It \cdot R[It]$ for the augmentation ideal of $R[It]$. Consider the local cohomology modules attached to the sequence (3):

$$0 \to H^d_N(R[It]) \longrightarrow H^d_N(R) \longrightarrow H^{d+1}_N(L) \longrightarrow H^{d+1}_N(R[It]) \to 0.$$

Because $H^d(R)$ is concentrated in degree 0, to show that $H^d_N(R[It])$ vanishes it suffices to show that it does not have any nonzero component in degree 0. From the sequence (2) we have the cohomology exact sequence

$$0 \to H^d_N(R[It]) \to H^d_N(\mathrm{gr}_I(R)) \to H^{d+1}_N(L[+1]) \to H^{d+1}_N(R[It]) \to 0,$$

and we are going to show that $H^d_N(\mathrm{gr}_I(R))$ has trivial components in non–negative degrees.

PROPOSITION 5.2. *Let $(R, \mathfrak{m})$ be a Cohen–Macaulay local ring, and let J be an ideal of codimension g, with sliding depth and satisfying the condition $\mathcal{F}_1$. Then $a(\mathrm{gr}_J(R)) \leq -g$.*

PROOF. Let $\nu(J) = t + g$. The hypotheses include depth $H_i(J) \geq \dim R - (g+t) + i$, which together with $\mathcal{F}_1$ implies that the approximation complex

$$0 \to H_t(J) \otimes B[-t] \xrightarrow{\varphi_t} \cdots \longrightarrow H_1(J) \otimes B[-1] \longrightarrow H_0 \otimes B \longrightarrow S(J/J^2) \to 0,$$

is acyclic and $S(J/J^2) = \mathrm{gr}_J(R)$ is Cohen–Macaulay.

We denote this algebra by G_0 and use the complex to estimate $a(G_0)$. (A similar calculation is carried out in [**10**] in case J is strongly Cohen–Macaulay.) Let $M_i = H_i \otimes B[-i]$ and let L_i be the image of φ_i. Note that M_i has dimension $d - g + (t + g) = d + t$ and depth at least $d - (t + g) + i + (t + g) = d + i$.

Decompose the approximation complex into short exact sequences

$$0 \to L_{i+1} \longrightarrow M_i \longrightarrow L_i \to 0.$$

As in [**10**], we obtain exact sequences

$$0 \to H_N^d(M_0)_j \to H_N^d(G_0)_j \to H_N^{d+1}(L_1)_j \tag{17}$$

$$0 \to H_N^{d+i}(M_i)_j \to H_N^{d+i}(L_i)_j \to H_N^{d+i+1}(L_{i+1})_j, \ 1 \leq i < t \tag{18}$$

$$H_N^{d+t}(M_t)_j \simeq H_N^{d+t}(L_t)_j, \tag{19}$$

from which we want to argue that $H_N^d(G_0)_j = 0$ for $j > -g$, where $n = t + g$. We first note that $H_N^{d+i}(M_i)_j = 0$ for $j > -n + i$. This a consequence of the duality theorem. Indeed, let $M = H \otimes R[T_1, \ldots, T_n]$, where H is an R-module of depth $\geq d - n + i$. Consider the exact sequence

$$0 \to M[-1] \longrightarrow M \longrightarrow M/T_1M \to 0.$$

Assume, by induction, that $H_N^{d+i-1}(M/T_1M)_j = 0$ for $j > -(n-1) + i$. In the cohomology sequence

$$0 \to H_N^{d+i-1}(M/T_1M)_j \longrightarrow H_N^{d+i}(M)_{j-1} \longrightarrow H_N^{d+i}(M)_j,$$

we obtain the inclusion

$$H_N^{d+i}(M)_{j-1} \hookrightarrow H_N^{d+i}(M)_j,$$

for $j > -(n-1) + i$. Since $H_N^{d+i}(M)_j = 0$ for $j \gg 0$, this implies our claim.

Now we use that

$$H_N^{d+i}(M_i)_j = 0 \text{ for } \ j > -n + i$$

in the sequences (17), (18) and (19) to complete the proof of the assertion. □

We arrive now at the core of the proof of the theorem. The approach is by connecting properties of the Sally module $S_J(I)$ with the associated graded $\mathrm{gr}_I(R)$. Consider the commutative diagram of exact sequences

$$\begin{array}{ccccccccc} 0 & \to & R[Jt] & \longrightarrow & R[It] & \longrightarrow & C & \to & 0 \\ & & \uparrow & & \uparrow & & \uparrow & & \\ 0 & \to & I\cdot R[Jt] & \longrightarrow & I\cdot R[It] & \longrightarrow & S_J(I) & \to & 0. \end{array}$$

From the snake lemma there is an acyclic complex

$$0 \to K_S \longrightarrow \mathrm{gr}_J(R)\otimes R/I \longrightarrow \mathrm{gr}_I(R) \longrightarrow K_C \to 0, \tag{20}$$

where K_S and K_C are defined from the natural mapping $S_J(I) \to C$:

$$0 \to K_S \longrightarrow S_J(I) \longrightarrow C \longrightarrow K_C \to 0. \tag{21}$$

PROPOSITION 5.3. *If $r_J(I) = 1$ there exist exact sequences of Cohen–Macaulay modules of dimension d*

$$0 \to gr_J(R)\otimes R/I \longrightarrow \mathrm{gr}_I(R) \longrightarrow C \to 0, \tag{22}$$

and

$$0 \to C[+1] \longrightarrow gr_J(R) \longrightarrow gr_J(R)\otimes R/I \to 0. \tag{23}$$

PROOF. It is clear from the sequences (20) and (21) and the proof of Corollary 4.13. □

We can now conclude the proof of the theorem. By Proposition 5.2, the a–invariant of of the mixed associated graded ring satisfies $a(\mathrm{gr}_J(R)) \leq -g \leq -2$. We take this into the sequence (23) and get that $H^d_N(C[+1])_j$ and $H^d_N(\mathrm{gr}_J(R)\otimes R/I)_j$ also vanish for $j > -g$. Finally, going over (22), we get that $H^d_N(\mathrm{gr}_I(R))_j = 0$ in non–negative degrees, which by the early part of the argument implies that $H^d(R[It])_0 = 0$, and therefore $R[It]$ is Cohen–Macaulay. □

5.2. Reduction number. We make some observations about what is required for the equation $r_J(I) = 1$ to hold.

THEOREM 5.4. *Let I and J be ideals as in* Theorem 4.11, *and suppose every associated prime ideal of I has codimension at most s. If for each prime ideal $\wp$ of codimension s the equality $I^2_\wp = (JI)_\wp$ holds then $I^2 = JI$.*

PROOF. It will be enough to show that the associated prime ideals of JI have codimension at most s. From the proofs of Corollaries 4.12 and 4.13 we get the exact sequence

$$0 \to H_1(J) \longrightarrow (R/I)^\ell \longrightarrow J/JI \to 0,$$

which will be combined with the sequence

$$0 \to J/JI \longrightarrow R/JI \longrightarrow R/J \to 0.$$

Since depth $H_1(J) \geq d-s+1$, it follows from the first sequence and the condition that $\dim R_\wp \leq s$ for every associated prime $\wp$ of I, that any prime in $\operatorname{Ass}(J/JI)$ has codimension at most s. The claim now follows from the second sequence, since the associated prime ideals of R/J have codimension at most s.

Alternatively, we can argue as follows to show the vanishing of the Sally module $S_J(I)$. In the exact sequence (1), as in Proposition 2.2, $I \cdot R[Jt]$ being a maximal Cohen–Macaulay module (and therefore an unmixed ideal of codimension one) implies that $S_J(I)$ either vanishes or has Krull dimension d. By induction on $\dim R$ we may assume that $I^2 = JI$ holds on the punctured spectrum of the local ring $(R, \mathfrak{m})$. This means that $S_J(I)$ is annihilated by some power of $\mathfrak{m}$ so that the dimension of $S_J(I)$ is at most $\nu(J) = s < d$, which is a contradiction unless $r_J(I) = 1$. □

5.3. Symbolic powers. The application of these techniques to the comparison of the ordinary and the symbolic powers of a prime ideal is straightforward:

THEOREM 5.5. *Let I and J be ideals as in* Theorem 4.11, *and assume further that $r_J(I) = 1$. Suppose I is a prime ideal and* depth $R/I \geq d - s$. *If for each prime ideal $\wp$, of codimension at most s, the powers $I_\wp^m$ are primary ideals then the ideals I^m are also primary.*

PROOF. This follows from the exact sequences (22) and (23): Together they say that the associated prime ideals of the conormal modules I^m/I^{m+1} have codimension at most s. This suffices, along with the hypothesis, to ensure that each such module is a torsion–free R/I–module. □

We shall consider another application (cf. [**36**, Theorem 4.1]):

THEOREM 5.6. *Let $(R, \mathfrak{m})$ be a Gorenstein local ring and let I be a Cohen–Macaulay ideal satisfying sliding depth which is generically a complete intersection. Suppose that for all integers m the conormal module I^m/I^{m+1} is torsion free as a R/I–module. Then I satisfies $\mathcal{F}_1$, it is strongly Cohen–Macaulay* (*and in particular its Rees algebra $R[It]$ is Cohen–Macaulay*), *and is generated by at most* $\dim R - 1$ *elements.*

PROOF. The main point is to show for each prime ideal $\wp$, which is not a minimal prime of I, the localization $I_\wp$ is generated by at most $\dim R_\wp - 1$ elements. We may assume that the assertion holds for the punctured spectrum of R. According to [**11**], I is generated by analytically independent elements on the punctured spectrum of R.

Harmlessly we may also assume that the residue field of R is infinite. Let J be a minimal reduction of I; by [**4**], $\ell(I) \leq \dim R - 1$, so that $\nu(J) \leq \dim R - 1$. By Theorem 4.11 however J satisfies sliding depth so that its associated primes have codimension at most $\nu(J)$ and therefore $\mathfrak{m}$ cannot be one of them. By the induction hypothesis this implies that $I = J$. Now we use [**14**, Theorem 1.4] to get that I is strongly Cohen–Macaulay. The assertion about the Rees algebra will then follow from [**11**]. □

REFERENCES

1. M. Artin and M. Nagata, Residual intersections in Cohen–Macaulay rings, J. Math. Kyoto Univ. **12** (1972), 307–323.
2. M. Brodmann, Rees rings and form rings of almost complete intersections, Nagoya Math. J. **88** (1982) 1–16.
3. P. Brumatti, A. Simis and W. V. Vasconcelos, Normal Rees algebras, J. Algebra **112** (1988), 26–48.
4. L. Burch, Codimension and analytic spread, Proc. Camb. Phil. Soc. **72** (1972), 369–373.
5. R. C. Cowsik and M. V. Nori, On the fibers of blowing up, J. Indian Math. Soc. **40** (1976), 217–222.
6. S. Goto and Y. Shimoda, On the Rees algebras of Cohen–Macaulay local rings, Lecture Notes in Pure & Applied Mathematics **68**, Marcel Dekker, New York, 1979, 201–231.
7. S. Goto and K. Watanabe, On graded rings I, J. Math. Soc. Japan **30** (1978), 179–213.
8. U. Grothe, M. Herrmann and U. Orbanz, Graded rings associated to equimultiple ideals, Math. Z. **186** (1984), 531–556.
9. A. Grothendieck, *Local Cohomology*, (Notes by R. Hartshorne) Lecture Notes in Mathematics **41**, Springer–Verlag, Berlin–Heidelberg–New York, 1967.
10. M. Herrmann, J. Ribbe and S. Zarzuela, On the Gorenstein property of Rees and form rings of powers of ideals, Trans. Amer. Math. Soc., to appear.
11. J. Herzog, A. Simis and W. V. Vasconcelos, Koszul homology and blowing–up rings, in *Commutative Algebra*, Proceedings: Trento 1981 (S. Greco and G. Valla, Eds.), Lecture Notes in Pure and Applied Math. **84**, Marcel Dekker, New York, 1983, 79–169.
12. J. Herzog, A. Simis and W. V. Vasconcelos, On the arithmetic and homology of algebras of linear type, Trans. Amer. Math. Soc. **283** (1984), 661–683.
13. J. Herzog, A. Simis and W. V. Vasconcelos, On the canonical module of the Rees algebra and the associated graded ring of an ideal, J. Algebra **105** (1987), 285–302.
14. J. Herzog, W. V. Vasconcelos and R. Villarreal, Ideals with sliding depth, Nagoya Math. J. **99** (1985), 159–172.
15. S. Huckaba, On graded rings associated to analytic deviation one ideals of small height, Preprint, 1991.
16. S. Huckaba and C. Huneke, Powers of ideals having small analytic deviation, American J. Math. **114** (1992), 367–403.
17. S. Huckaba and C. Huneke, Rees algebras of ideals having small analytic deviation, Trans. Amer. Math. Soc., to appear.
18. C. Huneke, The theory of d–sequences and powers of ideals, Advances in Math. **46** (1982), 249–279.
19. C. Huneke, On the associated graded ring of an ideal, Illinois J. Math. **26** (1982), 121-137.
20. C. Huneke, Linkage and Koszul homology of ideals, American J. Math. **104** (1982), 1043–1062.
21. C. Huneke, Strongly Cohen–Macaulay schemes and residual intersections, Trans. Amer. Math. Soc. **277** (1983), 739–763.
22. C. Huneke, Hilbert functions and symbolic powers, Michigan Math. J. **34** (1987), 293–318.
23. C. Huneke, Symbolic powers of prime ideals and special graded algebras, Comm. Algebra **9** (1981), 339–366.
24. C. Huneke and B. Ulrich, Residual intersections, J. reine angew. Math. **390** (1988), 1–20.
25. S. Ikeda and N. V. Trung, When is the Rees algebra Cohen–Macaulay?, Comm. Algebra **17** (1989), 2893–2922.
26. M. Narita, A note on the coefficients of Hilbert characteristic functions in semi–regular local rings, Proc. Camb. Phil. Soc. **59** (1963), 269–275.
27. D. G. Northcott, A note on the coefficients of the abstract Hilbert function, J. London Math. Soc. **35** (1960), 209–214.

28. D. G. Northcott and D. Rees, Reductions of ideals in local rings, Proc. Camb. Phil. Soc. **50** (1954), 145–158.
29. A. Ooishi, δ-genera and sectional genera of commutative rings, Hiroshima Math. J. **17** (1987), 361–372.
30. C. Peskine and L. Szpiro, Liaison des variétés algébriques, Invent. Math. **26** (1974), 271–302.
31. J. D. Sally, On the associated graded ring of a local Cohen–Macaulay ring, J. Math. Kyoto U. **17** (1977), 19–21.
32. J. D. Sally, Cohen–Macaulay local rings of maximal embedding dimension, J. Algebra **56** (1979), 168–183.
33. J. D. Sally, Tangent cones at Gorenstein singularities, Compositio Math. **40** (1980), 167–175.
34. J. D. Sally, Hilbert coefficients and reduction number 2, J. Alg. Geo. and Sing. **1** (1992), 325–333.
35. B. Ulrich, Remarks on residual intersections, in *Free Resolutions in Commutative Algebra and Algebraic Geometry*, Proceedings, Sundance 1990 (D. Eisenbud and C. Huneke, Eds.), Research Notes in Mathematics **2**, Jones and Bartlett Publishers, Boston–London, 1992, 133–138.
36. B. Ulrich and W. V. Vasconcelos, The equations of Rees algebras of ideals with linear presentation, Math. Z., to appear.
37. P. Valabrega and G. Valla, Form rings and regular sequences, Nagoya Math. J. **72** (1978), 91–101.

DEPARTMENT OF MATHEMATICS, RUTGERS UNIVERSITY, NEW BRUNSWICK, NEW JERSEY 08903

E-mail address: vasconce@rings.rutgers.edu

Contemporary Mathematics
Volume 159, 1994

Infinite Cyclic Covers of strongly F-regular Rings

KEI-ICHI WATANABE

ABSTRACT. We will show that the infinite anti-canonical cyclic cover (the symbolic Rees algebra of a divisorial ideal whose inverse is isomorphic to a canonical module) of a strongly F-regular (resp. F-pure) ring is F-regular and Gorenstein (resp. F-pure), if the Rees algebra is Noetherian. To prove this, we calculate the highest local cohomology group of this Rees algebra explicitly. We will give examples showing that this Rees algebra is not Cohen-Macaulay in general, even if the base ring is F-rational or a rational singularity.

Introduction

The concept of F-regular ring was introduced by M. Hochster and C. Huneke for rings of characteristic $p > 0$ with several variations (weakly, strongy F-regular rings) using the Frobenius endomorphism [6], [7]. Also, the concept of F-rational ring was introduced in [1] and F-rational rings are supposed to correspond with rational singularities in characteristic 0 [1], [10].

On the other hand, the concept of F-regular ring is strictly stronger than that of F-rational ring (or that of rational singularities) and is rather mysterious from the viewpoint of "classifying the singularities". One of the characteristic properties of F-regular (or F-pure) rings is that it is inherited by pure subrings. Also, for Gorenstein rings, the concepts F-regular (strongly or weakly) and F-rational are equivalent.

The concept of F-pure ring was introduced by M. Hochster and L. Roberts in [7] and has played an important role. It is known that F-pure rings form a class slightly wider than that of F-regular rings and includes certain "elliptic singularities". For example, in dimension 2, roughly speaking, "F-regular" is

1991 *Mathematics Subject Classification.* Primary 13H10, 14B05.
Key words and phrases. Infinite Cyclic Cover, Symbolic Rees Algebra, F-regular Ring.
This paper is in final form and no version of it will be submitted for publication elsewhere.

equivalent to "quotient singularity" and a F-pure and Gorenstein ring is either F-regular, a "simple elliptic singularity" or a "cusp singularity". (cf. [8], [14], [15] for precise statements.)

If a normal local ring A has a canonical module K_A and if its class $cl(K_A)$ has a finite order in the divisor class group $Cl(A)$ of A, we can take the canonical cover S of A. In [15], we have seen that if the order $\mathrm{ord}(cl(K_A)) = r$ of the canonical class is not divisible by p, then A is strongly F-regular (resp. F-pure) if and only if so is S. Thus if A is strongly F-regular, A is a pure subring of an F-regular Gorenstein ring since S is Gorenstein.

Then we would ask, "What if the canonical class has infinite order ?" and the aim of this paper is to answer this question under the assumption that the Rees algebra we are going to discuss is Noetherian. Namely, our main theorem in this paper is the following.

THEOREM 0.1. *Let $(A, \mathfrak{m})$ be a normal local ring (or a positively graded normal ring over a field) of characteristic $p > 0$ with canonical module K_A and let I be a reflexive ideal of A such that $I^{-1} \cong K_A$. Let $\mathcal{R} = \oplus_{n \geq 0} I^{(n)} t^n$ be the symbolic Rees algebra of I.*

If $\mathcal{R}$ is Noetherian and A is strongly F-regular (resp. F-pure), then $\mathcal{R}$ is strongly F-regular and Gorenstein (resp. F-pure).

We will call such $\mathcal{R}$ an *infinite anti-canonical cover of A*. It is shown that anti-canonical covers are always quasi-Gorenstein ($\mathcal{R}$ has a trivial canonical module but not necessarily Cohen-Macaulay) if it is Noetherian (cf. [11], [4], [12]).

To prove the theorem, we write down the highest local cohomology group $H^{d+1}_{\mathfrak{M}}(\mathcal{R})$ of $\mathcal{R}$ with respect to the unique graded maximal ideal $\mathfrak{M}$ of $\mathcal{R}$ explicitly, so that we can write down the action of Frobenius on $H^{d+1}_{\mathfrak{M}}(\mathcal{R})$ in terms of the action on the injective envelope of $A/\mathfrak{m}$. Since $H^{d+1}_{\mathfrak{M}}(\mathcal{R})$ is an injective module by our choice of I, we conclude that $\mathcal{R}$ is strongly F-regular (resp. F-pure) if so is A.

Finally, we will give examples of infinite anti-canonical cover $\mathcal{R}$ of A, where $\mathcal{R}$ is Gorenstein and A is not Cohen-Macaulay, or A is F-rational (or a rational singularity) and $\mathcal{R}$ is not Cohen-Macaulay without assuming F-regularity.

§1. Preliminary

(1.1) Let A be a Noetherian ring of characteristic $p > 0$ and $F : A \to A$ be the Frobenius map defined by $F(a) = a^p$. Throughout this paper, the letter q always means a power $q = p^e$ of p and we will assume that A is an *integral domain* since we always assume that A is a normal domain. We write the destination of F^e as ${}^eA$ or $A^{1/q}$.

Also, *we will always assume that F is a finite morphism.* (We say that A is *F-finite* in this case).

For an A-module M, we define the Frobenius action on M by

$$F^e : M = M \otimes_A A \to F^e(M) := M \otimes_A {}^eA \tag{1.1.1}$$

where $F^e(x) = x \otimes 1$. This corresponds to the canonical map

$$M = M \otimes_A A \to M \otimes_A A^{1/q}, \tag{1.1.2}$$

sending $x \in M$ to $x \otimes 1$. We define the A-module structure of $F^e(M)$ via ${}^eA$, so that we have $F^e(ax) = a^q.F^e(x)$. In the notation as in (1.1.2), $a.F^e(x)$ corresponds to $x \otimes a^{1/q}$.

When we write $(A, \mathfrak{m})$, we always mean either A is local with maximal ideal $\mathfrak{m}$ or $A = \bigoplus_{n\geq 0} A_n$ is a graded ring with $A_0 = k$ a field and $\mathfrak{m} = A_+ = \bigoplus_{n>0} A_n$. In the latter case, we will say that A is a *special graded ring.* If A is special graded, we think $A^{1/q}$ as a $\frac{1}{q}\mathbb{Z}$-graded ring. If M is a graded A-module, $F^e(M_n) \subseteq (F^e(M))_{qn}$ in (1.1.1), while in (1.1.2), the map preserves the grading.

In some cases we will only discuss the case where A is local and omit the case A is special graded since the arguments are quite similar.

DEFINITION 1.2. ([6], [7]) We say that A is strongly F-regular (resp. F-pure) if for every $c \neq 0$ in A, we can take q so that the A-homomorphism $A \to A^{1/q}$, sending 1 to $c^{1/q}$ (resp. the inclusion $A \to A^{1/p}$) splits as an A-module.

If $(A, \mathfrak{m})$ is local or special graded, we have the following criteria for strongly F-regular rings using the Frobenius action on the injective envelope $E = E_A(A/\mathfrak{m})$ of the residue field.

PROPOSITION 1.3. (cf. [5], 8.23[1], [16], 2.6) *Let $(A, \mathfrak{m})$ be an F-finite local ring, $E = E_A(A/\mathfrak{m})$ and z be the generator of the socle of E. Then A is strongly F-regular (resp. F-pure) if and only if for every $c \neq 0$ in A, we can take e so that $c.F^e(z) \neq 0$ in $F^e(E)$ (resp. $F(z) \neq 0$).*

(This fact is heavily used in [16]. So, we must replace "F-regular" (or "weakly F-regular") by "strongly F-regular" everywhere in [16]. I hope there will no differences between these concepts (under some mild conditions) in near future.)

The action of Frobenius on $E = E_A(A/\mathfrak{m})$ can be described as one on the highest local cohomology module of the canonical module K_A of A. Let us review some results from [16].

PROPOSITION 1.4. ([16], 2.4, 2.5) *Let $(A, \mathfrak{m})$ be a normal local ring with $\dim A = d$ and I be an ideal of A of pure height 1. Then*

$$F^e(H^d_{\mathfrak{m}}(I)) \cong H^d_{\mathfrak{m}}(I^{(q)}) \qquad (or,\ \ H^d_{\mathfrak{m}}(I) \otimes_A A^{1/q} \cong H^d_{\mathfrak{m}}((I^{(q)})^{1/q})),$$

where, $I^{(q)}$ is the reflexive hull of I^q and $(I^{(q)})^{1/q}$ means $I^{(q)}$ as an ideal of $A^{1/q}$.

[1]An analogous statement for weakly F-regular rings is proved in this reference. The proof of this statement is essentially the same as that one.

In particular, we have $F^e(E) \cong H^d_{\mathfrak{m}}(K_A^{(q)})$ *(or,* $H^d_{\mathfrak{m}}(K_A) \otimes_A A^{1/q} \cong H^d_{\mathfrak{m}}((K_A^{(q)})^{1/q})$*).*

The same conclusion holds if A *is a special graded ring and if we assume that* I *is homogeneous.*

If, moreover, A is strongly F-regular and the canonical class $cl(K_A)$ has a finite order r in the divisor class group $Cl(A)$, "canonical cover" and "anti-canonical cover" become the same thing and are F-regular (and Gorenstein) if $(p, r) = 1$. In this respect, we refer to a result about finite covers from [16].

PROPOSITION 1.5. ([16], 2.7, 2.9) *Let* $(A, \mathfrak{m})$ *be a strongly F-regular (resp. F-pure) local ring (or, a special graded ring) and* I *be a (resp. homogeneous) divisorial ideal of* A *with* $\operatorname{ord}(cl(I)) = r$ *in* $Cl(A)$. *If* p *does not divide* r, *then the cyclic cover*

$$B = \bigoplus_{0 \leq n < r} I^{(n)}$$

is strongly F-regular (resp. F-pure).

§2. Highest local cohomology module of an infinite cyclic cover

(2.1) In this section, let $(A, \mathfrak{m})$ be a universally catenary Noetherian normal local domain with $\dim A = d \geq 2$ and $I = A(D)$ be a divisorial ideal of A, where $D \in Div(A)$ is a divisor of A with $D \leq 0$. (We denote

$$A(D) = \{x \in K \mid div_A(x) + D \geq 0\}$$

for a divisor D of A.) The symbolic n-th power $I^{(n)}$ of I is the reflexive hull of I^n and we have $I^{(n)} = A(nD)$. We define

$$\mathcal{R} = \mathcal{R}(I) = \bigoplus_{n \geq 0} I^{(n)} t^n = \bigoplus_{n \geq 0} A(nD) t^n$$

and

$$\mathcal{R}' = \mathcal{R}'(I) = \bigoplus_{n \in \mathbb{Z}} I^{(n)} t^n = \bigoplus_{n \in \mathbb{Z}} A(nD) t^n.$$

(Note that the component of degree $n < 0$ of $\mathcal{R}'$ is *not* A as is defined in many literatures.) We denote by

$$\mathfrak{M} = \mathfrak{m}\mathcal{R} + \mathcal{R}_+$$

the unique graded maximal ideal of $\mathcal{R}$. The aim of this section is to calculate the highest local cohomology group $H^{d+1}_{\mathfrak{M}}(\mathcal{R})$ when $\mathcal{R}$ is Noetherian.

THEOREM 2.2. *If* $\mathcal{R} = \mathcal{R}(I)$ *is Noetherian, then*

$$H^{d+1}_{\mathfrak{M}}(\mathcal{R}) \cong \bigoplus_{n<0} H^d_{\mathfrak{m}}(I^{(n)}) t^n \cong \bigoplus_{n<0} H^d_{\mathfrak{m}}(A(nD)) t^n.$$

The following fact is crucial for this result.

PROPOSITION 2.3. ([4],3.2) *If $\mathcal{R}$ is Noetherian, then for every prime ideal $\mathfrak{p}$ with* $\mathrm{ht}(\mathfrak{p}) \geq 2$, *we have*

$$\dim((\mathcal{R}/\mathfrak{p}\mathcal{R})_{\mathfrak{p}}) < \mathrm{ht}(\mathfrak{p}).$$

Since (2.3) is not explicit in the reference, I will give some explanation. The reference asserts that if $\mathcal{R}$ is Noetherian, then there exists a positive integer r such that $I^{(rn)} = (I^{(r)})^n$ for every $n \geq 0$ and that

$$\lambda(I^{(r)}.A_{\mathfrak{p}}) < \mathrm{ht}(\mathfrak{p})$$

for every prime ideal $\mathfrak{p} \subseteq I$ of A, where $\lambda(I)$ means the analytic spread of I. Since

$$\lambda(I^{(r)}.A_{\mathfrak{p}}) = \dim(\mathcal{R}(I^{(r)}.A_{\mathfrak{p}})/\mathfrak{p}.\mathcal{R}(I^{(r)}.A_{\mathfrak{p}})),$$

and

$$\mathcal{R}(I^{(r)}) = \mathcal{R}(I)^{(r)} = \bigoplus_{n\geq 0}\mathcal{R}_{nr},$$

we have

$$\dim((\mathcal{R}/\mathfrak{p}\mathcal{R})_{\mathfrak{p}}) = \lambda(I^{(r)}.A_{\mathfrak{p}}) < \mathrm{ht}(\mathfrak{p}).$$

Now, we will prove (2.2) in several steps.

(2.4) Consider the exact sequence

(2.4.1) $$0 \to \mathcal{R}(I) \to \mathcal{R}'(I) \to \mathcal{R}'(I)/\mathcal{R}(I) \cong \bigoplus_{n<0} I^{(n)}t^n \to 0.$$

Since the action of $\mathcal{R}_+$ on $\mathcal{R}'(I)/\mathcal{R}(I)$ is elementwise nilpotent, we have

$$H^d_{\mathfrak{M}}(\mathcal{R}'(I)/\mathcal{R}(I)) \cong \bigoplus_{n<0} H^d_{\mathfrak{m}}(I^{(n)})t^n.$$

Thus it suffices to prove $H^d_{\mathfrak{M}}(\mathcal{R}') = H^{d+1}_{\mathfrak{M}}(\mathcal{R}') = 0$ to show (2.2).

(2.5) Next, we put $X = \mathrm{Proj}(\mathcal{R})$ and $O_X(n) = (\mathcal{R}(n))^{\sim}$, following the standard notation. Since $H^0_{\mathfrak{M}} = H^0_{\mathfrak{m}} \circ H^0_{\mathcal{R}_+}$ on the category of graded $\mathcal{R}$-modules, we have a spectral sequence for composite functors;

(2.5.1) $$E_2^{p,q} = H^p_{\mathfrak{m}}(H^q_{\mathcal{R}_+}(\mathcal{R}')) \Longrightarrow H^{p+q}_{\mathfrak{M}}(\mathcal{R}')$$

and we can divide our calculation into those of $H^q_{\mathcal{R}_+}$ and $H^p_{\mathfrak{m}}$. First, we will discuss $H^q_{\mathcal{R}_+}$.

From the exact sequence (2.4.1), we have the exact sequence

(2.5.2) $$0 \to H^0_{\mathcal{R}_+}(\mathcal{R}'/\mathcal{R}) = \mathcal{R}'/\mathcal{R} \to H^1_{\mathcal{R}_+}(\mathcal{R}) \to H^1_{\mathcal{R}_+}(\mathcal{R}') \to 0$$

and the isomorphisms

(2.5.3) $$H^p_{\mathcal{R}_+}(\mathcal{R}) \cong H^p_{\mathcal{R}_+}(\mathcal{R}') \qquad (p \geq 2)$$

since $H^p_{\mathcal{R}_+}(\mathcal{R}'/\mathcal{R}) = 0$ for $p \geq 1$. On the other hand, by the general theory of Proj of a graded ring, we have an exact sequence

$$(2.5.4) \qquad 0 = H^0_{\mathcal{R}_+}(\mathcal{R}) \to \mathcal{R} \to \bigoplus_{n\in\mathbb{Z}} H^0(X, O_X(n)) \to H^1_{\mathcal{R}_+}(\mathcal{R}) \to 0$$

and isomorphisms

$$(2.5.5) \qquad \bigoplus_{n\in\mathbb{Z}} H^p(X, O_X(n)) \cong H^{p+1}_{\mathcal{R}_+}(\mathcal{R}) \qquad (p \geq 1).$$

Since $\mathcal{R}$ (resp. $\mathcal{R}'$) is normal, $\mathcal{R} \cong \bigoplus_{n\geq 0} H^0(X, O_X(n))$ (resp. $\mathcal{R}' \cong \bigoplus_{n\in\mathbb{Z}} H^0(X, O_X(n))$) and then $H^1_{\mathcal{R}_+}(\mathcal{R}) \cong \bigoplus_{n<0} H^0(X, O_X(n)) \cong \mathcal{R}'/\mathcal{R}$. From this fact and (2.5.2) we conclude

$$(2.5.6) \qquad H^1_{\mathcal{R}_+}(\mathcal{R}') = 0$$

and from (2.5.3) and (2.5.5), we have

$$(2.5.7) \qquad H^p_{\mathcal{R}_+}(\mathcal{R}') \cong \bigoplus_{n\in\mathbb{Z}} H^{p-1}(X, O_X(n)) \qquad (p \geq 2).$$

(2.6) Now, we examine the canonical morphism $\phi : X \to Spec(A)$ of "symbolic blowing-up" of I. The fibre $\phi^{-1}(\mathfrak{p})$ of ϕ at $\mathfrak{p}$ is $\mathrm{Proj}(\mathcal{R} \otimes_A k(\mathfrak{p}))$ and $\dim\phi^{-1}(\mathfrak{p}) \leq \mathrm{ht}(\mathfrak{p}) - 2$ if $\mathrm{ht}(\mathfrak{p}) \geq 2$ by (2.3) (and ϕ is an isomorphism at $\mathfrak{p}$ if $ht(\mathfrak{p}) = 1$ since $I \cdot A_{\mathfrak{p}}$ is principal). Thus, if $\mathfrak{p} \in \mathrm{supp}(H^q(X, O_X(n)))$ for some $\mathfrak{p}, q > 0$ and $n \in \mathbb{Z}$, then $\dim\phi^{-1}(\mathfrak{p}) \geq q$, hence $\mathrm{ht}(\mathfrak{p}) \geq q + 2$. This implies

$$\dim(\mathrm{supp}(H^q(X, O_X(n)))) \leq \dim A - q - 2 \text{ for } q \geq 1 \text{ and for every } n \in \mathbb{Z}$$

(by our assumption, $\dim(\mathcal{R}/\mathfrak{P}) + \mathrm{ht}(\mathfrak{P}) = d+1$ for every prime ideal $\mathfrak{P}$ of $\mathcal{R}$), hence

$$(2.6.1) \qquad H^p_{\mathfrak{m}}(H^q(X, O_X(n))) = 0 \text{ for } p + q \geq d - 1 \text{ and for every } n \in \mathbb{Z}.$$

Comparing this and (2.5.7), we get $H^p_{\mathfrak{M}}(\mathcal{R}') = 0$ for $p = d, d+1$ by (2.5.1) and the proof of Theorem (2.2) is now complete.

§3. F-regularity of infinite anti-canonical cover

(3.1) The canonical class of a normal Rees algebra is computed in [11], 3.3 or [4], 4.5 and in the case of anti-canonical cover, we always have

$$(3.1.1) \qquad K_{\mathcal{R}} \cong \mathcal{R}(-1).$$

This implies that $H^{d+1}_{\mathfrak{M}}(\mathcal{R})$ is isomorphic to the injective envelope of $\mathcal{R}/\mathfrak{M}$. By (1.3), to prove that our anti-canonical cover $\mathcal{R}(I)$ is strongly F-regular, we have only to examine the action of Frobenius on $H^{d+1}_{\mathfrak{M}}(\mathcal{R})$.

(3.2) In the proof of (2.2), we have shown $H^i_{\mathfrak{M}}(\mathcal{R}') = 0$ for i = d, d+1. Hence we have the following isomorphism.

(3.2.1)

$$H^{d+1}_{\mathfrak{M}}(\mathcal{R}) \otimes_{\mathcal{R}} \mathcal{R}^{1/p} \cong H^{d+1}_{\mathfrak{M}}(\mathcal{R}^{1/p}) \cong H^d_{\mathfrak{M}}((\mathcal{R}')^{1/p}/\mathcal{R}^{1/p})$$
$$\cong \bigoplus_{n<0} H^d_{\mathfrak{m}}((I^{(n)})^{1/p})t^{n/p}.$$

(3.3) By (3.2.1), the Frobenius action on $H^{d+1}_{\mathfrak{M}}(\mathcal{R}) \cong \bigoplus_{n<0} H^d_{\mathfrak{m}}((I^{(n)})^{1/p})t^{n/p}$ is given by $F(\zeta \cdot t^n) = F(\zeta) \cdot t^{pn}$, where $\zeta \in H^d_{\mathfrak{m}}(I^{(n)})$ and $F(\zeta) \in H^d_{\mathfrak{m}}(I^{(n)}) \otimes_A A^{1/p} \cong H^d_{\mathfrak{m}}((I^{(pn)})^{1/p})$.

(3.4) (*Proof of* Theorem (0.1)) Let ζ be the generator of the socle of $H^d_{\mathfrak{m}}(I^{(-1)}) \cong H^d_{\mathfrak{m}}(K_A) \cong E_A(A/\mathfrak{m})$. Then the socle of $H^{d+1}_{\mathfrak{M}}(\mathcal{R})$ is generated by $\zeta \cdot t^{-1}$. If A is F-pure then $F(\zeta) \neq 0$ and $F(\zeta \cdot t^{-1}) = F(\zeta) \cdot t^{-p} \neq 0$. Then $\mathcal{R}$ is F-pure by (1.3).

If A is strongly F-regular, for every $c \neq 0$ in A, $c \cdot F^e(\zeta) \neq 0$ for some $e > 0$. (Note that if $c \cdot F^e(\zeta) \neq 0$, $c \cdot F^{e+1}(\zeta) \neq 0$ since $F(c \cdot F^e(\zeta)) = c^p \cdot F^{e+1}(\zeta) \neq 0$.) Take $c \cdot t^n \in \mathcal{R}(I)$ $(c \in I^{(n)})$ and $e > 0$ so that $c \cdot F^e(\zeta) \neq 0$ in $H^d_{\mathfrak{m}}(I^{(-q)})$ $(q = p^e)$ and $q > n$. Then $(c \cdot t^n) \cdot F^e(\zeta \cdot t^{-1}) = (c \cdot F^e(\zeta)) \cdot t^{n-q} \in H^d_{\mathfrak{m}}(I^{(n-q)}) \cdot t^{n-q}$ and $c \cdot F^e(\zeta) \neq 0$ in $H^d_{\mathfrak{m}}(I^{(n-q)})$ since the homothety by c on $I^{(-q)}$ is the composition of the multiplication map by c from $I^{(-q)}$ to $I^{(n-q)}$ and the inclusion map from $I^{(n-q)}$ to $I^{(-q)}$. Thus we have proved the condition for $\mathcal{R}(I)$ to be strongly F-regular given in (1.3).

§4. Examples

A typical example of infinite anti-canonical cover is

EXAMPLE 4.1. Let k be a field and S be a polynomial ring over k with variables $\{X_{ij} \mid 1 \leq i \leq n,\ 1 \leq j \leq n-1\}$. Let J be the ideal of S generated by all the 2 by 2 minors of $\{X_{ij}\}$ and put $A = S/J$. Then A is isomorphic to the "*Segre product*" of two polynomial rings $k[X_1, \cdots, X_n]$ and $k[Y_1, \cdots, Y_{n-1}]$. (The concept of Segre products is explained in (4.2).) In this case the canonical module K_A is generated by $\{1 \otimes Y_1, \cdots, 1 \otimes Y_{n-1}\}$ and we can take the ideal I generated by $\{X_1 \otimes 1, \cdots, X_n \otimes 1\}$ as an anti-canonical ideal. In this case, the anti-canonical cover $\mathcal{R}(I)$ is isomorphic to the Segre product of $k[X_1, \cdots, X_n]$ and $k[Y_1, \cdots, Y_{n-1}, t]$, which is Gorenstein and both A and $\mathcal{R}(I)$ are F-regular, being pure subrings of polynomial rings.

It turns out that the concept of the Segre product is very useful to make various examples of Cohen-Macaulay or non-Cohen-Macaulay anti-canonical covers. So, let us recall this concept briefly.

(4.2) Given two graded rings $R = \bigoplus_{n\geq 0} R_n$ and $R' = \bigoplus_{n\geq 0} R'_n$ with $R_0 = R'_0 = k$, a field, we define the *Segre product* $S = R \sharp R'$ of R and R' by

$$S = R \sharp R' = \bigoplus_{n \geq 0} (R_n \otimes_k R'_n).$$

We will denote the unique graded maximal ideal of R (resp. R', S) by $\mathfrak{m}$ (resp. $\mathfrak{m}'$, $\mathfrak{M}_S$). $\quad (\mathfrak{m} = R_+ = \bigoplus_{n>0} R_n$ etc.)

Also, we define the Segre product $M \sharp M'$ of a graded R-module M and a graded R'-module M' by

$$M \sharp M' = \bigoplus_{n \in \mathbb{Z}} (M_n \otimes_k M'_n).$$

The local cohomology group $H^p_{\mathfrak{M}_S}(S) \quad (p \geq 0)$ is given by

$$(4.2.1) \quad H^p_{\mathfrak{M}_S}(S) \cong (R \sharp H^p_{\mathfrak{m}'}(R')) \oplus (H^p_{\mathfrak{m}}(R) \sharp R') \oplus \Big(\bigoplus_{i+j=p+1} H^i_{\mathfrak{m}}(R) \sharp H^j_{\mathfrak{m}'}(R')\Big)$$

under the assumption $\operatorname{depth}(R) \geq 2$ and $\operatorname{depth}(R') \geq 2$. The canonical module K_S of S is given by

$$(4.2.2) \qquad K_S \cong K_R \sharp K_{R'}.$$

For the proof, we refer to [3], Chapter 4.

In particular, if R and R' are quasi-Gorenstein with canonical modules $R(a)$ and $R'(a')$, respectively, we have $K_S \cong R(a) \sharp R'(a')$ and we can take $R \sharp R'(n)$ with $n = a - a'$ as an anti-canonical ideal of S. If we put $I = R \sharp R'(n)$, then $I^{(m)} \cong R \sharp R'(mn)$. When $n = -1$, the symbolic Rees algebra $\mathcal{R}(I)$ (which we denote by T or $(R \otimes_k R')_{\geq}$) becomes

$$(4.2.3) \qquad T = (R \otimes_k R')_{\geq} = \mathcal{R}(I) = \bigoplus_{n \geq 0} \Big(\bigoplus_{n \geq m \geq 0} (R_n \otimes_k R'_m)\Big).$$

(4.3) The local cohomology modules of T with respect to the unique graded maximal ideal $\mathfrak{M}_T = \bigoplus_{n>0} \Big(\bigoplus_{0 \leq m \leq n} (R_n \otimes_k R'_m)\Big)$ is calculated in [9]. Namely, if $\operatorname{depth} R \geq 2$, we have an exact sequence of graded T-modules

$$(4.3.1) \qquad \begin{aligned} 0 &\to (H^p_{\mathfrak{m}}(R) \otimes_k R')_{\geq} \to H^p_{\mathfrak{M}_T}(T) \\ &\to \bigoplus_{i+j=p} (H^i_{\mathfrak{m}}(R) \otimes_k H^j_{\mathfrak{m}'}(R'))_{<} \to 0 \quad (p \geq 0), \end{aligned}$$

where, for a graded R-module M and graded R'-module M', we put

$$(M \otimes_k M')_{\geq} = \bigoplus_{n \in \mathbb{Z}} \Big(\bigoplus_{n \geq m} (M_n \otimes_k M'_m)\Big) \text{ and } (M \otimes_k M')_{<} = \bigoplus_{n \in \mathbb{Z}} \Big(\bigoplus_{n < m} (M_n \otimes_k M'_m)\Big)$$

so that the following is an exact sequence of graded T-modules:

$$0 \to (M \otimes_k M')_{\geq} \to M \otimes_k M' \to (M \otimes_k M')_{<} \to 0.$$

Taking various R and R', we can give several kinds of examples of $\mathcal{R}(I)$.

EXAMPLE 4.4. ("A is F-rational (or a rational singularity)" does not imply "$\mathcal{R}(I)$ is Cohen-Macaulay".) Let $A = S = R \sharp R'$, where

$$R = k[X, Y, Z]/(X^4 + Y^3 + Z^3) \text{ with } deg(X) = 3, deg(Y) = deg(Z) = 4$$

and

$$R' = k[X, Y, Z]/(X^3 + Y^4 + Z^4) \text{ with } deg(X) = 4, deg(Y) = deg(Z) = 3.$$

Then $a(R) = 12 - (3 + 4 + 4) = 1$ and $a(R') = 12 - (4 + 3 + 3) = 2$. (Recall that $a(R) = max\{n | (H^d_{\mathfrak{m}}(R))_n \neq 0\}$ if dim R = d. If K_R is free, then $K_R \cong R(a(R))$ as graded R-modules. cf. [3], Chapter 2 for more informations about $a(R)$.) Then by (4.2.1), S is Cohen-Macaulay and since $R(1)_n = 0$ for $n < -1, n = 0, 1$ and $R'(2)_n = 0$ for $n < -2, n = -1, 0$, $a(S) = -2$. If $char(k) = 0$, then S is a rational singularity by [14], 2.2 or [2], (3.1), since $Spec(S) - \{\mathfrak{M}_S\}$ has only "toric" singularities (cf. [13], 3.6). Also, we can check that S is F-rational if $char(k) \geq 5$. (We can check that R and R' are "F-injective in negative degree" if $char(k) \geq 5$. That is, if we take a non-zero element ζ from $(H^2_{\mathfrak{m}}(R))_n$ or $(H^2_{\mathfrak{m}'}(R'))_n$, $n < 0$, then $F(\zeta) \neq 0$.)

On the other hand, we can easily see from (4.3.1), that the anti-canonical cover $\mathcal{R}(I) = T$ is not Cohen-Macaulay. (In fact, we can easily read from (4.3.1) that if T is Cohen-Macaulay, then $a(R) < 0$ and R' is also Cohen-Macaulay.

EXAMPLE 4.5. (A case where A is not Cohen-Macaulay and $\mathcal{R}(I)$ is Gorenstein) Let $R = k[X, Y, Z]/(f)$, where R is a normal graded ring with $a(R) = -1$ (the most famous one is $f = X^2 + Y^3 + Z^5$) and $R' = k[X, Y, Z]/(X^3 + Y^3 + Z^3)$ with $a(R') = 0$. Again, put $A = S = R \sharp R'$. Then S is *not* Cohen-Macaulay by (4.2.1), while the anti-canonical cover $\mathcal{R}(I) = T$ is Cohen-Macaulay (hence is Gorenstein) by (4.3.1). In this example, S and T are F-pure if char$(k) \equiv 1$ $(mod\ 3)$ but not F-rational in any positive characteristic.

REMARK. The author does not know an example where A is F-regular, I is a divisorial ideal with infinite order (but *not* an anti-canonical ideal) with $\mathcal{R}(I)$ Noetherian and *not* Cohen-Macaulay.

REFERENCES

1. Fedder, R. and Watanabe, K.-i., *A characterization of F-regularity in terms of F-purity*, "Commutative Algebra" (M. Hochster, C. Huneke and J. Sally, eds.), Proc. Microprogram MSRI 1987, Publ., vol. 15, Springer, 1989, pp. 227-245.
2. Flenner, H., *Rationale quasihomogene Singlaritäten*, Archiv der Math. **36** (1981), 35-44.
3. Goto, S. and Watanabe, K.-i., *On graded rings, I*, J. Math. Soc. Japan **30** (1978), 179-213.
4. Goto, S., Herrmann, M., Nishida, K., and Villamayor, O., *On the structure of Noetherian symbolic Rees algebras*, Manuscripta Math. **67** (1990), 197-225.
5. Hochster, M. and Huneke, C., *Tight closure, invariant theory, and the Briançon-Skoda theorem*, J. of Amer. Math. Soc. **3** (1990), 31–116.
6. Hochster, M. and Huneke, C., *Tight closure and strong F-regularity*, Mém. Soc. Math. France **38** (1989), 119-133.
7. Hochster, M. and Roberts, J.L., *The purity of the Frobenius and local cohomology*, Adv. in Math. **21** (1976), 117–172.

8. Mehta, V.B. and Srinivas, V., *Normal F-pure surface singularites*, J. of Alg. **143** (1991), 130–143.
9. Nara, R. and Watanabe, K.-i., *Two "products" of graded rings, in preparation.*
10. Smith, K.E., *F-rational rings have rational singularities, preprint.*
11. Tomari, M. and Watanabe, K.-i., *Filtered rings, filtered blowing-ups and normal two-dimensional singularities with "star-shaped " resolution*, Publ. RIMS. Kyoto Univ. **25** (1989), 681–740.
12. Tomari, M. and Watanabe, K.-i., *Normal $\mathbb{Z}_r$-graded Rings and Normal Cyclic Covers*, Manuscripta Math. **76** (1992), 325–340.
13. Watanabe, K.-i., *Some remarks concerning Demazure's construction of normal graded rings*, Nagoya Math. J. **83** (1081), 203–211.
14. Watanabe, K.-i., *Rational singularities with k^*-action*, "Commutative Algebra" (S. Greco and G. Valla, eds.), Proc. Conf. Trento/Italy 1981, Lect. Notes in Pure, Appl. Math., vol. 84, Marcel Dekker, 1983, pp. 229–251.
15. Watanabe, K.-i., *Study of F-purity in dimension two*, "Algebraic Geometry and Commutative Algebra", in Honor of M. Nagata, 1987, pp. 791–800.
16. Watanabe, K.-i., *F-regular and F-pure normal graded rings*, J. of Pure and Appl. Alg. **71** (1991), 341–350.

Department of Mathematical Sciences, Tokai University, Hiratsuka, Kanagawa 259-12

E-mail address: watnbkei@ss.u-tokai.ac.jp

Contemporary Mathematics
Volume **159**, 1994

Torsion in Picard Groups of Affine Rings

ROGER WIEGAND

ABSTRACT. Let k be a field and let n be a positive integer prime to the characteristic of k. It is shown that $\operatorname{Pic}(A)$ has only finitely many elements of order n if A is a commutative finitely generated k-algebra of Krull dimension at most two. The one-dimensional case of the theorem has the following application to the multiplicative structure of field extensions: Let K/k be a finite separable extension, and let E and F be intermediate fields. Then K^*/E^*F^* has only finitely many elements of order n.

Let k be a field and A a finitely generated k-algebra. For each positive integer n, let ${}_n\operatorname{Pic}(A) = \{\alpha \in \operatorname{Pic}(R) | n\alpha = 0\}$. (We write the Picard group additively.) The goal of this note is a proof of the following theorem:

0.1. THEOREM. *Let A be a finitely generated algebra over the field k. If* $\dim A \leq 2$ *then* ${}_n\operatorname{Pic}(A)$ *is finite for every n prime to* $\operatorname{char}(k)$.

The restriction that n be prime to $\operatorname{char}(k)$ is always intended to be vacuous in characteristic zero. This restriction cannot be omitted. For example, the Picard group of the cusp $A = k[t^2, t^3]$ is isomorphic to the additive group of k, so ${}_p\operatorname{Pic}(A)$ is infinite if k is an infinite field of characteristic p.

The case $\dim(A) = 1$ already has an interesting application (2.4) to the multiplicative structure of field extensions. The proof in dimension 2 uses the one-dimensional case in what looks like a promising start of an inductive proof of the general theorem. Unfortunately, the method seems to break down at dimension three unless one puts further restrictions on the ring A.

David Saltman pointed out several years ago that when the ground field k is separably closed the theorem follows from **[SGA4$\frac{1}{2}$, p. 235, (1.10)]**. The one-dimensional case of the theorem (with arbitrary ground field) was obtained

1991 *Mathematics Subject Classification.* Primary 14C22; Secondary 13B22, 12F10.
Key words and phrases. Picard group, affine ring, group scheme, integral closure.
This research was partially supported by the National Science Foundation.
This paper is in final form and no version of it will be submitted for publication elsewhere.

in correspondence with Bob Guralnick during the spring of 1991, and the essential ideas of the two-dimensional case followed shortly. The final form of the argument resulted from several very enlightening conversations with David Jaffe, who found the crucial geometric results in [**SGA6**] and explained patiently what they meant.

Recently Jaffe and Wayne Raskind [**JR**] have proved the theorem without the restriction on dimension, using the general approach suggested by Saltman's observation. I have decided to publish the present note anyway, since the proof here is quite different from theirs and involves ideas that may be useful elsewhere.

It is tempting to conjecture a much stronger result if k is an algebraic number field, namely, that the torsion subgroup of Pic A should be finite. An important special case of this conjecture has recently been proved by Darren Holley, a student at the University of Nebraska. In the notation of Corollary 2.4, he has shown that the torsion subgroup of K^*/E^*F^* is finite if k is an algebraic number field.

1. preliminaries

Let k be a field and n a positive integer. We assume either that k has characteristic 0 or that k has prime characteristic p not dividing n. The proof of the Theorem is accomplished most easily if we ignore those parts of the Picard group that do not affect the conclusion. The formal way to do this is to work in the abelian category $\mathcal{A}/\mathcal{S}$, where $\mathcal{A}$ is the category of abelian groups and $\mathcal{S}$ is the Serre subcategory of $\mathcal{A}$ generated by finitely generated groups and (if $\mathrm{char}(k) = p$) bounded p-groups. (Recall that a full subcategory $\mathcal{S}$ of an abelian category $\mathcal{A}$ is a Serre subcategory provided for every short exact sequence $0 \to X \to Y \to Z \to 0$, one has $Y \in \mathcal{S} \Leftrightarrow X, Z \in \mathcal{S}$. We refer the reader to [**S, Part I**] for a discussion of Serre subcategories and the construction of the quotient category $\mathcal{A}/\mathcal{S}$.)

Since {finitely generated groups} and {bounded p-groups} are themselves Serre subcategories of $\mathcal{A}$, it is easy to see that a group is in $\mathcal{S}$ if and only if it has a finite filtration whose factors are either finitely generated groups or (if $\mathrm{char}(k) = p$) bounded p-groups. If G is a p-group or is finitely generated, clearly ${}_nG$ is finite (where ${}_nG$ denotes the subgroup of G killed by n). Since the functor taking G to ${}_nG$ is left exact, the class of groups G for which ${}_nG$ is finite is closed under extensions. Therefore ${}_nG$ is finite for every $G \in \mathcal{S}$. The following proposition will allow us to ignore finitely generated groups and bounded p-groups in our calculation of Pic(A):

1.1. PROPOSITION. *There is an exact functor $T : \mathcal{A} \to \mathcal{A}/\mathcal{S}$ satisfying $T(G) = 0 \Leftrightarrow G \in \mathcal{S}$. If G and H are abelian groups such that $T(G) \cong T(H)$, then ${}_nG$ is finite if and only if ${}_nH$ is finite.*

PROOF. The functor described in [**S, Part I, Theorem 2.1**] is exact and annihilates objects in $\mathcal{S}$. The fact that $T(G) = 0 \Rightarrow G \in \mathcal{S}$ follows from [**S,**

Part I, Lemma 2.10]. To prove the second assertion we need a lemma.

1.2. LEMMA. *Let H be a subgroup of the abelian group G. If ${}_nG$ is finite and $H \in \mathcal{S}$, then ${}_n(G/H)$ is finite.*

PROOF. By induction on the length of a suitable filtration for H, we may assume that H either is a bounded p-group or is finitely generated. Suppose $\{\gamma_i : i \geq 1\}$ is an infinite list of distinct elements of ${}_n(G/H)$. Writing $\gamma_i = g_i + H$, we have $ng_i \in H$ for each i. In the first case, assume $p^eH = 0$. Then $p^e g_i \in {}_nG$ for each i, so there exist $i < j$ such that $p^e g_i = p^e g_j$. Then $\gamma_i - \gamma_j$ is killed by both p^e and n, and since the killers are relatively prime $\gamma_i = \gamma_j$, contradiction.

If H is finitely generated, we can assume (by induction on the number of generators) that H is cyclic with generator h. Write $ng_i = a_i h, a_i \in \mathbf{Z}$. There are infinitely many subscripts i for which the integers a_i are all in the same congruence class modulo n, and by discarding the rest of the i's we may assume that $a_i \equiv a_1 \pmod n$ for all i. Writing $a_1 - a_i = nb_i$, we see that $g_1 - g_i - b_i h \in {}_nG$ for all i. There exist $i < j$ such that $g_1 - g_i - b_i h = g_1 - g_j - b_j h$, and again we have $\gamma_i = \gamma_j$, contradiction.

Returning to the proof of (1.1), we suppose first that there is a homomorphism $f : G \to H$ such that $T(f)$ is an isomorphism. By the first assertion of (1.1) this means that both the kernel and the cokernel of f are in $\mathcal{S}$. If ${}_nG$ is finite, (1.2) implies that that ${}_n \operatorname{Im}(f)$ is finite. Since ${}_n \operatorname{Coker}(f)$ is finite, so is ${}_nH$. Conversely, if ${}_nH$ is finite, so is ${}_n \operatorname{Im}(f)$, and since ${}_n \operatorname{Ker}(f)$ is finite, so is ${}_nG$.

In the general case, choose an isomorphism $\phi : T(G) \to T(H)$. By [**S, Part I, Lemma 2.9**] there exist an abelian group E and homomorphisms $g : E \to G$ and $h : E \to H$ such that $T(g)$ is an isomorphism and $\phi T(g) = T(h)$. Of course this means $T(h)$ is an isomorphism as well. Two applications of the special case treated above complete the proof.

Let k^i be the purely inseparable closure of k, that is, the set of elements in the algebraic closure of k that are purely inseparable over k. Then k^i is perfect, and we lose nothing by replacing k by k^i, thanks to the following lemma:

1.3. LEMMA. *Let K/k be a field extension, let A be a k-algebra, and put $B = A \otimes_k K$.*

(1) *If* $[K : k] = m < \infty$ *then* $\operatorname{Ker}(\operatorname{Pic}(A) \to \operatorname{Pic}(B)) \subseteq {}_m \operatorname{Pic}(A)$.

(2) *If K/k is either a finite extension of degree prime to n or a directed union of such extensions, then the natural map ${}_n \operatorname{Pic}(A) \to {}_n \operatorname{Pic}(B)$ is injective.*

PROOF. Since the functor ${}_n \operatorname{Pic}(\ )$ commutes with direct limits, (2) follows from (1). Let I be a rank-one projective A-module, and suppose $I \otimes_k K \cong B$. As A-modules, $I \otimes_k K \cong \oplus^m I$ and $B \cong \oplus^m A$. Taking m^{th} exterior powers we get $\otimes^m I \cong A$, which proves (1).

1.4. Remarks. The map in (1.3) is not necessarily surjective. For example, let $A = k[x, y]/(y^2 - x^3 - x^2)$, the affine coordinate ring of the nodal cubic. Then

$\mathrm{Pic}(A) \cong k^*$. If we take $k = \mathbf{Q}$ and $K = \mathbf{Q}(\omega)$, where ω is a primitive cube root of unity, then ${}_3\mathrm{Pic}(A) \to {}_3\mathrm{Pic}(A \otimes_k K)$ is not surjective.

The assumption that $(m,n) = 1$ cannot be omitted, even in characteristic 0: The ring $\mathbf{R}[x,y]/(x^2+y^2-1)$ has an element of order 2 in its Picard group (represented by the maximal ideal $(x, y-1)$) that becomes trivial over the complex numbers.

2. dimension one

2.1. THEOREM. *Let A be a finitely generated k-algebra of dimension one. Then ${}_n\mathrm{Pic}(A)$ is finite for every integer n prime to the characteristic of k.*

PROOF. We may assume k is perfect by (1.3) and that A is reduced. Our first task is to embed $X := \mathrm{Spec}(A)$ as an open subscheme of a projective scheme Y in such a way that Y has no singularities outside X. Write $A = k[x_1, \dots, x_m]/I$. For each $f \in I$ let $f^h \in \Lambda := k[x_0, \dots, x_m]$ be the homogenization of f with respect to x_0, and let J be the ideal of Λ generated by $\{f^h | f \in I\}$. Then X can be identified with an open subscheme of $W := \mathrm{Proj}(\Lambda/J)$. The projective scheme W is one-dimensional and reduced [**Ku, p. 38, 6**^{c}], so we just have to normalize the singularities of W without messing up X. This is easily done by repeatedly blowing up singular points outside X until we obtain a projective scheme Y and a finite birational morphism $\phi : Y \to W$ such that ϕ maps $\phi^{-1}(X)$ isomorphically to X and Y is smooth away from $\phi^{-1}(X)$. (Alternatively, let $v_1, \dots, v_s$ be the singular points of W outside X. Choose, for each j, an affine open neighborhood U_j of v_j missing every singular point of X, and let $\tilde{U}_j$ be the normalization of the affine scheme U_j. Then let Y be the scheme obtained by gluing the schemes $X, \tilde{U}_1, \dots, \tilde{U}_t$ in the obvious way.)

The restriction map $\rho : \mathrm{Pic}(Y) \to \mathrm{Pic}(X)$ is surjective. To see this, let $\mathcal{L}$ be any invertible sheaf on X. By [**H, Chap. II, Exer. 5.15**] there is a coherent $\mathcal{O}_Y$-module $\mathcal{M}$ extending $\mathcal{L}$. The double dual $\mathcal{M}^{\check{}\check{}}$ is then an invertible $\mathcal{O}_Y$-module restricting to $\mathcal{L}$.

Let $Y - X = \{z_1, \dots, z_t\}$. Each z_i determines a Cartier divisor ζ_i on Y as follows: Cover Y by the two open sets $U, V := Y - \{z_i\}$, where $U = \mathrm{Spec}(R)$ is an affine open neighborhood of z_i missing all the other z_j's. Let $f \in \Gamma(U, \mathcal{O}_Y)$ generate the maximal ideal of the discrete valuation ring $\mathcal{O}_{z_i}$. By shrinking U if necessary we may assume that f is a unit of $\mathcal{O}_z$ for every $z \in U - \{z_i\}$. Taking $g = 1 \in \Gamma(V, \mathcal{O}_Y)$, we see that the pair (f, g) defines a Cartier divisor ζ_i on Y. Identifying $\mathrm{Pic}(Y)$ with the Cartier divisor class group [**EGA, (21.3.4)**], one checks easily that $\zeta_1, \dots, \zeta_t$ generate $\mathrm{Ker}(\rho)$.

By (1.1) it suffices to show that ${}_n\mathrm{Pic}(Y)$ is finite. It is convenient to pass to the algebraic closure of k, which we can now do harmlessly, in view of the following result (proved in [**M, (6.2)**] under the additional assumption that Y is geometrically integral):

2.2. PROPOSITION. *Let Y be a projective scheme over a field k, and let K*

be the algebraic closure of k. Put $Z = Y \times_k \operatorname{Spec}(K)$, and let $\pi : Z \to Y$ be the projection. The induced map $\operatorname{Pic}(Y) \to \operatorname{Pic}(Z)$ is an injection.

PROOF. Since the extension K/k is a direct limit of finite extensions, it suffices to prove the following:

2.3. LEMMA. *Let Y be a projective scheme over a field k, and let K be a finite algebraic extension of k. Put $Z = Y \times_k \operatorname{Spec}(K)$, and let $\pi : Z \to Y$ be the projection. Let $\mathcal{M}$ and $\mathcal{N}$ be coherent $\mathcal{O}_Y$-modules such that $\pi^*(\mathcal{M})$ and $\pi^*(\mathcal{N})$ are isomorphic. Then $\mathcal{M} \cong \mathcal{N}$.*

PROOF. Choose a basis $\{\alpha_1, \ldots, \alpha_d\}$ for K over k. Let $\mathcal{A} = \pi_*(\pi^*(\mathcal{M}))$. Over any affine open set $U \subset Y$ we have $\mathcal{A}(U) = K \otimes_k \mathcal{M}(U)$, and there is a unique $\mathcal{O}_Y(U)$-module isomorphism f_U from $\mathcal{A}(U)$ to the direct sum of n copies of $\mathcal{M}(U)$, taking $\alpha_i \otimes m$ to the element with m in the ith coordinate and 0's elsewhere. These isomorphisms are compatible as U varies, so $\pi_*(\pi^*(\mathcal{M}))$ is isomorphic to the direct sum of n copies of $\mathcal{M}$. The analogous statement holds for $\mathcal{N}$, and since $\pi_*(\pi^*(\mathcal{M})) \cong \pi_*(\pi^*(\mathcal{N}))$ we conclude from the Krull-Schmidt Theorem [**S, Part II, Theorem 2.18**] that $\mathcal{M} \cong \mathcal{N}$. (The crucial fact is that the endomorphism ring of every coherent $\mathcal{O}_Y$-module $\mathcal{M}$ is a finite-dimensional algebra over k, since it is the module of global sections of the coherent $\mathcal{O}_Y$-module $\operatorname{Sheafhom}_{\mathcal{O}_X}(\mathcal{M}, \mathcal{M})$. See [**H, Chap. II, Theorem 5.19**].)

Returning to the proof of (2.1), we need only show that ${}_n\operatorname{Pic}(Z)$ is finite, where Z is as in (2.2). Now Z is a one-dimensional projective scheme over the algebraically closed field K. Moreover, Z is reduced, since we assumed the original ground field k was perfect. Let $\tilde{Z}$ be the normalization of Z. There is an exact sequence [**H, Chap. II, Exercise 6.9**] or [**EGA, (21.8.5.1)**]

$$0 \to \bigoplus_z \tilde{\mathcal{O}}_z^*/\mathcal{O}_z^* \to \operatorname{Pic}(Z) \to \operatorname{Pic}(\tilde{Z}) \to 0,$$

where z ranges over the singular points of X, and $(\ )^*$ denotes the group of units. (Caution: The sequence need not be exact for affine curves. See the comments after (3.4) of [**W**].) It will suffice to show that the groups ${}_n(\tilde{\mathcal{O}}_z^*/\mathcal{O}_z^*)$ and ${}_n\operatorname{Pic}(\tilde{Z})$ are finite. The proof of this is fairly standard, but we will give a sketch.

For the groups ${}_n(\tilde{\mathcal{O}}_z^*/\mathcal{O}_z^*)$, let $A = \mathcal{O}_z, B = \tilde{A}$, and let C be the seminormalization of A in B. The group C^*/A^* has a finite filtration by additive groups of (finite-dimensional) vector spaces over k. In characteristic p these can be ignored, in view of (1.1). In characteristic 0 the filtration splits (since the additive groups are divisible), and $B^*/A^* \cong (C^*/A^*) \oplus (B^*/C^*) = V \oplus (B^*/C^*)$, where V is a vector space. Therefore in either case all we need to worry about is B^*/C^*. But this is just a direct sum of $s - 1$ copies of K^*, where s is the number of maximal ideals of B.

Finally, we consider ${}_n\operatorname{Pic}(\tilde{Z})$. Working with one connected component at a time, we just have to show that ${}_n\operatorname{Pic}(W)$ is finite when W is a smooth connected projective curve over k. We have an exact sequence

$$0 \longrightarrow \operatorname{Pic}^0(W) \longrightarrow \operatorname{Pic}(W) \xrightarrow{\delta} \mathbf{Z} \longrightarrow 0,$$

where δ is the degree function. It is known that $\operatorname{Pic}^0(W)$ is the group of closed points of an abelian variety [**H, Chapter II, 6.10.3**], whence ${}_n\operatorname{Pic}^0(W)$ is finite [**M, Remark 8.4**].

2.4. COROLLARY. *Let K/k be a finite separable extension, and let E and F be intermediate fields. Then, for every n prime to the characteristic, ${}_n(K^*/E^*F^*)$ is finite.*

PROOF. Choose $\alpha \in K$ such that $K = E[\alpha]$, and let $A = \{f \in E[X] | f(\alpha) \in F\}$. We have a Cartesian square:

$$\begin{array}{ccc} A & \longrightarrow & E[X] \\ \downarrow & & \downarrow{\scriptstyle \epsilon} \\ F & \longrightarrow & K, \end{array}$$

where ϵ takes X to α. It follows from Milnor's Mayer-Vietoris exact sequence [**B, Chap. IX, (5.3)**] that $\operatorname{Pic}(A) \cong K^*/E^*F^*$, and we apply (2.1).

There are easier proofs of (2.4) if E and F are linearly disjoint over k, but I know of no other proof for the following very simple example: E and F are two distinct embeddings of $\mathbf{Q}(\sqrt[3]{2})$ in the splitting field K of $x^3 - 2$ over $k = \mathbf{Q}$. It is interesting to note that the analogous result fails for *three* intermediate fields. For example, let E_1, E_2, E_3 be the three proper intermediate fields for the splitting field K of $X^4 - 4$ over $\mathbf{Q}$. Then $K^*/E_1^*E_2^*E_3^*$ is an infinite elementary abelian 2-group, [**GW, (1.2)**].

In the next section we will need a stronger form of (2.4), in which the field K is replaced by a finite-dimensional separable algebra over k.

2.5. COROLLARY. *Let k be a field, and let $K_1, \ldots, K_t$ be finite separable field extensions of k. Put $K = K_1 \times \cdots \times K_t$, and let E and F be k-subalgebras of K. Then ${}_n(K^*/E^*F^*)$ is finite for every n prime to the characteristic.*

PROOF. If k is finite there is nothing to prove, so we assume that k is infinite. If we can find an element $\alpha \in K$ such that $K = k[\alpha]$, then $K = E[\alpha]$ as well, and the proof of (2.4) goes through word-for-word. The following lemma does the job:

2.6. LEMMA. *Let k and K be as in (2.5), with k infinite. Then $K = k[\alpha]$ for some element $\alpha \in K$.*

PROOF. Choose primitive elements α_i such that $K_i = k[\alpha_i]$ for each i, and let $f_i \in k[X]$ be the minimal polynomial for α_i. If all the polynomials f_i are distinct, we see that $K \cong k[X]/(f)$, where $f = f_1 \cdots f_t$. What if some of the f_i are the same? Assume inductively that $f_1, \ldots f_{t-1}$ are distinct. I claim that for each $i \leq t-1$ there are at most $\deg(f_t)$ constants $c \in k$ such that $f_t(X + c) = f_i(X)$.

For, if $f_t(X+c) = f_t(X+d)$, then $f_t(c) = f_t(d)$. Since k is infinite we can choose c outside the union of these $t-1$ finite sets and replace α_t by $\alpha_t - c$.

3. Dimension two

The proof in dimension two proceeds in two steps. First we prove the result for normal domains, following the same general outline as the proof in dimension one. Then we examine what happens to the Picard group of a not normal ring when one goes to the normalization. The second step works in any dimension, and uses the one-dimensional case of the theorem.

3.1. THEOREM. *Let k be a field and let A be a finitely generated k-algebra of dimension two. For every integer n prime to* $\mathrm{char}(k)$, ${}_n\,\mathrm{Pic}(A)$ *is finite.*

PROOF. We may assume k is perfect. Also, if k happens to be a finite field of order $q = p^e$ we can use (1.3) to replace k by the union of the fields of order $q^{p^m}, m \geq 1$. Therefore we assume from now on that k is an infinite perfect field.

Suppose first that A is a normal domain. Embed $X = \mathrm{Spec}(A)$ as an open subscheme of an integral scheme W_1 projective over k. Since passing to the normalization doesn't disturb X we may assume W_1 is normal. Again, we want to get rid of the singular points outside X. This is accomplished by means of Lipman's approach to resolution of singularities for surfaces, [**A**]: Let W_2 be the normalization of the scheme obtained by blowing up the singular points of W_1 outside X. We have a proper map $f_1 : W_2 \to W_1$, and we identify X with its isomorphic copy $f_1^{-1}(X) \subset W_2$. After repeating this process a finite number of times we arrive at an integral normal projective scheme Y containing X as an open subscheme, such that Y has no singular points outside X.

3.2. LEMMA. *Let F be any field, and let Y be an integral scheme of finite type over F. Assume Y is normal and two-dimensional, and let X be a non-empty open subset of Y containing all the singular points of Y. Then the restriction map* $\mathrm{Pic}(Y) \to \mathrm{Pic}(X)$ *is surjective, and its kernel is finitely generated.*

PROOF. The argument used to prove surjectivity in the one-dimensional case works in the current situation, since duals are free over a two-dimensional regular local ring. To see that the kernel is finitely generated, let $C_1, \ldots, C_t$ be the one-dimensional irreducible components of the reduced scheme $Y - X$. Since $\mathcal{O}_Y$ is factorial at each point of $Y - X$, each C_i is locally defined by a single equation; therefore each corresponds to a Cartier divisor ζ_i as before. Using [**EGA, (21.3.4)**] again to identify $\mathrm{Pic}(Y)$ with the Cartier divisor class group, we see that each ζ_i is in the kernel of the restriction map. To see that they generate the kernel, let η be a Cartier divisor whose restriction to X is principal. After multiplying by a suitable element of the function field, we may assume that $\eta(x) \equiv 1 (\mathrm{mod}\ \mathcal{O}_x^*)$ for every $x \in X$. (We are thinking of η as a global section of the sheaf $\mathcal{K}^*/\mathcal{O}_Y^*$.) Now let n_i be order of η along C_i (that is, its value for the discrete valuation of the local ring at the generic point of C_i). The Cartier divisor $\eta\zeta_1^{-n_1}\cdots\zeta_t^{-n_t}$ represents a unit of $\mathcal{O}_y$ at every non-closed point $y \in Y$. Since Y is normal, $\eta = \zeta_1^{n_1}\cdots\zeta_t^{n_t}$.

To prove (3.1) in the case of a normal domain A, it therefore suffices to prove that ${}_n\operatorname{Pic}(Y)$ is finite, where Y is an integral projective scheme over k of dimension two. By (2.2) we can replace Y by $Z := Y \times_k \operatorname{Spec}(K)$, where K is the algebraic closure. Of course Z may not be integral, but it is at least reduced, since k was assumed to be perfect. The following result therefore completes the proof in this case:

3.3. THEOREM. *Let Z be a reduced scheme, proper over an algebraically closed field k. Then ${}_nPic(Z)$ is finite, for every integer n prime to* $\operatorname{char}(k)$.

This result is well known to algebraic geometers and follows from the representability of the Pic functor. I will provide a brief guided tour of the results needed to piece together a proof. Let $\mathbf{Sch}_k$ denote the category of schemes (not necessarily of finite type) over k, and consider the functor $Pic_{Z/k} : \mathbf{Sch}_k \to \mathcal{A}$ (= abelian groups) defined (more or less) in **[SGA6, Exp. X]**. Since k is algebraically closed, one has $Pic_{Z/k}(\operatorname{Spec}(k)) = \operatorname{Pic}(Z)$, by **[SGA6, Exp. X, (2.4)]**. The big theorem is that $Pic_{Z/k}$ is representable, **[SGA6, Exp. XII, (1.5)]**. The representing object, which we denote by $\operatorname{Pic}_{Z/k}$, is a commutative group scheme **[O]** locally of finite type over k. We have $\operatorname{Pic}(Z) = \operatorname{Hom}_{\mathbf{Sch}_k}(\operatorname{Spec}(k), \operatorname{Pic}_{Z/k})$, that is, $\operatorname{Pic}(Z)$ can be identified with the group of closed points of $\operatorname{Pic}_{Z/k}$. The nth power map $\Phi_n : \operatorname{Pic}_{Z/k} \to \operatorname{Pic}_{Z/k}$ is a morphism of finite type, **[SGA6, Exp. XIII, (3.6)]**. (This much is true even if n is a multiple of the characteristic.) It follows that $G := \operatorname{Ker}(\Phi_n)$ is a commutative group scheme of finite type over k. Since ${}_n\operatorname{Pic}(Z)$ can be identified with the group of closed points of this scheme, all we need to do is show that G is a finite group scheme. By **[O, Chap. II, (6.1)]**, G has a finite filtration by subgroup schemes such that each factor is either the additive linear group G_a, the multiplicative linear group G_m, an abelian variety, or a finite group scheme. But since G is n-torsion and n is prime to the characteristic, each of the first three possiblities is ruled out. Therefore G is finite.

We now know that (3.1) is true if A is a normal domain. In the general case, we may assume A is reduced. Let $\tilde{A}$ be the normalization of A, that is, the integral closure of A in its total quotient ring. Since $\tilde{A}$ is a direct product of a finite number of normal domains, ${}_n\operatorname{Pic}(\tilde{A})$ is finite. Also, if $\mathbf{c} = (A : \tilde{A})$ denotes the conductor of A in $\tilde{A}$, then ${}_n\operatorname{Pic}(A/)$ is finite by (2.1). Therefore it will suffice to show that the kernel of the map $\operatorname{Pic}(A) \to \operatorname{Pic}(\tilde{A}) \times \operatorname{Pic}(A/\mathbf{c})$ has finite n-torsion. This is a special case of (3.8) below, but it will be helpful to establish a few preliminary results.

3.4. NOTATION. *Let $A \subseteq B$ be commutative rings, and let $\mathbf{a} = (A : B)$ be the conductor (the largest ideal of B contained in A). We denote by $\kappa(A, B)$ the kernel of the map* $\operatorname{Pic}(A) \to \operatorname{Pic}(B) \times \operatorname{Pic}(A/\mathbf{a})$.

3.5. LEMMA. *Let $A \subseteq B \subseteq C$, and let $\mathbf{a} = (A : B)$, $\mathbf{b} = (B : C)$ and $\mathbf{c} = (A : C)$ be the various conductors. The map* $\mathrm{Pic}(A) \to \mathrm{Pic}(B)$ *induces a map $\phi : \kappa(A, C) \to \kappa(B, C)$, and* $\mathrm{Ker}(\phi) \subseteq \kappa(A, B)$. *If ${}_n\kappa(A, B)$ and ${}_n\kappa(B, C)$ are finite, so is ${}_n\kappa(A, C)$.*

PROOF. Since $\mathbf{c} \subseteq \mathbf{b}$, the inclusion $A \to B$ induces a map $A/\mathbf{c} \to B/\mathbf{b}$. Therefore, if $\alpha \in \kappa(A, C)$, its image in $\mathrm{Pic}(B)$ is in $\kappa(B, C)$. Thus the map ϕ makes sense. The assertion regarding $\mathrm{Ker}(\phi)$ follows from the fact that $\mathbf{c} \subseteq \mathbf{a}$. Now we have an exact diagram

$$
\begin{array}{ccccccc}
 & & 0 & & & & \\
 & & \downarrow & & & & \\
0 & \longrightarrow & \mathrm{Ker}(\phi) & \longrightarrow & \kappa(A,C) & \longrightarrow & \kappa(B,C) \\
 & & \downarrow & & & & \\
 & & \kappa(A,B) & & & &
\end{array}
$$

Applying the left exact functor ${}_n(\ )$, we get the last statement.

It seems worthwhile to prove that ${}_n\kappa(A, \tilde{A})$ is finite when A is reduced and of finite type over an *arbitrary* field F, even though we will need this fact only for the infinite perfect field k.

3.6. LEMMA. *Let F be a field, and let $A \subseteq B$ be reduced F-algebras. Let K/F be a field extension, and consider the inclusion $C := (A \otimes_F K)_{red} \subseteq D := (B \otimes_F K)_{red}$, where $(\ )_{red}$ denotes reduction modulo nilpotents. If* ${}_n\mathrm{Pic}(A) \to {}_n\mathrm{Pic}(A \otimes_F K)$ *is an injection, so is ${}_n\kappa(A, B) \to {}_n\kappa(C, D)$.*

PROOF. Let $\mathbf{a} = (A : B)$ and $\mathbf{c} = (C : D)$ be the conductors. Since $\mathbf{a} \subseteq \mathbf{c}$, the injection $A \to C$ induces a map $A/\mathbf{a} \to C/\mathbf{c}$. Therefore the map $\psi : \mathrm{Pic}(A) \to \mathrm{Pic}(C)$ carries $\kappa(A, B)$ into $\kappa(C, D)$. Since killing nilpotents does not affect the Picard group, ψ induces an injection from ${}_n\mathrm{Pic}(A) \to_n \mathrm{Pic}(C)$, and the result follows.

Next, we recall the following well-known result, attributed to Roquette in [**Kr**]:

3.7. PROPOSITION. *Let F be a field and A a reduced ring of finite type over F. Let E be the integral closure of F in A. Then E is a direct product of finitely many fields, each finite over F; and A^*/E^* is a finitely generated group.*

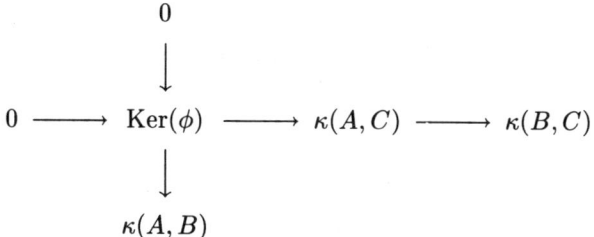

PROOF. Let B be the integral closure of A in its total quotient ring, and write $B = B_1 \times \cdots \times B_t$, where each B_i is a normal domain (of finite type over F). Let L_i be the quotient field of B_i. The integral closure C of F in B is clearly $K_1 \times \cdots \times K_t$, where K_i is the algebraic closure of F in L_i. Since each L_i is a finitely generated field extension of F, each K_i is finite over F, whence C is a finite-dimensional F-algebra. Since $E \subseteq C$, E is finite-dimensional as well, and since it is reduced, the first assertion follows. The usual formulation of Roquette's theorem (for domains) says that each B_i^*/K_i^* is finitely generated, whence B^*/C^* is finitely generated. Now $C \cap A = E$, and it follows that $C^* \cap A^* = E^*$. Therefore A^*/C^* embeds in B^*/E^*, and we are done.

At last we can complete the proof of (3.1). The following theorem, which holds with no restriction on the dimension, is the final step needed in the proof:

3.8. THEOREM. *Let F be any field, and let A be a reduced F-algebra of finite type. Let C be any ring between A and $\tilde{A}$. Then ${}_n\kappa(A, C)$ is finite for every integer n prime to the characteristic of k.*

PROOF. By (3.6) and (1.3) we may assume that F is an infinite perfect field. Also, it is convenient and harmless to assume that A is connected. Let B denote the seminormalization **[T]** of A in C. We will show that ${}_n\kappa(A, B)$ and ${}_n\kappa(B, C)$ are both finite, and then invoke (3.5). We will deal with the extension $A \subseteq B$ first.

Recall from **[T]** that $\operatorname{Spec}(B) \to \operatorname{Spec}(A)$ is bijective and that for each prime Q of B the residue field of B at Q is equal to the residue field of A at $Q \cap A$. By a result due to Jaffe **[J, (7.3)]** there is a chain of rings $A = A_0 \subset \cdots \subset A_t = B$ such that the conductor of A_{i-1} in A_i is a prime ideal of A_{i-1} for $1 \leq i \leq t$. Using (3.5), we may assume that the conductor of A in B is a prime ideal P of A. When viewed as an ideal of B, P has a unique minimal prime ideal Q, and we have $Q^m \subseteq P$ for some $m \geq 1$. Consider the following pullback diagram:

$$\begin{array}{ccc} A & \longrightarrow & B \\ \downarrow & & \downarrow \pi \\ A/P & \longrightarrow & B/P \end{array}$$

We know **[B, Chap. IX, (5.3)]** that $\kappa(A, B) = (B/P)^*/(A/P)^*\pi(B^*)$. Let E, F and K be the integral closures of k in B, A/P and B/Q, respectively. (Of course, F and K are fields, but in fact so is E, since B is connected.) By (3.7) and (1.1) it is enough to show that $(B/P)^*/E^*F^*$ has finite n-torsion. (We identify E and F with their images in B/P.) Consider the chain of surjections $B/P \to B/(P + Q^{m-1}) \cdots \to B/Q$. We need to know that E^*F^* is mapped isomorphically by each of these surjections, that is, E^*F^* is disjoint from Q/P. To prove this it is enough to show that $E \cup F$ is contained in a reduced subring of B/P. But $E \otimes_k F$ is reduced (since k is perfect) and zero-dimensional, so its image in B/P is also reduced.

By (2.4), ${}_n(K^*/E^*F^*)$ is finite; therefore ${}_n((B/Q)^*/E^*F^*)$ is finite, by (3.7) and (1.1). Assume inductively that ${}_n((B/(P + Q^e))^*/E^*F^*)$ is finite for some integer $e, 1 \leq e \leq m - 1$. We will show that ${}_n((B/(P + Q^{e+1}))^*/E^*F^*)$ is finite. When $e = m - 1$ this gives the desired result that ${}_n((B/P)^*/E^*F^*)$ is finite. Put $D = B/(P + Q^{e+1})$, and let $I = (P + Q^e/(P + Q^{e+1})$. We have an exact sequence

$$1 \to 1 + I \to D^* \to (D/I)^* \to 1.$$

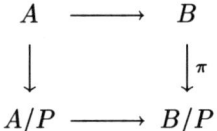

Let $H = E^*F^* \subseteq D^*$, and note that $H \cap (1+I) = 1$ (since $H \cap I = \emptyset$). Therefore we get another exact sequence

$$1 \to 1 + I \to D^*/H \to (D/I)^*/H \to 1.$$

Finally, we note that since $I^2 = 0$, $1 + I$ is isomorphic to the additive group of I, which is a k-vector space. Therefore ${}_n(1 + I) = 0$. Since ${}_n((D/I)^*/H)$ is finite by our inductive hypothesis, so is ${}_n(D^*/H)$. This completes the proof that ${}_n\kappa(A, B)$ is finite, where B is the seminormalization of A in C.

By (3.5) all that remains is to show that ${}_n\kappa(B, C)$ is finite. Since B is seminormal in C, the conductor $\mathbf{b}$ of B in C is a radical ideal of C, by [**T, Lemma 1.3**]. Therefore all the rings in the following pullback diagram are reduced:

$$\begin{array}{ccc} B & \longrightarrow & C \\ \downarrow & & \downarrow \pi \\ B/\mathbf{b} & \longrightarrow & C/\mathbf{b} \end{array}$$

We want to show that $(C/\mathbf{b})^*/(B/\mathbf{b})^*\pi(C^*)$ has finite n-torsion. Let E, F and K be the integral closures of k in C, $B/\mathbf{b}$ and $C/\mathbf{b}$, respectively. By (1.1) and (3.7) it is enough to show that ${}_n(K^*/E^*F^*)$ is finite. But this is exactly what (2.5) asserts!

This completes the proof of (3.8) and (3.1). The approach we have used in dimension two does not seem to carry over to higher dimensions. The problem is that a normal affine variety with infinitely many singularities can never be embedded in a projective variety without introducing new singularities (since the singular locus is closed). Of course, in certain situations there are no difficulties. For example, if A is three-dimensional and its normalization is smooth we can (assuming resolution of singularities for three-folds) easily deduce that ${}_n \operatorname{Pic}(A)$ is finite for n prime to the characteristic. In any case, we have the following result in dimension three:

3.9. COROLLARY. *Let k be any field and let A be a three-dimensional reduced k-algebra of finite type. Let B be any ring between A and the normalization $\tilde{A}$. Then $\delta(A, B) := \operatorname{Ker}(\operatorname{Pic}(A) \to \operatorname{Pic}(B))$ has finite n-torsion for n prime to* char(k).

PROOF. Let $\mathbf{a} = (A : B)$ be the conductor, and let H be the kernel of $\operatorname{Pic}(A) \to \operatorname{Pic}(A/\mathbf{a})$. By (3.8), ${}_nH \cap \delta(A, B)$ is finite. Also, ${}_nH$ is the kernel of the map from ${}_n \operatorname{Pic}(A)$ to the finite (by (2.1) or (3.1)) group ${}_n \operatorname{Pic}(A/\mathbf{a})$. Therefore ${}_nH$ has finite index in ${}_n \operatorname{Pic}(A)$, and the result follows.

REFERENCES

[A] M. Artin, *Lipman's proof of resolution of singularities for surfaces*, Arithmetic Geometry, Springer-Verlag, New York, 1986.

[B] H. Bass, *Algebraic K-Theory*, W. A. Benjamin, New York, 1968.

[EGA] A. Grothendieck and J. Dieudonné, *Éléments de Géométrie Algébrique IV, Quatrième Partie*, Publ. Math. I.H.E.S. **32** (1967).

[GW] R. Guralnick and R. Wiegand, *Galois groups and the multiplicative structure of field extensions*, Trans. Amer. Math. Soc. **331** (1992), 563–584.

[H] R. Hartshorne, *Algebraic Geometry*, Springer-Verlag, New York, 1977.

[J] D. Jaffe, *Coherent functors, with application to torsion in the Picard group*, (preprint).

[JR] D. Jaffe and W. Raskind, *Finiteness of the n-torsion of the Picard group of a variety over a field*, (preprint).

[Kr] J. Krempa, *Unit groups and commutative ring extensions*, Comm. Algebra **16** (1988), 2349–2361.

[Ku] E. Kunz, *Introduction to Commutative Algebra and Algebraic Geometry*, Birkhäuser, Boston, 1985.

[M] J. S. Milne, *Abelian Varieties*, Arithmetic Geometry, Springer-Verlag, New York, 1986.

[O] F. Oort, *Commutative Group Schemes*, Springer Lect. Notes in Math. **15** (1966).

[S] R. G. Swan, *Algebraic K-Theory*, Springer Lect. Notes in Math. **76** (1968).

[SGA4$\frac{1}{2}$] P. Deligne, *Cohomologie Étale*, Springer Lect. Notes in Math. **569** (1977).

[SGA6] P. Berthelot, A. Grothendieck and L. Illusie, *Théorie des Intersections et Théorème de Riemann-Roch*, Springer Lect. Notes in Math. **225** (1971).

[T] C. Traverso, *Seminormality and Picard group*, Ann. Scuola Norm. Sup. Pisa **24** (1970), 585–595.

[W] R. Wiegand, *Cancellation over commutative rings of dimension one and two*, J. Algebra **88** (1984), 438–459.

Department of Mathematics and Statistics, University of Nebraska, Lincoln, NE 68588-0323

E-mail address: rw@hoss.unl.edu

Recent Titles in This Series

(Continued from the front of this publication)

128 **Ridgley Lange and Shengwang Wang,** New approaches in spectral decomposition, 1992

127 **Vladimir Oliker and Andrejs Treibergs, Editors,** Geometry and nonlinear partial differential equations, 1992

126 **R. Keith Dennis, Claudio Pedrini, and Michael R. Stein, Editors,** Algebraic K-theory, commutative algebra, and algebraic geometry, 1992

125 **F. Thomas Bruss, Thomas S. Ferguson, and Stephen M. Samuels, Editors,** Strategies for sequential search and selection in real time, 1992

124 **Darrell Haile and James Osterburg, Editors,** Azumaya algebras, actions, and modules, 1992

123 **Steven L. Kleiman and Anders Thorup, Editors,** Enumerative algebraic geometry, 1991

122 **D. H. Sattinger, C. A. Tracy, and S. Venakides, Editors,** Inverse scattering and applications, 1991

121 **Alex J. Feingold, Igor B. Frenkel, and John F. X. Ries,** Spinor construction of vertex operator algebras, triality, and $E_8^{(1)}$, 1991

120 **Robert S. Doran, Editor,** Selfadjoint and nonselfadjoint operator algebras and operator theory, 1991

119 **Robert A. Melter, Azriel Rosenfeld, and Prabir Bhattacharya, Editors,** Vision geometry, 1991

118 **Yan Shi-Jian, Wang Jiagang, and Yang Chung-chun, Editors,** Probability theory and its applications in China, 1991

117 **Morton Brown, Editor,** Continuum theory and dynamical systems, 1991

116 **Brian Harbourne and Robert Speiser, Editors,** Algebraic geometry: Sundance 1988, 1991

115 **Nancy Flournoy and Robert K. Tsutakawa, Editors,** Statistical multiple integration, 1991

114 **Jeffrey C. Lagarias and Michael J. Todd, Editors,** Mathematical developments arising from linear programming, 1990

113 **Eric Grinberg and Eric Todd Quinto, Editors,** Integral geometry and tomography, 1990

112 **Philip J. Brown and Wayne A. Fuller, Editors,** Statistical analysis of measurement error models and applications, 1990

111 **Earl S. Kramer and Spyros S. Magliveras, Editors,** Finite geometries and combinatorial designs, 1990

110 **Georgia Benkart and J. Marshall Osborn, Editors,** Lie algebras and related topics, 1990

109 **Benjamin Fine, Anthony Gaglione, and Francis C. Y. Tang, Editors,** Combinatorial group theory, 1990

108 **Melvyn S. Berger, Editor,** Mathematics of nonlinear science, 1990

107 **Mario Milman and Tomas Schonbek, Editors,** Harmonic analysis and partial differential equations, 1990

106 **Wilfried Sieg, Editor,** Logic and computation, 1990

105 **Jerome Kaminker, Editor,** Geometric and topological invariants of elliptic operators, 1990

104 **Michael Makkai and Robert Paré,** Accessible categories: The foundations of categorical model theory, 1989

103 **Steve Fisk,** Coloring theories, 1989

102 **Stephen McAdam,** Primes associated to an ideal, 1989

101 **S.-Y. Cheng, H. Choi, and Robert E. Greene, Editors,** Recent developments in geometry, 1989

100 **W. Brent Lindquist, Editor,** Current progress in hyperbolic systems: Riemann problems and computations, 1989

(See the AMS catalog for earlier titles)